Mitigating Unsustainable Practices in Construction and Architecture

Roberto Alonso González-Lezcano
Universidad CEU San Pablo, Spain

Sunil Kumar Sansaniwal
NITI Aayog, India

IGI Global
Scientific Publishing
Publishing Tomorrow's Research Today

Vice President of Editorial	Melissa Wagner
Director of Acquisitions	Mikaela Felty
Director of Book Development	Jocelynn Hessler
Production Manager	Mike Brehm
Cover Design	Jose Rosado

Published in the United States of America by
IGI Global Scientific Publishing
701 East Chocolate Avenue
Hershey, PA, 17033, USA
Tel: 717-533-8845 | Fax: 717-533-7115
Website: https://www.igi-global.com E-mail: cust@igi-global.com

Library of Congress Cataloging-in-Publication Data

Names: González-Lezcano, Roberto Alonso, 1971- editor http://id.loc.gov/rwo/agents/n2021009547 | Sansaniwal, Sunil Kumar, 1989- editor http://id.loc.gov/rwo/agents/n2024040790

Title: Mitigating unsustainable practices in construction and architecture / edited by Roberto Alonso González-Lezcano and Sunil Kumar Sansaniwal.

Description: Hershey, PA : IGI Global Scientific Publishing, [2026] | Includes bibliographical references and index. | Summary: "This book aims to compile proposals, studies, research and concrete actions that can be widely implemented in the built environment for a more sustainable construction"-- Provided by publisher.

Identifiers: LCCN 2025013500 (print) | LCCN 2025013501 (ebook) | ISBN 9798337325552 hardcover | ISBN 9798337325569 paperback | ISBN 9798337325576 ebook

Subjects: LCSH: Sustainable construction http://id.loc.gov/authorities/subjects/sh2009003365 | Sustainable design http://id.loc.gov/authorities/subjects/sh2007001468

Classification: LCC TH880 .M58 2025 (print) | LCC TH880 (ebook) | DDC 624.028/6--dc23/eng/20250520

LC record available at https://lccn.loc.gov/2025013500

LC ebook record available at https://lccn.loc.gov/2025013501

British Cataloguing in Publication Data

A Cataloguing in Publication record for this book is available from the British Library.

Table of Contents

Detailed Table of Contents

This chapter explores emerging trends for addressing unsustainable practices in the construction process through the use of modern approaches to technology, regulations, and management. It offers an assessment of how digital technology, development and promotion of sustainable materials, and emphasis on the circular economy of industry practices limits (or seeks to mitigate) environmental and socio-economic impacts of the built environment. The role of regulations, government policies, international sustainability standards, and the capacity of all construction stakeholders to adopt sustainable practices are critically assessed. After identifying pathways toward action and key constraints to sustainable change, this research provides direction on how the construction sector address the opportunities and challenges of global sustainability in association with "smart" development and industry projects characterized by resilient practices that are more efficient and ecologically sound.

In-house engineering can be a powerful driver of innovation and sustainability in the construction industry. The authors examine the strategic choice between internalizing engineering services and outsourcing them and how it influences a firm's ability to innovate and meet sustainability goals. Integrating the four established theoretical perspectives (Transaction Cost Economics, Resource-Based View of the Firm, Dynamic Capabilities, and Absorptive Capacity), they develop a decision framework providing insights into the contexts and situations when in-house engineering adds the most value, becoming the preferred choice for construction firms. They start with an overview of industry trends, showing a clear shift from traditional outsourcing toward more integrated models as firms seek long-term competitive advantage. Then, they integrate decision criteria (cost, risk, innovation potential, sustainability, knowledge retention, and governance considerations) into a unified framework and illustrate them with case studies of major construction and engineering firms.

AI presents a paradox in business sustainability, addressing high-end environmental and social challenges while posing risks of algorithmic bias and ethical concerns. Sustainable AI requires balanced, risk-focused regulation that maintains technology neutrality and proportionality. Without oversight, AI can harm the economy, society, and the environment. Strong policies and enforcement are vital for ethical AI management, ensuring it serves as a common good. Integrating AI with blockchain enhances transparency, traceability, and accountability in supply chains, driving sustainable industrial transformation and carbon accounting. However, governance must align with environmental interests to mitigate socio-ecological risks. Inclusive policy norms counteract information asymmetry and promote equitable benefits.

This chapter reviews the integration of agricultural stubble, including rice straw, wheat straw, maize stalks, and sugarcane bagasse, into sustainable construction materials. It examines how stubble enhances mechanical, thermal, and acoustic properties in concrete, bricks, plaster, and particleboards, while addressing environmental

This chapter explores artificial intelligence (AI) and machine learning (ML)-driven processes in transactions that detect optimization, emphasizing forecast analysis, auto tomatoes, and data-based decisions. It investigates how AI/ML models increase the accuracy of planning, resource allocation, reducing risk and real-time monitoring of construction activities. By integrating history with intelligent algorithms, construction workflow can be trimmed for more efficiency, cost-effectiveness and durability. Case studies and emerging technologies highlight AI's transformative prospects in eliminating industry challenges such as delays, budget overruns, and safety risks. The chapter outlines practical structures and future directions for the implementation of AI and ML to optimize the construction project exhibition from the end to end.

Global trade considerably depends on maritime transportation, yet these activities generate substantial greenhouse gas (GHG) emissions, raising critical concerns among policymakers about climate change. In response, the International Maritime Organization (IMO) has implemented stringent regulatory measures towards the construction of energy-efficient ships and operations to reduce carbon emissions from both new and existing vessels. This chapter explores key IMO approaches, including the Energy Efficiency Design Index (EEDI), the Energy Efficiency Existing Ship Index (EEXI), the Ship Energy Efficiency Management Plan (SEEMP), the Energy Efficiency Operational Index (EEOI), and the Carbon Intensity Indicator

this chapter. The use of low-carbon cement, recycled composites, passive design, prefabrication, and AI is changing the construction field. At the same time, the high expenses, strict rules, and reserved attitude among investors delay the industry's progress. It points out that policy incentives, financial instruments, and effective regulations are the most important for a change to green and efficient construction.

This chapter will explore the dynamic intersection of nature-based design, cutting-edge technological solutions, and interdisciplinary collaboration in the pursuit of sustainable architecture. In an era marked by increasing environmental challenges, the integration of natural systems with advanced technologies offers a transformative approach to building design that can significantly reduce resource consumption, enhance resilience, and promote ecological stewardship. The chapter will provide a comprehensive overview of how these elements—nature, technology, and collaboration—interact in practice, offering case studies, theoretical insights, and practical recommendations for architects, engineers, and other design professionals working together to achieve sustainability goals in the built environment.

Cities are undergoing a profound transformation driven by technological advances such as artificial intelligence (AI), the Internet of Things (IoT), and digitalisation, which are redefining how people interact with their urban environment. These changes underscore the need to design more sustainable, inclusive, and resilient cities, capable of balancing technological innovation with social well-being and environmental protection. In this context, the chapter offers a comprehensive analysis of emerging urban design paradigms within the framework of smart and sustainable cities. Based on a systematic literature review and a multidisciplinary perspective, it explores the intersections between technology, sustainability and urbanism, with special attention to key issues such as cybersecurity, resilience, and the circular economy. It also identifies innovative strategies, including green and blue infrastructure, humanistic and inclusive urbanism, and hybrid and modular housing. Finally, it proposes actions to foster urban spaces that are resilient, accessible, and aligned with contemporary challenges.

Chapter 12

Lhousileno Shuya, School of Planning and Architecture, India
Manika Goel, School of Planning and Architecture, India

Streets are vital to a city's social, cultural, and economic life, serving as public spaces that enable movement and interaction. However, in many cities, including Delhi, streets have been overtaken by vehicular traffic, losing their public and social function. Chandni Chowk, once a vibrant heritage street, is now facing congestion, overcrowding, and a decline in its historic character. Increased commercial activity has further worsened traffic issues and reduced pedestrian safety. This study investigates the challenges faced by Chandni Chowk's streetscape and explores what urban streets should ideally offer. Using 25 evaluation parameters, supported by literature on street design and standards, the research involves on-site observations, interviews, and the application of design guidelines. The study concludes with an analysis of potential design benefits and offers recommendations to revive the street's livability and heritage value.

Chapter 13

Deeksha Khati, KR Mangalam University, India
Manika Goel, School of Planning and Architecture, India

Urban street design plays a vital role in shaping sustainable and inclusive cities by promoting walkability, connectivity, and placemaking. This study examines Vikas Marg in Delhi as a case of equitable road space allocation and effective urban design. Drawing on principles from The Image of the City, Theory of a Good City, and A Hidden Dimension, the research highlights the street's transformation toward legibility, inclusivity, and environmental quality. Through spatial analysis, stakeholder observations, and design evaluation, the study identifies initial challenges such as traffic congestion, inadequate pedestrian facilities, and lack of greenery. The redesigned street demonstrates how integrated design interventions can create vibrant, people-centric public spaces that enhance mobility, social interaction, and urban livability.

Preface

INTRODUCTION

As editors of *Mitigating Unsustainable Practices in Construction and Architecture*, we present this volume in response to one of the most pressing challenges of our time: the urgent need to transition the construction industry toward sustainability, resilience, and responsibility. Today, the global community is striving for safety, inclusion, and environmental stewardship by embracing circular economy principles and green technologies. Yet, construction—one of the most influential industries in shaping our societies—must take a leading role in facilitating these processes and supporting the green and digital recovery through the adoption of clean energy and sustainable practices.

The construction sector, with its complex web of activities, inevitably carries significant economic and environmental burdens. This diffuse unsustainability stems from multiple, interrelated factors: the necessity for new infrastructure, particularly in developing regions; the rehabilitation and renovation demands of ageing building stocks; energy inefficiency; and the scarcity of natural resources. As the world continues to urbanize, the demand for energy in built environments intensifies, resulting in substantial emissions and contributing to climate instability.

A shift toward energy-efficient and decarbonized built environments is therefore essential. Such a transformation underpins global efforts to achieve climate neutrality, enhance economic resilience, and prepare societies to withstand future scenarios shaped by climate change and energy crises. In recognition of this, governments and institutions worldwide are embedding urban energy efficiency, carbon neutrality, and sustainable construction within policies and regulatory frameworks to advance both mitigation and adaptation strategies.

This book compiles a diverse collection of proposals, studies, research insights, and practical actions that can be applied broadly across the built environment. Its thematic focus spans sustainable architectural and engineering design, innovative technologies, low-energy and energy-efficient buildings, indoor environmental quality, green construction practices, multifunctional and eco-friendly materials, nanotechnology applications, and life cycle assessment approaches.

We envision this volume as a resource for researchers, academics, professionals in engineering, architecture, environmental studies, and public health, as well as graduate and postgraduate students. By providing knowledge, methods, and real-world examples, this book seeks to inspire the adoption of sustainable practices in construction and architecture and contribute meaningfully to the global transition toward a cleaner, more resilient, and sustainable future.

Organization of the Book

Chapter 1, *A Study on Emerging Trends in Mitigating Unsustainable Practices in the Construction Process: Innovations, Regulations, and Strategic Pathways*, examines how modern approaches in technology, regulations, and strategic management are reshaping the construction industry toward sustainability. The chapter evaluates the role of digital tools, sustainable materials, and circular economy principles in reducing environmental and socio-economic impacts, while also addressing the influence of regulations, international standards, and stakeholder readiness. By identifying opportunities and constraints, this contribution highlights actionable pathways for advancing resilient, efficient, and ecologically sound practices in construction and development projects.

Chapter 2, *In-House Engineering as a Catalyst for Sustainable Innovation in Construction*, explores the critical role of internal engineering capacity as a driver for innovation and sustainable growth. Drawing upon Transaction Cost Economics, the Resource-Based View, Dynamic Capabilities, and Absorptive Capacity, the chapter introduces a decision-making framework to help firms evaluate when internal engineering offers greater value than outsourcing. Illustrated through case studies, this discussion demonstrates how in-house approaches enhance knowledge retention, innovation, and governance while reducing risks, positioning firms to achieve long-term competitive and sustainability advantages.

Chapter 3, *Role of AI in Mitigating Unsustainable Practices*, discusses the dual potential of artificial intelligence in addressing sustainability challenges while introducing new ethical concerns. The chapter stresses the importance of balanced regulatory oversight to ensure that AI applications serve the common good without exacerbating risks. By examining AI in combination with blockchain, the authors illustrate how transparency, traceability, and accountability can be strengthened across supply chains, enabling carbon accounting and sustainable industrial practices.

At the same time, this analysis warns that policies must align with environmental interests and promote inclusivity to counteract biases and socio-ecological risks.

Chapter 4, *Stubble-Reinforced Materials for Sustainable Construction: Advancing SDG 9, 11, and 12 Through Waste-to-Resource Innovations*, showcases how agricultural residues such as rice straw, wheat straw, maize stalks, and sugarcane bagasse can be transformed into valuable construction materials. The chapter details improvements in mechanical, thermal, and acoustic performance, while addressing critical issues of waste reduction and carbon emissions. By connecting these innovations to the United Nations Sustainable Development Goals, the discussion emphasizes both environmental and socio-economic benefits, including rural development and cost savings, while also acknowledging technical and regulatory challenges that must be addressed for broader adoption.

Chapter 5, *Energy Losses in Construction Parameters in Multi-Family Dwellings in Madrid, Spain*, investigates air infiltration and energy inefficiency in residential buildings. Through extensive Blower Door Tests, the study identifies key points of energy loss and quantifies renovation rates, highlighting that many existing dwellings fall short of updated technical building code requirements. By linking air permeability with energy consumption, the chapter offers practical insights into potential improvements, their amortization, and the pathways for enhancing energy efficiency in urban housing stock.

Chapter 6, *Prediction and Enhancing Construction Processes Through Machine Learning and Recent Optimization Techniques*, demonstrates how AI and machine learning can transform construction workflows by enhancing forecasting, planning, and resource allocation. Through case studies and emerging applications, the chapter illustrates how intelligent algorithms reduce risks, optimize real-time monitoring, and improve efficiency, cost management, and safety. By proposing future directions for AI/ML integration, this contribution highlights how technology can address chronic industry challenges such as delays and budget overruns, ultimately paving the way for smarter, more resilient construction processes.

Chapter 7, *CO2 Emission Calculations in Shipping for Sustainable Ship Design, Construction, and Operation: Assessing Strategies and Regulatory Tools for Low-Carbon Maritime Innovation*, extends the sustainability discourse to maritime construction and operations. With global trade heavily dependent on shipping, the chapter assesses the role of International Maritime Organization measures—such as EEDI, EEXI, SEEMP, EEOI, and CII—in reducing greenhouse gas emissions. By combining regulatory frameworks with quantitative analyses of real-world emissions, this contribution demonstrates the necessity of integrating sustainable design and operational practices to drive low-carbon innovation in the shipping industry, reinforcing maritime's role in global sustainability transitions.

Chapter 8, *Smart and Sustainable Strategies for Urban Energy Efficiency: An Integrated Framework*, emphasizes the growing importance of energy efficiency in cities as they become the focal points of consumption and carbon emissions. The chapter introduces a five-pillar framework—smart infrastructure and data, policy ecosystems, innovation platforms, governance coordination, and citizen engagement—illustrated through global case studies from Sweden, Germany, the UAE, Saudi Arabia, and the UK. With examples of emerging technologies such as AI, digital twins, edge computing, and vehicle-to-grid systems, the chapter identifies just energy transition values including equity, resilience, and digital inclusion. It concludes with practical planning recommendations and calls for future research in behavioral science, multisector integration, and adaptive governance, particularly for cities in the Global South.

Chapter 9, *Advancing Smart and Sustainable Construction: Integrating Circular Economy, Energy Efficiency, and Economic Viability in the Built Environment*, underscores the immense impact of the construction sector on global emissions and resource consumption. While the industry has begun adopting sustainable products, circular models, and digital applications, progress remains slow due to high costs, strict regulations, and investor hesitancy. Reviewing innovations such as low-carbon cement, recycled composites, prefabrication, and AI-driven tools, the chapter illustrates how energy-efficient practices can redefine the built environment. It emphasizes the critical role of targeted policy incentives, financial mechanisms, and clear regulations in accelerating the transition to a green, efficient, and economically viable construction industry.

Chapter 10, *Sustainability in Practice: The Intersection of Nature, Technology, and Collaborative Design*, explores how the convergence of nature-based systems, advanced technologies, and interdisciplinary collaboration can revolutionize sustainable architecture. By presenting case studies and theoretical perspectives, the chapter demonstrates how this triad reduces resource dependency, fosters resilience, and nurtures ecological stewardship. It positions collaborative design practices as central to realizing sustainability goals and offers practical recommendations for architects, engineers, and allied professionals to integrate ecological principles with cutting-edge solutions in real-world projects.

Chapter 11, *Sustainable Construction Strategies that Guarantee a Humanistic, Inclusive, and Technologically Innovative Urban Design*, reflects on how digitalization, artificial intelligence, and the Internet of Things are transforming cities into smart and interconnected spaces. Recognizing the risks of imbalance between technological progress, social inclusion, and environmental integrity, the chapter proposes urban design strategies that prioritize resilience and inclusivity. Through a multidisciplinary review, it discusses cybersecurity, circular economy, green and blue infrastructure, modular housing, and human-centered design. The chapter offers a vision of urban environments that are technologically innovative while remaining accessible, secure, and aligned with the broader goals of sustainability and human well-being.

Chapter 12, *Towards Inclusive and Sustainable Urban Streetscapes: Lessons from Chandni Chowk's Revitalization*, examines one of Delhi's most historic streets and its struggle with congestion, overcrowding, and loss of heritage character. Using 25 evaluation parameters alongside on-site observations and stakeholder interviews, the study identifies how commercial pressures and vehicular dominance have undermined livability and pedestrian safety. The chapter concludes by offering design recommendations to restore Chandni Chowk's identity as a vibrant public space while balancing heritage conservation, mobility, and social interaction, thus providing a blueprint for sustainable streetscape renewal in dense urban contexts.

Chapter 13, *Designing for All: A Road Space Equity Study of Vikas Marg, Delhi*, demonstrates the role of equitable road space allocation in advancing inclusive and sustainable cities. Drawing from spatial analysis, stakeholder perspectives, and design principles from seminal urban theories, the chapter documents challenges such as traffic congestion, inadequate pedestrian pathways, and insufficient greenery. The redesign of Vikas Marg is presented as evidence of how integrated interventions can transform a traffic-dominated corridor into a vibrant, people-oriented street that enhances mobility, environmental quality, and community life.

Chapter 14, *Integrating BGI into Urban Development: A Case of Indian Cities*, highlights the role of Blue-Green Infrastructure (BGI) in addressing the environmental pressures of rapid urbanization. By analyzing case studies from Bengaluru, Mumbai, London, and Copenhagen, the chapter identifies how BGI fosters resilience, sustainability, and public health in diverse contexts. It stresses the importance of local adaptability, regulatory support, public awareness, and multi-stakeholder collaboration for successful implementation. The findings advocate for Indian cities to embed BGI within urban policy and planning, enabling a balance between development needs and ecological conservation.

Chapter 15, *Sustainable Innovation for Pothole Detection and Road Maintenance*, introduces a frugal, technology-driven model for improving road conditions in developing regions. Using Arduino-based sensors and IoT applications, the system detects potholes, measures depth, and automates filler application for repairs. A complementary mobile application allows remote monitoring and configuration, ensuring efficient resource allocation. The chapter positions this approach as a sustainable innovation that enhances road safety, reduces economic losses, and improves infrastructure maintenance. By combining robotics, automation, and affordability, it demonstrates how context-sensitive technology can serve both social and environmental goals in transportation management.

CONCLUSION

This volume, *Mitigating Unsustainable Practices in Construction and Architecture*, brings together diverse perspectives, research findings, and innovative solutions that collectively highlight both the urgency and the opportunity of transforming the built environment. From strategies to reduce emissions and integrate circular economy principles, to the adoption of advanced digital technologies, nature-based solutions, and inclusive design practices, the chapters demonstrate how sustainability in construction and architecture is no longer a niche aspiration but a fundamental necessity.

What emerges throughout these contributions is a clear recognition that meaningful change depends on a convergence of factors: supportive regulatory frameworks, technological innovation, interdisciplinary collaboration, and a commitment to equity and resilience. The case studies, frameworks, and methodologies presented in this book underscore that sustainable transformation is not only technically possible but also economically and socially beneficial when pursued with foresight and cooperation.

We hope this collection inspires researchers, practitioners, policymakers, and students to continue advancing knowledge and implementing practices that reduce the environmental burden of construction and architecture while enhancing the quality of life for present and future generations. It is only through shared responsibility, bold innovation, and sustained commitment that we can truly reimagine the built environment as a driver of sustainability, resilience, and human well-being.

Chapter 1

A Study on Emerging Trends in Mitigating Unsustainable Practices in the Construction Process: Innovations, Regulations, and Strategic Pathways

Vaibhav Singh
https://orcid.org/0000-0003-4092-223X
Department of Civil Engineering, Shri Vaishnav Vidyapeeth Vishwavidyalaya, Indore, India

Ashok Kumar N. Rajanavar
https://orcid.org/0000-0002-0777-8566
Department of Civil Engineering, Walchand Institute of Technology, Solapur, India

R. Somasundaram
https://orcid.org/0000-0002-0377-9094
Department of Management Studies, Kongu Engineering College, Erode, India

Somu Chinnusamy
Department of Research and Development, RSP Science Hub, Coimbatore, India

Shruti Laxmikant Gore
https://orcid.org/0009-0006-5440-0538
Department of Engineering Sciences and Humanities, Vishwakarma Institute of Technology, Pune, India

M. Kumar
https://orcid.org/0009-0006-2989-9883
SRM University, India

DOI: 10.4018/979-8-3373-2555-2.ch001

ABSTRACT

This chapter explores emerging trends for addressing unsustainable practices in the construction process through the use of modern approaches to technology, regulations, and management. It offers an assessment of how digital technology, development and promotion of sustainable materials, and emphasis on the circular economy of industry practices limits (or seeks to mitigate) environmental and socio-economic impacts of the built environment. The role of regulations, government policies, international sustainability standards, and the capacity of all construction stakeholders to adopt sustainable practices are critically assessed. After identifying pathways toward action and key constraints to sustainable change, this research provides direction on how the construction sector address the opportunities and challenges of global sustainability in association with "smart" development and industry projects characterized by resilient practices that are more efficient and ecologically sound.

INTRODUCTION

The construction sector is a widely accepted economic powerhouse driving economic development, infrastructure development, resource users and waste producers. The construction industry is not only a key driver of economic development, but it is also resource-intensive and has a significant environmental impact. Globally, the construction industry accounts for a significant share of total energy consumption, natural resource use, greenhouse gas emissions and waste production. The United Nations Environment Program states that the construction and building sector accounts for approximately 37% of global CO_2 emissions (El-Hakim & AbouZeid, 2024). Technology and advocacy policies cannot seem to overcome the unsustainable practices in terms of inefficiencies in material use, energy use, poor waste disposal practices, and harm to the ecology present in the various phases in the construction life cycle. As the call for sustainable development grows louder in the context of climate change, resource depletion and social inequality, the trends that are advocating for a more sustainable building process must be examined.

The idea of sustainability in construction is not new, however, the ways in which it is addressed have changed dramatically in recent years. Responsiveness to unsustainability may look like: compliance with rules and regulations, environmental assessment; these types of responses are no longer enough on their own in isolation. The sector is progressing into more systemic and proactive strategies like lifecycle thinking, digitalization, decarbonization and circularity. The recently emerging trends where unsustainable practice is being mitigated, demonstrate this shift and show how construction stakeholders, for instance: developers, designers, contractors, regulators etc., are embracing innovation, re-thinking governance and resetting business models to become more sustainable. These trends demonstrate new technologies, but are also institutional and behavioral, merging engineering solutions with the will to change through policy instruments and cultural change (Agyemang et al., 2024).

Technology has become one of the primary disruptors advancing sustainable construction. With the advent of Building Information Modeling (BIM), Artificial intelligence (AI), digital twins, and Internet of Things (IoT)-based monitoring, are changing how we plan, construct, and manage a building post-occupation more efficiently and accurately. Information modeling and advanced monitoring allow stakeholders to simulate environmental damage, optimize material flows, analyze energy performance, and develop redundant options, substantially lowering emissions and resource waste. Additionally, many innovations in material science, such as low-carbon cement, recycled composite materials, self-healing concrete, and bio-based materials, are changing the environmental impact of the components used in construction (Musonda et al., 2025). These innovations increase performance and resource conservation while reducing the embedded energy in building systems.

In tandem with technical advances in sustainability, regulatory frameworks will become more comprehensive and harmonized towards sustainability objectives. For example, governments and intergovernmental organizations are creating ambitious carbon-neutral commitments, codifying green building certification, and providing fiscal incentives for sustainable practices (Satyam & Sharma, 2024). Earth-friendly policies, like extended producer responsibility (EPR), green procurement standards, mandates for energy use and adaptation to climate changes are altering the regulatory landscape. In particular, effective implementation of these policies by regulators, combined with compliance mechanisms for stakeholders and transparency, is essential for successfully reducing harmful environmental behavior. Voluntary standards such as LEED (Leadership in Energy and Environmental Design), BREEAM (Building Research Establishment Environmental Assessment Method), and EDGE (Excellence in Design for Greater Efficiencies) are already beginning to have an impact on project design and delivery processes and inform firms to implement best practices even when not required by regulation.

Along with innovations and regulations, strategic approaches such as sustainable delivery methods, integrated approaches, and collaboration are all becoming very significant. Early contractor involvement (ECI), integrated project delivery (IPD), and public and private partnerships (PPPs) are being utilized to encourage collaboration between stakeholders in the construction value chain, improve risk allocation amongst stakeholders, and create common sustainability goals or objectives (Omopariola et al., 2024). Firms are beginning to adopt Environmental, Social, and Governance

(ESG) criteria in their decision-making processes and reporting mechanisms when assessing their performance as sustainability leaders. They are realizing that their financial performance and reputation is linked directly back to their environmental stewardship. Sustainability mechanisms and tools including lifecycle cost analysis, sustainability performance metrics, and circular economy approaches centered on material reuse, modular building, and planning for deconstruction should be mandatory aspects of sustainability planning.

Notwithstanding these positive advancements, moving to sustainable construction is still met with barriers such as financial limitations, barriers related to technology, a fragmented supply chain, unskilled labor and resistance to change. In developing countries in particular, the upfront costs associated with sustainable technologies and lack of necessary regulations inhibiting mass take-up can act as a barrier (Musonda et al., 2025). Notably, the construction industry is also distracted by a largely short-term and linear production mindset, instead of a systems approach that intends to look at longer-term impacts and outcomes. It will take some considerable effort to navigate through all of these barriers which will include innovation and regulation, but also capacity building, stakeholder engagement, and a shift in perception of success that captures environmental and social outcomes as well as financial outcomes.

In order to comprehend the future path of sustainable construction, it is important to take into consideration how emerging trends interact and change in a broader socio-economic and technological context. For example, the connection between sustainability and digital transformation initiatives—including smart cities, green infrastructure, and energy-positive buildings—highlights the potential for new synergies and opportunities for scale. Similarly, the emergence of sustainability-focused financing mechanisms and climate-related financial disclosures impact investment decisions and project viability. These trends demonstrate the need for multi-dimensional strategies that transcend disciplines and institutions, and position the construction sector not as a barrier but a catalyst for sustainable development (Masyhur et al., 2024).

The goal of this chapter is to thoroughly investigate the evolving practices to reduce unsustainable practices, an organizational process. First, it sets out to provide a map of the current environmental crises linked to the industry, and explain how the sustainability field has developed over time. The chapter then delves into the principal technological, regulatory, and strategic shifts that are impacting the industry's response to these challenges. Further, cases and examples from around the world are provided to represent useful interventions, and lessons learned. Finally, gaps, opportunities, and recommendations for policy action are presented to accelerate the uptake of sustainable science and practice in construction (Oke et al., 2025).

This chapter situates the discussion in a larger framework of environmental responsibility, innovation, and governance and the hope is that it can contribute to conversations about sustainable urban development in both academic and practitioner settings. It aims to underscore the collective nature of this work, the importance of continual learning and adaptability, and that we can create a future where the construction industry meets human needs while respecting planetary boundaries.

Objectives

- To investigate cutting-edge technologies and innovative practices made to mitigate unsustainable practices in the construction process.
- To look at what emerging regulatory frameworks are doing to facilitate sustainable construction practices.
- To look at strategies which integrate environmental, economic, and social sustainability throughout the construction project life cycle.

BACKGROUND: UNSUSTAINABLE PRACTICES IN CONSTRUCTION

The construction sector has played, and will play, an important role in urbanization, and the development of infrastructure and economic growth; unfortunately, it has been associated, and often labeled, as unsustainable in its forms, and incompatible with the environmental, economic or social needs of its time. The demand for infrastructure in residential, commercial and industrial terms is increasing worldwide, particularly in rapidly urbanizing economies - this will accelerate the ecological footprint of construction projects (Naser & Kodur, 2025). Traditional approaches to construction and project delivery often prioritize short-term economic efficiency of long-lived projects may lead to building practices that conflict or are inconsistent with responsible development practices. To fully articulate the urgency and necessity of changing these unsustainable project delivery practices, requires a sophisticated grasp of the environmental implications, socio-economic impacts, and systemic origins. Fundamental elements of unsustainable practices in construction as shown in Figure 1.

Figure 1. Basic elements of unsustainable practices in construction

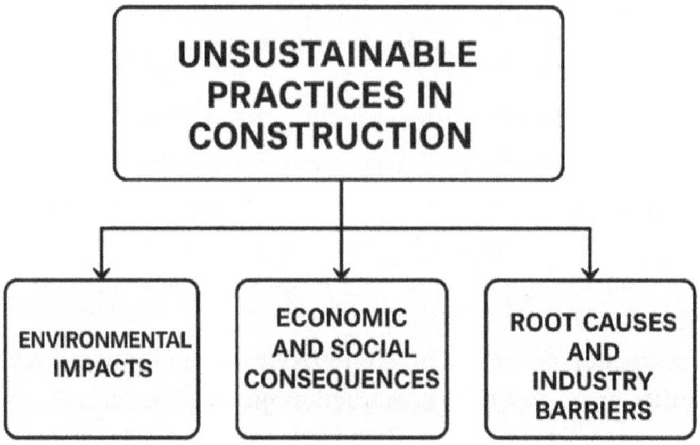

Environmental Impacts

One of the biggest issues with unsustainable building is the harmful effects sustainability-related actions on the environment. The construction industry is among the biggest users of raw materials – nearly 50% of global material extraction is attributed to the construction industry. The construction industry not only consumes enormous amounts of raw materials but is also responsible for vast amounts of carbon emissions and energy use. The International Energy Agency (IEA) estimates that buildings and construction together cause approximately 37% of energy-related CO_2 emissions globally. This represents CO_2 emitted by both the manufacture of construction materials (e.g., cement and steel), the process of building, and from the emissions produced through the lifecycle of a building's operation (Naser & Kodur, 2025; Yadav & Nikalje, 2024).

The overconsumption of non-renewable resources, such as sand, gravel, fossil fuels, and water, intensifies the current degradation of ecosystems. In addition, construction (C&D) waste, which generates up to 40% of the total urban solid waste in some countries, is often overproduced and poorly managed, leading to land pollution, loss of biodiversity, and the extinction of species. Fund objects to ecological balances through habitat destruction, while weak environmental site planning, soil erosion, deforestation, and poor stormwater management have long-lasting impacts. Impermeable surfaces disrupt natural hydrologic cycles, making flooding a reality and depleting groundwater sources.

Economic and Social Consequences

The economic and social costs of unsustainable construction practices are substantial in addition to the environmental damage caused. On the economic side, the inefficient use of materials, poor-quality control leading to rework, energy-inefficient buildings create high operational and maintenance costs. Further, construction pollution and emissions create outside costs to public health and local economies such as increased healthcare spending, decreased worker productivity and economic losses associated with climate-driven damages such as heat waves, flooding, and air pollution episodes (Masyhur et al., 2024).

From a social point of view, the negative externalities associated with construction are experienced disproportionately by marginalized communities. Projects that operate without an adequate social impact assessment have been associated with forced displacement, loss of livelihoods, and disruption to local cultures. Moreover, many construction sites in developing regions still function under unsafe work conditions for workers, poor labor conditions for workers, and wage exploitation, which negates the social sustainability of the industry. Likewise, informal labor in particular, is often excluded from social security benefits, skills development, and regulatory protections in low-income countries (Sharma et al., 2024).

Urban sprawl—a result of poorly planned development—exacerbates socio-economic inequality by developing disconnected, car-dependent neighborhoods with little access to services, employment or amenities. The unchecked lack of inclusive and universal design in construction makes it difficult for persons with disabilities, the elderly and other marginalized groups to access spaces and services.

Root Causes and Industry Barriers

There are a number of structural and operational factors that reinforce the persistence of unsustainable practices in the construction industry. The first is the construction value chain's fragmentation, which typically has multiple actors and actors—planners, architects, contractors, suppliers—often working in silos with conflicting incentives. This disjointed methodology stifles the potential for integrated sustainability strategies and obscures collaboration across project phases (Lucatello & Alcántara-Ayala, 2024; Masyhur et al., 2024).

Thirdly, limited awareness and technical capacity also still represent an issue. Many developers and contractors, particularly, from small and medium enterprises (SMEs), are not aware of sustainable materials, sustainable techniques, or existing certifications that demonstrate sustainability or sustainable practice. Even when they know there are options, technical and financial constraints may prevent developers from actually adopting, or implementing, a more sustainable alternative.

Fourthly, cost and time pressure often dominates decision-making. While sustainable construction may cost more up front—because of sourcing of green materials, engaging in certification or utilizing specialized labour—developers often revert to non-sustainable procurement options to maintain competitiveness and timelines. Upon reflection, these practices may reduce immediate costs, but in the long run, they can result in negative financial implications related to increasing energy use, ongoing maintenance, and environmental costs.

There is also limited enforcement of both regulatory methods to advance sustainability and inconsistent policies. While some countries have green building codes, or have standards for environmental performance, enforcement has ranged from ineffective to sporadic. In many instances (e.g., social procurement standards), the regulations for sustainability are voluntary, or exist at a municipal level leading to inconsistencies of their adoption across and/or within markets. Additionally, there are everyday gaps in regulations that affect sustainability, such as in construction waste management, lifecycle emissions reporting, and sustainable procurement standards; directly impacting the potential strength and effectiveness of sustainability-oriented regulations.

Resistance to innovation and to change also creates hurdles. The construction industry is traditionally risk-averse and thereby slow to implement new technologies or new business models. Cultural inertia, along with no incentives for innovation, limits the adoption of digital tools, renewable energy systems, and/or modular construction practices that could lead to sustainability (Naser & Kodur, 2025; Oke et al., 2025).

Finally, the dynamics of the market and investor expectations are also factors. Developers and investors focus largely on short-term returns, and investments associated with sustainability do not automatically translate to short term financial performance gains. Also, if there is not an explicit market demand for sustainable infrastructure or if clients do not pay a premium for green features, developers could continue to be hesitant to shift focus.

Unsustainable construction practices arise from a range of structural inefficiencies; regulatory failures; and economic pressures. The consequences of these behaviors risk environmental damage; greater economic costs; and socio-economic inequities. Their resolution and eventual removal requires a systemic shift in three areas: i) integrated planning and sustainability determination; ii) regulatory reform; iii) industry collaboration between stakeholders, and iv) innovative thinking. Realizing the challenges of these behaviors is the first stage in identifying and then scaling and sustaining behaviors, processes and materials that can reshape construction as a more sustainable, resilient and inclusive sector.

TECHNOLOGICAL INNOVATIONS FOR SUSTAINABLE CONSTRUCTION

The construction sector is in a major transformation due to the urgent need to reduce its impact on the environment and increase resource efficiency. Technological advances are essential to this shift, and key to greater sustainability is the ability of stakeholders to design, build, and operate buildings more sustainably as shown in Figure 2. By combining green materials, digital modeling technologies, and smart technologies, the construction industry is rethinking traditional approaches and aligning with global sustainability goals. This section considers the three major areas of technology relevant to sustainable construction, namely green building materials and smart technologies; Building Information Modeling (BIM) and digital twins; and automation, robotics and artificial intelligence (AI)(Hassan et al., 2024).

Figure 2. Technological innovations for sustainable construction

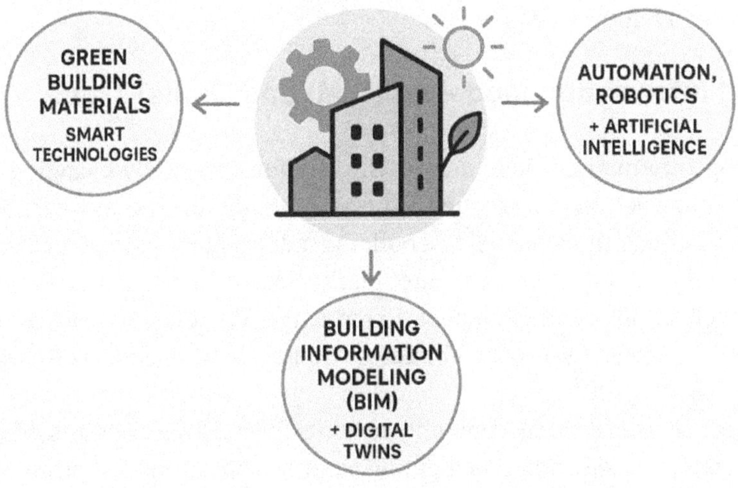

Green Building Materials and Smart Technologies

One of the clearest paths to sustainability in construction is through the use of green building materials. Green building materials are materials that are designed or selected to minimize the environmental impact over their entire lifecycle, from raw material extraction and manufacturing to consumption and disposal. Examples of green building materials include recycled steel, cross-laminated timber (CLT), low carbon concrete, fly ash cement, and bio-based materials, such as hempcrete or bamboo. Green building materials save resources and reduce embodied energy and embodied emissions (Zhang et al., 2024).

Low carbon concrete, for example, replaces a portion of Portland cement with SCMs like slag or fly ash, which results in a significant reduction in CO_2 emissions without sacrificing strength. Cross-laminated timber is renewable and carbon-sequestering, with structural strength similar to steel but fewer environmental impacts. Insulation products, like phase change materials (PCM) and aerogels, are also being used to improve temperature control and energy efficiency of building envelopes (Madan et al., 2024; Singh Madan et al., 2024).

Smart technologies (e.g., energy-efficient lighting, HVAC systems, intelligent building controls) add an additional layer of sustainability by leveraging real-time sensors and resulting adaptive algorithms to monitor and improve energy use, water use, and indoor environmental quality. Smart thermostats, for example, can modify heating or cooling based on occupancy patterns, considerably minimizing energy waste. Water-conserving fixtures and greywater recycling systems are effective sustainable resource management tools, especially in areas with limited water resources or high building density.

Building Information Modeling (BIM) and Digital Twins

Building Information Modeling, or BIM, is fundamentally changing how we design, plan, and deliver construction projects. It is a digital process where designers create a three-dimensional model of a building; additionally, project data in relation to the building's architecture, structure, mechanical systems, and other attributes, can be integrated into a single model. Essentially BIM offers increased visualization, planning, and analysis over a building's life cycle from concept to demolition (Yadav & Nikalje, 2024).

There are also tangible sustainability benefits from BIM as it assists with energy modeling, material selections, daylighting analysis and carbon footprints at the design level. It allows architects and engineers to predict building performance and to compute and refine imprecisions before implementation, leading to less errors and costly rework. Lastly it is a much stronger collaboration tool among team members since data is pushed to one place (the model), together with building performance analyses which improves communication, coordination and less material waste.

A digital twin is an extension of BIM. This provides a live, dynamic copy of the real building through collection of data from sensors, drones, monitoring, etc. Digital twins support real-time monitoring and predictive maintenance, so buildings run at maximum efficiency during their total lifecycle. For example, digital twins can identify energy-intensive systems, track equipment deterioration, or anticipate indoor air quality issues—all of which contribute to better sustainability outcomes.

The synergy between BIM and digital twins promotes data-driven decision-making and lifecycle thinking, encouraging the design of buildings that are not only efficient during construction but also resilient and adaptable during use and decommissioning.

Automation, Robotics, and AI in Sustainable Practices

Automation and robotics represent emerging key enabling technologies of accuracy, speed, and safety in sustainable construction. Robotic systems and devices can now be developed and deployed for many construction-associated activities, including: bricklaying, 3D printing concrete structures, autonomous operability of equipment and assembly of modular components. They can reduce human error, reduce material waste and increase efficiency—particularly in large-scale or repetitive projects (Naser & Kodur, 2025; Zhang et al., 2024).

A cutting-edge recent development pertains to construction 3D printing, which allows to build components geographically in layers using an optimized amount of material. Optimized operational processes with 3D printing also have the advantages of reducing time and labor costs and considerably reduce wastes as structures built with 3D printing would be built with accurate specifications. Robotic systems also operate in hazardous or remote environments and as such, help to remove the risk on-site injuries also creating a safer workplace.

Artificial Intelligence (AI) supplements automation by refining project planning, monitoring, and sustainability evaluations. AI algorithms have the ability to engage with very large datasets that are otherwise impossible to assess—such as those covering material orders, scheduling, and environmental performance aspects—to expose inefficiencies and identify potential solutions. For example, AI platforms are capable of predicting material needs; anticipated equipment downtime; and identifying historic carbon emissions in real time allowing interventions before problems come to fruition (Boopathi et al., 2023).

AI is also being incorporated into predictive maintenance for buildings and infrastructure, which will contribute to asset life cycling and reduce unnecessary replacements. When working with IoT devices, AI systems have the ability to control lighting, ventilation, or irrigation systems according to environmental conditions and occupancy automatically and improve energy savings. Given enough historical and geospatial data machine learning models are helping impact climate adaptation strategies by identifying risks from past events such as flooding, heat waves, or soil erosion—allowing construction teams to assess the most resilient construction materials and systems.

Technological innovations are transforming sustainable construction in various ways. Use of green materials and smart technologies have started to increase the sustainable impacts of construction inputs and operations. Building Information Modeling (BIM) and digital twins have increased construction accuracy, integration, and long-term performance management. Automation, robotics, and artificial intelligence (AI) also streamline and optimize construction efficiency, processes, and predictive capability. These advancements are creating a transition from reactive and fragmented processes towards integrated, intelligent, and sustainable construction ecosystems

Figure 3. Effective regulatory and policy frameworks are critical to guiding the construction industry

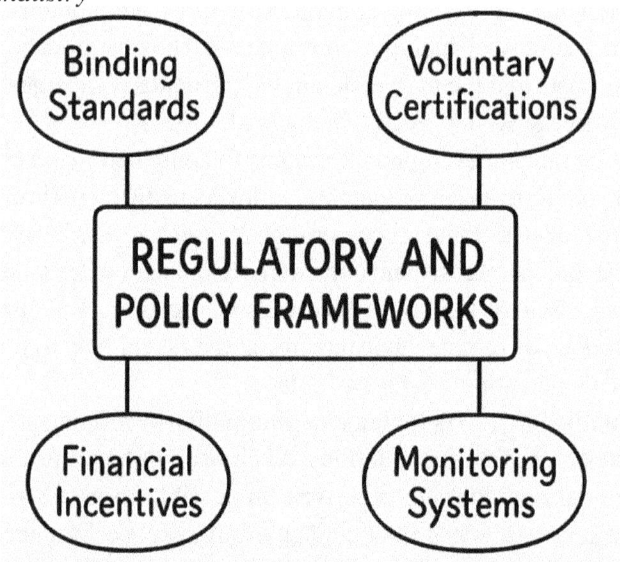

REGULATORY AND POLICY FRAMEWORKS

The existence of effective regulatory and policy frameworks is an important aspect of channeling the construction industry toward sustainable practices. Given the sector's significant environmental impacts, coherent and enforceable policies assist in establishing a process to reduce emissions, protect behaviors for responsible resource use, and improve environmental, social, and economic durability. Through the past 20 years, governmental organizations, international organizations, and industry bodies have sought to value sustainability through 'performance indicators' all while embracing a strategy of creating binding standards, voluntary standards/certifications, financial disincentives, and standards of monitoring. The availability of effective or even relevant regulatory and policy frameworks are heterogeneous and are, to some extent, dependent on local governance capacities, the stage of market maturity, and the prevalence of affordability development (Naser & Kodur, 2025; Yu et al., 2024). Figure 3 depicts effective regulatory and policy frameworks are critical to guiding the construction industry

National and International Standards

National and international standards play a foundational role in shaping sustainable construction. These standards set the minimum requirements or best practices for environmental performance, energy efficiency, material selection, safety, and lifecycle assessment. They provide a regulatory baseline and a reference point for benchmarking progress toward sustainability goals (Firoozi et al., 2024; Naser & Kodur, 2025).

Globally there are several standards and conventions related to sustainability in the built environment. Examples are the ISO 14000 series on environmental management systems, and ISO 21930 (Sustainability in building construction). There are guidance documents, too, for integrating environmental considerations during a build's lifecycle. If we are considering the differences of scale between built projects and politics, we would mention the United Nations Sustainable Development Goals (SDGs), wherein SDG 11 (Sustainable Cities and Communities) and SDG 13 (Climate Action) is intended to compel governments and industry to plan their construction and built environment in context of a wider development agenda and climate action framework.

Regional bodies like the European Union (EU) have been proactive in legislating sustainability in the built environment. The Energy Performance of Buildings Directive (EPBD) requires the establishment of near-zero energy buildings (NZEBs), energy performance certificates and audit requirements for buildings, putting pressure on the member states to raise the quality of buildings and reduce emissions during operations. Governments at the national level will create building codes, and environmental regulations based on their context. For example, the International Green Construction Code (IgCC) and ASHRAE standards in the United States will be examples of codes that make energy-efficient, high-most effective actions objectives in building design. Although not an extended list, India's Energy Conservation Building Code (ECBC) and Singapore's Green Mark Scheme are all examples of regional sustainability and policy approaches, but nonetheless, they serve an equivalent plan if not a strategy in approaches to support sustainability objectives globally - even in a functional category. The same can be said for the National Construction Code (NCC) in Australia.

Incentives, Certifications, and Compliance Mechanisms

Incentive-based strategies are becoming increasingly common in accelerating the adoption of sustainability practices in construction. These strategies include financial incentives such as the use of tax credits, subsidies, grants, low-interest loans, and green bonds to offset the high upfront costs of a green technology or sustainable materials. Many countries provide rebates for the installation of solar energy systems, energy-efficient HVAC systems, and green roofs, for example. Public-private partnerships (PPPs) are also leveraged to mobilize investment in green infrastructure projects (Musonda et al., 2025; Omopariola et al., 2024).

Market-based instruments of carbon pricing and emissions trading schemes (ETS) are increasing in popularity in the built environment, asking firms to either reduce emissions or buy offsets for those emissions. For example, established cap levels for emissions in regulated markets mean that buildings that exceed the established levels may have to pay a penalty and possibly buy carbon credits, this gives economic rationale to investment in sustainability (Narra et al., 2024; Vellingiri et al., 2024).

Certification schemes are key instrumental tools to mainstream sustainability and provide voluntary albeit important benchmarks for buildings. Programs such as:

- LEED (Leadership in Energy and Environmental Design)
- BREEAM (Building Research Establishment Environmental Assessment Method)
- EDGE (Excellence in Design for Greater Efficiencies)
- WELL Building Standard
- DGNB (German Sustainable Building Council)

They provide the framework for assessing the environmental, health and socially responsible behaviours of buildings. They enable transparency, foster innovation, and provide market value to sustainability credentials for developers with investors, tenants and regulators.

Compliance mechanisms typically consist of permitting processes, routine audits and inspections, and performance-based codes that require some verification in design, construction, and operational phases. Digitized compliance systems are also becoming popular alternatives to streamline processes, ensure transparency, maximize flexibility, and support real-time monitoring.

Policy Challenges in Developed vs. Developing Nations

While the number of standards and incentives are growing, there continues to be significant differences in effectiveness of sustainability policies and strategies between developed and developing countries as a result of varying priorities, institutional capacity, and economy. In developed countries, the presence of robust governance structures, mature markets, and greater public awareness of sustainability issues tends to produce a much greater ability to enforce standards for sustainability. Policy makers in those countries have progressed from simply ensuring compliance with building regulations, to more ambitious targets including net-zero emissions, circular construction, and smart cities, among others. Even so, developing the appropriate institutional capacity to ensure compliance with green building regulations, providing incentives for deep retrofits, and balancing between innovation and affordability, are some of the remaining challenges to sustainability policy even in developed countries (Babalola & Harinarain, 2024).

In developing countries, the list of challenges is much longer. One primary challenge in construction in low-income countries is the overwhelming momentum of urbanization, where governments are faced with the challenge of rectifying infrastructure deficits, often within a constricting public budget environment, despite their attempts at fiscal prudence. With limited public funds, many governments will always position quantity over quality. The relatively high upfront cost of design options, building materials, and building technologies associated with sustainable construction, and the lack of private and public sector financing, are challenges in ensuring implementation of green building codes and policies in developing countries. Limited institutional capacity and adequate training also limits the ability of governments to implement and enforce codes within the construction sector. The limited influence of regulatory bodies in low-income countries also relates to the potential size and significance of the informal construction sector, where construction often occurs without procedures or controls regulated by governmental authorities (Musonda et al., 2025).

Another regular issue learned in developing countries is the fragmentation of policy frameworks. Multiple ministries may be responsible for various aspects of construction (e.g. housing, energy, environment). As a result, they can have overlapping mandates with different standards. The lack of coordination from regulatory bodies and technician training means there is little opportunity to translate policy into practice. However, many countries in the Global South are making progress. For example, Rwanda and Kenya have developed green building guidelines appropriate for local climate and economic conditions. India's Smart Cities Mission aims to blend sustainable urban planning and smart city initiatives with digital forms of governance. Colombia and Mexico are even harnessing some of the ferocity of international climate finance to scale energy-efficient housing. Global support mechanisms such as the Green Climate Fund (GCF), UN Habitat, and the World Bank's EDGE program would service policy development, knowledge transfer and technical support for projects and sustainable construction in emerging markets.

The necessary regulatory and policy frameworks to assist the construction industry with the transition to sustainability are essential. Regulations and standards help create frameworks and consistency, incentives and certifications foster voluntary endorsement and innovation, and enforcement (or compliance) mechanisms help to hold the industry accountable. However, using regulations to bridge the policy-performance gap, particularly in the developing world, requires context and strategies tailored appropriately, institutional capacity, and international cooperation. As the climate crisis escalates, impacts increase, and resource supply becomes uncertain, developing policies that are smart, inclusive and enforceable will be critical to achieving sustainable built environments.

Figure 4. Strategic pathways for sustainability integration

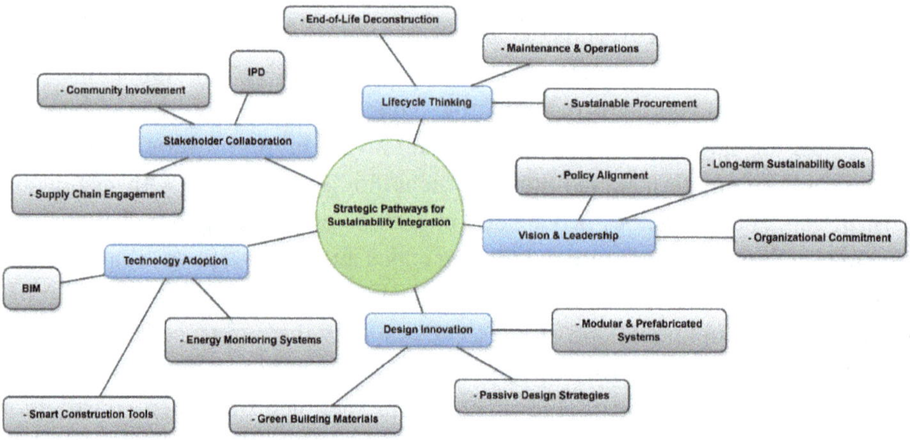

STRATEGIC PATHWAYS FOR SUSTAINABILITY INTEGRATION

Incorporating sustainability into construction must be more than technological upgrades or meeting regulations; it will require a radical departure from project inception, planning, delivery, and management. Environmental, social and economic considerations are interconnected throughout the value chain in construction, and new strategies are taking shape that allow for a more holistic and systems-based approach to sustainable development. In this regard, the circular economy, sustainable procurement and supply chain moves, and integrated project delivery supporting environmental, social and governance (ESG) frameworks have the potential to fundamentally change the way we design, manage, and operate built environments

(Nooradzhar et al., 2024). Together, these practices share a common goal of waste reduction, resource flow optimization, engagement of stakeholders, and ultimately delivering long-term value beyond the traditional dollar in the form of durability, maintainability and longevity over time. Strategic pathways for sustainability Integration as shown in Figure 4.

The circular economy is one of the most significant frameworks moving sustainable construction away from a traditional linear "take-make-dispose" model—toward a closed loop system where materials, energy, and resources are reused and regenerated. In construction, this means designing buildings that can be disassembled and assembling based on re-used or recycled materials, prolonging infrastructure life, and reducing waste over the entire life of the project. Circular construction also requires designs to be modular, prefabricated, and encourages design adaptations, which allows the structure to evolve over its life with minimum impacts to the environment. For example, prefabricated modular construction provides sound assembly and disassembly practices which can stop waste occurring on site (and reduce embodied energy). Material passports—digital agreements of building materials and components—are emerging as another prominent form of enabling reuse at the end of a building's life.

Construction and demolition waste represent a significant portion of municipal solid waste; it can constitute as much as 40% in some cases. Poor disposal of construction waste and demolition waste is detrimental to land, can lead to water pollution, and can also cause GHG emissions. When organizations add a circular approach to their project planning and implementing programs, they can effectively mitigate what goes to the landfill and use resources more efficiently. For example, when selective demolition is adopted and the time is taken to identify items that can be recovered for potential secondary use, steel, brick, and timber can be diverted from the waste stream. In addition, by adopting circularity, organizations can also use support climate impacts; as fewer carbon intensive virgin materials can be used, and, instead, the use of less-impact materials such as recycled aggregates and bio-based materials also gets highlighted. It's worth noting that the move to circular construction is also in response to growing investor and consumer demand for responsible environmental behaviors (Folorunso et al., 2024; Reddy et al., 2024).

Another strategic lever for sustainability is sustainable procurement and supply chain management. Sustainable procurement involves applying environmental and social performance criteria in the selection of materials, services, and vendors—not solely based on cost or convenience. Many times, much of the carbon footprint and environmental impact of a project is encapsulated in the supply chain—through the cycle of material extraction, processing and transport—therefore the procurement strategies will help actualize sustainability-related outcomes. These may include purchasing low-carbon materials, certified green products, locally sourced materials, and ethical suppliers who honor labor and environmental standards.

Sustainable procurement also implicates life-cycle thinking: Considering the environmental and social footprint of materials, services and products from cradle to grave—not just their cost and functionality. Procurement policies can set clear sustainability requirements for suppliers; require third-party certifications (such as FSC for timber or EPDs for building products); and even use performance-based contracts. Finally, the availability of digital tools—including supply chain mapping platforms and blockchain—activates transparency, traceability, and risk mitigation, which in turn facilitates more effective assessment and mitigation of sustainability features of procurement practices.

In addition to materials and logistics, shared delivery models and ESG programs can help embed sustainability through the entirety of the construction process. Traditional delivery mechanisms often result in fragmentation, competing motivations, and adversarial relationships between the different parties involved, leading to poor sustainability and inefficient project outcomes. Collaborative delivery models such as Integrated Project Delivery, Design-Build, and Early Contractor Involvement all encourage enhanced collaboration and transparency between project members, and shared responsibility toward the project outcomes. Collaborative delivery models allow sustainability to be embedded in the process as early as the design phase, where integrated decision-making, whole life costing, value engineering, and risk-sharing can both promote environmental goals and save costs.

Coupled with this delivery model, is the growing recognition of the role of Environmental, Social and Governance (ESG) performance as a measure of both organizational and project success. ESG frameworks provide a systematic way of measuring how construction firms are performing against their environmental obligations (e.g., emissions, water use, biodiversity), social performance (e.g., working conditions, community participation), and governance performance (e.g., ethics, compliance, accountability to stakeholders). ESG reporting is becoming mandatory for investors, regulators, and clients which does affect access to finance and market position. Construction firms adopting an ESG approach are likely to adopt operational practices which will attract sustainable investment, diminish reputational risk, and generate long-term value returns to shareholders and society (Babalola & Harinarain, 2024; Umana et al., 2024).

The inclusion of ESG in project planning also encourages firms to use tools such as sustainability reporting technologies (e.g., GRI, SASB), carbon disclosure, and materiality frameworks into project planning. These frameworks assist with data-leading decision-making, performance measures and allowing for continuous improvement. It is also important to highlight that ESG strategies are more than compliance but instead a striving culture of sustainability that can focus on how they are governed, designed, operated and interacted with wider stakeholders.

Finally, there are strategic pathways of adopting circular economy, sustainable procurement, and delivering projects in alignment with the ESG, which provide holistic and forward looking frameworks for embedding sustainability into construction practice. These strategies go further than just individual or ad-hoc relations, and instead can enable a transformational shift of the way construction projects are implemented and assessed. While there are obstacles and limitations to implementation, including cost constraints, capacity issues and resistance to change, the long-term outcomes are substantial, such as resource flow efficiency, regulatory compliance, stakeholder trust and competitive advantage.

CASE STUDIES AND GLOBAL BEST PRACTICES

The push to integrate sustainable practices in construction has gained exponential momentum globally, with increasing urban and infrastructure projects that showcase innovation, resilience, and environmental sustainability. There is no "one size fits all" approach to sustainability as approaches taken by governments, private developers, and urban planners are appropriate to the geography but also contribute to the ultimate goal of sustainability. As a result, they share best practices provide valuable insight related to what effective implementation looks like and what conditions allow this to occur (Ajirotutu et al., 2024; Likita et al., 2024).

Masdar City in the United Arab Emirates (UAE) was envisioned as one of the most sustainable cities globally and is recognized as an exemplar of sustainable urban development. It uses passive design, renewable energy and efficient water management to consider and subsequently impact its ecological footprint. The orientation, shading features, and smart grid manage their energy consumption, while usability of solar photovoltaic panels and wind towers is utilized to prioritize self-sufficiency. Masdar City was designed as a car-free development and relies on walking, as well as electric public transport, to provide accessibility features like all sustainable cities while espousing a car-free lifestyle. While Masdar City is still in development it acts as a living laboratory for green technologies and sustainable infrastructure in the desert (Moradi & Sormunen, 2024a; Orieno et al., 2024).

Over in Europe, Stockholm, Sweden illustrates how sustainability can be integrated into urban redevelopment. This eco-district utilizes closed-loop systems for energy, water, and waste treatment. The buildings are constructed from high-efficiency materials and linked to a central biofuel district heating facility. Wastewater is reclaimed, while food and organic waste is turned into biogas. Public transport and walking-friendly transitions help to further mitigate emissions while successfully engaging the community. The project is an example of how comprehensive sustainability outcomes can result from integrated urban planning, community engagement, and collaboration across sectors.

Germany has now reached the top of the leadership pyramid for the green construction market on a national scale, based on sound governance, technological advancement and significant industry-wide commitment to sustainability. The DGNB (German Sustainable Building Council) standard is now promoting lifecycle and engineered-based assessment and performance expansion in construction. Addressing the Germany energy transition (Energiewende) nationally to fulfill the broader sustainability objectives. Singapore's Green Mark Scheme has had a very positive exposure and impact on sustainable building design in the Southeast Asia region. The Green Mark Scheme, and similar performance rating tools, enable market transformation by establishing benchmarks and enabling performance analysis in energy efficiency, indoor air quality, and water. It has encouraged the government and the private sector across the region to adopt green building principles and green technologies.

Emerging economies are clearly moving towards context-specific solutions. The CII-Sohrabji Godrej Green Business Centre in the city of Hyderabad, India, is a LEED Platinum-certified building which employs natural ventilation, solar shading, rainwater harvesting, and energy-efficient lighting. This building type demonstrates that sustainable construction can be undertaken in low and resource-constrained settings through innovations, use of local materials, and by engaging and informing all key stakeholders (Chen et al., 2024a; Moradi & Sormunen, 2024a).

A comparison of the examples points to some significant enablers of successful implementation. Developed countries benefit from existing regulatory mechanisms, financial support, and research and development capacity, enabling them to create more technical, complex technologies and holistic policies. Conversely, developing countries have tended to rely on locally sourced materials using low-tech solutions, as well as the advantage of relying on global partners to meet their sustainability vision. Regardless of these distinctions, common success features included: engagement of stakeholders, early adoption of sustainability in design, facilitation through governance, and engagement.

Hence, global best practices demonstrate that sustainable construction is not a universal solution, rather a contextually dependent strategy requiring local innovations, committed leadership, and commitment to a longer-term outcome.

CHALLENGES AND LIMITATIONS

Although momentum for sustainable construction continues to increase, the continued transition is slowed down by a number of inter-connected barriers to implementation across a variety of contexts. These barriers - financial, technical, organizational and cultural - can lead to delayed uptake, piecemeal implementation, and uneven results as shown in Figure 5. Better understanding of these barriers, and supporting policies, training, and market incentives, would enable a shift toward sustainability in construction and the built environment (Likita et al., 2024; Moradi & Sormunen, 2024b).

Figure 5. Challenges and limitations in sustainable construction

Financial, technical, and organizational barriers are by far the greatest barriers. Sustainable construction typically involves higher upfront capital expenditure for certified green material, energy-efficient materials, and design consultation. Although these costs usually lead to lower operational costs in the long run, the size of the expenditures can be prohibitive—especially for SMEs or public projects that have tight budgets. While access to green financing is still limited in some areas, ROI is sometimes unclear or under-quantified, lessening the interest of other stakeholders to pursue sustainable solutions.

From a technical perspective, the materials and technologies have significant variation depending on the geography. Sustainable construction products such as low-carbon concrete, prefabricated modular units, or smart building systems may simply not be available, may be very expensive, or may not fit within the established supply chain for much of the developing world. Organizationally, many construction companies do not control sustainability on an integrated basis and there is often ineffective accountability due to a lack of relevant performance and measurement systems. Firms typically view sustainability solely as a compliance issue, rather than a (governed) strategic approach (Chen et al., 2024b).

Another significant challenge involves the skills gap and employability. Practicing sustainable construction necessitates specialized knowledge in many areas like; green building certification and energy modeling as well as lifecycle assessment and digital construction technology such as Building Information Modeling (BIM). These resources are often not available to construction workers, project managers, engineers, and architects in these parts of the world. The skills gap is even more pronounced in emerging economies where technical education systems may not have evolved to address the rapidly-changing needs of a modern-day sustainable construction industry.

Moreover, the workforce is usually stratified, and subcontractors, temporary workers, and informal laborers may never receive the same opportunities concerning exposure to sustainability standards or best practices—without regular training, assessment of capacity, and supervision, even the best-intentioned project will not achieve its environmental objectives in such cases. Closing the skills gap will require a collaborative effort amongst government, industry associations, educational institutions, and private enterprise to develop curriculum, certification pathways, and continuing professional development for green construction (Moradi & Sormunen, 2024b).

One of the more nuanced but equally important barriers that we can not overlook is the resistance to change and delays in adoption. The construction industry is primarily conservative and generally risk-averse. It mostly depends on known processes, known materials, and known vendor relationships. New approaches, such as circular design, off-site construction, or performance-based contracting, generally meet with skepticism based on perceived risks, lack of precedent, or unfamiliarity with issues the key actors face. A lot of the cultural inertia faced—especially among senior decision-makers—can also mean a lack of appetite for adopting innovation even when evidence shows that there are benefits.

Furthermore, regulatory uncertainty, lack of enforcement, and inconsistent requirements for sustainability policies can infer even more delays in adoption. Without clear policy direction or demand for the service from the market, sustainability initiatives may be deprioritized to focus on immediate cost and schedule. These systemic delays can result in stagnation, but perhaps worse is, that they undermine our trust and continue to weaken momentum to enable systemic shift across the industry.

Thus, the streamlining to sustainable construction has several constraining challenges that require both strategies and tactical portraits. To overcome these challenges strategic to sustainability must address the financial barriers through incentives and financing models, develop technical capacity, enable skilled worker, and create a culture of innovation and collaboration. The construction industry can only unlock the full spectrum of sustainability through a collective, collaborative, and proactive vision. With these factors, the construction industry can significantly move toward contributing to global environmental and developmental aspirations.

FUTURE OUTLOOK AND EMERGING TRENDS

As the construction sector addresses continued challenges with environmental degradation, urbanization pressure, and economic uncertainty, it will continue to be an important contributor to an innovative and resilient future. The path to sustainable construction is becoming continually molded by new technology, funding, and strategic resourcing that go beyond existing concepts of green. The highlight of what is on the horizon is a much modernized, collaborative and proactive construction space, empowered by smart cities and climate-resilient infrastructure, sustainability-linked financing and new types of materials and constructs (Abasova & Mamedov, 2025).

One of the most disruptive trend is the emergence of smart cities and resilient infrastructure. Smart cities are built utilizing a combination of digital technology such as the Internet of Things (IoT), big data analytics, and real-time monitoring to create increased urban efficiency, sustainability and livability. In the construction sector, smart buildings and infrastructure deploy sensors and potentially digital controls, and embedding data capture through all stages of construction, maintain information related to energy use, occupancy, environmental data, and structural health. This available data-driven information can firstly align contractors adopting predictive maintenance strategies, and secondly, respond and adapt operations to the changes in preference as represented by the data.

Additionally, with the rise in climate-induced disruptions (floods, heatwaves, and extreme storms for example), resilience is becoming a forefront consideration in design philosophy. Infrastructure of the future needs to not just be sustainable but also climate-adaptive, built to withstand, respond, and transform in reaction to environmental pressures. This may include elements like flood defences, green roofs, permeable surfaces, and flexible utility systems as essential components of urban planning. Globally, cities like Copenhagen, Singapore, and Rotterdam are exemplifying developments in resilient thinking, weaving it into building codes and infrastructure futures, creating urban environments that are efficient and resilient (and adaptable) to uncertain futures related to climate change (Alam Bhuiyan & Hammad, 2024).

In parallel to these developments of technological transformation related to built form and climate adaptation, sustainability linked financing is transforming as well. Financial markets are increasingly clear that sustainability is not just an ethical consideration; principles of sustainability should be regarded as material risk factors which can alter asset value, regulatory scrutiny, and operational continuity. As a result, the construction and real estate sector is responding to investor demand and potential negative implications of sustainability within construction leverage options such as green bonds, sustainability-linked loans, and investment portfolios focusing on ESG, with structures tied to the performance of environmental engagement, such as reducing overall carbon intensity, energy use, or water consumption.

As required by the changes in the capital markets, green building certifications and ESG ratings are on the rise for institutional investors by providing evidence to back capital requests for bigger developments. Investors want more transparency and accountability around how construction firms are managing the environmental risk, social impact, and governance frameworks of their investments. This newly demanded accountability pushes rollout of sustainability metrics in the business model of project developers to allow access to capital, get brand value, meet regulatory expectations, and build competitive advantage. Climate disclosure frameworks based on TCFD requirements are also translating the sustainability into decision making at both project and portfolio levels.

Lastly, the construction sector of the future will inevitably be shaped by next-generation materials and construction methods. Innovations in materials science are extending new opportunities for greener, more efficient, and even regenerative building products. Carbon-sequestering concrete, self-healing materials, mycelium-based insulation, and transparent solar panels are examples of these materials that can produce more favorable performance characteristics, with less embodied carbon and a chance for circularity. Furthermore, additive manufacturing - or 3D printing - is changing the way builds are made with the opportunity for on-demand and precise manufacture of building components, using far less waste and energy (Chen et al., 2024c).

At the same time, new paradigms such as marketed/modular construction, industrialized construction, and design for disassembly (DfD), are gaining momentum. These models achieve speed, scalability, and other sustainable outcomes via component standardization, construction disruption mitigation, and future reuse/recycling. Paired with digital twins and AI-based design applications, these models achieve levels of control, customization, and performance optimization that are unprecedented across the lifecycle of a building.

Thus, the future of sustainable construction is defined by convergence—of technologies, financial products, and approaches that emphasize long-term value instead of short-term cost. Data-driven smart, resilient cities; source-based investments generating climate and social outcomes; regenerative and adaptive materials, all signal a construction industry that is smarter, greener, and more resilient. As these trends accelerate, the real challenge will be about equity, policy coherence, and the capacity to scale—to ensure that sustainable construction benefits all neighborhoods, working sectors, and communities.

CONCLUSION

The construction industry is at a crossroad to rapidly urbanize and combat significant environmental, economic and social impacts. This chapter articulated the various unsustainable dimensions of the many socio-ecological complications in the construction life cycle, and the creative approaches to intervention that can drive progress towards unsustainable development in the construction industry. This includes the environmental damage to land and communities through resource heavy practices, and the extensive barriers for construction producers as a result of chronic organizational and policy complexities. The implication is for coordinated, scalable, and contextual action.

New trends, such as green building materials, smart building technologies, and data-driven design methodologies (as with Building Information Modeling (BIM) and digital twins) are reconstituting the possibilities for optimizing construction processes for sustainability. Automation, robotics, and artificial intelligence are improving accuracy and minimizing waste while improving site performance. Regulatory structures at the national- and international-level are crucial for institutionalizing sustainability, and there are still gaps between developed and developing contexts that need to be addressed. Certifications, incentives, and compliance are moving the industry forward to achieve greater environmental performance; however, performance is still dependent on governance and collaboration among stakeholders.

Hence, reducing unsustainable work practices in construction is not just an afterthought, it's a key strategic concern. It is a key strategic path forward that requires not just innovation but visionary leadership, cooperative governance and a collective commitment to building for people and the planet. Sustainability is no longer a hope; it is a must. The construction industry has a key part to play in developing urban resilience, environmental integrity and inclusive economic landscape as sustainability transitions from preference to need.

REFERENCES

Abasova, A., & Mamedov, E. (2025). *Analysis of modern development trends in the construction sector*. Academic Press.

Agyemang, E., Ofori-Dua, K., Dwumah, P., & Forkuor, J. B. (2024). Towards responsible resource utilization: A review of sustainable vs. Unsustainable reuse of wood waste. *PLoS One*, *19*(12), e0312527. PMID: 39715227

Ajirotutu, R. O., Adeyemi, A. B., Ifechukwu, G.-O., Iwuanyanwu, O., Ohakawa, T. C., & Garba, B. M. P. (2024). Future cities and sustainable development: Integrating renewable energy, advanced materials, and civil engineering for urban resilience. *International Journal of Sustainable Urban Development, 3*.

Alam Bhuiyan, M. M., & Hammad, A. (2024). Engineering and Design for Sustainable Construction: A Bibliometric Analysis of Current Status and Future Trends. *Sustainability*, *16*(7), 2959.

Babalola, A., & Harinarain, N. (2024). Policy barriers to sustainable construction practice in the Nigerian construction industry: An exploratory factor analysis. *Journal of Engineering. Defence Technology*, *22*(1), 214–234.

Boopathi, S., Umareddy, M., & Elangovan, M. (2023). Applications of Nano-Cutting Fluids in Advanced Machining Processes. In *Sustainable Utilization of Nanoparticles and Nanofluids in Engineering Applications* (pp. 211–234). IGI Global.

Chen, L., Hu, Y., Wang, R., Li, X., Chen, Z., Hua, J., Osman, A. I., Farghali, M., Huang, L., & Li, J. (2024). Green building practices to integrate renewable energy in the construction sector: A review. *Environmental Chemistry Letters*, *22*(2), 751–784.

El-Hakim, Y., & AbouZeid, M. N. (2024). Towards mitigating climate change negative impact: The role of regulations and governance in the construction industry. *Sustainability*, *16*(16), 6822.

Firoozi, A. A., Firoozi, A. A., Oyejobi, D., Avudaiappan, S., & Flores, E. S. (2024). Emerging trends in sustainable building materials: Technological innovations, enhanced performance, and future directions. *Results in Engineering*, *24*, 103521.

Folorunso, A., Olanipekun, K., Adewumi, T., & Samuel, B. (2024). A policy framework on AI usage in developing countries and its impact. *Global Journal of Engineering and Technology Advances*, *21*(01), 154–166.

Hassan, H., Rodriguez-Ubinas, E., Al Tamimi, A., Trepci, E., Mansouri, A., & Al-mehairbi, K. (2024). Towards innovative and sustainable buildings: A comprehensive review of 3D printing in construction. *Automation in Construction, 163*, 105417.

Likita, A. J., Jelodar, M. B., Vishnupriya, V., & Rotimi, J. O. B. (2024). Lean and BIM integration benefits construction management practices in New Zealand. *Construction Innovation, 24*(1), 106–133.

Lucatello, S., & Alcántara-Ayala, I. (2024). Sustainable Synergy: Strengthening disaster risk reduction in Latin America and the Caribbean through nature-based solutions. *International Journal of Disaster Risk Reduction, 113*, 104860.

Madan, B. S., Viraj Dravid, S., Deenadayalan, G., Rathod, P., Arumugam, G., & Boopathi, S. (2024). Innovations and Applications of Engineered Cementitious Composites and Alternative Materials: Reimagining Construction. In *Advances in Chemical and Materials Engineering* (pp. 53–86). IGI Global. DOI: 10.4018/978-1-6684-8182-0.ch003

Masyhur, R. T., Alias, A. H., Haron, N. A., & Zulkafli, Z. (2024). A systematic review on green practices in the Malaysian construction industry: Status, challenges, key motivations, rating systems and technology evolution. *Energy and Building, 320*, 114550.

Moradi, S., & Sormunen, P. (2024a). Integrating lean construction with BIM and sustainability: A comparative study of challenges, enablers, techniques, and benefits. *Construction Innovation, 24*(7), 188–203.

Moradi, S., & Sormunen, P. (2024b). Integrating lean construction with BIM and sustainability: A comparative study of challenges, enablers, techniques, and benefits. *Construction Innovation, 24*(7), 188–203.

Musonda, J., Ma, L., Liu, T. Y., Shahzad, M., & Ali, A. (2025). Megaprojects' environmental impact mitigation by sustainable construction mediated by sustainable practices and regulations. *Cleaner and Responsible Consumption*, 100281.

Narra, M., Vij, M., Shreenidhi, K. S., Anandaram, H., Hari, B. S., & Boopathi, S. (2024). Redefining Sustainability in Building Practices in Circular Construction Process. In *Innovations in Energy Efficient Construction Through Sustainable Materials* (pp. 293–324). IGI Global. DOI: 10.4018/979-8-3693-3398-3.ch011

Naser, M., & Kodur, V. (2025). Vulnerability of structures and infrastructure to wildfires: A perspective into assessment and mitigation strategies. *Natural Hazards, 121*(8), 9995–10015.

Nooradzhar, F. S., Abdullah, M. N., & Asmoni, M. (2024). Strategic Pathways to Sustainable Development: A Comprehensive Review of the Benefits and Barriers of Green Buildings. *International Journal of Real Estate Studies*, *18*(2), 91–102.

Oke, O. S., Aliu, J. O., Oke, A. E., Ekundayo, D., & Duduyegbe, O. M. (2025). Green Building Practices: Fuzzy Synthetic Evaluation of the Drivers of Deforestation and Forest Degradation in a Developing Economy. *Sustainability*, *17*(4), 1538.

Omopariola, E. D., Olanrewaju, O. I., Albert, I., Oke, A. E., & Ibiyemi, S. B. (2024). Sustainable construction in the Nigerian construction industry: Unsustainable practices, barriers and strategies. *Journal of Engineering. Defence Technology*, *22*(4), 1158–1184.

Orieno, O. H., Ndubuisi, N. L., Eyo-Udo, N. L., Ilojianya, V. I., & Biu, P. W. (2024). Sustainability in project management: A comprehensive review. *World Journal of Advanced Research and Reviews*, *21*(1), 656–677.

Reddy, V. J., Hariram, N., Ghazali, M. F., & Kumarasamy, S. (2024). Pathway to sustainability: An overview of renewable energy integration in building systems. *Sustainability*, *16*(2), 638.

Satyam, V., & Sharma, P. (2024). Unsustainable construction material and process leading to severe environmental degradation. *AIP Conference Proceedings*, *3100*(1), 040018.

Sharma, M., Sharma, M., Sharma, N., & Boopathi, S. (2024). Building Sustainable Smart Cities Through Cloud and Intelligent Parking System. In *Handbook of Research on AI and ML for Intelligent Machines and Systems* (pp. 195–222). IGI Global.

Singh Madan, B., Logeswaran, S., Veerapathran, M., Rai, P., Kumar Singh, S., & B., S. (2024). Engineered Cementitious Composites (ECC) Through Nano-Composite Integration: Enhancing Construction. In *Advances in Chemical and Materials Engineering* (pp. 210–243). IGI Global. https://doi.org/DOI: 10.4018/978-1-6684-8182-0.ch008

Umana, A. U., Garba, B. M. P., Ologun, A., Olu, J. S., & Umar, M. O. (2024). The role of government policies in promoting social housing: A comparative study between Nigeria and other developing nations. *World Journal of Advanced Research and Reviews*, *23*(03), 371–382.

Vellingiri, A., Kokila, R., Nisha, P., Kumar, M., Chinnusamy, S., & Boopathi, S. (2024). Harnessing GPS, Sensors, and Drones to Minimize Environmental Impact: Precision Agriculture. In *Designing Sustainable Internet of Things Solutions for Smart Industries* (pp. 77–108). IGI Global. https://doi.org/DOI: 10.4018/979-8-3693-5498-8.ch004

Yadav, K., & Nikalje, G. C. (2024). Comprehensive analysis of bioplastics: Life cycle assessment, waste management, biodiversity impact, and sustainable mitigation strategies. *PeerJ*, *12*, e18013. PMID: 39282116

Yu, H., Zahidi, I., Chow, M. F., Liang, D., & Madsen, D. Ø. (2024). Reimagining resources policy: Synergizing mining waste utilization for sustainable construction practices. *Journal of Cleaner Production*, *464*, 142795.

Zhang, L., Sun, H., Pu, T., Sun, H., & Chen, Z. (2024). Do green finance and hi-tech innovation facilitate sustainable development? Evidence from the Yangtze River Economic Belt. *Economic Analysis and Policy*, *81*, 1430–1442.

Chapter 2
In-House Engineering as a Catalyst for Sustainable Innovation in Construction

Mitchell Roy Woodworth
https://orcid.org/0009-0004-5041-2676
University of Calgary, Canada

Oleksiy Osiyevskyy
https://orcid.org/0000-0002-5765-6161
University of Calgary, Canada

ABSTRACT

In-house engineering can be a powerful driver of innovation and sustainability in the construction industry. The authors examine the strategic choice between internalizing engineering services and outsourcing them and how it influences a firm's ability to innovate and meet sustainability goals. Integrating the four established theoretical perspectives (Transaction Cost Economics, Resource-Based View of the Firm, Dynamic Capabilities, and Absorptive Capacity), they develop a decision framework providing insights into the contexts and situations when in-house engineering adds the most value, becoming the preferred choice for construction firms. They start with an overview of industry trends, showing a clear shift from traditional outsourcing toward more integrated models as firms seek long-term competitive advantage. Then, they integrate decision criteria (cost, risk, innovation potential, sustainability, knowledge retention, and governance considerations) into a unified framework and illustrate them with case studies of major construction and engineering firms.

DOI: 10.4018/979-8-3373-2555-2.ch002

INTRODUCTION

The construction industry faces mounting pressure to improve sustainability and innovation in project delivery. Construction and infrastructure projects contribute significantly to carbon emissions and resource consumption, making sustainable practices a growing priority (Oakland & Marosszeky, 2017). At the same time, innovation is increasingly recognized as vital to a company's competitiveness, yet the construction sector has historically lagged behind other industries in innovation performance (Blayse & Manley, 2004). Fragmented project delivery, characterized by siloed responsibilities and heavy outsourcing, has often been blamed for this lag (Gann & Salter, 2000). Many construction firms traditionally outsource substantial portions of design and engineering scope to external consultants, a practice rooted in specialization and risk aversion (Abdel-Rahman, Dokhan, & El-Dash, 2023). This "make-or-buy" decision, whether to perform engineering in-house or contract it out, is not merely operational; it is strategic and has far-reaching implications for a firm's innovative capacity and sustainability outcomes (Sinha, Osiyevskyy, & Radnejad, 2022).

In-house engineering ("make" approach) refers to a firm directly employing engineers to handle project design, planning, and technical tasks, whereas outsourced engineering ("buy" approach) refers to hiring external engineering consultants to perform those services. Historically, outsourcing was pursued for perceived cost efficiency and access to specialized expertise not available internally. By transferring design work to external parties, contractors could reduce fixed staffing costs and tap into a competitive market of engineering firms. However, over time, drawbacks of excessive outsourcing became apparent within the industry. Coordination difficulties, misaligned incentives, and loss of institutional knowledge often accompanied the outsourcing model (Blayse & Manley, 2004; Gann & Salter, 2000). Projects with complex, interdependent components often suffered from fragmented roles, designs optimized in isolation by external consultants that did not translate well to on-site construction realities, leading to costly changes or project delays. Additionally, outsourcing can incur hidden transaction costs in managing contracts, communications, and quality control that offset the apparent savings. As project complexity and the need for sustainable innovation have risen, many firms have begun to reconsider the balance between external and internal engineering capabilities.

Currently, an emerging shift is underway: forward-thinking construction companies are investing in in-house engineering teams as strategic assets to drive innovation and embed sustainability into projects from the outset. Integrating engineers within the contractor's organization can break down silos between design and construction, creating feedback loops for continuous improvement. When designers and builders work in tandem under one roof, sustainable design principles can be woven into projects from conception through execution. This contrasts with the traditional approach where sustainability might be an afterthought or value-engineering item. The integrated approach of in-house engineering aims to achieve better outcomes for projects, aligning technical decisions with long-term environmental and operational performance goals.

Despite these potential benefits, the decision to internalize engineering is complex. Construction executives face a strategic dilemma: should they invest in building and maintaining internal engineering capabilities, or continue relying on external engineering consultants? This dilemma involves balancing short-term cost efficiency against long-term innovation and sustainability benefits. Outsourcing can offer flexibility and lower immediate costs, but it may limit a firm's ability to innovate or coordinate complex work. In-house teams require upfront investment and overhead, but promise greater control, knowledge retention, and the ability to develop proprietary innovations. There is no one-size-fits-all answer, and many firms adopt a hybrid approach. However, clear frameworks to guide this decision have been lacking, leaving companies to rely on improvised judgements or traditional models.

The chapter addresses this gap by examining how in-house engineering can act as a catalyst for sustainable innovation in construction, and under what conditions this approach yields superior results compared to outsourcing. We integrate insights from several theoretical lenses to analyze the make-or-buy decision for engineering services. Transaction Cost Economics, Resource-Based View of the Firm, Dynamic Capabilities, and Absorptive Capacity theory each shed light on different facets of the choice. Using these theories, we propose a structured decision framework for practitioners. We also review key industry trends, from historical outsourcing practices to recent shifts toward integration, to frame the decision. Through multiple case studies of both successful and cautionary examples, including large construction contractors and project owners, we illustrate the real-world consequences of different sourcing strategies on innovation and sustainability outcomes.

In doing so, the chapter provides both analytical and practical contributions. First, it forms a theoretical foundation for viewing engineering sourcing as a strategic driver of innovation. Second, it derives a decision criteria framework that companies can use to evaluate when internal engineering is most beneficial. Third, it offers evidence-based lessons and actionable recommendations for organizations aiming to enhance sustainable innovation in construction. The overarching argument is that leveraging in-house engineering is not just a technical staffing choice, but a strategic move that can strengthen a firm's innovation strategy and long-term competitiveness in an industry facing unavoidable change and sustainability demands.

THEORETICAL FOUNDATIONS

Multiple theoretical perspectives help explain the trade-offs in the in-house versus outsourced engineering decision. In this section, we outline key theories that form the chapter's analytical foundation: Transaction Cost Economics (TCE), the Resource-Based View (RBV) of the firm, Dynamic Capabilities, and the concept of Absorptive Capacity, as well as relevant behavioral considerations. Together, these provide an understanding of how firms can strategically manage engineering services to drive innovation and sustainable outcomes.

Transaction Cost Economics (TCE) provides a framework for evaluating the "make-or-buy" decision by comparing the costs of using the market (outsourcing) with the costs of internal organization (Williamson, 1996; Walker & Kwong Wing, 1999). According to TCE, firms will internalize an activity when the transaction costs of outsourcing (such as the costs of finding vendors, negotiating and enforcing contracts, coordinating schedules, and safeguarding against opportunism) exceed the administrative costs of performing the activity in-house. Several factors increase transaction costs when outsourcing engineering. High uncertainty in project requirements or external environments can lead to contract disputes or unexpected changes that are costly to handle with external parties. High asset specificity, for example, project-specific knowledge or specialized methods, makes it risky to rely on the market, since external providers might not recoup investments specific to one project, or they may hold up the firm knowing it cannot easily switch suppliers. Under such conditions, TCE predicts that "make" (in-house) is preferable to "buy."

Empirical work supports these ideas. In a seminal study on make-or-buy decisions, Walker and Weber (1984) found that comparative production cost was a primary factor, but when demand uncertainty was high, firms tended to internalize production to avoid dependence on suppliers (i.e., to reduce transaction risks). In construction, a comparable scenario is a company with a steady, predictable pipeline of projects (ensuring high utilization for an internal engineering team), such a firm may find it cost-effective to maintain in-house engineers rather than repeatedly contracting out. Conversely, a company with very sporadic projects might lean toward outsourcing to avoid paying full-time staff during downtime. TCE also highlights opportunism risk: if an engineering service is hard to fully specify in a contract, an external provider might cut corners or demand higher fees later. In-house teams, being managed from within, are easier to realign if priorities change. Thus, TCE suggests that as projects become more complex and uncertain, and as coordination needs increase (all of which raise transaction costs of outsourcing), the case for in-house engineering strengthens.

The Resource-Based View (RBV) theorises that a firm's sustainable competitive advantage derives from resources and capabilities that are valuable, rare, imperfectly imitable, and non-substitutable (VRIN) (Barney, 1991; Goh & Loosemore, 2017). Under RBV, an internal engineering department can be seen as a strategic resource. If a company develops unique engineering expertise, proprietary design methods, or innovative processes in-house, these can be sources of differentiation that competitors cannot easily copy. For example, a contractor that pioneers a novel construction technique (say, a new modular assembly method or a custom project management software integrated with BIM) using its internal R&D could gain a time and cost advantage on projects. Such internally developed innovations may be rare (not widely available on the market) and valuable in improving project outcomes. If protected effectively (treated as trade secrets or patented), they are inimitable by competitors. In contrast, by outsourcing engineering, the knowledge and processes reside with external firms, who can offer similar services to the contractor's competitors, any innovation introduced by an external consultant is not exclusive to one contractor. Thus, outsourcing by itself usually cannot create unique advantages, since the same pool of consulting expertise is open to all contractors. By internalizing, a firm can accumulate organization-specific, non-substitutable knowledge over time, leading to continuous improvement and a know-how that competitors who outsource lack.

An important aspect of RBV is knowledge protection. Firms worry that outsourcing critical design work could leak intellectual property or erode their distinctive competencies. A recent study on R&D governance regimes emphasizes that beyond cost, companies consider knowledge spillovers and behavioral uncertainties when deciding governance structures (Sinha et al., 2022). In construction, keeping innovative engineering tasks in-house helps protect intellectual property and build up organizational learning, factors RBV suggests are crucial for long-term advantage. However, RBV also acknowledges that not all resources need to be internal, only those that meet VRIN and align with the firm's strategy. Routine engineering tasks that do not qualify as unique, might be safely outsourced without hindering a company's competitive advantage. The challenge is identifying which engineering functions are truly strategic resources.

While RBV focuses on valuable static resources, Dynamic Capabilities theory extends this view by examining a firm's ability to continuously adapt, integrate, and reconfigure resources in response to changing environments (Teece, 1997; Adam & Lindahl, 2017; Aghimien et al., 2023). In fast-changing industries, the capacity to innovate and respond, often referred to as organizational agility, is more critical than any particular asset. For construction firms, dynamic capabilities include the ability to sense emerging opportunities or threats (e.g. new sustainable technologies, shifting client demands), seize those opportunities by assembling the right resources, and reconfigure organizational processes to stay ahead of change. Strong internal engineering teams can enhance all three of these aspects. Because in-house engineers are embedded in the firm, they are in a good position to sense opportunities on projects, for instance, noticing a chance to implement a novel green design solution or a digital tool that improves efficiency. They can then seize the opportunity by developing and piloting the solution internally, without waiting for lengthy external procurements. Finally, they help the firm reconfigure its operations by spreading new practices across projects and departments.

A dynamic capabilities perspective suggests that the speed and effectiveness with which a construction firm can innovate are partly determined by its internal technical capacity. Firms with strong internal engineering can modify processes and adopt new technologies more quickly, because they do not have to renegotiate contracts or convince an external party of the value, they can simply do it. As an illustration, consider how EPC (Engineering-Procurement-Construction) companies like Bechtel and Fluor managed the rise of modular construction techniques. Bechtel leveraged its internal engineering capabilities to pivot toward modular design/fabrication in response to market demand for faster project delivery; it was able to redeploy its in-house experts across projects to implement modular methods swiftly. This kind of rapid reconfiguration is a dynamic capability that stems from having the knowledge in-house. Meanwhile, a contractor relying solely on external engineering might miss such opportunities or adopt them much later, due to inertia or lack of internal champions.

Dynamic capabilities also emphasize learning mechanisms, organizations improve through iterative learning, and retaining key functions internally provides a better environment for experiential learning. Internal teams accumulate knowledge from project to project, refining their approaches. Over time, this learning improves the firm's ability to handle new challenges. For example, if an internal team experiments with a cutting-edge sustainable material on one project, the lessons remain within the company and can be applied to future projects. If that experiment were outsourced, much of the learning would leave with the consultant. In essence, in-house engineering contributes to an agile, learning-oriented organization that can innovate continuously. This aligns with the dynamic capability view that it's not just the resources you have, but how quickly you can transform them into new strategies that matters.

An often overlooked but critical concept explaining the firm success in the construction industry is Absorptive Capacity (Gann, 2001), an organization's ability to recognize the value of new external information, integrate it, and apply it. Even firms that outsource a lot can innovate if they are good at absorbing knowledge from their partners; conversely, firms with internal teams can stagnate if they are insular. Absorptive capacity is largely a function of the firm's prior related knowledge and its internal technical competence (Cohen & Levinthal, 1990). In the construction context, having some in-house engineering "critical mass" greatly enhances a firm's ability to benefit from external innovations. When a construction firm employs qualified engineers internally, those engineers can interface effectively with outside experts, understand new technologies or research findings, and integrate those ideas into the firm's practices. Without in-house technical expertise, a contractor might receive a novel idea from a consultant but fail to exploit it fully, essentially, the knowledge passes through without sticking.

Studies have noted that project-based firms often struggle to learn from project to project, partly because knowledge tends to remain tacit and project-specific unless there are deliberate efforts to capture it (Gann & Salter, 2000; Blayse & Manley, 2004). Building absorptive capacity through internal teams is one way to address this. Blayse and Manley (2004) emphasize that to take full advantage of knowledge transfer, construction organizations need sufficient in-house technical staff who can interpret and act on new information. These internal professionals serve as knowledge receptors and translators, they can collect external knowledge and integrate it with internal processes and translate the firm's project experiences into lessons learned. A firm with high absorptive capacity might, for example, quickly adopt a cutting-edge software tool introduced by a consulting engineer, because its own engineers rapidly understand the tool's capabilities and tailor it to the company's needs. In contrast, a firm lacking absorptive capacity might ignore or misuse the same tool. Thus, even when leveraging external sources of innovation, having in-house engineering talent determines how well those external ideas are absorbed and turned into business improvements.

Behavioral and Organizational Considerations

Beyond economic and resource considerations, there are human and organizational factors influencing the in-house vs. outsource decision. TCE itself originates from behavioral assumptions like bounded rationality (the idea that decision-makers do not have perfect information or unlimited cognitive capacity) and opportunism (the risk that parties may shirk or cheat if not properly monitored). These imply that managers might misjudge the costs and benefits of outsourcing if they underestimate coordination difficulties or overestimate their suppliers' reliability. Behavioral economics suggests that even well-intentioned external consultants can suffer from divided attention or misaligned goals, for instance, an external engineer juggling multiple clients might not prioritize a contractor's project sufficiently, leading to underperformance. In-house engineers, by contrast, are directly invested in the firm's success and culture, providing greater assurance of effort and alignment.

Bounded reliability is another concept (from behavioral strategy and transaction cost economics) stressing that even when parties intend to fulfill obligations, real-world distractions or conflicts can limit reliability (Verbeke & Greidanus, 2009). Sourcing decisions should account for these additional factors. Trust and communication ease innovation: if a firm's internal engineers and field managers trust each other, they can communicate openly about problems and experiment with solutions. When using external teams, establishing that trust and seamless communication is more difficult, often requiring formal mechanisms. Some research (Sinha et al., 2022) argues for governance regimes that consider these behavioral aspects, for example, cultivating long-term relationships with a select group of engineering partners to mimic the benefits of internal teams in terms of trust and shared understanding. In summary, a purely economic analysis might suggest outsourcing to the cheapest bidder, but wise managers also weigh culture, communication, and the ability to coordinate complex, innovative tasks when making sourcing decisions.

In conclusion, these theoretical foundations collectively suggest that internalizing engineering services can provide significant strategic benefits when innovation, integration, and long-term knowledge development are important to the firm. TCE urges caution about outsourcing under high uncertainty or complexity due to high transaction costs. RBV and Dynamic Capabilities highlight the competitive advantages of building unique capabilities and being able to adapt rapidly; two features often tied to having key expertise in-house. Absorptive capacity underscores that even an open innovation approach requires internal experts to capture value from external ideas. None of this means outsourcing is inherently bad; rather, it means that the decision should hinge on the nature of the work and the firm's context. Many firms will find an optimal balance, e.g., in keeping core innovative engineering tasks internal, while outsourcing standardized or peripheral tasks to trusted partners. The following sections build on these insights to examine industry trends and to develop a practical framework for making these decisions in the construction industry.

INDUSTRY OVERVIEW AND TRENDS

The balance between in-house and outsourced engineering in construction has evolved over time, influenced by changes in project delivery models, market pressures, and strategic thinking within firms. This section provides an overview of industry trends, from the historical dominance of outsourcing in certain segments, to the recent shifts towards more integrated approaches. We examine the traditional rationale for outsourcing, the benefits and drawbacks as understood *then and now*, and emerging trends such as digitalization and sustainability that are reshaping sourcing decisions.

cause the external designers had limited incentive or feedback about construction efficiencies. Firms realized that by integrating engineering with construction, they could achieve breakthroughs in how projects are delivered. A notable example is the use of Building Information Modeling (BIM) and other digital technologies. BIM requires constant iteration between design and construction planning; contractors that developed in-house BIM and virtual design teams found they could reduce errors and clashes during construction much better than if they relied on an external party for BIM modeling. As the UK government mandated BIM on public projects by 2016, firms globally saw the writing on the wall that digital integration would be key (Shibeika & Harty, 2015). Many contractors responded by hiring BIM specialists and engineers, effectively internalizing these roles to ensure digital innovations were embedded across their operations.

Knowledge retention and continuous improvement have also factored into the shift. Project after project, outsourcing can lead to a "brain drain", when the lessons learned on one job walk out the door when the consulting engineer finishes their contract. In contrast, when engineers are employed by the contractor, each project's learnings (good or bad) can be institutionalized. Firms aiming for long-term improvement recognized that having a stable core of engineering staff allows them to build up institutional memory, standardize best practices, and progressively push the boundaries of efficiency and quality. Fewings and Henjewele (2019) note that closer integration of design and construction teams is vital to drive future improvements in project delivery. This integration fosters a culture of problem-solving and innovation inside the firm.

Sustainability and Life-Cycle Focus

In recent years, the construction industry's move toward sustainability has reinforced the value of integration. Achieving green building certifications, low-carbon construction, or simply better life-cycle performance requires tight collaboration between design and execution. Sustainability is not just about choosing greener materials; it often involves iterative design optimizations (to reduce material quantities, improve energy modeling, etc.) and considering construction methodologies that reduce waste and emissions. These iterative loops benefit from having designers and builders on the same team, aligning goals. Contractors with in-house design capabilities can integrate sustainable design choices early and ensure they are feasible and cost-effective, rather than treating sustainability as an external imposition to be engineered by someone else. In essence, sustainable innovation in construction is pushing firms toward more integrated project teams, and many are choosing to build those teams internally.

Hybrid Approaches and Outsourcing 2.0

It is important to note that the trend is not a wholesale rejection of outsourcing; rather, it is about smarter sourcing. Many leading firms use a hybrid model: they maintain a core in-house engineering group for critical functions and project integration but continue to outsource specialized tasks or overflow work to external consultants. The difference now is that these decisions are made strategically. For instance, a firm might keep all means-and-methods engineering (temporary works, construction planning) in-house because it directly impacts their field productivity and risk management, while outsourcing the bulk of the design to an outside firm. The key difference is that many modern firms utilize their in-house capabilities to oversee the designs produced by others to ensure they are in alignment with project objectives. Even when outsourcing, firms today often seek long-term partnerships and deeper collaboration (sometimes co-locating external designers with their team or using integrated digital platforms to work together). This "Outsourcing 2.0" is more relationship-based and outcome-focused, as opposed to purely transactional, to mitigate the downsides of traditional outsourcing.

Industry Surveys and Evolving Attitudes

Recent industry surveys echo the idea that while cost remains a factor, firms are now considering innovation and long-term value impacts in their sourcing decisions. According to a Deloitte (2024) global outsourcing survey, many organizations in engineering-intensive sectors reported that they evaluate providers not just on price but on their ability to contribute new ideas and improvements. In the same vein, a KPMG (2023) report on the future of outsourcing noted that companies increasingly pursue a *"multi-dimensional sourcing"* strategy, meaning they mix in-house, outsourcing, and partnerships in a way that optimizes not just cost, but also agility and innovation potential. In construction, this is evidenced by some contractors bringing on joint venture design partners to blend the benefits of internal and external sourcing.

In summary, the industry is moving beyond the simplistic view of outsourcing as purely a cost-saving tactic. There is a growing recognition that engineering sourcing choices are strategic. Firms that historically outsourced by default are re-evaluating which capabilities they should own to remain competitive, especially in delivering complex, innovative, and sustainable projects. The next section will introduce a framework of decision criteria that firms can use to systematically assess the in-house vs. outsource choice, building on both theory and industry lessons. Before diving into specific case studies, this framework sets the stage for understanding when internalizing engineering is most advantageous.

DECISION CRITERIA AND FRAMEWORK

Construction firms must weigh a complex set of criteria when addressing the make-or-buy decision. This section presents a structured framework that integrates insights from the four major perspectives discussed (TCE, RBV, Dynamic Capabilities, and Absorptive Capacity) and aligns them with practical criteria: cost, risk/quality, innovation potential, sustainability, knowledge retention, and governance/strategic alignment. For each category, we outline key considerations. The section culminates in a practical decision framework that decision makers can use as a guide, including a visual decision matrix (Figure 1).

Cost Considerations

Cost is often the first factor examined in sourcing decisions. On the surface, outsourcing engineering can provide direct cost benefits: it converts fixed salaries into variable project expenses. When workloads fluctuate, hiring external engineers only as needed can prevent under-utilization of staff during lulls. Many contractors use this tactic to remain lean. In addition, outsourcing can tap into economies of scale that large consulting firms provide. In theory, competitive bidding among consultants drives engineering fees toward an efficient market rate, whereas internal costs might be higher due to inefficiencies or overhead.

TCE frames the cost decision as a trade-off between production costs and transaction costs. The production cost of outsourcing includes the consultant's fees (which incorporate their labor, overhead, and profit), while the production cost of in-house includes salaries, benefits, training, and overhead for the engineering department. Many early outsourcing decisions were driven by seeing a lower sticker price for external services compared to maintaining an internal staff. However, transaction costs, the costs of managing and coordinating the outsourced relationship, must be added to the equation. These include the effort to draw up detailed contracts (since work scope and expectations need to be clearly defined for an external party), ongoing communication to manage interfaces, and monitoring work quality and progress. If a design is outsourced, the contractor might need to have employees (e.g., project managers or coordinators) spend significant time liaising with the designer, reviewing submittals, and ensuring the design meets practical needs. These costs, while not always tracked explicitly, can be substantial.

Empirical evidence shows that outsourcing's financial merits depend on the nature of the work. The Texas DOT study mentioned earlier found that routine, low-complexity tasks were cheaper to outsource, likely because they are easy to specify and require less coordination (Radhakrishnan, 2010). On the other hand, complex or innovative tasks incurred high coordination costs when outsourced, sometimes making in-house execution more cost-effective in the end. In-house engineers, while a fixed cost, can generate long-term savings by improving other aspects of project performance. For instance, internal teams often enhance bid accuracy during tendering, as a contractor with in-house design capability can produce better estimates and value-engineer the project in the proposal stage, potentially winning more work or avoiding underbidding errors. These benefits translate to financial gains. Companies like PCL have cited the role of their internal engineering in refining bids and reducing costly change orders downstream, thus adding value that doesn't show up simply as reduced design cost, but in overall project profitability.

There are also opportunity costs to consider. Outsourcing might save money on a given project's design, but what opportunities might the firm be missing by not having that expertise internally? An internal engineering group can potentially be a revenue generator: some contractors offer their engineering services externally or use them to expand into design-build-operate realms. If engineering is entirely outsourced, those opportunities remain untapped. Additionally, if poor-quality external engineering work leads to site delays or failures, the contractor may incur costs (penalties, rework expenses, reputation damage) that dwarf the original design fee savings. For example, if an outsourced design is flawed and causes a significant change during construction, the contractor might bear costs that far exceed what an in-house team's salary would have been, plus the project takes a schedule hit.

Surveys of industry practice consistently reveal that while cost is a leading driver of outsourcing, it is not the only one. Firms also cite concerns about quality and strategic capability loss from too much outsourcing (KPMG, 2023). Many successful firms adopt a blended approach to cost: they maintain a core internal team for critical tasks or oversight, ensuring those roles continuously add value across projects, and outsource only the overflow or truly one-off needs. This way, internal fixed costs are kept manageable, and external resources are used as a flexible supplement. The core team prevents the company from becoming completely price-taker in the engineering market, they have enough internal knowledge to roughly estimate what things should cost and to negotiate intelligently with consultants. This helps avoid being overcharged or relying on low bids that might be unrealistically low, leading to poor service.

In summary, the cost criterion involves more than just comparing hourly rates. Contractors must account for transaction costs, downstream impacts, and the value-add that internal engineering can provide. The cheapest short-term option may not yield the lowest total cost of ownership for the project or the firm's sustained competitiveness. A prudent approach is to outsource commoditized tasks where ample competition ensures low cost and minimal coordination needs, but to insource tasks where the firm can ill afford errors or where an internal capability will pay off over multiple projects.

Risk and Quality Management

Risk management is a pivotal criterion in the make-or-buy decision for engineering. Construction projects are fraught with technical risks (design errors, safety issues, unforeseen conditions) and managing these risks is central to a contractor's success and reputation. When deciding on in-house versus outsourced engineering, firms consider how each option affects their control over quality and their exposure to various risks.

Outsourcing engineering introduces certain risks largely stemming from the separation between the contractor and the designer. Quality risk is paramount: the contractor must rely on the external engineer to produce designs that are safe, code-compliant, and buildable. If the external designer underperforms, the contractor might catch issues late or not at all until problems occur on site. TCE argues that when uncertainty is high and performance is hard to fully specify in a contract, internalizing can reduce exposure to opportunistic behavior or negligence. In a complex project, it is nearly impossible to specify every requirement and scenario in the initial design contract, things will evolve, and an external party might not proactively manage changes unless incentivized. There's also the risk of moral hazard: the external engineer might prioritize their profit (limiting hours spent, for example) over thoroughness, unless closely managed.

Quality risk goes beyond outright failures; it also includes rework and performance shortfalls. Designs that are incomplete or impractical lead to change orders and disputes, a risk to budget and schedule. When using external designers, contractors mitigate this by conducting thorough plan reviews, but those require internal experts. Many firms thus maintain at least a small in-house engineering group specifically for quality assurance of outsourced designs. These internal engineers act as an "intelligent customer," reviewing plans, checking calculations, and ensuring the design suits the firm's construction approach. Without such oversight, a contractor is essentially flying blind with someone else's design, which can be risky.

In contrast, in-house engineering strengthens quality assurance and control. Internal engineers are part of the contractor's organization and culture, so they have a deep understanding of the company's standards, preferred methods, and lessons from past projects. They can design with execution in mind, reducing the risk of downstream issues. If a mistake or change is needed, it can be addressed quickly through internal coordination rather than through formal external change processes. For example, leading Canadian contractors like EllisDon and PCL have dedicated in-house units for crucial construction engineering tasks to ensure those are done to their internal standards. This tight alignment between design decisions and field conditions means fewer surprises. Site personnel can coordinate easily with in-house engineers to discuss a problem and solve it in real time, rather than issuing an RFI and waiting days for an external consultant's response.

Schedule risk is also relevant. On fast-track projects, waiting on external designers for revisions can delay construction. In-house teams, being integrated, can turn around changes faster and prioritize as needed. The risk of miscommunication (a common cause of delay) is reduced when teams sit under one roof. There is a behavioral element here: since the external consultants juggle multiple clients, your project might not always get top priority. Internal engineers, however, focus only on their employer's projects, their incentives are aligned to get the project done on time. In line with the bounded reliability perspective, even without ill intent, a consultant might have shifting priorities that conflict with your schedule. In-house teams are fully incentivized by the firm's success (often sharing in its rewards), which can enhance commitment and responsiveness.

Conversely, insourcing concentrates responsibility and risk within the firm. If something is designed in-house and fails, the firm cannot shift blame; it bears it directly. This is why not every firm insources; some prefer the risk transfer of having a third party to hold accountable. But transferring risk is only effective if the third party is truly competent and if the contract covers the scenario, which is not always the case.

In summary, risk management considerations often tip the scales in favor of at least some degree of insourcing. High-risk, safety-critical, or mission-critical engineering activities are often best done or closely monitored in-house to ensure quality and responsiveness. Outsourcing is better suited for lower-risk tasks or when the firm's internal resources are overstretched, and even then, partnerships and oversight are key. The ideal is to blend internal control with external support wherever possible. This way, the firm maintains central quality control and risk management, while still benefiting from outside expertise where needed.

Innovation and Competitive Advantage

Innovation is at the heart of long-term competitiveness in any industry, and construction is no exception. The ability to innovate in construction can manifest as developing new construction methods, adopting novel materials, implementing advanced technologies, or improving processes for productivity and safety. The question is: does having in-house engineering help a construction firm innovate more effectively? The evidence and theory suggest so, since in-house engineering fosters an environment where new ideas can be generated, tested, and refined as an ongoing cycle within the company.

From an RBV perspective, we discussed that proprietary knowledge and capabilities can be key competitive differentiators. In-house engineering is a direct enabler of such proprietary innovation. When engineers are on the payroll, a company can direct them to pursue R&D or to spend time on continuous improvement initiatives that may not be project specific. In-house teams embedded in operations tend to have a problem-solving culture that continuously seeks better solutions. They get direct feedback from project teams about what works and what doesn't, spurring iterative improvements. This tight feedback loop is an engine for process innovation. For instance, if field crews encounter an issue, the internal design group can collaboratively brainstorm a solution and implement it on the next project, a cycle that gradually pushes the efficiency frontier.

Dynamic Capabilities theory emphasizes not just one-time innovation, but the capacity to keep adapting and innovating. In-house engineering contributes to this dynamic capability by making the organization more nimble. Consider how Bechtel responded to changes in the energy market: when Liquefied Natural Gas (LNG) projects boomed, they needed to accelerate construction through modularization. With a large internal engineering and fabrication planning force, Bechtel quickly developed standard modular designs and redeployed engineers to refine modular construction techniques across multiple projects. They didn't have to wait to hire external experts; they retrained and refocused their own. External consultants might introduce cutting-edge ideas, but if the contractor lacks internal capability, those ideas might lose momentum after the consultant's involvement ends, an issue of absorptive capacity as discussed. Large EPC firms often not only develop their own innovations, but also actively absorb external innovations and institutionalize them. They use their internal teams to evaluate new technologies presented by vendors or partners and figure out how to roll them into standard practices company wide.

It is important to note that outsourcing can certainly provide fresh perspectives and specialized innovation. Collaboration with external firms can inject new ideas that an insular in-house team might miss. However, to convert those ideas into a competitive advantage, the contracting firm needs to internalize the knowledge, again coming back to absorptive capacity. Without a strong internal technical base, the firm risks being just a platform where other entities innovate, without capturing unique value. Competitors could hire the same innovative consultant and get a similar outcome, eroding any advantage.

Another point is speed to market with innovations. If a company relies on external engineering for innovative concepts, it might face delays in negotiating contracts for experimental work or finding the right partner who is willing to innovate (since many consultants are often conservative, sticking to tried solutions to avoid liability). In-house teams, being part of the firm, can be directed to try something new as a strategic initiative. The firm can tolerate some trial and error internally more readily than a failure by an external consultant. In other words, in-house engineering can create a safe sandbox for innovation.

From a competitive strategy standpoint, firms like Fluor have long viewed their in-house engineering as part of their brand identity and value proposition. Fluor has a reputation for engineering excellence and innovative project execution, backed by its decades of internal know. This reputation helps win mega-projects where clients want a contractor who can solve complex problems. This illustrates how in-house engineering, when properly cultivated, can translate into market advantage and not just execution capability.

It must be noted that not all innovation comes from within. There are instances where external innovation brokers can drive change that an in-house team might overlook. This is especially relevant in the current sustainability era. For example, a specialized sustainability consultant might introduce a novel carbon reduction technique that the contractor's own team had never encountered. If the contractor is open and has some internal expertise to learn from that consultant, they gain a new skill. The key is the partnership; some firms establish long-term collaborations with universities or specialty firms to continually refresh their innovation pipeline while still building internal capacity.

In summary, the innovation criterion often favors having a strong in-house capability if the firm aspires to be an industry leader or differentiate itself. In-house teams are engines for continuous improvement, allow for proprietary developments, and enable quicker and more reliable implementation of new ideas. Outsourcing may still play a role, especially to bring in cutting-edge knowledge, but it works best if paired with high absorptive capacity internally. Firms should insource the domains where they want to distinguish themselves and innovate (making those their core competencies) and consider outsourcing in areas where innovation is less critical or can be accessed through partners without losing competitive ground.

Sustainability Objectives

Sustainability has moved from a buzzword to a central concern in construction. Governments, clients, and society at large are increasingly mandating or expecting that infrastructure and building projects minimize environmental impact, reduce carbon emissions, and enhance resilience. This adds another layer to the engineering sourcing decision: which approach better supports a firm's sustainability goals and its ability to deliver sustainable solutions?

Integrating sustainability effectively into construction projects requires early and holistic consideration of environmental factors in design, long before shovels hit the ground. Decisions made during planning and design, such as site orientation, structural system choice, material selection, façade design for energy efficiency, etc., largely determine a project's eventual sustainability performance. Contractors who outsource all engineering often have little influence until the design is mostly complete. If sustainability was not deeply embedded by the external designer, the contractor is limited to smaller tweaks or improvements during construction. In contrast, if the contractor has in-house capabilities and is involved from the conceptual phase, they can embed sustainability principles from day one. An internal team may propose a design alteration that uses less material or integrates a construction method that lowers emissions, such as modular prefabrication to reduce waste, or integration of low carbon concrete mixes to reduce a project's carbon footprint. Internal engineering enables a feedback loop where sustainability measures are considered in tandem with cost and schedule. Rather than viewing green features as costly add-ons, an integrated team can find balanced solutions that achieve sustainability targets efficiently. This is harder to do when sustainability input comes from an external consultant who might not fully grasp the construction implications, or when the contractor is not involved until after design.

Dynamic capabilities come into play with sustainability as well. The regulatory landscape for sustainability (building codes, carbon pricing, client expectations like net-zero goals) is rapidly evolving. Firms that can quickly adapt to new sustainability requirements have a competitive edge. Internal engineering helps because the firm can rapidly train and refocus its team on the latest standards. For example, if a city introduces a new green building requirement, a contractor with an internal design team can immediately start updating their design approaches across all projects to comply or exceed the required standards, rather than relying on disparate external designers who each need to be briefed or might have varying levels of commitment to the standards. In-house teams create a unified approach, management can set a company-wide sustainability policy and internal engineers can implement it across projects consistently. In an outsourcing model, it's harder to enforce such consistency because external designers work for many clients and often follow only contract specifications or internal design standards.

Furthermore, companies with in-house capabilities are better poised to set ambitious internal sustainability goals. For instance, they might invest in researching alternative materials or construction methods (like exploring carbon-fiber reinforcement or mass timber construction) through their engineering department. The knowledge gained can then differentiate them in proposals. An external firm typically would not conduct such research for the contractor's exclusive benefit without being paid specifically for it, whereas an internal team might pursue it as part of the continuous improvement ethos.

Another angle with which to view sustainability is governance and accountability. Sustainability often requires going above minimum requirements, something that needs commitment. If a firm has in-house designers, it can decide to pursue higher standards (LEED Gold, for example) and mobilize its team to achieve it, even if the client did not explicitly require it, as a value-adding option. The firm can absorb some effort in design optimization to deliver a better product for the client as a competitive strategy. If design is outsourced, the consultant might not be willing to do extra work for higher sustainability unless the client pays for it or requests it. Thus, internal teams provide more control over meeting or exceeding sustainability objectives.

That said, external consultants, especially those specializing in sustainability (energy modelers, LEED consultants, etc.), are valuable. Many contractors without large in-house design groups do hire sustainability consultants to advise on projects. But again, having internal competence is crucial to fully utilizing those consultants. If a sustainability consultant recommends a novel greywater recycling system, an internal engineering team is in a better position to incorporate that into the project design and justify it in terms of cost/benefit. Without internal expertise, a contractor might not know how to implement or might doubt the feasibility, potentially missing an innovation.

In conclusion, achieving sustainability goals strongly aligns with integration. In-house engineering fosters an environmentally conscious culture within the firm, where sustainability is a shared value rather than an external imposition. It allows continuous improvement in sustainable practices across projects (learning what worked for one green project and applying to the next). On the other hand, outsourcing puts more onus on contract specifications to drive sustainability; one only gets what you explicitly contract for. A firm truly committed to sustainability will likely internalize key capabilities or at least maintain very close collaborative relationships with designers, effectively blurring the line between in-house and outsource through integrated teams. The decision framework should therefore consider sustainability-critical functions as candidates for insourcing or close integration, especially if the firm's strategy is to lead in sustainable construction.

Knowledge Retention and Learning

Construction should be considered a "knowledge industry" as much as it is a manufacturing one, since each project yields lessons, and firms that capture and reuse this knowledge can improve over time. However, the project-based nature of construction, where teams often disband after each project, makes knowledge retention challenging (Gann & Salter, 2000). The sourcing of engineering can significantly affect how knowledge is managed.

When engineering is kept in-house, the firm has a far greater ability to retain knowledge from project to project. Internal engineers carry tacit knowledge, the reasoning behind design decisions, the nuances of what was tried and what succeeded or failed, which can inform future designs (Nonaka, 1994). They can develop standard details, checklists, and best practices based on accumulated experience. Many large firms have internal knowledge management systems that their engineering staff contribute to (Argote & Miron-Spektor, 2011). This helps to ensure mistakes are not repeated and successful innovations are replicated. In effect, an in-house engineering department acts as a repository of institutional knowledge.

In contrast, if engineering is outsourced, a lot of the knowledge leaves with the consultants. The contractor might only retain the final drawings and documents, not the thought process or intellectual property behind the design. Also, external designers may not be inclined to highlight things that went wrong in a project to the contractor, since that could expose them to criticism; whereas internal teams can more openly discuss "what we could do better next time" (Blayse & Manley, 2004). Absorptive capacity theory ties in here, as without an internal knowledge base, firms struggle to even recognize what they could learn from an external partner (Cohen & Levinthal, 1990). A contractor might get innovative designs from different consultants on different jobs, but without an internal comparative analysis, they might not notice patterns or integrate those innovations into a coherent company knowledge base. Organizational learning literature emphasizes that for learning to stick, organizations often need to internalize processes and build cumulative expertise (Argote & Miron-Spektor, 2011).

Maintaining in-house teams requires an investment in training and development to keep knowledge updated. The benefit, though, is that training yields long-term returns directly to the firm. A contractor might, for instance, sponsor their engineers to get LEED accreditation or advanced structural certifications, which then benefits all upcoming projects. If that skill was only accessed via outsourcing, the firm could be continuously paying a premium for it rather than building it once. In practice, many firms use a hybrid knowledge strategy: they keep some senior experts in-house who act as stewards of knowledge, even if they outsource a lot of production work. These senior in-house engineers might not design everything themselves, but they review and compile lessons, updating company standards. This model allows cost flexibility while still retaining core know-how (Gann & Salter, 2000).

Another knowledge aspect is innovation diffusion within the firm. Under the standard outsourcing format, if one project team tries something new and it works well, how does that get communicated to other project teams within the company? With an insourcing model, this could be captured through internal training or updates to internal standards (Argote & Miron-Spektor, 2011). Without it, the only communication might be informal or reliant on individual project managers to disseminate the information within the company. Formalizing learning is easier with an internal structure. In conclusion, for firms that view continuous improvement as essential, having an internal mechanism for knowledge capture is crucial. In-house engineering provides that mechanism in the realm of design and technical knowledge. While it is possible to enforce knowledge capture from outsourced projects, it is more difficult and often less effective. The best practice emerging is a balanced approach: maintain a robust internal knowledge core and supplement it with external capacity, ensuring that external contributions are absorbed. In the decision framework, activities that are key to future learning and improvement might be worth keeping in-house to build long-term capability (Cohen & Levinthal, 1990; Nonaka, 1994).

Governance, Control, and Strategic Alignment

Finally, firms must consider how the choice of in-house vs. outsourced engineering aligns with their governance structure and strategic objectives. Governance refers to how decisions are made and monitored within and across organizational boundaries, and strategic alignment refers to how well an activity supports the company's overall strategy and goals.

Internal Governance and Control

Having engineering in-house gives a firm direct control over technical decisions and priorities. The engineers are part of the internal chain-of-command and subject to the firm's management, culture, and incentives. This means management can more easily steer engineering efforts to align with company strategy, whether that's prioritizing cost-saving innovations, focusing on client-specific customizations, or accelerating schedules. It also means informal coordination is possible: issues can be discussed in internal meetings or impromptu conversations, rather than always through formal correspondence. This agility can be crucial when projects need quick decisions. In-house teams can shift resources between projects as needed under the direction of company leadership, which is something much harder to negotiate with independent firms.

By contrast, outsourcing creates inter-organizational governance challenges. The contractor must establish contracts, communication protocols, and oversight mechanisms to ensure the external engineer's work meets its needs. TCE warns of dependency risks here, if a firm outsources a critical function, it becomes dependent on the supplier, which can be risky if the relationship sours or if the supplier underperforms. Companies often put in place extensive monitoring (progress meetings, design reviews, etc.), which increases overhead. Also, decision-making can slow down because of contractual boundaries, e.g., making a change might require a formal change order with the external designer, whereas an internal team might implement it following a managerial directive.

Strategic alignment is another consideration. A company's strategy might emphasize being a leader in certain types of projects or delivering a unique value proposition. If internal engineering is key to that, then it should be controlled internally. If the strategy is to be a low-cost bidder, the firm might outsource more to keep overhead lean, but then it sacrifices some control over quality unless mitigated. Many top contractors see themselves not just as builders, but as integrators of complex systems. For them, internal engineering units could act as innovation hubs that connect with other departments.

There's also the matter of intellectual property (IP) and proprietary methods. If a firm's engineers develop a novel solution in-house, the firm owns that IP and can reuse or even patent it. If engineering is outsourced, typically the design firm might own the IP or at least the know-how resides with them unless contracted otherwise. Internalizing engineering thus allows companies to build a library of proprietary designs, software tools, and processes that add to their competitive assets. Many firms now see data as strategic assets, if created externally, it's important to ensure they are contractually owned by the contractor. In-house creation avoids that ambiguity.

From a strategic agility standpoint, firms that insource can pivot more quickly. Say the market shifts and there is a sudden opportunity in a new type of project (like renewable energy facilities). A contractor with a versatile internal engineering team can reorient them, train if needed, and pursue that market. A firm without that would have to hire new staff or partner with a new design firm, taking more time and perhaps not capturing as much value.

Importantly, with internal control comes the need for good internal governance, having in-house teams requires strong internal management processes. Some firms struggle with under-utilization of internal staff in slow periods, which can create pressure to find them work (sometimes leading to taking projects just to feed the machine). Outsourcing offers simplicity: if you don't have work, you simply do not contract it out. This is where a hybrid governance model can shine.

In terms of corporate strategy, companies that view engineering as central to who they are will obviously keep it in-house. Others who see themselves more as construction managers might outsource design and focus on coordination skills. Both can be valid strategies, but they lead to different organizational structures. It's crucial that a firm's approach to engineering sourcing is consistent with its identity and value proposition. Misalignment, for example, a firm that sells itself as an innovator but outsources all design, could lead to strategic failure.

In summary, governance and strategic alignment often favor a model where core strategic engineering functions are kept in-house, ensuring the company retains control, flexibility, and alignment with its mission, whereas non-core or supplementary functions can be outsourced under robust management. Each firm must define what is "core. The decision framework will prompt executives to ask: *Does this engineering activity underpin our competitive advantage or strategic goals?* If yes, lean towards insourcing. If not, the decision can be based more on cost and resource considerations.

Practical Decision Framework

Bringing together the above criteria, we propose a structured decision framework to guide whether a construction firm should internalize an engineering function or outsource it. The framework involves evaluating the function along three key dimensions derived from our analysis: Strategic Importance, Risk/Complexity, and Cost/Resource Fit. Each dimension influences the make-or-buy decision:

- **Strategic Importance Assessment:** Determine how central the engineering function is to the firm's competitive advantage, innovation goals, or sustainability commitments. If the function directly contributes to the firm's unique value proposition or long-term capability, it has high strategic importance and favors insourcing. Conversely, if it is a generic support function with low strategic impact, it may be suitable for outsourcing.
- **Risk and Complexity Evaluation:** Evaluate the level of uncertainty, complexity, and integration required for the task. If the engineering task is tightly integrated with construction means and methods, or carries significant safety/quality risk, it likely favors internal delivery or at least a high degree of internal oversight. High-risk activities require close control and fast coordination. If the task is low-risk, well-understood, and modular, outsourcing is less risky and can be considered. Complex tasks might be handled by a hybrid model to balance expertise needs and oversight, outsourcing complex design tasks while maintaining a high degree of internal oversight using in-house expertise.
- **Cost and Resource Analysis:** Consider the availability and efficiency of internal resources relative to the market. If the firm has or can develop internal expertise at a reasonable cost, and the task would keep those resources effectively utilized, insourcing can be justified. If outsourcing offers genuine cost savings even after accounting for transaction/coordination costs, and the task is not strategic or high-risk, then outsourcing may be prudent, especially for routine, surge, or one-off needs. Also consider resource flexibility: if internal staff are at capacity, outsourcing additional workload can be a relief valve.

In practice, a company can map its engineering activities using these dimensions using a matrix with Strategic Importance on one axis and Risk/Complexity on the other and then considering Cost as a filter or third dimension (see Figure 1). Activities that are high in strategic importance and high in risk/complexity (upper right quadrant) align best with in-house execution. This is where the firm wants control and has most to gain by investing internally. On the opposite end, activities low in strategic importance and low in complexity (bottom left quadrant) are prime candidates to outsource to efficient external providers. The other quadrants represent intermediate cases:

- High strategic importance but low complexity (bottom right) might call for selective insourcing, the firm keeps core aspects in-house to retain knowledge/advantage but could outsource some parts once the overall approach is set.
- Low strategic importance but high complexity (upper left) often suggests a hybrid model, these tasks are not core to the firm's identity but are risky enough that pure outsourcing is dangerous. A solution is to outsource to a specialist but maintain strong internal oversight or co-management.

By applying this framework, firms can make more nuanced decisions than a one-size-fits-all policy. It helps avoid the extremes of outsourcing everything or internalizing everything, neither of which is optimal universally. Instead, it points to a portfolio approach to sourcing: internalize where it counts most, outsource where it makes sense, and manage the interfaces carefully.

Figure 1 below illustrates a simplified decision matrix mapping engineering activities by their strategic importance and risk/complexity. It indicates the general recommendation for each quadrant, assuming cost considerations are not prohibitive in the in-house cases:

Figure 1. Engineering sourcing decision matrix

		Strategic Importance	
		Low	**High**
Risk/Complexity	**High**	Hybrid Model or Oversight-Heavy Outsourcing	In-house Delivery
	Low	Outsource	Selective Insourcing

Cost/Resource Fit as a Filtering Layer
-If in-house resources are available and cost-effective: →Favour Insourcing -If outsourcing yields substantial savings and internal capacity is limited: →Favour Outsourcing -If cost difference is marginal: →Default to strategic alignment and risk control in decision

By systematically analyzing engineering functions through these lenses, construction firms can develop sourcing strategies that bolster their innovative capacity and sustainability performance while managing cost and risk. The following case studies will provide concrete illustrations of how different companies have navigated these decisions and the outcomes they experienced.

CASE STUDIES

To ground the discussion in real-world contexts, this section examines several case studies of construction and engineering organizations. These cases illustrate a range of approaches to in-house versus outsourced engineering and highlight lessons regarding innovation, sustainability, and strategic outcomes. We cover both positive examples where internal engineering drove success, and cautionary tales where missteps in sourcing led to problems. The cases include major Canadian contractors (EllisDon, PCL, SNC-Lavalin) and international firms (Fluor, Bechtel).

EllisDon: Strategic Insourcing and Innovation

EllisDon is one of Canada's largest construction services companies, known for its entrepreneurial culture and willingness to adopt new practices. Over the past decade, EllisDon deliberately insourced key engineering and technological functions to drive innovation. It established internal divisions such as Construction Engineering Services (to handle complex structural design, sequencing, and temporary works in-house) and Building Digital (a tech-focused unit developing digital tools and BIM capabilities for construction). This strategic insourcing was a departure from the traditional general contractor model and was driven by a belief that integrating design and construction would yield better project solutions and even new revenue streams.

According to EllisDon's own statements, being "fully integrated" and "relentlessly innovative" allows the company to mitigate risk and offer unmatched value to clients (EllisDon, 2025). They see their internal engineering and tech units as central to delivering one-stop solutions. A practical example is EllisDon's work on complex healthcare and infrastructure projects. With in-house engineers, they have innovated in modular construction and even experimented with 3D printing of building components for hospitals, which helped reduce on-site construction time. During the COVID-19 pandemic, these capabilities paid off: when hospitals needed reconfiguration for isolation rooms and airflow changes, EllisDon's in-house design/detailing team quickly re-engineered layouts and systems to meet new health requirements without having to wait on external consultants, enabling fast adjustments to projects in progress.

The EllisDon case illustrates that insourcing engineering fostered a proactive, problem-solving culture. The company even launched an internal innovation accelerator program, where their engineers collaborate with startups to pilot new construction technologies on EllisDon projects. By having internal experts engaged, they can tailor these new technologies to real project needs and rapidly implement them. This synergy between internal knowledge and external innovation opportunities is a hallmark of their strategy. The result has been an enhanced market reputation: EllisDon is explicitly sought after for design-build and P3 (public-private partnership) projects because they can handle both design and build in-house, providing single-point responsibility. In competitive bids, this integrated capability often becomes a decisive factor in winning contracts.

Key takeaway from EllisDon: A deliberate strategy of insourcing critical engineering and embracing technology has enabled greater innovation and adaptability.

It required investment in talent and creation of new internal departments, but it positioned the firm as a leader in delivering complex, fast-track, and innovative projects. The company's experience shows that in-house engineering can not only solve current project challenges but also open new business avenues.

SNC-Lavalin: A Hybrid Model Approach

SNC-Lavalin is a long-established Canadian engineering firm that expanded into construction contracting, making it an example of a hybrid model. Historically an engineering powerhouse (especially in sectors like civil infrastructure and energy), SNC-Lavalin in the 2000s acquired construction companies to offer full EPC (Engineering, Procurement, Construction) services. Essentially, SNC moved from pure engineering into also executing construction, thereby insourcing construction capability into an engineering firm. However, by the late 2010s, after experiencing significant project losses on some turnkey construction projects (due to cost overruns and execution difficulties), SNC-Lavalin re-evaluated this strategy.

The company decided to reduce its exposure to self-performing large-scale construction and refocus on its core strength: engineering and project management. It adopted what could be called an "asset-light" approach to construction. In practice, SNC-Lavalin now often acts as an EPCM (Engineering, Procurement, and Construction Management) contractor. It keeps engineering and design in-house, which has always been its forte and a source of innovation. However, it outsources much of the on-site construction work to subcontractors or partners. This model attempts to capture the best of both worlds: maintain in-house the high-value, innovative work (design, IP, specialized engineering) while outsourcing labor-intensive or lower-margin execution work to avoid the fixed costs and risks of directly employing large construction crews.

In terms of our framework, SNC-Lavalin treats engineering as high strategic importance and construction work as lower strategic importance to them. The risk of outsourcing construction is mitigated by SNC's strong project management processes and oversight; they have had to develop rigorous systems to manage and integrate the outsourced construction activities to ensure quality and schedule are maintained. This includes clear contractual frameworks, on-site SNC managers, and often using joint ventures to align incentives. Behaviorally, trust between SNC's internal engineers and the external contractors is crucial; they work on fostering collaboration so that outsourcing doesn't lead to finger-pointing.

The outcomes for SNC-Lavalin have been mixed. On one hand, refocusing on engineering has stabilized the company after a tumultuous period, and it aligns with what they historically do best. They've secured contracts where clients want SNC's design expertise but are okay with them not self-performing all construction. On the other hand, some projects still encountered issues, indicating that just keeping engineering in-house isn't a cure-all if the governance of outsourced segments isn't excellent.

The SNC-Lavalin case demonstrates a different side of the coin: even a firm with deep in-house engineering may choose to outsource other portions of work to reduce risk. It underscores that defining what is core is a strategic choice. SNC concluded that detailed construction execution was not core to their identity or advantage, whereas cutting-edge engineering and project integration is. So they realigned accordingly. For other firms, the line might be drawn elsewhere. The key is that SNC's approach still values in-house engineering highly, it did not outsource that, reaffirming that they see it as integral to innovation, while outsourcing what they view as commoditized or risk-heavy construction work.

PCL Construction: In-House Capabilities Fueling Integration

PCL Construction is another major Canadian contractor, notable for being employee-owned and for its broad portfolio (buildings, civil infrastructure, industrial projects). Traditionally, PCL, like many large general contractors, did not do most design in-house, it worked with external architects and engineers, especially in design-bid-build contexts. However, over time PCL has built substantial in-house technical capabilities in areas that interface closely with construction, essentially insourcing for integration rather than for offering standalone design services.

PCL employs engineers for functions like construction planning (means and methods), temporary works design (shoring, formwork, crane foundations), BIM/VDC coordination, and quality management. They have an "Integrated Construction Services" group that includes structural and civil engineers who, for example, design crane bases, formwork systems, and sequencing plans internally. The reason is that these elements directly affect field productivity and safety, and PCL found that having in-house experts to handle them ensures they fit their construction approach perfectly.

While PCL still typically partners with external design firms for the permanent works, its internal engineers collaborate during preconstruction to influence those designs. They embed with design teams to guide constructability. A case in point: on large oil sands projects in Alberta, PCL experienced costly delays when late design changes by external engineers disrupted construction. In response, PCL began embedding its own engineers earlier in the design phase to work with the designers so that the plans would be construction-friendly and to catch any issues early.

Moreover, PCL's in-house efforts extend to R&D in sustainability and new construction methods. This includes researching low-carbon construction methods and then applying them across projects, which gives them an edge as clients increasingly prioritize green construction. In essence, PCL leverages in-house engineering not to replace external design consultants, but to augment and challenge them, ensuring the final product is both innovative and buildable. This synergy often comes to fruition in design-build projects, where PCL has won numerous awards. They credit their ability to integrate design and construction (through processes and internal expertise) as a factor in that success.

PCL's model shows that a contractor does nt need to do all design internally to benefit from internal engineering. By focusing internal talent on the connection between design and construction, they improve outcomes. This aligns with the decision framework: PCL identified which parts of engineering were strategically important for them and insourced. Permanent facility design they still consider outside their core, so they partner for that, but guided by their internal standards and oversight.

In terms of innovation, PCL's internal technical team has allowed them to adopt technologies like BIM enterprise-wide effectively, and they've developed internal software tools to manage field operations that tie into design models, an integration that would be harder relying purely on external data from designers.

In summary, PCL illustrates an integration-focused insourcing: keep inside the company those engineering tasks that ensure plans and reality line up, and capture lessons from construction back into design. The result is improved efficiency and fewer issues on site, which in turn solidifies PCL's reputation for delivering complex projects on time. Clients may not see PCL as a "designer," but they see them as a highly technically competent builder, which is directly related to PCL's in-house engineering strength in construction technology.

Fluor Corporation: Managing Innovation with In-House Engineering

Fluor is a US-based global contractor frequently active in oil, gas, and mining sectors. Fluor traditionally offers fully integrated services (hence the common label EPC for its projects). It has long maintained large multidisciplinary engineering teams in-house as part of its corporate structure. Fluor's case sheds light on how a heavy in-house model competes and the challenges it faces.

Because Fluor retains such broad engineering capacity internally, it has a rich corporate culture of innovation in project delivery. Historically, Fluor developed its own methodologies, for example, it was a pioneer in 3D model-based design. Today, Fluor continues this with initiatives like a "Technology Leadership" program, which encourages internal experts to create and share new practices across the company. They have proprietary project management platforms that integrate engineering, procurement, and construction data, which stem from decades of internal development. All this is possible because they have the scale and commitment to keep R&D largely in-house.

On a project level, Fluor's integrated approach has enabled feats like novel modularization strategies in industrial projects. One example includes the Athabasca Oil Sands project where Fluor's in-house engineers designed large modules that could be built off-site and shipped to the project, significantly cutting on-site construction time. Achieving this required close interplay between design and construction planning, exactly the scenario where in-house capability shines. They had to consider transport limits, lifting techniques, etc., all in the design phase. Fluor could do that internally without handing off between separate companies. The result was an innovation in execution that gave them a competitive advantage in delivering mega-plants efficiently.

However, Fluor's story also highlights the financial risk that comes with doing everything in-house on a lump-sum basis. An example of this is a gas-fired power plant project in 2017 where Fluor faced major cost overruns. Fluor took a hit because, with full in-house scope, they had essentially assumed all design and construction risk. The problems on the project (a series of U.S. gas power projects) led Fluor to re-assess how it manages risk and to be more selective in the fixed-price projects it chases. Notably, Fluor did not respond by outsourcing engineering, instead, it tightened internal processes and project selection criteria (Korman, 2017). This indicates confidence that their integrated model is still their strength, and the issue was how they executed and governed it, not the model itself. Fluor's belief (aligned with RBV thinking) is that their integrated expertise is their differentiator, so rather than break it apart when a project fails, they improve it internally.

Thus, the Fluor case demonstrates both the upside and the challenge of the in-house-heavy strategy. The upside: strong capacity to innovate in methods and technology, enabling tackling of extremely complex projects with less reliance on others (and thus capturing more value). The challenge: the firm bears the full brunt of any shortcomings, there's no external party to pass risk to, and overhead is high. Fluor mitigates the latter by only going after projects that can support that overhead and by constantly trying to improve productivity through innovation.

For other firms, Fluor's experience is a lesson that if you choose to internalize broadly, you must invest correspondingly in risk management and process discipline to make it work and be strategic about the types of projects you pursue.

Bechtel Corporation: Integrated EPC With Self-Performing Capabilities

Bechtel is another global giant, often considered in the same league as Fluor in terms of providing end-to-end project delivery. Historically, Bechtel has an extensive in-house workforce for engineering and construction management, but like most, it subcontracted many portions of physical construction (civil works, equipment erection, etc.). Recently, Bechtel made strategic moves to self-perform more construction work that it used to outsource, effectively internalizing some of the execution labor in addition to engineering.

On some large LNG projects on the U.S. Gulf Coast, Bechtel decided to bring certain previously subcontracted scopes (like earthworks, marine construction for LNG terminals) in-house. They found that by directly employing craft labor and supervisors for these critical scopes, they could better control the schedule and maintain a steady workforce across projects. Shaun Kenny of Bechtel noted that doing more work with their own forces gave "much more flexibility and certainty of outcome," and allowed them to provide continuous employment for skilled crews between projects. Essentially, Bechtel concluded that for their pipeline of projects, insourcing more construction was beneficial, a vertical integration move mirroring what many contractors consider.

This vertical integration by Bechtel signals confidence in its internal processes to manage additional risk. They weighed that the benefits of control (schedule reliability, direct quality oversight, keeping talent employed) outweighed the risks of taking on more labor and safety responsibility. It also aligns with dynamic capabilities: they can move crews from one project to another, maintaining institutional knowledge, rather than losing those crews when a subcontract ends.

The competitive advantage angle for Bechtel is clear: if they can deliver projects faster or more reliably by insourcing, they will gain an edge in winning future contracts (clients often care about schedule certainty and single-point responsibility). Also, employing more of the workforce can have social and economic benefits for the firm (it builds loyalty and a strong culture, etc.). Many construction firms face the decision of which trades or specialties to self-perform; Bechtel's choice in LNG projects is representative of how a firm might carefully analyze that on a market-by-market basis. They found steady enough work to justify self-performing, which many firms can not unless they have backlog to keep crews busy.

At the same time, Bechtel's example and others reveal a common insight: whichever approach a firm takes, governance and oversight are crucial. Companies like EllisDon and PCL that heavily insource still hire many subcontractors, but their strong internal teams allow them to manage those subs better. Likewise, companies that outsource heavily would be wise to maintain at least a basic in-house team to oversee and intelligently manage those contracts. Industry guidelines often recommend that owners or contractors who outsource design should have an internal "intelligent owner" function, essentially a small internal engineering group, to review and direct the outsourced work.

Collectively, these case studies yield several insights:

- Internal engineering can drive innovation and value (EllisDon, PCL examples).
- Deciding what is core to keep in-house is an identity-level strategic choice (SNC-Lavalin and Fluor define it differently, but both succeed when they align with their strengths).
- A unifying theme is the importance of governance and oversight: whether a firm outsources heavily or insources, success hinges on managing the interfaces and processes well. Those with strong internal engineering can even manage outsourced segments better since they possess the technical insight to supervise effectively. Conversely, firms that outsource extensively should still maintain a nucleus of internal expertise to be smart clients.

In the next section, we translate these lessons into broader strategic implications for construction organizations, addressing how they can structure themselves and make decisions to maximize innovation and sustainability performance.

STRATEGIC IMPLICATIONS

The analysis and cases above carry several strategic implications for construction firms weighing in-house engineering against outsourcing. These implications span organizational structure, financial performance, risk management, and long-term competitiveness. Firms must consider these when formulating their sourcing strategy to ensure it supports their overall business objectives.

Organizational and Cultural Implications

Choosing to internalize engineering fundamentally transforms a firm's organization. It shifts the culture toward integration and can break down silos between design and construction departments. Companies that have traditionally seen themselves purely as builders may need to evolve into design-build organizations, which involves retraining staff, redefining roles, and possibly adopting new project management approaches. An implication is the need to invest in leadership and management that understand both design and construction, integrated teams require integrated management. Firms might reorganize into multi-disciplinary project teams rather than sequential hand-offs. Culturally, fostering collaboration and continuous learning becomes paramount. If a company's culture is supporting these, in-house engineering will flourish; if not, simply hiring engineers will not yield benefits until the culture shifts. Organizations also need to consider career development paths for engineers internally, something consulting firms handle as their core, but contractors need to emulate to attract and retain top talent.

Financial and Competitive Implications

From a financial perspective, internalizing engineering turns what might be variable costs into fixed costs. This can pressure margins in downturns but can pay off in booms through higher productivity and value capture. Strategically, firms should align their in-house capacity with their expected volume of work. Those that anticipate steady demand in their niche can afford more internal staff; those in volatile markets might limit insourcing to avoid under-utilization. A potential competitive advantage of insourcing is the ability to offer clients more for the same price, e.g., value engineering, faster schedule, innovative options, which can win work even if the internal overhead slightly increases costs. Over time, a reputation for delivering superior results (fewer change orders, better sustainability, etc.) can allow a firm to command higher bid prices or preferred contractor status, offsetting the cost of internal teams.

However, firms must be mindful of not over-extending. The Fluor example shows that even giants had to become more selective on lump-sum projects because doing everything in-house meant they were exposed to full project risk. Strategically, a firm with in-house engineering might stick to delivery models that reward integration (like design-build or cost-plus projects) and be cautious with models that don't (like low-bid build-only jobs where their extra capability isn't valued).

For outsource-heavy firms, the financial benefit is flexibility and often a lighter balance sheet. However, the strategic risk is lack of differentiation. If everyone is outsourcing to the same pool of engineering firms, it's hard to stand out. These firms compete mostly on price and project management efficiency. They might find themselves in a race to the bottom unless they carve a niche or excel in some other differentiator. Strategically, they may ally closely with particular consultants in long-term relationships to quasi-internalize some benefits without formal integration.

Risk Management Implications

Strategically, firms must decide how much risk they are willing and able to carry. Insourcing brings accountability in-house. Firms with strong risk management processes, insurance coverage, and technical rigor can handle this and benefit from control. Others might wisely avoid internalizing high-risk functions if they lack confidence in managing them, it might be better to pay a premium for a highly qualified external firm that carries professional liability for design. A key implication is that any function that is outsourced but carries significant risk must be tightly managed through oversight or contractual protections. If these safeguards are not in place, the contractor is essentially carrying the risk without the benefit of control, a worst-of-both-worlds scenario.

In-house teams can proactively reduce risk via better quality control and faster problem resolution. This can result in better safety records and fewer disputes, important risk outcomes that also have strategic value.

Innovation Trajectory

Strategically, a firm needs to align its sourcing model with its desired innovation trajectory. A company aiming to be at the forefront of Construction 4.0, sustainability, or any emerging trend likely needs internal capacity to experiment and implement new technologies across projects. This suggests a more insourced model, or at least a strong internal R&D unit. A company content to be a fast follower might rely on external partners to bring innovations and adopt them a bit later. Both are viable strategies; the former can lead to market leadership but requires investment, while the latter saves cost but can risk falling behind if partnerships aren't sufficient. This kind of responsiveness, which includes the ability to reconfigure technical strategy, workflows, and talent quickly, is a hallmark of agility. In-house engineering plays a foundational role in supporting that capability.

It's also worth noting the absorptive capacity implication: a firm with too little internal expertise may one day find itself unable to pivot when an industry shift happens (e.g., widespread BIM mandates or sustainability standards). Building up a base level of internal engineering knowledge is like strategic insurance for adaptability.

Talent Implications

A strategic side effect of in-house engineering is becoming a more attractive employer for technical talent who want to see projects through the entire lifecycle. Many engineers enjoy working for contractors that allow them to be involved in design and then see their designs built in real time, as opposed to a pure design office environment. This can help firms recruit top young talent who are interested in integrated project delivery and innovation. On the flip side, maintaining that talent requires providing growth opportunities.

In conclusion, the strategic implications underscore that the in-house vs outsource decision is not just a cost or operational issue; it is a long-term strategic positioning choice. Firms need to consider where they want to be on the innovation curve, how they will differentiate, how they will manage risk and knowledge, and what culture they want to foster. The decision should flow from these considerations. Many successful firms blend models, but even in blending, one can see which parts they consider strategic by what they keep inside versus contract out. The next section provides recommendations for executives and managers on how to implement these strategic insights in practice.

RECOMMENDATIONS

Based on the analysis of theory, industry trends, and case studies, we offer the following actionable recommendations for construction firms seeking to leverage in-house engineering as a catalyst for sustainable innovation, while judiciously using outsourcing where appropriate. These recommendations aim to help organizations achieve the best of both worlds: maximizing innovation, quality, and knowledge retention, and maintaining flexibility and cost-effectiveness.

Identify and Internalize Core Competencies

Conduct a strategic assessment of which engineering and technical functions provide your firm with the greatest competitive advantage, risk control, or innovation potential. Internalize those core functions by building or strengthening in-house teams around them. This ensures that key knowledge and skills remain in-house and continuously improve. Conversely, do not feel compelled to internalize functions that are truly commodity-like or not differentiators for your business, those can remain outsourced with proper management.

Maintain an "Intelligent Client" Team

Even if you outsource significant engineering work, always retain a capable internal engineering oversight team. This team should be involved in setting scopes, reviewing deliverables, and liaising with external consultants. They act as the guardian of your interests and the bridge between external designs and internal construction execution. This team also preserves institutional memory and absorptive capacity, ensuring your organization learns from every outsourced project.

Foster Cross-Functional Collaboration

If you bring engineering in-house, actively integrate engineers with construction and other departments. Break down silos by forming multidisciplinary project teams that include engineers, construction managers, estimators, and sustainability experts working together from project inception. Encourage practices like co-location of design and construction staff and regular coordination meetings. This collaboration accelerates problem-solving and innovation. A collaborative culture will maximize the benefits of in-house expertise.

Invest in Knowledge Management Systems

To capitalize on in-house engineering, implement formal knowledge capture and sharing processes. Develop a lessons-learned database where engineers and project managers document insights from each project. Set up internal technical forums or communities of practice for engineers to share innovations and improvements. Ensure that any new technique or solution developed on one project is disseminated across the company. This institutionalizes learning and prevents "reinventing the wheel."

Leverage In-House Teams for Sustainability Goals

Use your internal engineering capacity to embed sustainability into your projects at every stage. For firms with sustainability objectives, task in-house engineers with early-stage evaluations of energy efficiency, carbon footprint, and materials options on each project. Set internal sustainability targets and empower your engineering team to achieve them through design optimizations. By doing this in-house, you maintain control and are able iterate quickly. Additionally, consider creating roles such as "innovation or sustainability champions" within your team. These individuals can drive sustainable innovation and ensure it aligns with construction practices.

Develop Strategic Partnerships for Non-Core Areas

For engineering areas you choose to keep outsourced, build long-term partnerships with reputable consulting firms or specialists. Treat them as extensions of your team: share your standards, involve them in debriefs, and possibly enter alliance agreements that incentivize mutual success beyond one-off contracts. This creates a more integrated approach even in outsourcing and secures priority and consistency. This approach mitigates some disadvantages of outsourcing by building trust and better communication.

Right-size and Train Your Internal Team

Aim for an internal engineering team sized to handle a baseline workload plus some flexibility for innovation tasks, but not so large that they'll be idle in slow periods. Use outsourcing as a buffer for peak loads. Continuously train your in-house team in new technologies, software, and methods, invest in their development so they bring cutting-edge ideas into the firm. This keeps your internal capability sharp and up-to-date with external industry advances, improving your absorptive capacity.

Employ Rigorous Risk Management and QA Processes

If you insource engineering, implement strict internal quality assurance and peer review processes to avoid complacency. It can be valuable to sometimes use external peer reviewers for critical designs even if done in-house, to get independent validation. Ensure that internal designs are thoroughly checked and that lessons from any errors are captured non-punitively to improve future work. Conversely, for outsourced designs, conduct rigorous internal reviews before construction and don't assume the external stamp guarantees constructability.

Utilize the Decision Framework Regularly

Make the make-or-buy evaluation a standard step in planning new projects or new business lines. Use the criteria of strategic importance, risk, and cost to decide on a case-by-case basis how to approach engineering. For instance, during project kickoff, convene a meeting to discuss: "Which parts of this project's engineering will we do internally vs. outsource? What is the rationale?" This deliberate approach ensures consistency with your strategic goals and prevents defaulting to habit. Over time, refine your framework thresholds based on outcomes.

Communicate Value to Clients

If you do build strong in-house capabilities, market their value to clients. Highlight in proposals how your integrated team will reduce risk, enhance innovation, or speed up delivery. Provide tangible examples. By educating clients on the benefits of your model, you can often negotiate collaborative contracts that fully utilize your strengths. Similarly, if you maintain key partnerships for some engineering aspects, reassure clients how you manage those seamlessly by showing a track record of successful collaboration.

Implementing these recommendations requires commitment from top management and alignment across the company. Change may be gradual, but the payoff is a more resilient, innovative organization. By thoughtfully blending in-house excellence with smart outsourcing, firms can achieve superior project outcomes and long-term sustainable improvement.

CONCLUSION

In the construction industry's pursuit of sustainable innovation and improved project delivery, the sourcing decision for engineering services emerges as a critical strategic lever. This chapter has explored how in-house engineering, when aligned with a firm's goals and managed effectively, can serve as a powerful catalyst for innovation, enabling construction companies to integrate new ideas, maintain high quality, and meet sustainability targets more effectively than traditional fragmented approaches. Conversely, it acknowledges that outsourcing, in the right contexts, provides flexibility and access to external expertise, and remains an important tool in a firm's strategy. The key is finding the optimal balance and management approach.

Through theoretical lenses like Transaction Cost Economics, Resource-Based View, Dynamic Capabilities, and absorptive capacity, we established that the value of in-house engineering goes beyond just comparing salary costs to consulting fees. It influences a firm's ability to create and protect competitive advantages, to learn and adapt over time, and to coordinate complex, uncertain project elements with agility. Industry trends show a gradual shift from treating engineering as a commodity to be purchased, toward viewing it as an integral part of a contractor's value proposition, especially under delivery models that reward integration and amid rising demands for sustainable, technologically advanced construction solutions.

The Decision Framework presented encapsulates the multifaceted criteria a firm should consider: strategic importance of the function, risk and complexity, and cost/resource availability. Applying this framework helps avoid one-dimensional decisions and instead tailors the sourcing strategy to the specific situation. High strategic, high-risk items generally warrant insourcing, whereas low strategic, low-risk items can be efficiently outsourced; intermediate cases benefit from hybrid approaches. This structured thinking, combined with the illustrative Figure 1, offers a practical guide for decision-making.

Strategically, the chapter highlights that embracing in-house engineering is not an all-or-nothing proposition, but rather about making conscious choices regarding what capabilities to build internally. Those choices should align with a firm's identity and future aspirations. A contractor aiming to lead in sustainable construction might internalize engineering, for instance, whereas one focusing on cost leadership might internalize efficient design-management processes but outsource actual design. There is room in the industry for varied models; what separates high performers is their clarity on when and why they choose one approach over the other, and their ability to manage the chosen approach well.

For firms that do decide to invest in in-house engineering, the chapter's recommendations provide a roadmap for implementation: from integrating teams and managing knowledge to balancing fixed and flexible resources and forging strategic external partnerships. For firms maintaining significant outsourcing, recommendations emphasize retaining internal oversight and knowledge functions to remain a savvy client and to capture lessons for continuous improvement.

In conclusion, the chapter asserts that in-house engineering, wielded strategically, is a strong enabler of sustainable innovation in construction. By embedding engineers within the project lifecycle, companies can more effectively iterate on designs, prevent issues, incorporate sustainable practices, and respond to changing requirements, leading to better project and business outcomes. Yet, the chapter also pragmatically acknowledges that outsourcing will continue to play a role and can complement internal efforts when used wisely. The ultimate message is one of balance and intentionality: construction firms should neither outsource by default nor internalize by ideology but instead should carefully architect their sourcing strategy to support their long-term vision of innovation, excellence, and sustainability in the modern construction landscape.

REFERENCES

Abdel-Rahman, S., Dokhan, S., & El-Dash, K. (2023). Outsourcing engineering tasks by construction firms: State-of-the-art. *Engineering Research Journal, 52*(4), 157–168.

Adam, A., & Lindahl, G. (2017). Applying the dynamic capabilities framework in the case of a large public construction client. *Construction Management and Economics, 35*(7), 420–431. DOI: 10.1080/01446193.2016.1274452

Aghimien, D., Aigbavboa, C., & Matabane, K. (2023). Dynamic capabilities for construction organizations in the Fourth Industrial Revolution era. *International Journal of Construction Management, 23*(5), 855–864. DOI: 10.1080/15623599.2020.1813832

Argote, L., & Miron-Spektor, E. (2011). Organizational learning: From experience to knowledge. *Organization Science, 22*(5), 1123–1137.

Barney, J. (1991). Firm resources and sustained competitive advantage. *Journal of Management, 17*(1), 99–120. DOI: 10.1177/014920639101700108

Bechtel Corporation. (2025). *Dream, Design, Deliver.* Retrieved from https://www.bechtel.com/approach/

Blayse, A. M., & Manley, K. (2004). Key influences on construction innovation. *Construction Innovation, 4*(3), 143–154.

Cohen, W. M., & Levinthal, D. A. (1990). Absorptive capacity: A new perspective on learning and innovation. *Administrative Science Quarterly, 35*(1), 128–152.

Construction, P. C. L. (2024). *Company website.* Retrieved from https://www.pcl.com

Deloitte. (2024). *U.S. global outsourcing survey 2024 report: Multidimensional sourcing.* Deloitte Consulting LLP.

EllisDon. (2025). *EllisDon furniture, equipment and design division more than meets the eye.* Retrieved from https://www.ellisdon.com/news/ellisdon-furniture-equipment-and-design-division-more-than-meets-the-eye

Engineering.com. (2016). *Outsourcing CAD services.* Retrieved from https://www.engineering.com/outsourcing-cad-good-bad/

ENR. (2009). *State Engineers Union Claims Private Engineers "Cost Double".* Retrieved from https://www.enr.com/blogs/12-california-views/post/15255-state-engineers-union-claims-private-engineers-cost-double

Fewings, P., & Henjewele, C. (2019). *Construction project management: An integrated approach* (3rd ed.). Routledge.

Fluor Corporation. (2016, October 18). *Fluor completes module fabrication for oil sands project.* Retrieved from https://newsroom.fluor.com/news-releases/news-details/2016/Fluor-Completes-Module-Fabrication-for-Oil-Sands-Project/default.aspx

FR-504. (2022). *Final report: Transportation infrastructure evaluation and performance assessment* (Report No. FR-504). Louisiana Transportation Research Center, Louisiana State University.

Gann, D. M. (2001). Putting academic ideas into practice: Technological progress and the absorptive capacity of construction organizations. *Construction Management and Economics, 19*(3), 321–330. DOI: 10.1080/01446190010022632

Gann, D. M., & Salter, A. J. (2000). Innovation in project-based, service-enhanced firms: The construction of complex products and systems. *Research Policy, 29*(7-8), 955–972.

Goh, E., & Loosemore, M. (2017). The impacts of industrialization on construction subcontractors: A resource-based view. *Construction Management and Economics, 35*(5), 288–304. DOI: 10.1080/01446193.2016.1248985

Kenny, S. (2022, February). The innovation function: An interview with Shaun Kenny of Bechtel. *McKinsey & Company.* Retrieved from https://www.mckinsey.com/industries/infrastructure/our-insights/the-innovation-function-an-interview-with-shaun-kenny-of-bechtel

Korman, R. (2017, August 8). *Fluor stumbles on fixed-price power projects.* Engineering News-Record. Retrieved from https://www.enr.com/articles/42478-fluor-stumbles-on-fixed-price-power-projects

KPMG. (2023). *Future of outsourcing.* KPMG LLP.

Nonaka, I. (1994). A dynamic theory of organizational knowledge creation. *Organization Science, 5*(1), 14–37.

Oakland, J. S., & Marosszeky, M. (2017). *Total construction management: Lean quality in construction project delivery.* Routledge.

Radhakrishnan, K. (2010). *Development of a decision-making procedure for outsourcing engineering services in TxDOT.* Texas Department of Transportation.

Shibeika, A., & Harty, C. (2015). Diffusion of digital innovation in construction: A case study of a UK engineering firm. *Construction Management and Economics, 33*(5-6), 453–466.

Sinha, K. K., Osiyevskyy, O., & Radnejad, A. B. (2022). Propositions for R&D governance regimes: A behavioral perspective. *International Journal of Innovation and Technology Management, 19*(6), 2250017.

SNC-Lavalin. (2022, May). *Investor relations overview: May 2022*. SNC-Lavalin. Retrieved from https://s29.q4cdn.com/301499865/files/doc_presentations/2022/06/SNC_Investor-Relations-Overview-May-2022_FINAL.pdf

Verbeke, A., & Greidanus, N. S. (2009). The end of the opportunism versus trust debate: Bounded reliability as a new envelope concept in research on MNE governance. *Journal of International Business Studies, 40*(9), 1471–1495. DOI: 10.1057/jibs.2009.44

Walker, A., & Kwong Wing, C. (1999). The relationship between construction project management theory and transaction cost economics. *Engineering, Construction, and Architectural Management, 6*(2), 166–176. DOI: 10.1108/eb021111

Walker, G., & Weber, D. (1984). A transaction cost approach to make-or-buy decisions. *Administrative Science Quarterly, 29*(3), 373–391.

Williamson, O. E. (1996). *The mechanisms of governance*. Oxford University Press.

KEY TERMS AND DEFINITIONS

Absorptive Capacity: The ability of an organization to recognize the value of new external information or knowledge, assimilate it, and apply it to commercial ends. Absorptive capacity in construction determines how effectively a contractor can learn from external sources and use that knowledge to innovate or improve. Higher absorptive capacity often requires a base of relevant prior knowledge and skilled internal staff who can interpret and integrate new insights.

Dynamic Capabilities: The abilities of an organization to integrate, build, and reconfigure internal and external competences to address rapidly changing environments. Dynamic capabilities enable firms to sense opportunities or threats, seize opportunities by mobilizing resources, and transform their operations accordingly.

In-House Engineering: The practice of performing engineering design and related technical services within a construction firm's own organization, using its direct employees. In-house engineering involves integrating design capabilities under the same corporate roof as construction management, allowing for closer coordination and alignment with the company's goals.

Outsourced Engineering: The practice of contracting external engineering consultants or firms to perform design and technical services for a project. Outsourcing transfers responsibilities that could be done internally to outside entities, typically through formal agreements. It is often used to access specialized expertise or adjust to workload fluctuations but requires managing external parties through contracts and communication.

Resource-Based View (RBV): A management perspective that sees a firm's resources and capabilities as key to achieving competitive advantage. Under RBV, resources that are Valuable, Rare, Inimitable, and Non-Substitutable (VRIN) can lead to sustained competitive advantage. In the context of construction, a strong in-house engineering team can be viewed as a strategic resource if it develops unique competencies or innovations that competitors cannot easily replicate.

Sustainable Innovation: Innovations in processes, technologies, or designs that improve environmental, social, or economic sustainability in construction. Sustainable innovation might include new methods to reduce carbon emissions, increase energy efficiency of buildings, minimize waste, or enhance the resiliency and life-cycle performance of infrastructure. It implies an advancement that not only yields business benefits but also positive sustainability outcomes.

Transaction Cost Economics (TCE): A theoretical framework used to understand and predict whether a firm should "make" (insource) or "buy" (outsource) a given activity. TCE considers the costs of using the market (e.g., negotiating and enforcing contracts, potential opportunism by suppliers) versus the costs of organizing the activity within the firm (administrative and coordination costs). The theory suggests firms internalize activities when external transaction costs are high.

Chapter 3
Role of AI in Mitigating Unsustainable Practices

Kirti Dilip Singh
https://orcid.org/0009-0003-4161-2871
Amity University, India

Meenakshi Deshwal
Amity University, India

Vijay Bhutani
Amity University, India

Manjeet Singh
https://orcid.org/0009-0002-7049-5937
Amity University, India

Sunil Kumar Sansaniwal
https://orcid.org/0000-0002-0997-5096
NITI Aayog, New Delhi, India

ABSTRACT

AI presents a paradox in business sustainability, addressing high-end environmental and social challenges while posing risks of algorithmic bias and ethical concerns. Sustainable AI requires balanced, risk-focused regulation that maintains technology neutrality and proportionality. Without oversight, AI can harm the economy, society, and the environment. Strong policies and enforcement are vital for ethical AI management, ensuring it serves as a common good. Integrating AI with blockchain enhances transparency, traceability, and accountability in supply chains, driving

DOI: 10.4018/979-8-3373-2555-2.ch003

sustainable industrial transformation and carbon accounting. However, governance must align with environmental interests to mitigate socio-ecological risks. Inclusive policy norms counteract information asymmetry and promote equitable benefits.

INTRODUCTION

AI is emerging as an essential component in almost all sectors, but the impact on the environment is contradictory. Its extensive computing resources increase carbon emissions. Additionally, mining the necessary minerals often exploits marginalized groups. As climate change reaches crisis levels, the incorporation of AI as an infrastructure that increases carbon emissions needs scrutiny. Sustainable AI provides a way to manage environmental justice throughout its life cycle, (Chen et al., 2023). To some degree, these measures will ensure a responsible, ethical approach to Deployed AI in an ecological collapsed world. The intangible environmental impacts of AI are primarily exceptional sustainability threats. While AI is extensively applied in response to climate change, its use and environmental footprint are comparatively less studied. Inadequate disclosure of the environmental impacts of AI impedes quality decision-making, (Onyeaka et al., 2023).

Figure 1. Data foundation of AI

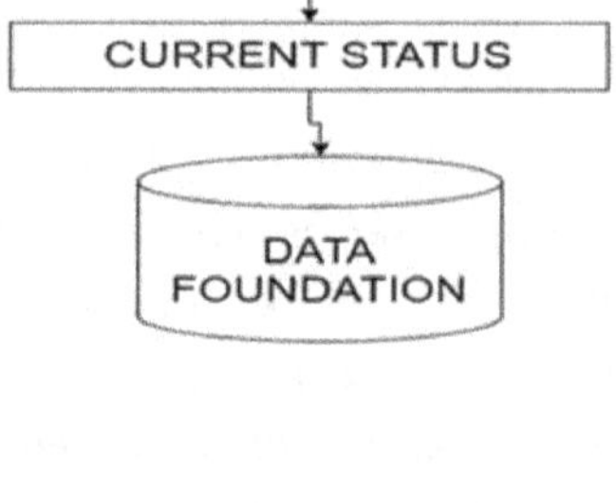

Incorporating AI into infrastructure requires ethical consciousness since what is being created in AI today will determine the future's sustainability. Since AI is being applied aggressively, its sustainability must be a research and development area to counter long-term sustainability risks and enable effective technological development. Unsustainable behavior is widespread across all industries and has catastrophic social and environmental impacts. Monoculture farming, excessive use of pesticides, and deforestation for agriculture result in land degradation, loss of biological diversity, and climate change, (Onyeaka et al., 2023). The manufacturing process of fossil fuel-based industries generates high carbon emissions and toxic waste. Fast fashion in the apparel industry generates high textile waste and contaminates water bodies. The technology industry has electronic waste and energy-guzzling data centers resulting in excessive resource consumption. Transport is the largest polluter as it is fossil fuel based. Systemic change implies that there must be policy change, technological innovation, and circular economy shift, (Onyeaka et al., 2023; Popescu et al., 2024).

The consequences for the environment are clear, particularly with the current day overexploitation of resources. There are advanced measures to develop to control climate change. These sustainable developmental farming practices also clear forests, practice monoculture, and use high amounts of chemical fertilizers and pesticides. These practices result in the destruction of soil, loss of biodiversity, and the pollution of water. The reliance on fossil fuels in industrials production results in the discharge of greenhouse gases along with toxic water and air pollutants, (Srivastava & Maity, 2023). The fashion industry promotes fast fashion, which results in overproduction, ginormous textile waste, and chemical dye water pollution. The technology industry also produces ginormous electronic waste, while data centers, being energy-intensive, produce ginormous electricity, resulting in further carbon emissions. The transport industry is also a severe polluter, where overuse of fossil fuels produces greenhouse gas emissions and air pollution in urban areas. Even the construction industry pollutes the environment due to overuse of resources, high energy usage, and waste generation. These unsustainable operations not only ruin ecosystems but also endanger human health and economic stability. Systemic reforms are required to address these issues, such as stricter regulations, technology, and transition to sustainable alternatives like renewable energy, circular economies, and sustainable consumption. These issues need to be addressed through collaborative efforts by governments, industries, and individuals to develop a more sustainable future, (Olawade et al., 2024; Haleem et al., 2023).

Harnessing Artificial Intelligence for Sustainable Development Goals

The SDGs provide a useful structure to assess the effects of AI on Society, Economy, and Environment. According to the expert review AI contributes positively to 79% of SDG targets and the most affected sector is Environment (93% of targets positive). AI is, nevertheless, deleterious to Society as 38% of the targets negatively interacted with AI.AI facilitates SDGs 8 and 9 by generating employment and improving productivity, while SDG 1 is facilitated by poverty monitoring with satellite-based technologies. AI facilitates SDGs 7 and 13 by improving energy efficiency in smart grids and air pollution management, (Zhang et al., 2024). AI facilitates SDGs 14 and 15 by improving biodiversity conservation and natural resource management. AI is a major threat to the achievement of SDGs in the Economy and Society. Loss of jobs is one of the main concerns, though several studies have estimated that the net impact could not be negative. Asymmetric access to computing resources for AI can also increase economic inequalities, making societies more unequal. Algorithmic discrimination and social polarization by AI can also occur, particularly in data-intensive use such as contact tracing of COVID-19. Abuse of such technology can also increase polarization of societies, impacting peace and stability, which are the heart of SDGs 16, and another pressing issue is the growing energy demand of AI. ICT electrical demand globally would rise from 1% at present to potentially up to 20%, (Banerjee et al., 2022). Such remarkable growth could pose additional strains on energy resources as well as trigger environmental degradation. Though there are numerous benefits of AI, its unintended consequences need to be dealt with by ethical engineering and equitable access. To avoid economic and social inequalities and retard its environmental footprint, a judicious use of AI must be promoted. Policy makers and stakeholders must implement long-farsighted policies to balance the pros and cons of AI so that technology can further evolve to benefit, not jeopardize, overall sustainable development in every aspect, (Kaya, 2025).

Artificial Intelligence for a Thriving 21st Century

Some issues such as ethical ones, problems of security or even policies may derive from the immeasurable difficulties AI can solve. The key concern of AI is the advancement of security, which can be achieved through the minimization of human errors. As an example, self-driving cars are less prone to accidents than regular vehicles. As well, most hospital systems powered by AI are efficient in managing patient care while minimizing medical errors, (Singh et al., 2021; Chavan et al., 2024). AI supports rapid detection of diseases like lung cancer and blindness in diagnosis, and speeds up medical research and the creation of drugs. AI also revolutionizes education by personalizing learning and delivering it globally. But AI can be disastrous if misused. Data mining errors such as the Cambridge Analytica scandal show the potential of AI to control human behavior and sway elections. AI-guided drone attacks caused casualties that raise the moral concerns of autonomous weapons. Financial markets are not spared either, such as the Knight Capital trading system failure lost $440 million. And security vulnerabilities in AI-controlled air systems, such as Boeing's code, show the vulnerability to cyber-attacks. The rapid evolution of AI is difficult to ensure its ethical, fair, and sustainable use, (Ataş &

Talay, 2022). The most significant challenge is that the policymakers lack a proper vision of the possibilities of AI, and thus it cannot be regulated and enforced. The most difficult task is to gain the greatest benefits of AI while simultaneously halting its misuse. With the pace with which AI is developing, it is an urgent requirement to formulate rules and regulations, (Boyer et al., 2016).

One of the possibilities is to create a clear line of demarcation between acceptable and unacceptable uses of AI. It is not the case that it is impossible to achieve world consensus over the regulation of AI, but there is a precedent. The Asilomar Conference did manage to elicit the world's agreement to ban bioweapons, showing that world consensus over the governance of technology is achievable. The Asilomar AI Principles also provided ethical guidelines for the design of AI that are being signed up to by leading researchers, business leaders, and policymakers. A shared vision for AI globally, rooted in the Sustainable Development Goals (SDGs), can make AI work for humanity, (Brevinni, 2020). AI can be employed to fuel prosperity, accelerate scientific breakthroughs, assist in averting climate change, revolutionize healthcare, and enhance global happiness. Collaboration, transparency, and responsible use of AI can enable the world to reap the benefits of AI for the greater good.

AI IN ENVIRONMENTAL SUSTAINABILITY

The Broadband Commission for Sustainable Development emphasized the potential of technology to accelerate the pace of achievement towards global goals, notably the narrowing of the digital divide. With the complementarities of digitalization and sustainability, the Commission aimed to make ICT access global. Melinda Gates and Jack Ma then co-authored The Age of Digital Interdependence, advocating five main proposals in a bid to actualize the potential of digital technology without danger.AI, in contrast to previous ICTs, is self-developing and self-learning, and therefore accountability and governance are required. If left unregulated, AI could result in digital pollution, e.g., privacy infringement, bias, and discrimination, (El Etaiby, 2023; Popkova, 2023). AI is value-neutral by itself, but the use imparts value implications. Misuses can exacerbate inequality, erode trust, and result in defective decisions. Proper governance of AI is the key to how technology can be a blessing to humankind. Environmental sustainability can be enhanced by Artificial Intelligence (AI) at a point where the utilization of resources is maximized, the carbon footprint is reduced, and conservation processes are accelerated. Of all the applications of AI, climate modeling is especially potent; large volumes of data are sifted through by advanced machine learning algorithms to predict weather patterns more precisely, monitor climate change, and predict natural disasters. These projections enable policymakers and scientists to form preventive measures of climate resilience, disaster preparedness, and climate mitigation. Artificial intelligence enhances the utilization of renewable sources of energy like wind, solar, and hydroelectricity by predicting changes in demand, improving the efficiency of the grid, and controlling energy storage units, (Tokhayeva et al., 2020). AI-supported smart grids do away with energy loss and therefore guarantee the supply of stable and reliable electricity. Further, AI-supported automation of industrial production processes leads to processes that are optimized, minimized resource utilization, less material wastage, and shorter supply chains.

But with all its benefits, artificial intelligence is at risk from sustainability due to the energy needs of large machine learning models and data centers. To address the challenge, innovation in energy-efficient AI algorithms, green hardware, and increased reliance on renewable energy sources for AI use is necessary. Researchers are developing solutions such as neuromorphic computing and federated learning to make AI low carbon without sacrificing its performance, (Son, Chen, & Syu, 2024).

With the incorporation of AI in environmental policy, societies will be able to make wiser, smarter choices to address pressing ecological issues. In climate action, energy conservation and efficiency, agriculture, or urban planning, AI is a great tool for sustainability. The future is in the future development of technology, as the key is to find a balance between the potential of AI and environmental and ethical issues to integrate it into a greener, more sustainable future for generations to come.

AI for Climate Change Prediction and Mitigation

Artificial Intelligence (AI) is revolutionizing climate change prediction and mitigation by facilitating accurate prediction, informed decision, and new means of environmental footprint minimization. AI-based climate models analyze enormous amounts of real-time and historical data to generate climate pattern forecasts, predict extreme climatic events, and improve long-term climate pattern estimates.AI-driven knowledge helps governments, scientists, and policymakers to strategize effective disaster preparedness programs, coastline protection, and resource management. AI also helps prevent climate change by optimizing renewable resources like wind, sun, and water energy. Machine learning algorithms maximize energy efficiency by forecasting demand, grid management optimization, and minimizing fossil fuel use. Moreover, manufacturing of AI-based automation saves emissions through optimal manufacturing, wastage minimization, and optimal logistics. With water and fertilizer optimization, AI-driven precision agriculture allows farmers to yield more with less.

AI Powered Renewable Energy Optimization

The rise in demand for sustainable energy resources has benefitted significantly from AI technology thanks to its application in enhancing efficiency, cost, and overall system reliability. Use of AI with renewable energy is changing the entire landscape of clean energy generation, storage and distribution. AI alongside Machine Learning and modern analytics make it possible to forecast energy production and optimize the management of electrical grids and storage systems.

A big challenge to renewable energy is the reliability of solar and wind power which is dependent on specific weather conditions. Maintenance of solar and wind powered energy generators becomes more convenient because anticipating system failure usually takes less time, (Saadi & Azdimousa, 2023). Better predictions result in managers being able to reduce the amount of fossil fuel-based generators in service. AI absorbs these issues by analyzing vast meteorological databases in conjunction with historical energy production to create sound predictions. More accurate assumptions lead to better control of supply and demand balance on the grid resulting in a higher, moreover, AI is now simplifying the decentralization of energy systems by enabling small producers such as solar panel homeowners to sell their excess energy back to the grid. The use of AI in demand response Systems enables more rational distribution of energy that minimizes waste, making it possible to promote sustainability. The integration of AI within renewable electricity systems not only enhances efficiency level but also moves towards achieving a more sustainable and cleaner environment. AI has the potential to positively shift the energy paradigm from skepticism over fossil fuels to embracing renewable energies over the next few decades, (Rashid & Rasheed, 2022; Rojek et al., 2024).

AI Applications in Carbon Footprint Reduction and Emission Control

Fifth Industrial Revolution is evolving with astonishingly new ideas and technologies. Artificial Intelligence is one such peak of innovation and modern-day technology. AI is used in virtually all aspects of life, from construction to agriculture, healthcare, business, military, entertainment, and education. It's not surprising that AI is undergoing such widespread applications, given how advanced AI development technologies have become. We must recognize, though, that they do not come at no cost. Everything needs massive material resources and energy, including human labor. Their existence creates, apart from their computational power and possibilities, countless ethical problems that need closer scrutiny particularly from sustainability perspective, (Challoumis, 2024). Environmental sustainability of AI Practically all the technologies which we use today have their source from the internet. It doesn't matter if it's a human or machine doing it, the process will always need energy. With the push for creating an AI, there is also a need for power, which presently comes from fossil fuel. There is always a huge relationship between concerns regarding the environment and the digital infrastructure Vinessa and other authors in their research on AI technology in accomplishing SDGs treated AI functions as simple instruments, paying no attention to the economic implications stemming from their deployment respectively, (Vinuesa et al., 2020). The first attempts have been undertaken to check whether these hopes are likely to come to fruition. AI is significant in emission reduction and the mitigation of carbon footprints in different sectors. AI helps to extract sense out of the abundant data available to assist business and governmental authorities in monitoring, forecasting, and minimizing their ecological footprint. One of the most prevalent uses these days is saving energy, saving energy is very important and it can be done through AI systems supervise energy expenditure within buildings, factories, and transport systems to detect operational waste and recommend corrective actions.AI systems augment emission control in plants by streamlining the processes and monitoring for emissions and pollutants in real time. By using smart sensors and machine learning, it is possible to control excessive emissions by optimizing the burning of fuel and the execution of various operational activities.AI also helps in lowering carbon footprint by waste elimination and improving logistics in the supply chains as well as recommending alternative sustainable materials, (Slater & Cooper, 2015; Lv & Xie, 2024). AI powered climate models assists policymaking in coming up with working strategies for carbon reduction. Business and governmental authorities stand a better chance of making more intelligent decisions regarding environmental matters through the implementation of AI, which will help ensure global efforts at sustainability.

AI IN SUSTAINABLE AGRICULTURE AND FOOD SECURITY

The use of AI technologies automates the measuring and analyzing of plant traits during plant phenotyping research, AI is very beneficial in plant growth research which is now being implemented in the study of plant growth and adaptation.

AI Driven Precision Farming

New approaches to crop management and research are possible with the advent of AI-powered monitoring. Even though AIs has greatly contributed to the plant sciences, further development is needed in the model's robustness and interpretability. Other issues such as privacy and bias algorithms have also been taken care of. It is imperative that plant scientists, data scientists, and engineers work together to harness the full power of AI in agriculture. The progression of AI heralds important changes in species identification, disease management, breeding, and sustainability for a more secure and environmentally sustainable future.

The studies conducted utilized keywords like "Technology in Agriculture," "Agriculture 5.0," "Artificial Intelligence in Agriculture," "The Digital Divide," and "Food Safety and Security" to extract pertinent literature. The analysis of search returns showed "The Digital Divide" as the most frequently searched term with records, while "Agricultural technology" came second in Scopus and WOS databases, (Lin & Huang, 2012). The term "AI in Agriculture" received 17 Scopus and 6 WOS outputs, while "Agriculture 5.0" was poorly represented in both databases. Google scholar was not included due to being less accurate despite their large volume of records. The research was split into categories which included: research articles, review papers, books, and general articles for everything else. A trend analysis was conducted over three terms before the year 2000, 2000-2019, and 2020-2023. The last decade increased focus on AI as well as food security and agricultural technology. To maintain relevance towards the findings a clean set of filters was employed that encompassed studies older than five years and those focusing on significant issues within the agricultural field, (Jean, 2024; Diabat & Govindan, 2011).

Leveraging Digitalization for Sustainable Development in Fisheries and Aquaculture

Digital technologies have found a place in nearly every part of human life, including the fishing industry. Observe that even common activities of people are included in everyday living, in the case caring for the socio-economic, environment, and sustainability work, which makes digitalization relevant. Perceives it as the process of taking physical information and converting it into a computable form for the web of things, this is very important as it acquires a different computable form. The Internet of Things is the interconnection of tangible objects using embedded devices in them for real-time data transportation and analysis, (Vineusa et al., 2020). With the other digital technologies' advancement comes the incorporation of Artificial Intelligence, machine learning, and big data, increasing the effectiveness and sustainability of multiple domains. Digitalization in fisheries and aquaculture enables complete control of production and shipment processes and improves management with the usage of cloud-edge computing and artificial intelligence systems. There is also progress in the application of digital technologies in recirculating aquaculture systems (RAS), submersible cages, and vertical aquaponics. These technologies include deep-water fish farming and vessel-based aquaculture, which require further development. These tools further improve resource efficiency and sustainability, (Slater & Cooper, 2015).

Figure 2. Smart AI-driven agriculture technology

AI FOR SUSTAINABLE BUSINESS AND CIRCULAR ECONOMY

According to, Artificial Intelligence (AI) is the science and engineering of constructing intelligent machines. It is one of the top five transformational technologies of our age. Advances in AI technology can be explained by three factors: the availability of significant amounts of data, improved algorithms, and extraordinary gains in computation hardware capabilities. These developments have caused a surge in interest and funding from leading technology firms like Google, Amazon, and Microsoft, which have started to offer AI and machine learning platforms on their clouds. These include machine learning, neural networks, deep learning, and fuzzy logic which enables machines to mimic the human ability to make decisions and solve problems AI. These AI systems can enhance decision-making processes, increase productivity and offer intelligent automation services to different sectors of the economy. In addition, AI enables machines to examine and anticipate results in even the most uncertain of situations and set the most optimal or likely achievable outcomes.

Artificial Intelligence as a Catalyst for Circular Economy Transformation

Organizations are undergoing a change with the advent of AI technologies which replicate human thought processes, aid in decision making, and perform manual work. Management, innovation, business strategizing, and creating value are some of the areas impacted, (Lv & Xie, 2024). Recently, the emergence of ChatGPT and other transformer-based models have increased conversations about AI and its implementation in companies. Searl referred to strong AI as which literally does what a human does while weak AI functions as augmenting human intelligence with its tools and thus changes how people think and perform ideas. The arguments around strong AI are still considered hypothetical while in case of weak AI, it has been proved to increase efficiency in several areas of a business including speed, flexibility, customization, scaling, and even decision making.

Table 1. Various AI applications and their impact on different stages of the circular economy.

SR.NO	AI Application	Circular Economy Stage	Description	Impact on CE Transformation
1	Predictive Analytics	Circular Design	AI predicts product lifecycle and design improvements	Reduces waste and enhances product longevity
2	Machine Learning	Circular Production	AI optimizes manufacturing processes	Increases efficiency and reduces resource consumption
3	Computer Vision	Circular Consumption	AI monitors product usage and maintenance	Extends product life and improves user experience
4	Natural Language Processing	Circular Reuse and Repair	AI assists in diagnosing and repairing products	Facilitate product reuse and reduces need for new products
5	Robotics	Circular Recycling	AI-powered robots sort and recycle materials	Enhances recycling efficiency and reduces landfill waste

These networks are organized in clusters, which represent distinct themes, with each cluster given a specific color to enhance data interpretation. With VOS viewers challenging bibliometric data is visualized in a simple way, revealing the role of AI in various steps taken in the circular economy. AI has emerged as an important factor for the transition into the circular economy (CE) as it offers more than piecemeal solutions, (Lin & Huang, 2012). Unlike digital technologies in general, AI facilitates paradigm changes in business model innovation towards circularity. Although there are existing studies that indicate the possibility of AI integration, most apply single case studies instead of a comprehensive approach. This dissertation aims to fill that gap by conducting systematic reviews of 63 articles using PRISMA protocol and content analysis. The purpose is to construct a model that shows the role of AI in different CE activities instead of providing fragments of information and capturing the full impact of the AI and CE interaction, (Jean, 2024).

AI FOR ECOFRIENDLY PRODUCT DESIGN AND MANUFACTURING

AI is also having an impact on the design and manufacturing of sustainable products. AI Algorithms can also be used to conduct sensitivity analysis on variables based on the impacts of product design, to minimize and obtain the product with the least environmental footprint from a pool of options by processing large data sets related to materials, their properties, environmental impacts and key consumer behaviors It can include identifying sustainable material options, predicting product life cycles, and simulating how certain design decisions might affect resource use and waste generation, (Zhu, Sarkis, & Lai, 2008). AI systems in manufacturing can help optimize energy use and increase the power of resources, minimize material waste by controlling processes accurately, and deliver predictive maintenance that extends tool life. 4 AI also facilitates efficient reclamation of scarce resources via closed-loop systems, where products are designed for disassembly and recycling. Such integration of AI not only makes the environmental sustainability more rational.

Figure 3. AI-powered circularity: a conceptual framework

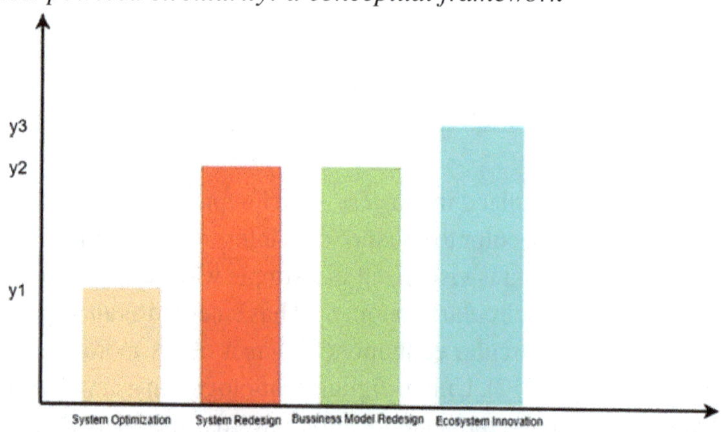

AI circularity is a new concept that integrates circular economy and artificial intelligence to maximize efficiency in the consumption of resources, minimize wastage, and increase sustainability. With the application of AI-enabling analytics, machine learning, and automation, organizations can maximize utilization of efficiency in recycling products, managing the product life cycle, and minimizing wastage. Employing AI can also predict demand, enhance supply chain management, and enable intelligent recycling systems to be operational, which have improved capability to detect, sort, and recycle, (Zhu, Sarkis, & Lai, 2008). Creating products from scratch that can be designed to last longer, be easily repaired, and easily recycled is also possible with design platforms on AI. At the manufacturing level, AI can track manufacturing processes in real time, minimize energy and material consumption, and produce faults. AI-powered robots and computer vision technologies can sort waste more efficiently, redirecting more material away from landfills and into recycling streams. AI also makes sharing economy models more sustainable by maximizing the use of products, anticipating when maintenance is required, and maximizing asset life. The confluence of circularity initiatives and artificial intelligence helps policymakers and businesses transition from a linear "take-make-dispose" approach to a regenerative framework that favors sustainability. With evolving artificial intelligence, its capability to promote circularity will be the key driver that will enable sectors to extract the maximum value and reduce their imprint towards a circular future, (Di Vaio et al., 2020).

AI-Driven Sustainable Product Design

AI sustainable product design minimizes environmental footprints through optimizing materials, energy consumption, and recyclability. AI LCA examines the complete lifecycle of a product, making designers opt for biodegradable materials, modular products, and low-energy production. AI simulations also maximize product longevity to reduce waste. AI fosters circular economy practices where products are created for reuse, refurbishment, or recycling, (Toll et al., 2020). Autodesk and Siemens are already adopting AI-based sustainability solutions. This strategy also is cost-saving while building brand credibility. With AI evolving increasingly, it will be a decisive driver for green innovation. The study focuses on the manufacturing industry, electricity, gas, steam, and air conditioning supply, supply of water, sewerage, waste management and remediation, in 26 EU Member States, excluding Bulgaria as data for this country were not available, (Zhou & Halli, 2021). It only encompasses companies with more than 10 employees since application of AI among micro-enterprises (less than or equal to 9 workers) is assumed to be irrelevant. Percentages in terms of enterprises look at the adoption of one or more AI technologies, such as text mining, speech recognition, natural language generation, image recognition, machine learning for data analysis, workflow automation, and autonomous movement of equipment. Quantification of AI adoption in these sectors and their trends and gaps is the intention of the study.

AI in Green Manufacturing and Waste Reduction

Artificial intelligence is revolutionizing green manufacturing along the lines of optimized operations, minimized waste, and energy savings. With predictive analytics, machine learning, and automation, AI assists industries to reduce their impact on the environment and optimize operations productivity. Another key application is predictive maintenance, in which AI keeps monitoring equipment real-time and identifies in advance failures that will take place later. It reduces downtime, avoids unnecessary replacement, and prolongs equipment lifespan, consequently eliminating industrial waste, (Toll et al., 2020). Smart manufacturing systems powered by AI have a significant impact on resource efficiency through data-driven monitoring of energy use and material consumption. This capability to offer insights enables manufacturers to identify wasteful processes, cut down on excess raw materials, and lower carbon emissions. These AI-driven quality control systems employ computer vision and deep learning to spot defects in production, which leads to less waste from faulty products. AI plays a key role in boosting recycling across many aspects of waste management. In the recycling industry, sorting technologies benefit from AI. The teamwork of AI robots and computer vision helps to correctly identify separate and sort recyclables, which improves recycling productivity and cuts down on landfill waste. AI is also crucial for closed-loop manufacturing where waste materials become new products, (Toll et al., 2020; McNally, 2020).

ETHICAL IMPLICATIONS IN AI-DRIVEN SUSTAINABILITY

Algorithmic bias poses a major ethical issue in AI-powered sustainable design. When training data mirrors past inequalities, AI might strengthen these disparities, which hurts fairness in sustainability efforts. This can increase social friction in mixed communities and due to implementation of such bias it will grow the strength of sustainability. To tackle bias, designers and policy experts should apply data cleanup, model checks, and input from various stakeholders, so that algorithmic bias can be easily handled. Fairness measures can evaluate AI results to make sure they don't harm or help specific groups too much. Spotting and reducing bias at the start of the design process allows for step-by-step upgrades ensuring AI-driven sustainability matches up with inclusion and social fairness, it is equally important in cultivating and strengthening.

AI Sustainability Standards and Regulations

To keep high-level designs using AI for sustainable habitats consistent with issues of quality, safety, and accountability, there must be clear standards and rules. Conventional building codes and environmental legislation often fail to address the complex interconnections of an AI-enhanced system, given the depth of linkages between the many attributes of design, which are constantly evolving, (Leshinsky

& Schatz, 2018). This suggests the collaboration of government organizations and industry groups to come together and prepare the specific guidelines; these involve both process requirements-AI algorithms need to be tested—and technical specifications, such as how much prediction models shall have increasingly shifted different minima. Those standards should also force AI-based decision-making to be transparent, allowing architects, engineers, and others to trace how inputs (like weather data or how humans use buildings) led to specific design decisions. Opaque models of artificial intelligence foster a lack of responsibility for their unintended aftermath: bias or harm to the ecosystem. Keeping either proper documentation or conducting protocol validations and recording error rates enables an inspector to notice or take criticism against whatever faulty details there may be. Aside from transparency, the imposition of performance benchmarks-for example, energy efficiency and emission reduction-further serves as a form of regulatory oversight, (Moorthy & Nathiya, 2023). Lawmakers may require AI-driven designs to adhere to sustainability standards, imposing fines for non-compliance while rewarding innovation with incentives. Such regulations will not only provide feedback on input but will also foster advancement in AI-driven sustainability. Setting clear thresholds and rewarding excellence as per the regulatory frameworks will lead to further integration of AI technologies in sustainable design, while instilling responsible and green innovation, (Mitra et al., 2023).

Financial Incentives and Investment in AI Sustainability

A financial mechanism will yield added value to extending the coverage of AI-centered paradigms to sustainable habitat designs. Tax exemptions, subsidies, or low-interest loans could take a form acting as an incentive regarding investment in AI-motivated applications for energy conservation, zero waste, and circular economy investment. Such financial mechanisms diminish some spans of capital burdens while persuading firms to bear budgetary constraints to upgrade into advanced AI technology, (Talwar & Koury, 2017). The PPPs enhance the feasibility of AI in sustainability even more so due to their promotion of collective enterprise across government institutions, research bodies, and the private sector. Such partnerships assure equitable risk sharing and are, therefore, enabling innovations in the long-term investment of AI; including advanced life-cycle assessment tools and multi-objective optimization platforms that enable sound urban planning decisions. Moreover, funding programs should be structured with adaptive incentive frameworks to promote accountability as well as continuous advancement toward sustainability targets. Most importantly, by creating tiers for support payment disbursement into phased milestones dependent on sustainability indicators-declines in energy consumption, CO2 emissions, or waste in resources-governments and donors can be assured that their investments into AI yield some benefits for the public. Such evolving funding paradigms highly expand the sustainability culture of innovation and refinement by pointing toward the widespread deployment of AI-driven approaches for attaining more sustainable and resilient built environments, (Mesjasz-Lech, 2017; Bartosiewicz, 1992).

Algorithmic Interpretability and Justification

Several AI algorithms are inherently opaque and, thus, problematic in guaranteeing stakeholder trust and informed decision-making, particularly in sustainable habitat design. In this space, XAI models are critical for ensuring that the adoption of AI in providing recommendations delivers true environmental and social benefits rather than just short-term efficiency gains. On top of that, for the sake of transparency, these AI systems would have to lay down some degree of traceability so that stakeholders--including policymakers, architects, and controllers--could understand how input variables determinant, such as climate data or user behavior, translated into final design parameters. Attractive, user-friendly interfaces that would elucidate everything from decision trees to layers and weights in neural networks to design components would illustrate whatever internal decision-making might be going on, (Zawada, 2009). Narratives that complement these may further contextualize what framing design alternatives in those specific terms-measure sustainability, for example, or resilience or cost-effectiveness-means to one another. Such an approach would widen accessibility and support inclusive, data-driven collaboration between differently staged stakeholders. External independent ethical audits could also be established that could periodically confirm whether AI systems abide by principles of fairness, data integrity, and practicality in the sustainability domain, (Li et al., 2022). When exposed, these audit findings provide a certain degree of accountability that ensures AI decisions can meet the broader public interest goals. This leads all the way back to establishing transparency: various levels of this spectrum-from explainability tools and external validation assessments of AI systems to the actual outcomes of AI decisions all establishing accountability of AI-driven sustainable design processes, (Mesjasz-Lech, 2018).

Figure 4. Strategic considerations and ethical frameworks in AI-enhanced sustainable development

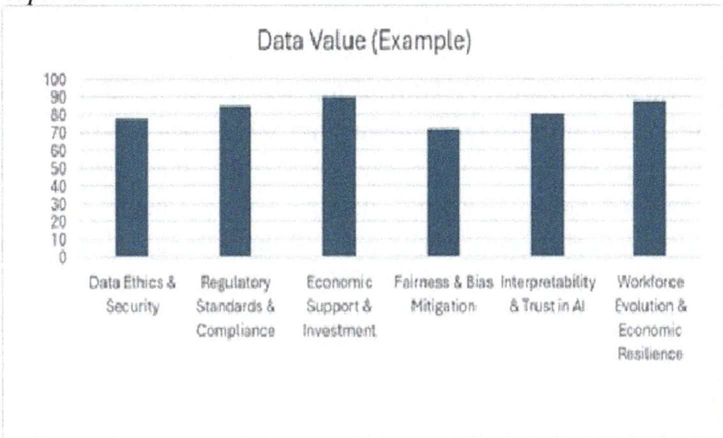

CONCLUSION

Research suggests that various AI-enabled mechanisms contribute to sustainability in construction. Machine learning, generative design, and IoT systems optimize resource utilization, regulating thermal comfort and enhancing circular economic principles. Such innovations support the global climate goals while enabling painless improvements in the wellbeing of numerous societies. Through the infusion of AI, planners, architects, and policymakers may now place themselves in a position to redefine and completely revolutionize the design of sustainable habitats so that through smarter and more advanced interventions at both urban and building levels, the environmental footprint could be reduced while enabling increased long-term resilience and sustainability. Research suggests that various AI-enabled mechanisms contribute to sustainability in construction. Machine learning, generative design, and IoT systems optimize resource utilization, regulating thermal comfort and enhancing circular economic principles. Such innovations support the global climate goals while enabling painless improvements in the wellbeing of numerous societies. Through the infusion of AI, planners, architects, and policymakers may now place themselves in a position to redefine and completely revolutionize the design of sustainable habitats so that through smarter and more advanced interventions at both urban and building levels, the environmental footprint could be reduced while enabling increased long-term resilience and sustainability. The ethical debate is about distinguishing AI applications that materially contribute towards achieving sustainability goals from the sustainability of the technologies themselves. This explicit conceptual distinction has significance in the contexts of both environmental and social sustainability. The discourse on "sustainable AI" often lends itself into tones that do not illuminate the issue fully; rather, it-an aspect set on the sustainability spectrum in "(fill in a particular PCI dimension or milestone)"-labels technology as sustainable in deference to other more ambiguous analyses of it. Selective assessments of this flavor could amount to a kind of green washing aligned to other economic or political interests, revealing a potential "dark side of sustainability." In countering the iteration, a far more specked-out analysis of these areas may assist in carving out "sustainable AI" concerning social and environmental impact and cost. Such a dimension scope, known as "matrix consideration," offers various ways to assess comprehensively the real sustainability question the AI poses; such considerations afford an introduction of a measured complexity to discussions of sustainability. The assessments avoid grand simplifications while giving further clarity about the sustainability of AI.

REFERENCES

Ataş, M., & Talay, A. C. (2022). Development of automatic tree counting software from UAV based aerial images with machine learning. arXiv preprint . 2201. 02698. DOI: 10. 48550/ arXiv

Banerjee, K., Vats, I., Akhtar, N., Harsha, K. G., Vinooth, P., Kumar, A., . . . Gautam, P. (2022, December). A review on artificial intelligence based sign language recognition techniques. In *2022 5th International Conference on Contemporary Computing and Informatics (IC3I)* (pp. 2195-2201). IEEE.

Bartosiewicz, S. (1992). *Econometrics with elements of mathematical programming and comparative analysis*. Wrocław University of Economics Press.

Bernstein, M. S., Levi, M., Magnus, D., Rajala, B., Satz, D., & Waeiss, C. (2021). ESR: ethics and society review of artificial intelligence research. arXiv preprint https:// arxiv. org/ pdf/ 2106. 11521.pdf

Boyer, R., Peterson, N., Arora, P., & Caldwell, K. (2016). Five approaches to social sustainability and an integrated way forward. *Sustainability*. Advance online publication. DOI: 10. 3390/ su809 0878

Brevini, B. (2020). Black boxes, not green: Mythologizing artificial intelligence and omitting the environment. *Big Data & Society*, *7*(2). Advance online publication. DOI: 10. 1177/ 20539 51720 935141

Challoumis, C. (2024, October). Building a sustainable economy-how ai can optimize resource allocation. In *XVI International Scientific Conference* (pp. 190-224).

Chavan, L., Kumbhakar, R., Agrawal, P., Rakesh, N., Parvathi, R., & Shahid, M. (2024, November). AI in Agriculture: Yield Optimization and Food Security. In *2024 4th International Conference on Technological Advancements in Computational Sciences (ICTACS)* (pp. 536-541). IEEE.

Chen, L., Chen, Z., Zhang, Y., Liu, Y., Osman, A. I., Farghali, M., & Yap, P. S. (2023). Artificial intelligence-based solutions for climate change: A review. *Environmental Chemistry Letters*, *21*(5), 2525–2557.

Di Vaio, A., Palladino, R., Hassan, R., & Escobar, O. (2020). Artificial intelligence and business models in the sustainable development goals perspective: A systematic literature review. [CrossRef]. *Journal of Business Research*, *121*, 283–314.

Diabat, A., & Govindan, K. (2011). An analysis of the drivers affecting the implementation of green supply chain management. *Resources, Conservation and Recycling*, *55*(6), 659–667.

Fatima, S., Desouza, K., Dawson, G., & Denford, J. (2021). *Analyzing Artificial Intelligence Plans in 34 Countries.* Available online: https://www.brookings.edu/blog/techtank/2021/05/13/analyzing-artificial-intelligence-plans-in-34-countries

Haleem, A., Javaid, M., Khan, I. H., & Mohan, S. (2023). Significant applications of artificial intelligence towards attaining sustainability. *Journal of Industrial Integration and Management*, *8*(04), 489–520.

Jean, G. G. (2024). Green Financing and Green Product Innovation. *Green Financing and Green Product Innovation. Green Financing in India: Insights from NIFTY 100's top 5 Companies and their Role in Financial and Sustainable Growth, 285.*

Kaya, C. (2025). Intelligent Environmental Control in Plant Factories: Integrating Sensors, Automation, and AI for Optimal Crop Production. *Food and Energy Security*, *14*(1), e70026.

Leshinsky, R., & Schatz, L. (2018). I don0t think my landlord will find out: Airbnb and the challenges of enforcement. *Urban Policy and Research*, *36*, 417–428.

Li, X., Ruan, T., Hou, K., & Qu, R. (2022). The configuring pathways of green technology advance, organizational strategy and policy environment for realizing low-carbon manufacturing from the perspective of simmelian tie: A qualitative comparative analysis of listed companies in China. *Journal of Cleaner Production*, *382*, 135149. DOI: 10.1016/j.jclepro.2022.135149

Lin, P. C., & Huang, Y. H. (2012). The influence factors on choice behavior regarding green products based on the theory of consumption values. *Journal of Cleaner Production*, *22*(1), 11–18.

Lv, Z., & Xie, S. (2024). Artificial intelligence in the digital twins: State of the art, challenges, and future research topics. *Digital Twin*, *1*(1), 12.

McNally, A. (2020). *Creating Trustworthy AI for the Environment: Transparency, Bias, and Beneficial Use*. Available online: https://www.scu.edu/environmental-ethics/resources/creating-trustworthy-ai-for-the-environment-transparency-biasand-beneficial-use

Mesjasz-Lech A. (2017), Factors Conducive to the Development of Business Cooperation in a Cluster Initiative – a Statistical Analysis, Zeszyty Naukowe. *Organizacja i Zarządzanie / Politechnika Śląska, 1984*(109), 65-75.

Mesjasz-Lech, A. (2018). Educational and research potential of regions and the innovativeness of businesses - an analysis of spatial concentration. IA TED Academy.

Mitra, A., Bera, B., Das Ashok, K., Jamal Sajjad, S., & You, I. (2023). Impact on blockchain-based AI/ML-enabled big data analytics for Cognitive Internet of Things environment. *Computer Communications, 197*, 173–185. DOI: 10.1016/j.comcom.2022.10.010

Olawade, D. B., Wada, O. Z., David-Olawade, A. C., Fapohunda, O., Ige, A. O., & Ling, J. (2024). Artificial intelligence potential for net zero sustainability: Current evidence and prospects. *Next Sustainability, 4*, 100041.

Onyeaka, H., Tamasiga, P., Nwauzoma, U. M., Miri, T., Juliet, U. C., Nwaiwu, O., & Akinsemolu, A. A. (2023). Using artificial intelligence to tackle food waste and enhance the circular economy: Maximising resource efficiency and minimising environmental impact: A review. *Sustainability, 15*(13), 10482.

Popescu, S. M., Mansoor, S., Wani, O. A., Kumar, S. S., Sharma, V., Sharma, A., & Chung, Y. S. (2024). Artificial intelligence and IoT driven technologies for environmental pollution monitoring and management. *Frontiers in Environmental Science, 12*, 1336088.

Popkova, E. G. (2023). Sustainable development of agro-industrial complex and food security through digital agriculture: Contribution of robots, big data, and AI. In *Digital agriculture for food security and sustainable development of the agro-industrial complex* (pp. 3–9). Springer International Publishing.

Rashid, A., & Rasheed, R. (2022). A Paradigm for measuring sustainable performance through big data analytics–artificial intelligence in manufacturing firms. *Available at SSRN* 4087758.

Rojek, I., Mikołajewski, D., Mroziński, A., & Macko, M. (2024). Green Energy Management in Manufacturing Based on Demand Prediction by Artificial Intelligence—A Review. *Electronics (Basel), 13*(16), 3338.

Saadi, S., & Azdimousa, H. (2023, October). Artificial Intelligence and Green Marketing: What Link? In *International Conference on Advanced Intelligent Systems for Sustainable Development* (pp. 104-113). Cham: Springer Nature Switzerland.

Saeed El Etaiby, Y. (2023). Smart agriculture and future technologies for achieving food security. *International Journal of Family Studies. Food Science and Nutrition Health, 4*(1), 56–71.

Singh, A., Rajput, V. D., Rawat, S., Sharma, R., Kumar, S. A., Kumar, S. A., & Tomar, R. S. (2021). Emerging tools for sustainable agriculture and food security: An introduction. *Emerging Tools for Sustainable Agriculture and Food Security, 1*(1), 3–16.

Slater, A. G., & Cooper, A. I. (2015). Function-led design of new porous materials. *Science, 348*(6238), aaa8075. PMID: 26023142

Son, N., Chen, C. R., & Syu, C. H. (2024). Towards artificial intelligence applications in precision and sustainable agriculture. *Agronomy (Basel), 14*(2), 239.

Sri Skandha Moorthy, R., & Nathiya, N. (2023). Botnet Detection Using Artificial Intelligence. *Procedia Computer Science, 218*, 1405–1413. DOI: 10.1016/j.procs.2023.01.119

Srivastava, A., & Maity, R. (2023). Assessing the potential of AI–ML in urban climate change adaptation and sustainable development. *Sustainability, 15*(23), 16461.

Talwar, R., & Koury, A. (2017). Artificial intelligence – the next frontier in IT security? *Network Security, 2017*(4), 14–17. DOI: 10.1016/S1353-4858(17)30039-9

Tokhayeva, Z. O., Almukhambetova, B. Z., Keneshbayev, B., & Akhmetova, K. (2020). Innovative processes' management in agriculture and food security: Development opportunities. *Entrepreneurship and Sustainability Issues, 7*(3), 1565.

Toll, D., Lindgren, I., Melin, U., & Madsen, C.Ø. (2020). Values, benefits, considerations and risks of AI in government: A study of AI policies in Sweden. *JeDEM-eJournal democracy Open Gov.*, 12, 40–60.

Vinuesa, R., Azizpour, H., Leite, I., Balaam, M., Dignum, V., Domisch, S., & Fuso Nerini, F. (2020). The role of artificial intelligence in achieving the Sustainable Development Goals. *Nature Communications, 11*(1), 233. PMID: 31932590

Zawada, M. (2009). Application of selected methods of multivariate statistical analysis in the investment process on the capital market in Poland. In A. Mesjasz-Lech (Ed.), *Modern management instruments* (pp. 78). Publishing Section of the Faculty of Management, Częstochowa University of Technology.

Zhang, H., Xu, J. J., Cui, H. W., Li, L., Yang, Y., Tang, C. S., & Boers, N. (2024). When Geoscience Meets Foundation Models: Toward a general geoscience artificial intelligence system. *IEEE Geoscience and Remote Sensing Magazine.*

Zhou, Y., & Kankan Halli, A. (2021). AI regulation for smart cities: Challenges and principles. In Estevez, E., Pardo, T., & Scholl, H. (Eds.), *Smart Cities and Smart Governance: Towards the 22nd Century Sustainable City* (pp. 101–118). Springer.

Zhu, Q., Sarkis, J., & Lai, K. H. (2008). Confirmation of a measurement model for green supply chain management practices implementation. *International Journal of Production Economics*, *111*(2), 261–273.

Chapter 4

Stubble-Reinforced Materials for Sustainable Construction: Advancing SDG 9, 11, and 12 Through Waste-to-Resource Innovations

Jagdeep Walia
https://orcid.org/0009-0002-2905-2739
Rayat Bahra University, Punjab, India

Rita Sharma
https://orcid.org/0009-0003-5119-8731
Rayat Bahra University, Punjab, India

Shalom Akhai
https://orcid.org/0000-0002-7533-457X
Maharishi Markandeshwar Engineering College, MM(DU), Mullana, India

ABSTRACT

This chapter reviews the integration of agricultural stubble, including rice straw, wheat straw, maize stalks, and sugarcane bagasse, into sustainable construction materials. It examines how stubble enhances mechanical, thermal, and acoustic properties in concrete, bricks, plaster, and particleboards, while addressing environmental challenges like carbon emissions and agricultural waste. The chapter highlights economic benefits, such as cost savings and rural development, alongside technical challenges like durability, scalability, and regulatory barriers. Future research directions focus on optimizing stubble-matrix bonding, fire resistance,

DOI: 10.4018/979-8-3373-2555-2.ch004

and large-scale production techniques. By transforming agricultural waste into eco-friendly building solutions, this chapter underscores the potential of stubble-reinforced materials to advance sustainable construction practices and contribute to global sustainability goals.

CHAPTER HIGHLIGHTS

- The chapter provides a detailed analysis of agricultural stubble types, including rice straw, wheat straw, maize stalks, and sugarcane bagasse, and their integration into construction materials, focusing on their physical and chemical properties such as cellulose content, lignin concentration, and tensile strength.
- Significant enhancements in mechanical, thermal, and acoustic properties are documented, including a 5-15% increase in compressive strength of concrete, improved thermal insulation in bricks, and up to 60% reduction in crack width in plaster and mortar through stubble reinforcement.
- The chapter underscores the environmental advantages of stubble-reinforced materials, such as reduced carbon emissions, minimized cement usage, and effective agricultural waste management, contributing to circular economy principles and sustainable development goals.
- Economic benefits are highlighted, including material cost reductions, energy savings in production, and the creation of rural employment opportunities through the development of stubble-based construction industries.
- Key challenges are addressed, including variability in stubble properties, long-term durability concerns, compatibility with existing construction practices, and the need for standardized testing methods and building code updates to facilitate widespread adoption.
- The chapter outlines critical research directions, such as optimizing stubble-matrix bonding, enhancing fire resistance, developing hybrid reinforcement systems, and advancing large-scale production techniques to enable the commercialization and scalability of stubble-reinforced materials.

INTRODUCTION

Agricultural stubble, the leftover plant material after crop harvesting, has long been considered a waste product in many parts of the world. It is defined as the basal part of herbaceous plants, especially cereals, left attached to the soil after harvesting. Stubble is abundantly available in agricultural regions, with global production estimated at over 800 million tons annually (Ashour, 2017). The traditional method of disposing of stubble through burning has become a significant environmental concern. Stubble burning releases large amounts of particulate matter, carbon dioxide, and other harmful gases into the atmosphere, contributing to air pollution and climate change. In regions like Northern India, stubble burning has led to severe air quality issues, affecting millions of people. In recent years, there has been a growing emphasis on sustainable construction practices to address environmental concerns and resource scarcity (Bhuyan et al., 2020). The construction industry, being one of the largest consumers of raw materials and energy, has been exploring alternative materials to reduce its environmental footprint. This has led to increased interest in using agricultural by-products, including stubble, as reinforcement materials in construction (Binici et al., 2013; Thirunavukkarasu et al., 2018; Pandey, 2004).

This review explores the use of agricultural stubble in building materials, focusing on its potential to enhance performance attributes. Various forms of stubble, including rice straw, wheat straw, maize stalks, and sugarcane bagasse, have been studied for their integration into building materials. The effectiveness of stubble as a reinforcing material depends on its physical and chemical characteristics, including cellulose concentration, lignin concentration, tensile strength, water absorption, and resistance to degradation. Optimized stubble fiber content in concrete can increase compressive strength by 5-15%, while bricks and blocks show improvements in thermal insulation, sound insulation, weight reduction, and fracture resistance. However, water absorption characteristics may need treatment. Future research should focus on optimizing stubble-reinforced materials for improved performance, including enhanced bonding between stubble and matrix, surface treatments, fire resistance, hybrid reinforcement systems, scalability, and market penetration strategies.

TYPES OF STUBBLE AND THEIR PROPERTIES

Several types of agricultural stubble have been investigated for use in construction materials. The most common types include:

- **Rice Straw:** One of the most abundant types of stubble, rice straw is widely available in Asia and other rice-producing regions. It is characterized by its high silica content, which can contribute to improved durability in construction materials (Garay et al., 2009).
- **Wheat Straw:** Widely available in temperate regions, wheat straw has been extensively studied for use in construction materials. It has a relatively high cellulose content and good mechanical properties (Li et al., 2007).

- **Corn Stalks:** Also known as maize stover, corn stalks are abundant in many parts of the world. They have a lower silica content compared to rice straw but offer good mechanical properties (Logeswaran et al., 2020).
- **Sugarcane Bagasse:** A by-product of sugar production, bagasse is rich in cellulose and has been successfully used in various construction applications, particularly in regions with significant sugarcane industries (Fang et al., 2017)

Table 1 discusses the properties and applications of agricultural studbble in sustainable construction materials, compares studbble types for reinforcement, highlights key characteristics and uses in eco-friendly construction, and discusses the performance and limitations of studbble-reinforced materials.

Table 1. Comparative analysis of stubble types for reinforcement in building materials

Stubble Type	Key Properties	Applications in Construction	Advantages	Limitations
Rice Straw	High silica content (improves durability), cellulose content (35-40%), water absorption (150-200% of its weight).	Used in bricks, concrete, and particleboards for improved durability and thermal insulation.	Enhances durability, reduces thermal conductivity, and provides good microbial resistance.	High water absorption can affect workability and setting time of cementitious materials.
Wheat Straw	High cellulose content (40-50%), tensile strength (50-315 MPa), moderate lignin content.	Reinforces concrete, plaster, and fiberboards for improved mechanical strength and crack resistance.	High tensile strength and cellulose content improve mechanical properties and reduce cracking.	Requires surface treatments to mitigate water absorption and compatibility issues with matrices.
Corn Stalks (Maize Stover)	Moderate cellulose content, high lignin content (15-20%), good mechanical properties.	Used in lightweight bricks, blocks, and insulation panels for thermal and acoustic properties.	Lightweight, good mechanical properties, and potential for thermal and acoustic insulation.	Lower silica content compared to rice straw may reduce durability in certain applications.
Sugarcane Bagasse	High cellulose content (40-50%), low lignin content, fibrous structure.	Incorporated into particleboards, fiberboards, and concrete for enhanced strength and insulation.	High cellulose content improves mechanical properties; abundant by-product of sugar production.	Requires processing to reduce water absorption and improve compatibility with binding materials.

The effectiveness of stubble as a reinforcement material in construction is largely dependent on its physical and chemical properties. Key properties include:

- **Cellulose Content:** Cellulose is the primary structural component of plant cell walls. Higher cellulose content generally correlates with better mechanical properties. Wheat straw and sugarcane bagasse typically have higher cellulose content (40-50%) compared to rice straw (35-40%) (Javaid et al., 2024)
- **Lignin Content:** Lignin acts as a natural binder and contributes to the stiffness of the stubble. It also affects the stubble's resistance to degradation. Lignin content varies among stubble types, with corn stalks having relatively high lignin content (15-20%) compared to rice straw (10-15%) (Ashour, 2017)
- **Tensile Strength:** The tensile strength of stubble fibers is crucial for their reinforcing effect. Wheat straw fibers typically exhibit tensile strengths ranging from 50-315 MPa, while rice straw fibers show values between 220-550 MPa (Athira et al., 2021)
- **Water Absorption:** Stubble fibers generally have high water absorption capacities, which can affect the workability and setting time of cementitious materials. Rice straw, for instance, can absorb water up to 150-200% of its own weight (Murali & Wong, 2024)
- **Degradation Resistance:** The natural resistance of stubble to degradation varies among types and affects the long-term durability of reinforced materials. Silica content, particularly high in rice straw, can contribute to improved resistance against microbial attack (Ayub et al., 2020)

Understanding these properties is crucial for optimizing the use of stubble in construction materials and addressing potential challenges in their application.

Table 2. Key properties of stubble and their impact on construction material performance

Cellulose Content: Primary structural component; higher content improves mechanical properties.	**Reinforcement in concrete, bricks, plaster, and composite panels.**	**Enhances compressive strength, flexural strength, and crack resistance.**	**Variability in properties requires standardized processing and quality control.**
Lignin Content: Natural binder; contributes to stiffness and degradation resistance.	Improves durability and resistance to microbial attack in construction materials.	Increases stiffness and resistance to environmental degradation.	High lignin content may reduce flexibility in some applications.
Tensile Strength: Critical for reinforcing effect; varies by stubble type (50-550 MPa).	Used in fiber-reinforced composites for structural applications.	Improves tensile and flexural strength of construction materials.	Lower tensile strength in some stubble types may limit their use in high-stress applications.
Water Absorption: Affects workability and setting time of cementitious materials.	Requires treatment for use in plaster, mortar, and concrete.	Can be managed with surface treatments and additives like superplasticizers.	High water absorption can delay setting time and reduce workability.
Degradation Resistance: Silica content (e.g., in rice straw) improves resistance to microbial attack.	Enhances long-term durability of stubble-reinforced materials.	Improves resistance to environmental factors like moisture and microbial activity.	Lower silica content in some stubble types may reduce durability.

Key Insights from the Table 1 and 2:

1. **Rice Straw**: High silica content makes it ideal for durable construction materials, but its high water absorption requires treatment.
2. **Wheat Straw**: High cellulose and tensile strength improve mechanical properties, but surface treatments are needed for compatibility.
3. **Corn Stalks**: Lightweight and good for insulation, but lower silica content may limit durability.
4. **Sugarcane Bagasse**: High cellulose content enhances strength, but processing is needed to address water absorption.
5. **General Properties**: Cellulose and lignin content, tensile strength, water absorption, and degradation resistance are critical for optimizing stubble use in construction.

STUBBLE REINFORCEMENT IN VARIOUS CONSTRUCTION MATERIALS

Stubble reinforcement has shown promising results in various construction materials, including concrete, bricks and blocks, plaster and mortar, particleboards, and fiberboards. Concrete can increase its compressive strength by 5-15% with optimal stubble fiber content (0.5-2% by volume), while bricks and blocks show significant improvements in thermal insulation, sound insulation, weight reduction, and crack resistance (Saini et al., 2024). Adding 1% wheat straw fibers to cement mortar reduced crack width by up to 60% under shrinkage conditions. However, stubble fibers can affect the setting time of plaster and mortar due to their water absorption properties, which requires proper treatment. Alterations in workability can be managed through proper mix design and the use of superplasticizers. Stubble-based boards can be produced with a wide range of densities, from low-density insulation boards to high-density structural panels (Ruiz Trejo, 2018; Asim et al., 2020). Proper treatment and processing of stubble fibers can lead to boards with good dimensional stability, as demonstrated by wheat straw fiberboards treated with silane coupling agents. Fire resistance properties have also been shown to improve in stubble-based boards, with rice husk-based panels showing better fire resistance compared to conventional wood-based panels due to the high silica content in rice husks. Concrete has shown compressive strength increases by 5-15% with optimal stubble fiber content, while bricks and blocks show significant improvements in thermal insulation, sound insulation, weight reduction, and crack resistant (Saini et al., 2024; Ruiz Trejo, 2018). Long-term studies are needed to fully assess the durability impacts of stubble-reinforced concrete and other construction materials. **Table 3** presents the enhancements made to construction materials through Stubble fiber reinforcement (Asim et al., 2020; Sangmesh et al., 2023; Qamar et al., 2018; Deraman et al., 2022).

Table 3. Construction material enhancements through Stubble fiber reinforcement

Construction Material	Property Enhanced	Observations	Example/Study Findings
Concrete	Compressive Strength	Increase by 5-15% with optimal stubble fiber content (0.5-2% by volume).	12% increase in compressive strength with 1% wheat straw fiber addition.
	Flexural Strength	Significant improvement, with potential increases of up to 25%.	25% increase in flexural strength with 1.5% rice straw fibers.
	Durability	Enhanced resistance to crack propagation, freeze-thaw cycles, and reduced permeability.	Results vary; long-term studies needed for full assessment.

continued on following page

Table 3. Continued

Construction Material	Property Enhanced	Observations	Example/Study Findings
Bricks and Blocks	Thermal Properties	Improved thermal insulation properties, reducing thermal conductivity.	5% rice straw addition reduced thermal conductivity by up to 30%.
	Sound Insulation	Enhanced sound absorption capabilities.	15% improvement in sound absorption coefficient with 3% corn stalk fibers.
	Weight Reduction	Results in lighter bricks and blocks, reducing transportation costs.	10-15% weight reduction in wheat straw-reinforced earth blocks.
Plaster and Mortar	Crack Resistance	Significant improvement in crack resistance.	1% rice straw fibers reduced crack width by up to 60% under shrinkage conditions.
	Setting Time	Generally increases due to water absorption by stubble fibers, requiring proper treatment.	Proper fiber treatment is crucial to mitigate increased setting time.
	Workability	Reduced workability, manageable through mix design and additives like superplasticizers.	Requires proper mix design for effective workability.
Particleboard sand fiberboards	Density Variations	Can be produced in a range of densities, suitable for various applications.	Rice straw particleboards with densities ranging from 500 to 800 kg/m^3.
	Dimensional Stability	Properly treated stubble fibers yield good dimensional stability, comparable to wood-based panels.	Wheat straw fiberboards treated with silane coupling agents demonstrated improved stability.
	Fire Resistance	Improved fire resistance due to the high silica content in stubble, particularly in rice husk-based boards.	Rice husk-based panels showed better fire resistance compared to conventional wood-based panels.

PERFORMANCE IMPROVEMENTS

Stubby fibers can significantly improve the mechanical properties of building materials, such as tensile strength, flexural strength, and impact resistance. The use of 1% wheat straw fibers can increase concrete's tensile strength by 30%, cement mortars by 40%, and concrete panels by 50%. Stubby reinforcement also offers heat insulation, sound absorption, and dampness resistance (Jiang et al., 2018; Akhil et al., 2023). For instance, mud bricks reinforced with 5% rice straw showed a 25% decrease in heat conductivity. Additionally, silane-treated rice straw fibers reduced water absorption by 40%. The durability of stubble reinforcement depends on the material and intended use. Research has shown enhanced weathering resilience in materials reinforced with stubble, with concrete blocks showing a 15% reduction in mass loss during accelerated weathering testing. Additionally, stubble fibers in concrete beams reinforced with 1.5% rice straw fibers may improve their performance under cyclic loading, with a 25% greater fatigue life than raw beams. However, the long-term stability of stubble-reinforced materials remains a concern, necessitating further durability research to evaluate their performance under different environmental conditions (Kamble & Behera, 2021).

ENVIRONMENTAL AND ECONOMIC BENEFITS

- Stubby construction materials are environmentally and economically beneficial, as they burn or decompose carbon, minimize cement use, and save energy. Rice straw particleboards use 30% less energy than wood-based ones and reduce stubble burning and agricultural waste. This promotes a circular economy in agriculture and construction, with a closed-loop system where one industry's trash benefits another. In India, using 10% rice straw reduced cement block material prices by 8%. Stubby-based insulation panels use less energy than foam insulation, and a straw-based building material sector has produced 500 rural jobs over three years (Aly, 2012).
- **Reduction in Carbon Footprint:** The use of stubble in construction materials can significantly reduce the carbon footprint of industry. It can partially replace cement in certain applications, reducing cement consumption. A study showed that replacing 20% of cement with rice straw ash in concrete production led to a 15% reduction in CO_2 emissions (Javaid et al., 2024). Additionally, stubble-based materials often require less energy than conventional materials, with rice straw particleboards consuming 30% less energy compared to wood-based particleboards. Furthermore, stubble can sequester carbon that would otherwise be released through burning or decomposition, potentially reducing the carbon footprint by up to 1.2 tons of CO2 equivalent per ton of straw used (Javaid et al., 2024; Puri et al., 2022).
- **Waste Management:** The use of stubble in construction materials can significantly reduce agricultural waste management by creating a market for it. A case study in Punjab, India, showed that promoting stubble use in local brick production reduced stubble burning by 40%. The use of agricultural by-products in construction also creates value from waste, aligning with the principles of a circular economy and resource efficiency (Maji et al., 2020). This integration of stubble into construction materials represents a step towards a more circular economy, creating a closed-loop system where waste from one industry becomes a resource for another (Singh & Brar, 2021).

- **Cost-effectiveness:** Stubby in construction materials can lead to long-term cost savings, including reduced material costs in regions with abundant stubble, as seen in India where 10% rice straw in cement blocks reduced overall material costs by 8%. Lower energy requirements in production can also result in cost savings, as seen in a 20% reduction in straw-based insulation panels compared to conventional foam insulation. Additionally, stubble can create new economic opportunities in agricultural regions, as demonstrated by a socio-economic study that several new jobs were created in a rural area over a three-year period (Maji et al., 2020; Singh & Brar, 2021).

CHALLENGES AND LIMITATIONS

Technical issues in stubble-reinforced composites include varying stubble characteristics, long-term durability, and compatibility with current building processes. Standardizing procedures and quality control are necessary to maintain product consistency. Economic constraints include processing equipment investment, market perception, and supply chain growth. Small-scale manufacturers may face financial challenges, while market adoption may be slow due to conservative building industry attitudes. Establishing a consistent stubble supply chain may be challenging in areas lacking infrastructure (Chaganti, 2008). Regulatory difficulties include standardization, building code compliance, and quality control. Stubby-reinforced material testing and performance requirements must be standardized to improve quality control and market acceptability. Obtaining permissions for stubble-reinforced materials is challenging due to their lack of inclusion in building standards (Kumar et al., 2024). A recommended set of standards for stubble-based goods is not widely embraced. Table 4 presents the challenges and barriers to studbble reinforcement in construction materials.

Table 4. Challenges and barriers to stubble reinforcement in construction materials

Section	Challenge/Barrier	Description	Potential Solutions	Impact	Priority	Responsible Party
Technical Challenges	Variability in Stubble Properties	Inconsistent stubble properties affect material performance	Standardized processing and quality control	High	High	Industry/Regulators
	Long-term Durability Concerns	Uncertainty about long-term durability under various conditions	Extensive long-term durability studies	Medium	Medium	Research Institutions
	Compatibility with Existing Practices	Need for specialized equipment and training	Specialized training programs for construction professionals	Low	Low	Industry/Training Providers
Economic Barriers	Initial Investment in Equipment	High upfront costs for specialized equipment	Financial incentives or subsidies for small producers	High	High	Government/Financial Institutions
	Market Acceptance and Perception	Reluctance to adopt new materials without proven performance	Large-scale demonstration projects and public education	Medium	Medium	Industry/Marketing Professionals
	Supply Chain Development	Complexities in establishing reliable supply chains	Localized supply chains to reduce transportation costs	Low	Low	Industry/Logistics Providers

continued on following page

Table 4. Continued

Section	Challenge/Barrier	Description	Potential Solutions	Impact	Priority	Responsible Party
Regulatory Issues	Lack of Standardization	Absence of standardized testing methods and performance criteria	International collaboration to develop standards	High	High	Regulators/Industry Associations
	Building Code Compliance	Difficulty in gaining approval due to outdated building codes	Accelerated updates to building codes	Medium	Medium	Regulators/Industry Associations
	Quality Control Measures	Ensuring consistent quality in stubble-reinforced materials	Strict quality control protocols specific to stubble-reinforced materials	Low	Low	Industry/Regulators

FUTURE RESEARCH DIRECTIONS

Future research should focus on optimizing stubble-reinforced materials for enhanced performance using proven techniques like Taguchi and RSM etc. (Kumar et al., 2024; Akhai & Rana, 2022; Akhai et al., 2021; Rana & Akhai, 2022). Enhancing studbble-matrix bonding is crucial for improving mechanical properties, and surface treatments and compatibility with different matrices are needed. Fire resistance is essential for building applications, and a combination of fire-retardant coatings and nano-additives can achieve comparable fire ratings. Hybrid reinforcement systems, combining stubble with other materials, could lead to superior performance. Scalability and commercialization are essential for widespread adoption. Large-scale production techniques, such as continuous straw-based insulation panels, are essential for commercialization (Wang et al., 2022). Automated Stubble Processing can improve efficiency and reduce costs, as seen in robotic stubble harvesting. Market penetration strategies, targeting niche markets like eco-friendly construction, are also crucial. Sustainable urban planning and execution, climate adaption measures, and intelligent approaches are essential for attaining resilient futures (Abbass et al., 2025; Akhai and Abbass 2025). Further research into consumer preferences, marketing strategies, and policy incentives is needed to drive adoption (Saha et al., 2024). Table 5 presents key research areas and future directions for stubble-based construction materials, outlining their potential applications in various construction applications (Saha et al., 2024; Barczewski et al., 2018; Rana et al., 2019; Serra-Parareda et al., 2020).

Table 5. Key future research areas in stubble-reinforced materials and the future directions for stubble-based construction materials

Research Direction	Focus Area	Key Points	Future Research Needs
Material Optimization	Enhancing Stubble-Matrix Bonding	Improving the bonding between stubble fibers and matrix materials can enhance mechanical properties.	Further studies on surface treatments and compatibility with various matrices.
	Improving Fire Resistance	Fire resistance is crucial for building applications; promising results have been achieved using fire-retardant coatings and nano-additives.	Development of cost-effective fire-resistant formulations.
	Developing Hybrid Reinforcement Systems	Combining stubble with synthetic fibers or nanoparticles can improve tensile strength and overall performance.	Exploration of synergistic combinations with other reinforcing materials.
Scalability and Commercialization	Large-scale Production Techniques	Efficient large-scale production methods, such as continuous production lines for straw-based panels, are essential for commercialization.	Research into scaling up production while maintaining quality and cost-effectiveness.
	Automation in Stubble Processing	Automation in the collection, processing, and incorporation of stubble can improve efficiency and reduce costs, as seen in robotic stubble harvesting.	Development of automated systems throughout the production chain.
Market Penetration Strategies	Market Entry and Expansion	Targeting niche markets like eco-friendly construction can facilitate market entry.	Research into consumer preferences, marketing strategies, and policy incentives.

CONCLUSION

This review highlights the potential of stubble-reinforced construction materials in improving performance and addressing environmental concerns. The use of agricultural stubble in construction materials enhances mechanical properties, thermal and acoustic insulation, and overall sustainability. It offers significant environmental benefits, such as reduced carbon emissions and effective waste management. Economic advantages include potential cost savings and rural economic development. However, challenges remain, such as technical issues, economic barriers to adoption, and regulatory hurdles. These issues necessitate continued research and development in key areas such as material optimization, scalable and automated production techniques, comprehensive life cycle assessments, and standard and building codes development. Stubble-reinforced materials could contribute to sustainable construction practices by reducing the environmental footprint of the construction industry and providing new economic opportunities in agricultural regions. As sustainable practices become increasingly important, stubble-reinforced construction materials represent a promising avenue for innovation, contributing significantly to circular economy principles, waste reduction, and overall sustainability. Collaboration between researchers, industry stakeholders, and policymakers is essential to realize the full potential of these materials and overcome existing barriers to their widespread use.

CHAPTER SUMMARY

This chapter comprehensively reviews the potential of agricultural stubble as a sustainable reinforcement material in construction, addressing both its technical and environmental implications. Various types of stubble, including rice straw, wheat straw, maize stalks, and sugarcane bagasse, are examined for their physical and chemical properties, such as cellulose and lignin content, tensile strength, and water absorption, which influence their performance in building materials. The integration of stubble into concrete, bricks, plaster, and particleboards has demonstrated significant improvements in compressive strength, thermal and acoustic insulation, crack resistance, and overall durability. Environmental benefits, such as reduced carbon emissions, minimized cement usage, and effective agricultural waste management, are highlighted, alongside economic advantages like cost savings and rural development opportunities. However, challenges related to material variability, long-term durability, scalability, and regulatory barriers remain critical hurdles. The chapter concludes by outlining future research directions, emphasizing the need for optimized stubble-matrix bonding, enhanced fire resistance, hybrid reinforcement systems, and large-scale production techniques. By transforming agricultural waste into eco-friendly building solutions, stubble-reinforced materials offer a promising pathway toward sustainable construction practices, aligning with global sustainability goals and circular economy principles. This chapter serves as a foundational resource for researchers, industry stakeholders, and policymakers aiming to advance sustainable construction innovations.

Table 6 provides a clear and structured summary of the chapter's key points.

Table 6. Summary of chapter

Aspect	*Summary*
Focus	Review of agricultural stubble (rice straw, wheat straw, maize stalks, sugarcane bagasse) as sustainable reinforcement in construction materials.
Key Properties	Examines cellulose and lignin content, tensile strength, water absorption, and degradation resistance of stubble for material performance.
Performance Improvements	Enhanced compressive strength (5-15%), thermal and acoustic insulation, crack resistance, and durability in concrete, bricks, plaster, and particleboards.
Environmental Benefits	Reduces carbon emissions, minimizes cement usage, and promotes agricultural waste management, aligning with circular economy principles.
Economic Advantages	Cost savings, energy efficiency, and rural economic development through stubble-based construction industries.
Challenges	Material variability, long-term durability concerns, scalability issues, and regulatory barriers hinder widespread adoption.
Future Research	Focus on stubble-matrix bonding, fire resistance, hybrid reinforcement systems, and large-scale production techniques for commercialization.
Conclusion	Stubble-reinforced materials offer a sustainable pathway for construction, contributing to global sustainability goals and fostering innovation in the industry.

REFERENCES

Abbas, M., Akhai, S., Abbas, U., Jafri, R., & Arif, S. M. (2025). Ai-enabled sustainable urban planning and management. In *Real-World Applications of AI Innovation* (pp. 233–260). IGI Global Scientific Publishing.

Abbass, M., Abbas, U., Jafri, R., Arif, S. M., & Akhai, S. (2025). AI and Machine Learning Applications in Sustainable Smart Cities. In *Sustainable Smart Cities and the Future of Urban Development* (pp. 1–32). IGI Global Scientific Publishing.

Akhai, S., & Abbass, M. (2025). Toward Resilient Futures: The Role of AI-Driven Strategies in Climate Adaptation. In *Nexus of AI, Climatology, and Urbanism for Smart Cities* (pp. 325-340). IGI Global Scientific Publishing.

Akhai, S., & Rana, M. (2022). Taguchi-based grey relational analysis of abrasive water jet machining of Al-6061. *Materials Today: Proceedings, 65*, 3165–3169.

Akhai, S., Srivastava, P., Sharma, V., & Bhatia, A. (2021, August). Investigating weld strength of AA8011-6062 alloys joined via friction-stir welding using the RSM approach. [). IOP Publishing.]. *Journal of Physics: Conference Series, 1950*(1), 012016.

Akhil, U. V., Radhika, N., Saleh, B., Aravind Krishna, S., Noble, N., & Rajeshkumar, L. (2023). A comprehensive review on plant-based natural fiber reinforced polymer composites: Fabrication, properties, and applications. *Polymer Composites, 44*(5), 2598–2633.

Aly, M. (2012). *Development of an eco-friendly composite material for engineering applications* (Doctoral dissertation, Dublin City University).

Ashour, T. (2017). Composites using agricultural wastes. *Handbook of composites from renewable materials.* Wiley.

Asim, M., Uddin, G. M., Jamshaid, H., Raza, A., Hussain, U., Satti, A. N., & Arafat, S. M. (2020). Comparative experimental investigation of natural fibers reinforced light weight concrete as thermally efficient building materials. *Journal of Building Engineering, 31*, 101411.

Athira, G., Bahurudeen, A., & Appari, S. (2021). Rice-straw ash as a potential supplementary cementitious material: Influence of thermochemical conversion on its properties. *Journal of Materials in Civil Engineering, 33*(6), 04021123.

Ayub, M. A., Usman, M., Faiz, T., Umair, M., ul Haq, M. A., Rizwan, M., & Zia ur Rehman, M. (2020). Restoration of degraded soil for sustainable agriculture. *Soil health restoration and management*, 31-81.

Barczewski, M., Matykiewicz, D., Piasecki, A., & Szostak, M. (2018). Polyethylene green composites modified with post agricultural waste filler: Thermo-mechanical and damping properties. *Composite Interfaces*, *25*(4), 287–299.

Bhuyan, N., Narzari, R., Gogoi, L., Bordoloi, N., Hiloidhari, M., Palsaniya, D. R., & Kataki, R. (2020). Valorization of agricultural wastes for multidimensional use. In *Current developments in biotechnology and bioengineering* (pp. 41–78). Elsevier.

Binici, H., Eken, M., Kara, M., & Dolaz, M. (2013, October). An environment-friendly thermal insulation material from sunflower stalk, textile waste and stubble fibers. In *2013 International conference on renewable energy research and applications (ICRERA)* (pp. 833-846). IEEE.

Chaganti, V. N. (2008). *Reclamation potential of amendments for soils irrigated with saline-sodic drainage water*. California State University.

Deraman, R., Nawi, M. N. M., Mydin, M. A. O., Ismail, M. H., Nordin, N. D. M., Sari, M. W., & Mohd-Danuri, M. S. (2022). Production of roof board insulation using agricultural wastes towards sustainable building material. *Journal of Advanced Research in Fluid Mechanics and Thermal Sciences*, *99*(1), 66–89.

Fang, Z., Deng, W., Zhang, Y., Ding, X., Tang, M., Liu, T., & Wang, X. (2017). Open burning of rice, corn and wheat straws: Primary emissions, photochemical aging, and secondary organic aerosol formation. *Atmospheric Chemistry and Physics*, *17*(24), 14821–14839.

Garay, R. M. M., Rallo, M. D. L. B., Carmona, R. C., & Araya, J. C. (2009). Characterization of anatomical, chemical, and biodegradable properties of fibers from corn, wheat, and rice residues. *Chilean Journal of Agricultural Research*, *69*(3), 406–415.

Javaid, A., Jain, D., & Kwatra, N. (2024). Pre-treatment and its effect on the thermal conductivity of natural lignocellulosic rice straw stubble waste boards: Analysis vis a vis potential for the civil engineering applications. *Construction & Building Materials*, *445*, 137903.

Jiang, D., An, P., Cui, S., Xu, F., Tuo, T., Zhang, J., & Jiang, H. (2018). Effect of leaf fiber modification methods on mechanical and heat-insulating properties of leaf fiber cement-based composite materials. *Journal of Building Engineering*, *19*, 573–583.

Kamble, Z., & Behera, B. K. (2021). Sustainable hybrid composites reinforced with textile waste for construction and building applications. *Construction & Building Materials*, *284*, 122800.

Kumar, H., Wadhwa, A. S., Akhai, S., & Kaushik, A. (2024). Parametric analysis, modeling and optimization of the process parameters in electric discharge machining of aluminium metal matrix composite. *Engineering Research Express*, *6*(2), 025542.

Kumar, H., Wadhwa, A. S., Akhai, S., & Kaushik, A. (2024). Parametric optimization of the machining performance of Al-SiCp composite using combination of response surface methodology and desirability function. *Engineering Research Express*, *6*(2), 025505.

Li, X., Wang, S., Duan, L., Hao, J., Li, C., Chen, Y., & Yang, L. (2007). Particulate and trace gas emissions from open burning of wheat straw and corn stover in China. *Environmental Science & Technology*, *41*(17), 6052–6058. PMID: 17937281

Logeswaran, J., Shamsuddin, A. H., Silitonga, A. S., & Mahlia, T. M. I. (2020). Prospect of using rice straw for power generation: A review. *Environmental Science and Pollution Research International*, *27*, 25956–25969. PMID: 32378100

Maji, S., Dwivedi, D. H., Singh, N., Kishor, S., & Gond, M. (2020). Agricultural waste: Its impact on environment and management approaches. *Emerging eco-friendly green technologies for wastewater treatment*, 329-351.

Murali, G., & Wong, L. S. (2024). A comprehensive review of biochar-modified concrete: Mechanical performance and microstructural insights. *Construction & Building Materials*, *425*, 135986.

Pandey, M. M. (2004). Present status and future requirements of farm equipment for crop production. *Central Institute of Agricultural Engineering, Bhopal, 24*, 69-113.

Puri, V., Kumar, S., Grover, K., & Sharma, M. (2022, July). Development of Eco-Friendly Bricks for Sustainable Construction. *IOP Conference Series. Materials Science and Engineering*, *1248*(1), 012109.

Qamar, F., Thomas, T., & Ali, M. (2018). Use of natural fibrous plaster for improving the out of plane lateral resistance of mortarless interlocked masonry walling. *Construction & Building Materials*, *174*, 320–329.

Rana, M., & Akhai, S. (2022). Multi-objective optimization of Abrasive water jet Machining parameters for Inconel 625 alloy using TGRA. *Materials Today: Proceedings*, *65*, 3205–3210.

Rana, M. N., Islam, M. N., Nath, S. K., Das, A. K., Ashaduzzaman, M., & Shams, M. I. (2019). Properties of low-density cement-bonded composite panels manufactured from polystyrene and jute stick particles. *Journal of Wood Science*, *65*, 1–10.

Ruiz Trejo, C. E. (2018). *Feasability study on maize husk as resource for a novel composite material* (Doctoral dissertation).

Saha, A., Kulkarni, N. D., & Kumari, P. (2024). Development of Bambusa tulda fiber-micro particle reinforced hybrid green composite: A sustainable solution for tomorrow's challenges in construction and building engineering. *Construction & Building Materials*, *441*, 137486.

Saini, K., Matsagar, V. A., & Kodur, V. R. (2024). Recent advances in the use of natural fibers in civil engineering structures. *Construction & Building Materials*, *411*, 134364.

Sangmesh, B., Patil, N., Jaiswal, K. K., Gowrishankar, T. P., Selvakumar, K. K., Jyothi, M. S., & Kumar, S. (2023). Development of sustainable alternative materials for the construction of green buildings using agricultural residues: A review. *Construction & Building Materials*, *368*, 130457.

Serra-Parareda, F., Julián, F., Espinosa, E., Rodríguez, A., Espinach, F. X., & Vilaseca, F. (2020). Feasibility of barley straw fibers as reinforcement in fully biobased polyethylene composites: Macro and micro mechanics of the flexural strength. *Molecules (Basel, Switzerland)*, *25*(9), 2242. PMID: 32397611

Singh, L., & Brar, B. S. (2021). A review on rice straw management strategies. *Nat Environ Pollut Technol*, *20*(4), 1485–1493.

Thirunavukkarasu, A., Nithya, R., Sivashankar, R., & Sathya, A. B. (2018). Bio-based building materials for a green and sustainable environment. *Bioprocess engineering for a green environment*, 47-65.

Wang, B., Gu, F., Wu, F., Ma, J., Yang, H., & Hu, Z. (2022). Analysis of influencing factors and operation quality evaluation strategy of straw crushing and scattering system. *Agriculture*, *12*(4), 508.

Chapter 5
Energy Losses in Construction Parameters in Multi-Family Dwellings in Madrid, Spain

Roberto Alonso González-Lezcano
https://orcid.org/0000-0002-6185-4929
San Pablo CEU University, Spain

Mar Hernández Martín
San Pablo CEU University, Spain

Sunil Kumar Sansaniwal
https://orcid.org/0000-0002-0997-5096
NITI Aayog, New Delhi, India

Sonia Cesteros García
San Pablo CEU University, Spain

ABSTRACT

This paper focuses on the characterization of the residential building stock in the Madrid area. 151 Blower Door Tests have been carried out in multi-family dwellings built in different periods whose execution has been regulated by the UNE-EN 13829 standard. Through its quantification by an n50 value, average values of 5.8 renovations per hour have been obtained; in addition to detecting the main points where air infiltration occurs. These results are intended to lay the groundwork for finding solutions for infiltration, focusing on energy savings. By establishing a relationship between hourly renovations and the energy expenditure of the different

DOI: 10.4018/979-8-3373-2555-2.ch005

dwellings, the amortization of these improvements could be quantified. Many of the results obtained from current dwellings would not comply with the new requirements of the technical building code. In addition, the influence of several construction characteristics on permeability results was assessed.

INTRODUCTION

Building energy demand has become one of the most important concerns in the construction sector. The European Energy Performance of Buildings Directive (EPBD) is committed to achieve a highly efficient and decarbonized building stock, considering that almost 50% of the final energy consumption is used for heating and cooling, of which 80% is used in buildings, (European Parliament, 2018).

Concern about the airtightness of dwellings has been considered in the United States since before 1995, (Sherman, 1995). It is not surprising EU Member States have committed to reduce primary energy consumption by 20% by 2020. as energy consumption in residential and commercial buildings accounts for approximately 40% of total final energy consumption (Darvish et al., 2020), and 52% of the main heat losses in dwellings are due to ventilation, and therefore infiltration, becoming clear that improving airtightness is essential for enhancing energy efficiency in buildings, (Almarzouq & Sakhrieh, 2018).

Most of the energy is consumed by buildings. Buildings aimed at providing user comfort conditions are in an important position in terms of consuming energy. Considering the building life cycle, energy is an indispensable element for the building. There are differences in the increase in energy use in buildings and its global distribution. Recently, buildings that are self-sufficient within the framework of the building life cycle and independent buildings for energy have been designed and constructed, (Oh et al., 2017.

Energy efficiency regulations for buildings have been radically tightened over the last decade in Europe, and increased insulation requirements have greatly reduced heat losses through the thermal envelope. As a result, the impact of heat losses through the air has become proportionally larger in the energy demand of buildings, (Hong & Kim, 2022).

A recent study (Martín-Garín et al., 2020) puts the incidence of infiltration at between 20% and 50% of the total energy demand and warns of its impact on the airtightness of buildings. In the case of existing buildings, the wide range of energy renovation measures has increasingly included airtightness improvements, especially in cases where other actions are not feasible, such as in historic buildings. Many practitioners are incorporating airtightness analysis in their energy assessments due to its high impact on these types of buildings.

While there is abundant documentation on air infiltration in cold-climate European, there is little to almost nothing for warm countries such as Spain. This knowledge gap highlights the need to investigate the primary causes of these both controlled and uncontrolled infiltration in warm climates, through by studying dwellings in Madrid, (Goubran et al., 2017).

Despite the lack of regulations on airtightness or infiltrations in European countries with warm climates (Spain, Portugal, Italy...), tests have been carried out in different countries (Fernández-Agüera et al., 2016) that offer various results that help us to get an idea of the situation in which they find themselves. In Portugal, we observe that they focus on social housing, with average n50 values between 6 and 8 h-1, depending on whether they are occupied or unoccupied (Oliveira et al., 2021), and in dwellings built in periods between 70-80, they reach values of up to 9 h-1, which are reduced to values of between 4 and 5, when they have been rehabilitated or due to the behaviour of the users, (Ramos et al., 2015). In Greece and Italy, the results of the 20 tests carried out in each study, in dwellings, and although they show some unevenness, more than half of the cases exceed the value of 7 h-1, (Sfakianaki et al., 2008; Alfano et al., 2012). In Spain, the values range between 6 and 9 h-1, (Fernández-Agüera et al., 2016), and estimates of these values have even been presented based on extrapolations from other climatic zones, (Guillén-Lambea, Rodríguez-Soria, & Marín, 2019). However, this is not sufficient to determine a definitive value, and it is necessary to show the importance of airtightness in these countries to help reduce energy demand.

In countries such as Spain and Italy, multi-family dwellings prevail over other northern European countries, many of which are social housing, i.e. they receive government aid and are therefore of low construction standards, as well as being similar on the Mediterranean coast, (Fernández-Agüera et al., 2016). For all these reasons, it is more than ever necessary for these countries (those on the Mediterranean coast) to improve their airtightness standards. Indoor thermal conditions in these areas, especially in social housing, have usually been conditioned by economic considerations, the prolonged absence of thermal systems, both cooling and heating, in dwellings and the lack of adequate capacities in these buildings, (Sendra et al., 2013).

Social housing, in the form of collective buildings, is one of the fundamental pillars in the configuration of today's cities in southern Europe. It is estimated that air leakage through the envelopes of southern European cities is one of the most important factors impacting energy consumption, representing between 20% and 50% of a household's energy demand, depending on the climate zone, (Ferdyn-Grygierek et al., 2018). Therefore, they are important for the control of energy consumption and the residential environment, as well as for air quality design, (Alfano et al., 2016).

Leakage is a current of air that circulates through cracks, interstices and unwanted openings, which is produced by a pressure difference, due to wind, temperature differences or the mechanical ventilation system itself. This uncontrolled flow of air can occur from the outside to the inside (infiltration, which is the common form in buildings) or in the opposite direction (exfiltration, which is rare, due to unique external and internal conditions), (Jokisalo et al., 2009).

In this field of study, airtightness is understood as that quality of the building envelope capable of preventing the entry of air or other elements such as water, noise, pollutants, etc., from an external environment to an internal one. In order to achieve good acoustic and thermal insulation, it is considered necessary for the building envelope to be airtight. This is one of the necessary conditions for energy savings, (Sherman, 1987; Bhandari et al., 2018; González-Lezcano, & Hormigos-Jiménez, 2016).

In addition to its impact on energy efficiency and energy cost, inadequate airtightness of the dwelling also affects indoor air quality, among others causes, which determines that these infiltrations cannot be considered natural ventilation due to the quality of the surrounding air, (Fernandez-Antolin et al., 2019). According to Sherman (2017), the causes of leakage in a dwelling are height, area and age. He sets the height coefficient at 1.15, i.e. adding one storey in height increases leakage rates by 15%. The greater the area, the higher the ratio of openings to the total volume of the building. The older they get, the higher the leakage will be, at a rate of 1% per year.

This paper focuses on the characterization of the residential building stock in the Madrid area. Despite the fact that no evidence was found to justify that climate is a significant variable in terms of airtightness, (Hamby, 1994; Villi et al., 2013; Bramiana, Entrop, & Halman, 2016), it seems clear that there are different aspects associated to the region where the building is located such as differences in construction quality, dwelling design or materials, or due to differences in building size or age, (Hamby, 1994; Villi et al., 2013; Bramiana, Entrop, & Halman, 2016; Sadineni, Madala, & Boehm, 2011; Fernandez-Antolin, del-Rio, & Gonzalez-Lezcano, 2019). Through 151 tests in the Community of Madrid, the aim is to delve into the main causes of the lack of airtightness of multi-family dwellings.

MATERIALS AND METHODS

Blower-Door is the measurement technique used to determine the level of infiltration in the dwellings under study. It consists of a fan placed on an exterior door or window that generates a pressure difference, either by pressurizing or depressurizing the room, by injecting or extracting a controlled flow of air. The result of dividing the flow rate at a pressure of 50 Pa by the volume of indoor air gives the renovations/hour of the building, as a way of measuring the airtightness.

Once the leakage rate for a building has been measured, it is useful to estimate the total size of all leaks or openings in the building envelope. The TECTITE Express airtightness test analysis software calculates two separate leakage areas, based on different assumptions about the physical shape of the hole. The equivalent leakage area (EqLA) is defined by Canadian researchers at the National Research Council of Canada as the area of a sharp-edged opening through which the same amount of air will leak as that measured at a pressure difference of 10 Pascals. The effective leakage area (ELA) was developed by the Lawrence Berkeley Laboratory (LBL) and is used in their infiltration model. The effective leakage area is defined as the area of a rounded edge opening through which the same amount of air will leak as that measured at a pressure difference of 4 Pascals, (Synnefa, Santamouris, & Akbari, 2007; Fernandez-Antolin, del Rio, & Gonzalez-Lezcano, 2021; Lee, Kim, & Ryou, 2011; Sfakianaki et al., 2008; Fernandez-Antolin et al., 2020).

The Blower Door airtightness test is mandatory in many European countries for new buildings and to obtain energy efficiency and sustainable certifications such as Pasisvhaus or BREEAM and LEED. In other countries, it is not compulsory but practically essential to ensure compliance with current regulations and to avoid surprises on completion of the building work. In Spain, however, the Technical Building Code still considers it as an optional test.

Figure 1 shows the equipment with which all measurements were carried out:

Figure 1. a) and b) Blower door equipment used in the test, c) Infiltration by capitalization d) Infiltration due to poorly sealed ducts e) Infiltration through woodwork

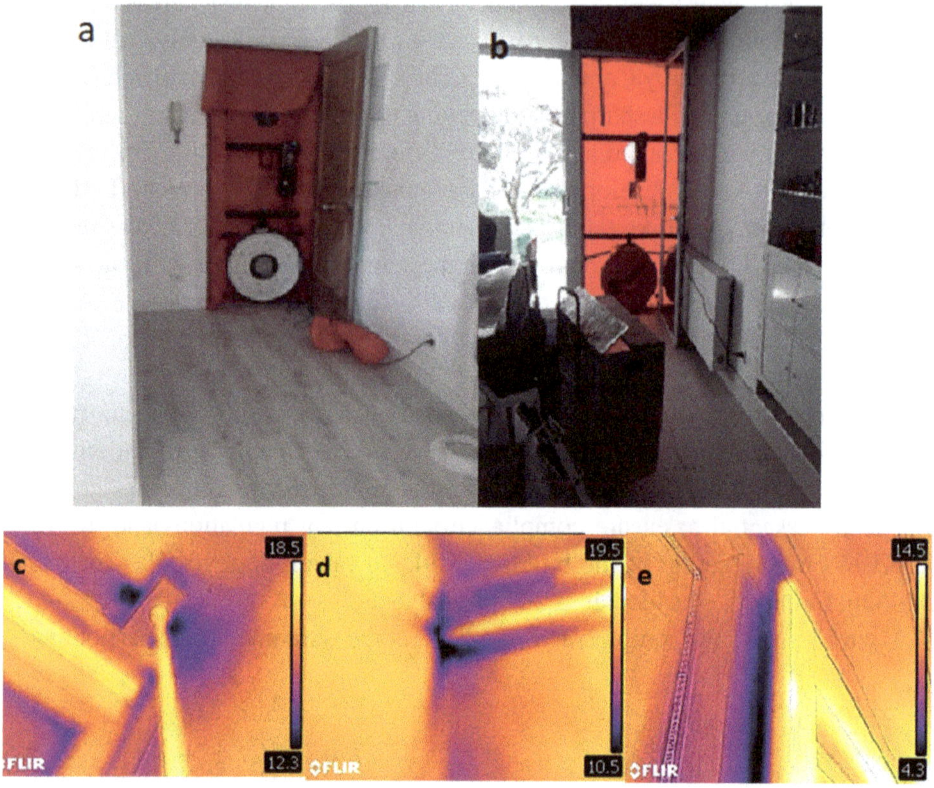

It is necessary to carry out an experimental analysis by means of tests, which make it possible to locate the openings or cracks through which the infiltrations take place. A widely used determination system is the thermographic analysis, the infrared radiation emitted by the bodies is captured thanks to a thermographic camera. During the depressurization process, data is taken through the thermography camera as shown in Figure 1, thus locating the points where the greatest infiltrations occur in the dwelling, for which it is necessary that there is a thermal gap between the interior and exterior.

RESULTS

By taking data from 151 measurements in collective dwellings in the Community of Madrid, from different years, the following results was drawn up (Figure 2-3-4)

Figure 2. Summarizing the results of the airtightness tests (Own elaboration)

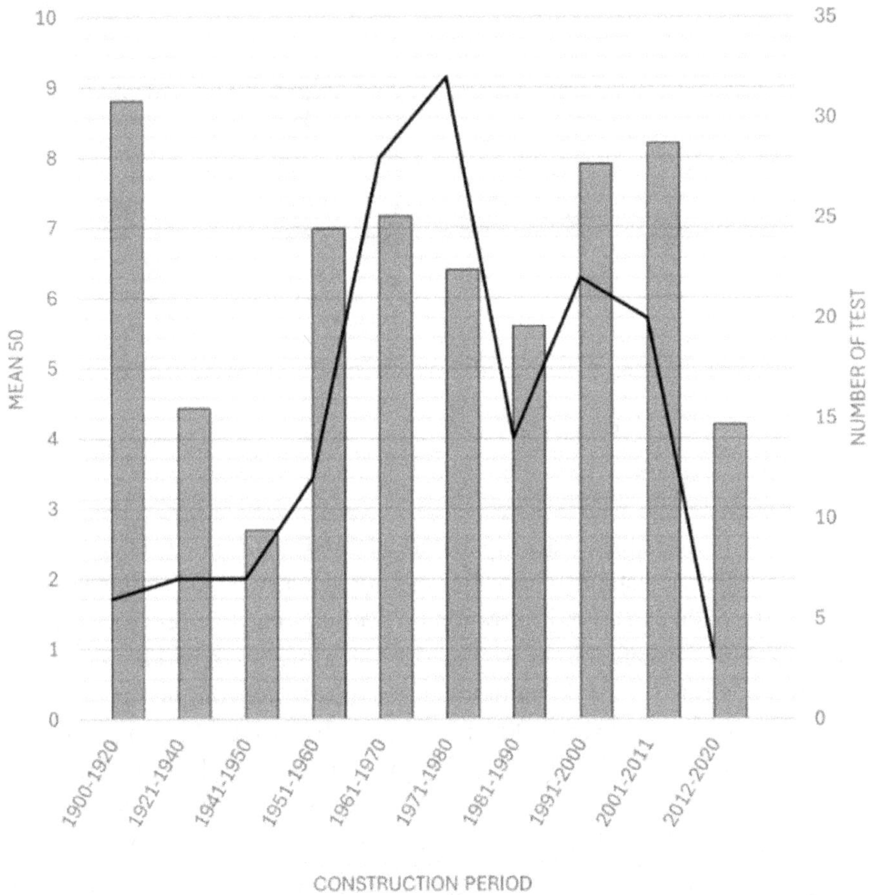

Figure 2 shows the evolution of the average number of renovations per hour (n50) as a function of the period of construction of the dwellings, contrasted with the number of tests carried out in each group. It can be seen that the highest infiltration values are concentrated in dwellings built between 1900-1920, despite the few tests carried out. Although there is a decrease in infiltration afterwards, it increases again in the following years, reaching values between 8 and 9 renovations per hour in the years 2001-2011, which is quite high despite the improvement in construction techniques. It is in more recent dwellings where an improvement in watertightness is shown, although with a lower number of tests carried out, with values between 4 and 5 renovations per hour, which may be due to slightly more demanding regulations.

Figure 3. Summarizing the cause of infiltration (Own elaboration)

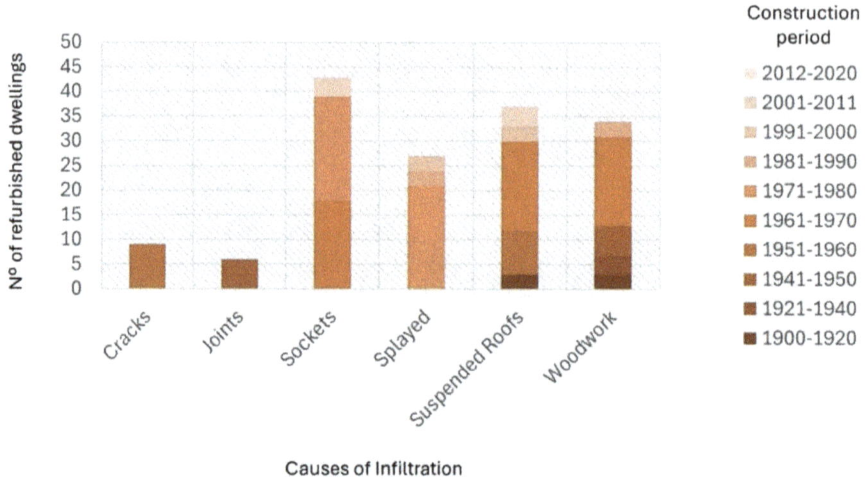

In Figure 3, it is possible to identify historical trends in the occurrence of infiltration, which can help to establish priorities in future energy rehabilitation strategies. In this case, the main causes of infiltration in dwellings can be observed, according to their period of construction, as well as their relationship to dwellings that were previously refurbished. Certain pathologies such as woodwork are more frequent in older buildings (e.g. between 1900-1960), probably due to the construction materials used, which also coincides with a higher number of refurbished dwellings. Pathologies such as suspended roofs appear in almost all periods, although more or less frequently in dwellings where interventions have been carried out. Other emerging pathologies in recent buildings are sockets or splayed, of which fewer have been refurbished.

Figure 4. Classification of infiltration standards in different European countries (Fernandez-Antolin et al., 2020)

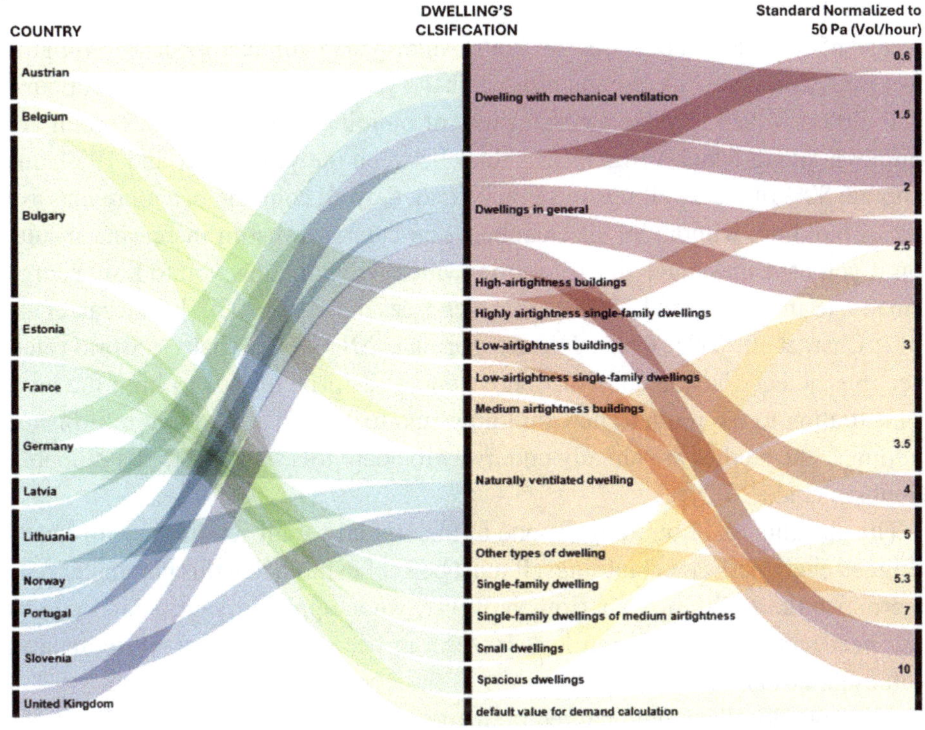

Figure 4 shows the categories of dwellings according to European countries that have standard reference values varying from 0.6 to 10 volume per hour for infiltration. It can be seen that countries such as Austria, Germany, Norway and Slovenia provide that dwellings with mechanical ventilation should achieve the lowest infiltration rate, while dwellings that use natural ventilation, as reflected in the countries of Germany, Lithuania and Slovenia, the reference value increases to 3, which reflects less control over heat exchange. Dwellings where higher infiltration is allowed correspond to the regulations in the UK and Belgium (for dwellings in general), and in Bulgaria (for low-airtightness buildings).

In addition, it is considered that the reference values of renovations/hour vary depending on the classification of housing by ventilation, size, use... and reflects the importance of having a regulation that addresses infiltrations adapted to the real built conditions of each country, as is the case in these European countries, and which is lacking in many others, such as Spain. In this way, overestimates or underestimates can be avoided in the design of heating, ventilation or cooling systems.

DISCUSSIONS

With the results obtained in the tests carried out in Madrid, we can conclude an average n50 of 5.80 renovations per hour. Among the main factors observed in the dwellings, as in the USA, the main factor is the year of construction, (Santamouris et al., 2001), which leads to a certain way of building and therefore a certain airtightness, giving better airtightness values between the years 1921 and 1950, and although 80% of the dwellings were renovated, they all coincide in a more massive way of building, without an air chamber. You cannot invest in increasing insulation if you have unwanted air infiltration, which will not allow you to limit energy demand, (Santamaria et al., 2020; Gonzales-Lezcano et al., 2020; Frontczak et al., 2012; Chan et al., 2005; Chan, Joh, & Sherman, 2013). Passivhaus, a world reference in efficient housing construction, as one of its principles is to limit the level of infiltration to 0.6 ren/h in new buildings and to 1 ren/h at 50 Pa in refurbished buildings, and studies are already underway to adapt this standard to hot climates, (Goubran et al., 2017).

The airtightness of buildings is one of the crucial parameters for new or newly designed energy efficient buildings. It must be kept in mind that this parameter only determines the infiltration air flow, however, when applying ventilation systems with heat recovery, infiltration heat losses can become dominant, (Friedman & Matheson, 2017).

Although, local and national requirements are generally diversified all over the world, there is no relation between local climatic conditions and the required airtightness levels. In addition, the requirements do not distinguish between buildings that are well-exposed to environmental conditions and those that are more sheltered, (Lezcano & Burgos, 2021; Gonzalez-Lezcano et al., 2023).

A methodology has been developed to investigate the airtightness of the architectural envelope of the existing residential park in Madrid by means of an experimental study using the blower door test. Considering a proportional quota sampling scheme, a representative number of 151 cases has been selected. In this way it is possible to extrapolate the results to other residential buildings in Madrid.

Building construction techniques using good insulation do not have the desired effect without good execution and supervision, this can lead to unwanted air infiltration due to various construction defects. At the same time, special care must be taken with the natural ventilation of the dwelling, as this can account for more than 50% of the heat losses, (Gonzalo et al., 2022).

CONCLUSIONS AND IMPLICATIONS

In Spain there are no mandatory regulations regarding the level of airtightness of residential buildings. In the latest update of the Technical Building Code (CTE) in its Basic Document on Energy Saving (DB-HE) of 2018 (Chan, Joh, & Sherman, 2013), it is indicated that to calculate the energy demand of buildings, losses due to ventilation and unwanted infiltrations should be considered, but no limit is established, (Mutschler et al., 2021; Miszczuk & Heim, 2021; de la Edificacion, 2018; Isaac & van Buuren, 2009; Burgos et al., 2020).

In the year 2019, the Technical Building Code in Spain (Agee, Nikdel, & Roberts, 2021), introduced modifications concerning the air permeability of the building envelope. These requirement are limited to new buildings for private residential use. In addition, it also limits the useful surface area to a minimum of 120 m^2 for this condition to be enforceable. What is determined is that the permeability of a new building for residential use (including single-family houses) with a useful surface area greater than 120 m^2 does not exceed the tabulated limit value that depends on the Compacity (V/A): Ratio between the volume enclosed by the thermal envelope (V) of the building (or part of the building) and the sum of the heat exchange surfaces of this envelope (A= $\sum$Ai). It is expressed in m^3/m^2.

> V/A less than or equal to 2; n_{50} less than 6
> V/A greater than or equal to 4; n_{50} greater than 3
> The average values of the intermediate compactnesses would be obtained by interpolation.

The standard does not indicate values for dwellings of less than 120 m^2; moreover, these values are quite high compared to most northern European standards. Many of the results obtained from current dwellings would not comply with the new requirements of the technical building code, (Miszczuk & Heim, 2021).

In countries with warm climates, such as Spain, Italy, Portugal or Greece (Fernández-Agüera et al., 2016; Guillén-Lambea, Rodríguez-Soria, & Marín, 2019), the airtightness of residential buildings is limited, depending on whether it is natural or mechanical ventilation, as the latter will be more demanding in terms of airtightness. It is important to lay the foundations in Spain for these requirements, but always with a link to energy demand, setting the guidelines to achieve these savings, not just demanding a value of renovations.

In addition, the influence of several construction characteristics on permeability results was assessed. General trends have been identified. Nevertheless, no statistically significant results could be obtained, in part due to the reduced sample size for some categories and also because of the difficulty of isolating the variables.

Extrapolations by means of energy simulations would be highly recommended for any analysis of air infiltration in the context of the energy performance of buildings; and thus the effect on energy consumption at different pressures can be analysed.

It has also been concluded that the effect of improving the indoor comfort conditions of the building by controlling the outdoor environmental conditions with the construction of different façade systems is very important in the architectural process. With the development of technology and intelligent systems, the impact of façades on climate analysis and energy efficient design strategies will become much more important in the future.

While ventilation is a necessary feature of buildings to guarantee good indoor air quality (IAQ), infiltration is an unintended and uncontrolled effect. Therefore, despite the extra fresh air it provides, infiltration doesn't assure proper IAQ and it can increase considerably the heating and cooling needs of buildings.

Practical Implications: Despite having identified some interesting correlations for airtightness and building type, they should be applied with caution, due to the limited number of cases analyzed. Further research is advised to complement the present results.

FUNDING

The authors wish to thank CEU San Pablo University Foundation for the funds dedicated to the ARIE Research Group, through the Project Ref. G20/6-06 provided by the CEU San Pablo University.

ACKNOWLEDGMENT

The authors wish to thank CEU San Pablo University Foundation for the funds dedicated to the ARIE Research Group, Project Ref. G20/6-06 by the CEU San Pablo University.

REFERENCES

Agee, P., Nikdel, L., & Roberts, S. (2021). A measured energy use, solar production, and building air leakage dataset for a zero energy commercial building. *Scientific Data*, *8*(1). Advance online publication. DOI: 10.1038/s41597-021-01082-8

Alfano, D. F. R., Dell'Isola, M., Ficco, G., & Tassini, F. (2012). Experimental analysis of air tightness in Mediterranean buildings using the fan pressurization method. *Building and Environment*, *53*, 16–25. DOI: 10.1016/J.BUILDENV.2011.12.017

Alfano, F. R. d., Dell'Isola, M., Ficco, G., Palella, B. I., & Riccio, G. (2016). Experimental air-tightness analysis in mediterranean buildings after windows retrofit. *Sustainability*, *8*, 991.

Almarzouq, A., & Sakhrieh, A. (2018). Effects of glazing design and infiltration rate on energy consumption and thermal comfort in residential buildings. *Thermal Science*, *2018*, 2951–2960. DOI: 10.2298/TSCI170910073A

Bhandari, M., Hun, D., Shrestha, S., Pallin, S., & Lapsa, M. (2018). A simplified methodology to estimate energy savings in commercial buildings from improvements in airtightness. *Energies*, *11*(12). Advance online publication. DOI: 10.3390/en11123322

Bramiana, C. N., Entrop, A. G., & Halman, J. I. M. (2016). Relationships between Building Characteristics and Airtightness of Dutch Dwellings. In *Energy Procedia* (Vol. 96, pp. 580–591). Elsevier Ltd. DOI: 10.1016/j.egypro.2016.09.103

Burgos, M. J. M., de Toledo, H. S. Á., Lezcano, R. A. G., & de Mera, A. G. (2020). The sedentary process and the evolution of energy consumption in eight native American dwellings: Analyzing sustainability in traditional architecture. *Sustainability (Switzerland)*, *12*(5), 1–28. DOI: 10.3390/su12051810

Chan, , W. R., Joh, J., & Sherman, M. H. (2013). Analysis of air leakage measurements of US houses. *Energy and Building*, *66*, 616–625. DOI: 10.1016/j.enbuild.2013.07.047

Chan, W. R., Nazaroff, W. W., Price, P. N., Sohn, M. D., & Gadgil, A. J. (2005). Analyzing a database of residential air leakage in the United States. *Atmospheric Environment, 39*(19), 3445–3455. DOI: 10.1016/j.atmosenv.2005.01.062

Darvish, A., Eghbali, S. R., Eghbali, G., & Mahlabani, Y. G. (2020). The effects of building glass facade geometry on wind infiltration and heating and cooling energy consumption. *International Journal of Technology, 11*(2), 235–247. DOI: 10.14716/ijtech.v11i2.3201

de la Edificación, C. T. (2018). *Documento Básico he Ahorro de energía. CTE.* DB-HE.

de la Edificación, C. T. (2019). Documento Básico Ahorro de Energía HE1: Limitación de la Demanda Energética. *Madrid. December.*

European Parliament, European Directive 2018/844 Amending Directive 2010/31/EU on the Energy Performance of Buildings Directive 2012/27, EU on energy efficiency, 2018.

Ferdyn-Grygierek, J., Bartosz, D., Specjał, A., & Grygierek, K. (2018). Analysis of Accuracy Determination of the Seasonal Heat Demand in Buildings Based on Short Measurement Periods. *Energies, 11*, 2734.

Fernández-Agüera, J., Domínguez-Amarillo, S., Sendra, J. J., & Suárez, R. (2016). An approach to modelling envelope airtightness in multi-family social housing in Mediterranean Europe based on the situation in Spain. *Energy and Building, 128*, 236–253. DOI: 10.1016/j.enbuild.2016.06.074

Fernandez-Antolin, M. M., del Río, J. M., Costanzo, V., Nocera, F., & Gonzalez-Lezcano, R. A. (2019). Passive design strategies for residential buildings in different Spanish climate zones. *Sustainability (Switzerland), 11*(18). Advance online publication. DOI: 10.3390/su11184816

Fernandez-Antolin, M. M., Del-Río, J. M., Del Ama Gonzalo, F., & Gonzalez-Lezcano, R. A. (2020). The relationship between the use of building performance simulation tools by recent graduate architects and the deficiencies in architectural education. *Energies, 13*(5). Advance online publication. DOI: 10.3390/en13051134

Fernandez-Antolin, M. M., del-Río, J. M., & Gonzalez-Lezcano, R. A. (2019). Influence of solar reflectance and renewable energies on residential heating and cooling demand in sustainable architecture: A case study in different climate zones in Spain considering their urban contexts. *Sustainability (Switzerland), 11*(23). Advance online publication. DOI: 10.3390/su11236782

Fernandez-Antolin, M. M., del Río, J. M., & Gonzalez-Lezcano, R. A. (2021). The use of gamification in higher technical education: Perception of university students on innovative teaching materials. *International Journal of Technology and Design Education*, *31*(5), 1019–1038. DOI: 10.1007/s10798-020-09583-0

Friedman, A., & Matheson, M. (2017). Selecting and Installing Energy-Efficient Windows to Improve Dwelling Sustainability. VITRUVIO - International Journal of Architectural Technology and Sustainability, 2(2), 2. https://doi.org/DOI: 10.4995/vitruvio-ijats.2017.7687

Frontczak, M., Schiavon, S., Goins, J., Arens, E., Zhang, H., & Wargocki, P. (2012). Quantitative relationships between occupant satisfaction and satisfaction aspects of indoor environmental quality and building design. *Indoor Air*, *22*(2), 119–131. DOI: 10.1111/j.1600-0668.2011.00745.x

González-Lezcano, R. A., & Hormigos-Jiménez, S. (2016). Energy saving due to natural ventilation in housing blocks in Madrid. In IOP Conference Series: Materials Science and Engineering (Vol. 138). Institute of Physics Publishing. DOI: 10.1088/1757-899X/138/1/012002

González-Lezcano, R. A., López-Fernández, E., Cesteros-García, S., & Sanglier-Contreras, G. (2020). Evaluation of energy-saving strategies according to spanish regulations. *International Journal of Energy and Environmental Engineering*, *26*(3), 197–217.

González-Lezcano, R. A., Sanglier Contreras, G., Iglesias Sanz, C. M., Sancho Alambillaga, R., & López Fernández, E. J. (2023). *Construction parameters that affect the air leaks of the envelope in dwellings in Madrid.* Frontiers in Built Environment. DOI: 10.3389/fbuil.2023.1220559

Gonzalo, F. D. A., Griffin, M., Laskosky, J., Yost, P., & González-lezcano, R. A. (2022). Assessment of Indoor Air Quality in Residential Buildings of New England through Actual Data. *Sustainability (Switzerland)*, *14*(2). Advance online publication. DOI: 10.3390/su14020739

Goubran, S., Qi, D., Saleh, W. F., & Wang, L. (Leon). (2017). Comparing methods of modeling air infiltration through building entrances and their impact on building energy simulations. *Energy and Building*, *138*, 579–590. DOI: 10.1016/j.enbuild.2016.12.071

Guillén-Lambea, S., Rodríguez-Soria, B., & Marín, J. M. (2019). Air infiltrations and energy demand for residential low energy buildings in warm climates. *Renewable & Sustainable Energy Reviews*, *116*, 109469. DOI: 10.1016/J.RSER.2019.109469

Hamby, D. M. (1994). A review of techniques for parameter sensitivity analysis of environmental models. *Environmental Monitoring and Assessment, 32*(2), 135–154. DOI: 10.1007/BF00547132

Hong, G., & Kim, C. (2022). Experimental analysis of airtightness performance in high-rise residential buildings for improved code-compliant simulations. *Energy and Building, 261*. Advance online publication. DOI: 10.1016/j.enbuild.2022.111980

Isaac, M., & van Vuuren, D. P. (2009). Modeling global residential sector energy demand for heating and air conditioning in the context of climate change. *Energy Policy, 37*(2), 507–521. DOI: 10.1016/j.enpol.2008.09.051

Jokisalo, J., Kurnitski, J., Korpi, M., Kalamees, T., & Vinha, J. (2009). Building leakage, infiltration, and energy performance analyses for Finnish detached houses. *Building and Environment, 44*(2), 377–387. DOI: 10.1016/j.buildenv.2008.03.014

Lee, M. J., Kim, N. I., & Ryou, H. S. (2011). Air tightness measurement with transient methods using sudden expansion from a compressed chamber. *Building and Environment, 46*(10), 1937–1945. DOI: 10.1016/j.buildenv.2011.04.001

Lezcano, R. A. G., & Burgos, M. J. M. (2021). Airflow analysis of the haida plank house, a breathing envelope. *Energies, 14*(16). Advance online publication. DOI: 10.3390/en14164871

Lezcano, R. A. G., Fernández, E. J. L., García, S. C., & Contreras, G. S. (2020). Optimisation of Heat Loss in Cities in Different Climatic Zones in a Mediterranean Climate.

Lozinsky, C. H., & Touchie, M. F. (2018). Improving energy model calibration of multi-unit residential buildings through component air infiltration testing. *Building and Environment, 134*, 218–229. DOI: 10.1016/j.buildenv.2018.02.040

Martín-Garín, A., Millán-García, J. A., Hidalgo-Betanzos, J. M., Hernández-Minguillón, R. J., & Baïri, A. (2020). Airtightness analysis of the built heritage⇓field measurements of nineteenth century buildings through blower door tests. *Energies, 13*(24). Advance online publication. DOI: 10.3390/en13246727

Miszczuk, A., & Heim, D. (2021). Parametric study of air infiltration in residential buildings—The effect of local conditions on energy demand. *Energies, 14*(1). Advance online publication. DOI: 10.3390/en14010127

Mutschler, R., Rüdisüli, M., Heer, P., & Eggimann, S. (2021). Benchmarking cooling and heating energy demands considering climate change, population growth and cooling device uptake. *Applied Energy, 288*. Advance online publication. DOI: 10.1016/j.apenergy.2021.116636

Oh, J., Hong, T., Hakpyeong, K., An, J., Kwangbok, J., & Koo, C. (2017). Advanced Strategies for Net-Zero Energy Building: Focused on the Early Phase and Usage Phase of a Building's Life Cycle. *Sustainability, 9*, 2272. DOI: 10.3390/su9122272

Oliveira, R., Vicente, R., Almeida, R. M. S. F., & Figueiredo, A. (2021). The importance of in situ characterisation for the mitigation of poor indoor environmental conditions in social housing. *Sustainability (Switzerland), 13*(17). Advance online publication. DOI: 10.3390/su13179836

Ramos, N. M. M., Almeida, R. M. S. F., Curado, A., Pereira, P. F., Manuel, S., & Maia, J. (2015). Airtightness and ventilation in a mild climate country rehabilitated social housing buildings – What users want and what they get. *Building and Environment, 92*, 97–110. DOI: 10.1016/J.BUILDENV.2015.04.016

Sadineni, S. B., Madala, S., & Boehm, R. F. (2011). *Passive building energy savings: A review ofbuilding envelope components. Renewable and Sustainable Energy Reviews*. Elsevier Ltd. DOI: 10.1016/j.rser.2011.07.014

Santamaria, B. M., Del Ama Gonzalo, F., Pinette, D., Gonzalez-Lezcano, R. A., Aguirregabiria, B. L., & Ramos, J. A. H. (2020). Application and validation of a dynamic energy simulation tool: A case study with water flow glazing envelope. *Energies, 13*(12). Advance online publication. DOI: 10.3390/en13123203

Santamouris, M., Papanikolaou, N., Livada, I., Koronakis, I., Georgakis, C., Argiriou, A., & Assimakopoulos, D. N. (2001). On the impact of urban climate on the energy consuption of building. *Solar Energy, 70*(3), 201–216. DOI: 10.1016/S0038-092X(00)00095-5

Sendra, J. J., Domínguez-Amarillo, S., Rojas, P. B., & Leon, A. L. (2013). Energy Intervention in the residential sector in the south of Spain: Current challenges. *Informes de la Construcción, 65*, 457–464.

Sfakianaki, A., Pavlou, K., Santamouris, M., Livada, I., Assimakopoulos, M. N., Mantas, P., & Christakopoulos, A. (2008). Air tightness measurements of residential houses in Athens, Greece. *Building and Environment, 43*(4), 398–405. DOI: 10.1016/J.BUILDENV.2007.01.006

Sherman, M. (1995). The Use of Blower-Door Data. *Indoor Air, 5*(3), 215–224. DOI: 10.1111/j.1600-0668.1995.t01-1-00008.x

Sherman, M. H. (1987). Estimation of infiltration from leakage and climate indicators. *Energy and Building, 10*(1), 81–86. DOI: 10.1016/0378-7788(87)90008-9

Shrestha, P. M., Humphrey, J. L., Barton, K. E., Carlton, E. J., Adgate, J. L., Root, E. D., & Miller, S. L. (2019). Impact of low-income home energy-efficiency retrofits on building air tightness and healthy home indicators. *Sustainability (Switzerland), 11*(9). Advance online publication. DOI: 10.3390/su11092667

Synnefa, A., Santamouris, M., & Akbari, H. (2007). Estimating the effect of using cool coatings on energy loads and thermal comfort in residential buildings in various climatic conditions. *Energy and Building*, *39*(11), 1167–1174. DOI: 10.1016/j.enbuild.2007.01.004

Villi, G., Peretti, C., Graci, S., & De Carli, M. (2013). Building leakage analysis and infiltration modelling for an Italian multi-family building. *Journal of Building Performance Simulation*, *2*(6), 98–118. DOI: 10.1080/19401493.2012.699981

Chapter 6

Prediction and Enhancing Construction Processes Through Machine Learning and Recent Optimization Techniques

M. Veerapathran
https://orcid.org/0000-0002-5515-247X
Department of Civil Engineering, Dr. N.G.P. Institute of Technology, Coimbatore, India

Somu Chinnusamy
https://orcid.org/0009-0006-5764-1660
Department of Research and Development, RSP Science Hub, Coimbatore, India

Himanshu Singh
Department of Civil Engineering, Uttranchal University, Dehradun, India

Amit Chawla
School of Emerging Media and Creator Economy, K.R. Mangalam University, Gurugram, India

Harkirat Kaur
https://orcid.org/0000-0002-1281-5608
Department of Civil Engineering, Gautam Buddha University, Greater Noida, India

DOI: 10.4018/979-8-3373-2555-2.ch006

ABSTRACT

This chapter explores artificial intelligence (AI) and machine learning (ML)-driven processes in transactions that detect optimization, emphasizing forecast analysis, auto tomatoes, and data-based decisions. It investigates how AI/ML models increase the accuracy of planning, resource allocation, reducing risk and real-time monitoring of construction activities. By integrating history with intelligent algorithms, construction workflow can be trimmed for more efficiency, cost-effectiveness and durability. Case studies and emerging technologies highlight AI's transformative prospects in eliminating industry challenges such as delays, budget overruns, and safety risks. The chapter outlines practical structures and future directions for the implementation of AI and ML to optimize the construction project exhibition from the end to end.

INTRODUCTION

In the construction industry, which has long been considered by the adoption of new technologies, has been taking deeply with the integration of artificial intelligence (AI) and machine learning (ML). These technologies are re -defining how construction projects are conceived, planned, implemented and managed. Among the increasing pressure for timely delivery, cost control, durability and safety, presents a powerful set of tools for processes throughout the life cycle of AI and ML construction projects. Their forecasts to adopt and mark the main shift from the reactive in prescription management approaches, enabling stakeholders to make informed decisions on the basis of data-based insights rather than inward or stable planning models (Kazeem et al., 2023).

At its root, AI refers to the simulation of human intelligence processes by machines, especially computer systems. These processes include learning, logic and self-improvement. ML, a subset of AI, focuses on the development of algorithms that allow the computer to learn from predictions or decisions based on data based on data. When applicable to construction management, these technologies provide unprecedented capabilities in the process of large and complex datasets to expose patterns, prediction results, and automatic routine and complex tasks equally (Baduge et al., 2022a).

The most attractive advantage of AI and ML in construction is predictive analysis. Construction Projects Produces a large amount of data through building information modeling (BIM), sensors, drones, project management. Software and Historical Tiachaic Records. M.L. Algorithms can analyze these data sources such as potential project risks such as cost overrun, scheduled delays, labor shortages and failure of devices. For example, observed education techniques can be applied to the history of the historical rhythmic project data to predict the duration of specific tasks or the possibility of certain risks, enabling project managers to allocate resources and adjust the timetable accordingly. This transition from traditional to prediction management increases efficiency and reduces uncertainty, improving overall project results (Sharma et al., 2022a).

Construction projects include numerous mutual dependent activities that must be carefully coordinated. AI and ML tools can imitate various construction sequences for cost, time or environmental influence. Reinforcement education, for example, can be applied to develop dynamic scheduled models that are convenient to change the situation of the project in real-time. In addition, the optimization algorithms can tune well with the allocation of resources such as labor, materials and devices, learning from past project data and improving continuous recommendations. The result is a crisp and adaptive project management framework that effectively responds to disruptions and benefits the opportunities for process improvement (Waqar, 2024).

AI enhances automatic and real-time monitoring, which significantly improves the productivity and safety of the site. Technologies such as computer vision and IOT-enabled sensors are being linked to AI algorithms to monitor construction sites, find unsafe behavior, track devices and ensure quality control. M.L. Models can process real-time video feeds and sensor data to trigger automatic actions such as alert supervisors and machinery or re-operate workflow. This not only increases operational efficiency but also reduces human error and improves the compliance with safety rules, which is important in high -risk environments such as construction sites (Abioye et al., 2021).

The implementation of AI and ML also presents advanced decision support systems. These systems integrate multiple data sources and provide active insights through dashboards, warnings and view simulations to project managers. Instead of relying on fragmented information, the decision -making can cause access to the widespread views of the project exhibition and imitate the influence of various decisions. For example, AI may imitate how a sub -contractor or material supplier can affect project costs and timelines, helping managers choose the best way to action. In this way, AI acts as a partner, increasing human judgment with machine intelligence (Pan & Zhang, 2023a).

Despite these advantages, the integration of AI and ML in construction is not without challenges. The quality and availability of the data remains the main obstacles. Construction data is often unstructured, sill, or incomplete, which limits the effectiveness of AI models. In order to maximize the value of AI, it is necessary to ensure data interoperability and standardization in the platform and stakeholders. In addition, the construction industry related to data vigil, AI and analytics is a distance of significant skills. Target training, interdisciplinary cooperation and investments in digital infrastructure are required to overcome this distance (Tapeh & Naser, 2023).

Moral and regime concerns must also be taken into account. The use of AI raises questions about responsibility, data privacy and algorithmic bias. For example, decisions taken by AI systems - such as labor display assessments or risk priority - are explained to transparent and stakeholders. Strong governance framework must be developed in the construction industry that ensures that AI systems are reliable, incorporated and aligned with extensive social and regulatory expectations. This includes clarity to adopt AI approaches and establish liability methods for automatic decisions (Tien et al., 2022).

Looking forward, the role of AI and ML in construction will only increase. Emerging trends such as digital twins, generating design, autonomous construction equipment and blockchain -based supply chain integration are ready to enhance AI's capabilities in construction. Virtual replicas of digital twin-physical construction projects-design, performance and maintenance can be continuously updated with real-time data of sensors to Ze pt. Generative design algorithms can look for thousands of design options based on definitive barriers and objectives, helping engineers more effectively on innovative solutions (Zhang et al., 2022).

The strategic integration of AI and ML in construction adjusts with widespread goals of stability, elasticity and smart urban development. AI contributes to more durable construction methods, enabling smart use of materials, reducing waste and increasing energy efficiency. In the context of global challenges such as climate change, urbanization and structural renewal, AI-celebrated construction management offers a forward-looking settlement that balances efficiency with liability (Soori et al., 2023).

Hence, the construction of AI and ML represents inspiration, planning projects, execute and shifting. These technologies empower stakeholders with predictions and prescription tools that increase decision -making, reduce disqualification and reduce risks. While challenges related to data, skills and morality must be carefully managed, the long -term benefits of adopting AI and ML are significant. As the industry accepts digital transformation, the AI-operated process will be part of the built environment and the foundation of a competitive advantage and sustainable project delivery.

Objectives

This chapter aims to explore the integration of artificial intelligence (AI) and machine learning (ML) for optimizing construction management processes. It tries to analyze how AI/ML can increase project planning, resource allocation and risk predicting when improving efficiency, safety and durability. The main objectives include practical applications and construction workflows of forecast analyzes and evaluation, real-time monitoring systems and data-based framework. This chapter also addresses the challenges of the implementation of the data, including the quality and moral concerns of the data, and proposes strategic solutions for successful adoption. After all, it provides a roadmap for the benefit of AI and ML to transform the influence of the construction project.

BACKGROUND: AI AND MACHINE LEARNING

Modern data -based modern decisions in industries, including artificial intelligence (AI) and machine learning (ML) construction, creates a technical backbone of modern decision-making processes. AI widely refers to the development of computer systems that usually require human intelligence, such as logic, education, problem solving and understanding of natural language. Machine learning, AI subset, especially focuses on enabling computers to learn from data patterns and improve their performance over time, without clearly programmed for each task. Together, AI and ML provide powerful tools for driving efficiency and innovation in construction, to analyze huge amounts of construction data, produce functional insights, and automate complex workflows (Glady et al., 2024; Kalaiselvi et al., 2024; Reddy et al., 2024).

The root of machine learning has different types of algorithms designed to handle different data challenges and objectives. These algorithms are usually classified in three main types: observed education, unhealthy education and reinforcement education. The observed education involves training a model on a labeled dataset, where input data is attached to the appropriate output. The goal is to learn the model mapping from the inputs to the output and make accurate predictions on new, invisible data. This category includes linear regression, decision trees, support vector machines and neural networks in the general algorithms. In construction, inspection education is used for analytical functions of predictions such as prediction project delay, cost estimates, or learning from the history of the historical project data to find potential safety risks(Gunasekaran et al., 2024; Madan et al., 2024).

Different from unhealthy education that the data provided to the model is without label, that is, there is no predefined output. Instead, the algorithm tries to identify the patterns or groups hidden inside the data. Clustering and association algorithms, such as K-Min clustering and major component analysis (PCA), are popular unhealthy techniques. In the context of construction, unaffected education helps to divide construction activities, grouping similar project risks or analyzing sensor data to find inconsistencies in the operation of devices without pre -knowledge.

Reinforcement education is a dynamic and interactive form of machine learning where an agent learns to decide by performing actions within the environment to maximize the accumulated rewards. Unlike observed and unsafe education, reinforcement education depends on the response of actions rather than stable datasets. This approach is especially useful for optimization problems, where the system should adapt to constant changing conditions. In construction management, reinforcement education can be operated to develop adaptive scheduled systems that respond to disruptions such as weather change or supply delays in real-time, where there is a resource allocation and to the index of functions(Mageswari et al., 2024; Narra et al., 2024).

The consistency of AI and ML with the construction industry is multilateral and more serious in view of the complexity and efficiency of the field. Construction projects include uncertainty, changes and mutual dependence, making traditional management approaches less effective. AI and ML enhanced construction professionals to benefit the historical accommodation and real-time data for advanced forecasts, risk management and decision support. For example, the predictive models can estimate the possibility of excessive or delay of cost depending on the Dells, project parameters and external factors, enabling active interventions (Muthukumar et al., 2024).

Moreover, construction sites are becoming more digitized by the use of building information modeling (BIM), Internet Th Things (IoT) sensors, drones and other techniques that produce a large amount of data. AI and ML algorithms process this data to monitor progress, find deviations, and automate observations. Computer vision techniques, a branch of AI enables the automatic identity of defects, safety compliance and materials by analyzing the images and videos captured on the site. Machine learning of construction processes leads to a better scheduled, reduced waste and the use of improved resources. For example, ML-powered process can recommend the most efficient order of construction activities, or the ideal allocation of workforce and machinery, which can be considered as barriers to the deadline, budget limit and site status. This increases productivity and reduces operational costs.

In addition, AI promotes innovation in the design and plan phase by integrating generating design algorithms that explore multiple design options based on influence measures such as structural integrity, cost and durability. Such data -based design tools improve the decision making and promote durable building methods. Despite these benefits, the construction industry faces the challenges in fully realizing the AI and ML benefits. Data fragmentation, inconsistent data quality and lack of standardization model accuracy and scalability. Moreover, traditional mentality and skill in the workforce can slow down. A unified effort is needed in digital infrastructure development, training and cross-disciplinary collaboration to overcome these obstacles(Singh Madan et al., 2024).

In summary, the base of AI and ML consists of various algorithms corresponding to different types of learning functions, each of which has a separate application in construction management. Their relevance is underscored by the complex nature of the increased digitalization and construction projects of construction workflows. Using AI and ML, the construction industry can achieve high efficiency, safety and durability, change how projects are operated and operated in the digital age.

APPLICATIONS IN CONSTRUCTION MANAGEMENT

Adoption of artificial intelligence (AI) and machine learning (ML) in construction management is revolutionizing traditional project workflows by making data -based decisions and increasing operational efficiency. The two most effective applications for risk and delay management are predictive analysis, and allotment of resource with timetable optimization. These capabilities take into account some constant challenges in construction: uncertainty, inability and excessive cost (Manchana, 2022). Figure 1 illustrates the utilization of AI and ML in the field of construction management.

Figure 1. Applications of AI and ML in construction management

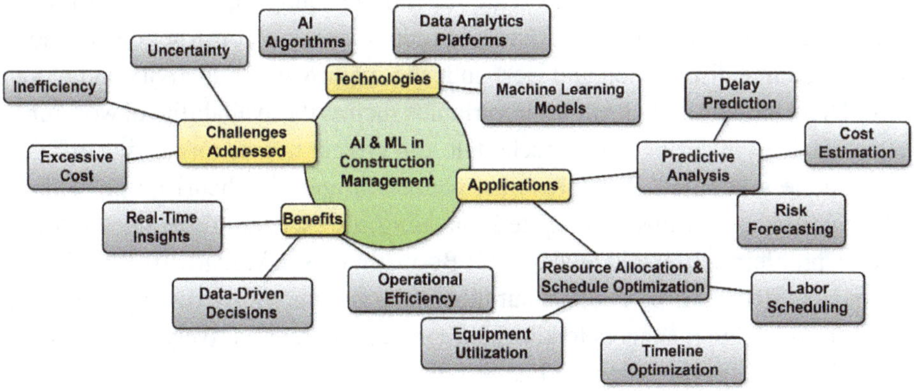

Predictive Analytics for Risk and Delay Management

Predictable analyzes for risk and delay management are a transformative application of AI and ML that empower construction managers to expect and reduce the risks of the project before completing. Construction projects are naturally complex, including unexpected external factors such as multiple stakeholders, various functions and weather, supply chain disturbances and regulatory changes. By giving the benefit of Historical Project data, sensor inputs and external datasets, M.L. Algorithms can analyze patterns associated with project delays, expenses, safety events, and quality problems. For example, the probability of adverse events based on observed learning models, risk factors and project parameters such as scope, timetable, budget and site conditions is trained (Baduge et al., 2022b).

The capacity of this prediction enables an active risk management strategy. For example, if a model predicts a high probability of delay in a particular construction phase due to a lack of labor or failure of the equipment, managers can take prioritizing measures, such as rearranging resources, adjusting timeline, or sourcing alternative suppliers. Moreover, AI-propelled risk analysts improve safety by identifying dangerous work conditions through real-time monitoring of the site data, preventing accidents and compliance violations. The result is a decrease in unexpected disruptions, reliability of the project and a decrease in the confidence of the enhanced stakeholder.

Resource Allocation and Scheduling Optimization

Resources allocation and timetable represents another critical field where AI and ML offer significant benefits. Careful coordination of labor, machinery, materials and subcontractors is required to meet the deadline and budget limits for construction projects. Machine learning algorithms are revolutionizing construction schedules through resource distribution and work index to maximize productivity and reduce waste. These models include multiple variables such as the availability of workforce, rates of devices, material delivery schedule and project targets for making efficient and adaptive schedules. Reinforcement education and metaheuristic algorithms, such as genetic algorithms or simulated annealing, are especially effective in solving complex problems(Boopathi et al., 2021; Boopathi & Sivakumar, 2013; Myilsamy & Sampath, 2017). AI procedures automatically and purify scheduling processes, enable construction managers to allocate resources more effectively, reduce downtime and avoid expensive delays (Pan & Zhang, 2023b).

AI -operated forecasts for risk and delay management, with state -of -the -art resource allocation and timetable are resorting the construction industry. These applications provide competitive benefits by improving the accuracy, response and efficiency, determining new standards for project excellence in the most complex and demanding environment. Construction of artificial intelligence (AI) and machine learning (ML) in construction management has produced significant progress, especially in areas such as quality control, safety monitoring and real-time project inspection. These applications not only improve operational efficiency but also increase safety standards and project results through data -based insights and Auto tomatoes.

Quality Control and Safety Monitoring

In safety inspection, AI increases the ability to predict and prevent accidents by analyzing unsafe conditions or data patterns associated with behavioral conditions. Wearable sensors and IOT devices can make the use of worker's activities, environmental conditions and equipment in real time.

Real-Time Monitoring with IoT and Computer Vision

Real-time monitoring with IoT and computer vision is another crucial application where AI and ML are running innovation. Construction projects produce huge trends of data through IoT-enabled devices, including environmental sensors, GPS trackers, drones and mobile equipment. Integrating these data sources with AI analytics facilitates the progress of the project, the use of resources and continuous monitoring of environmental factors. For example, an IOT sensor can monitor temperature, humidity, noise levels and vibrations to ensure the best functional conditions and prevent the failure of the devices. A.I. Models analyze this data to provide active insights, enable managers to adjust timely adjustments that reduce delays and costs (Kor et al., 2023).

Computer vision connected to drone technology allows wide site surveillance and progress tracking. Drone equipped with a high-resolution camera captures a detailed air image, which processes the AI algorithms to create 3D models and compare the actual progress against the planned schedule. This releases real-time comparison discrepancies, identifies obstacles, and supports active decision. Moreover, computer vision systems can make track and use of machinery movements and use, reducing theft and incompetence. The fusion of IoT, computer vision and AI not only enhances transparency, but also enables the maintenance of machinery by predicting potential rupture based on sensor data trends. They enable project teams to transition to active management by solving reactive problems, and to ensure data intelligence to ensure the success of the project. As the construction industry is increasingly adopted by these techniques, the benefits go beyond individual projects such as durability, workforce and overall industry modernization (Woschank et al., 2020).

Figure 2. Process optimization using AI and ML in construction management

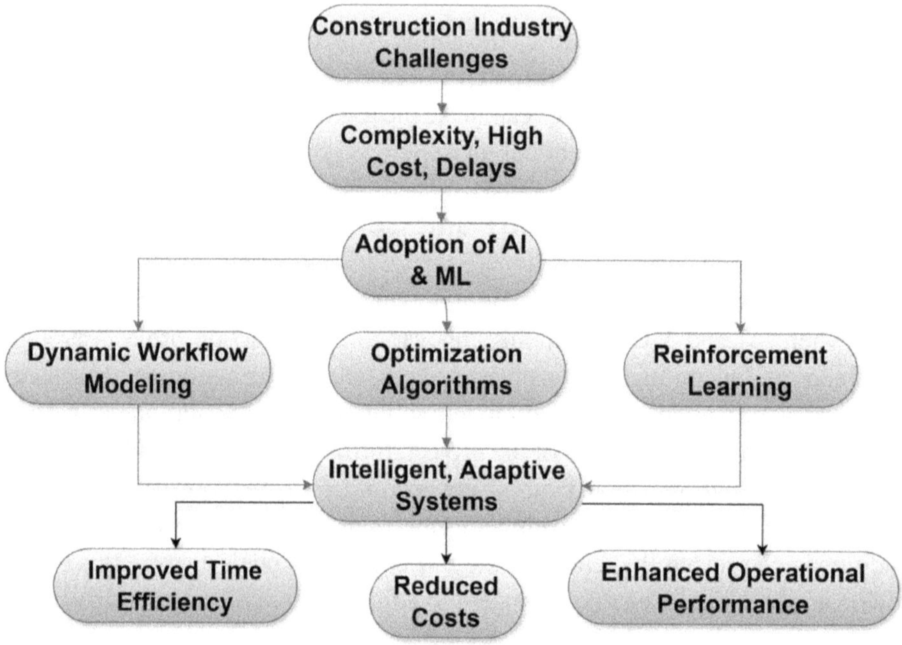

PROCESS OPTIMIZATION USING AI AND ML

The construction industry, characterized by its complexity, multifaceted workflow and dynamic environment, faces constant challenges related to incapacity, expenses and scheduled delays. To overcome these challenges, AI and machine learning (ML) process has emerged as a transformative tool, enabling construction management to develop from stable, manual methods intelligent, adaptive systems. The process using AI and ML includes dynamic workflow modeling, sophisticated optimization algorithms and reinforcement education, collectively time, cost and operational efficiency (Sharma et al., 2022b). Figure 2 depicts the implementation of AI and ML in process optimization in the construction management sector

Dynamic Workflow Modelling

Traditional construction workflows are linear and rigid, often failing to include real-time changes or unexpected disruptions such as weather conditions, labor shortages or supply chain delays. Dynamic workflow modeling operated by AI and

ML shifts this pattern by enabling adaptive and flexible project management that constantly adjusts new data and changes conditions. Dynamic workflow modeling involves the use of data-based algorithms that imitate and analyze construction processes as they manifest. By integrating historical history data, IoT devices and project management can map the complex dependence between real-time inputs, AI systems, functions, resources and timeline from Software. This model allows managers to imagine potential obstacles and dynamic work sequencing process(Pan & Zhang, 2023c).

For example, graph-based ML models can represent construction activities and their dependence, predicting how a task is delayed by the project timeline. These models simplify view planning, where alternative workflow is imitated to identify the most efficient ways to complete the project. The AI system learns from current project data, improves its recommendations, and provides insights to make the decisions responsible for the uncertainty and change. Moreover, the dynamic workflow project teams empowered to respond quickly to modify the timetable to reduce the recovery resources or downtime. This dexterity is important in large -scale or complex projects where rigid schemes can lead to expensive overruns. By accepting dynamic workflow modeling, construction management can transform stable gent charts and timetable of traditional CPM (complex path method) into living systems that continuous optimize execution.

Optimization Algorithms for Time, Cost, and Efficiency

The optimization algorithms form the main part of the AI-based process improvement in construction management. These algorithms use mathematical and heuristic approaches to find the best potential remedy of complex problems associated with multiple obstacles and objectives, such as reducing time and cost when maximizing resource use and quality. In construction, optimization challenges often include work schedules, resource allocation, procurement planning and logistics coordination. AI algorithms-or genetic algorithms, particle swarm optimization and simulated annealing solution spaces are suitable for exploring where traditional methods are calculated(Rane, 2023).

For example, AI-enabled scheduling tools optimize project timeline by adjusting the task dependence, labor availability, the limits of the equipment and the deadline. These tools evaluate numerous potential sequences and choose one that reduces the total project period without compromising the quality. By integrating cost data, algorithms can also propose a timetable that reduces costs, such as overtime or expensive devices. Resource allocation optimization is another important application. Construction projects are subject to various resources - workers, machinery, materials - requirement of integrated deployment, availability, skill level and barriers like delivery time. AI-Runner models analyze these factors with project priority to recommend the best resource assignments that maximize productivity and reduce idle time.

In addition to time and cost efficiency is focused on optimization waste, energy consumption and reducing environmental effects. Algorithms can analyze material consumption patterns and recommend adjustments that reduce excessive ingredient or site deterioration. Such optimization not only improves profitability, but also adjusts with goals of increasing stability in the construction sector. The main advantage of the AI optimization algorithms is their ability to learn and improve over time. By incorporating response loops from project results, these models become increasingly accurate and effective, enabling continuous process improvement in projects and institutions.

Reinforcement Learning in Construction Planning

Reinforcement Learning (RL) represents a cutting edge in AI for optimization of the process, especially for a dynamic, uncertain environment like construction. Unlike traditional inspection education, where models learn from fixed datasets, RLs include an agent communicating with the environment to learn the best actions through a trial and error, which guides by rewards or penalties. In construction planning, RL can be applied to develop adaptive schedules and decision -making systems that develop with the progress of the project. The RL agent gets real-time data, such as the functioning of the function, resource availability and external interruption-long-term project, such as re-scheduled tasks or rearrangement resources to maximize success (Sharma et al., 2022b).

For example, RL-based system activities can optimize which actions lead to rapid completion or low cost in different situations. If a particular schedule is delayed or prevents the devices from lazy, the agent receives a positive reward, strengthening the decision. On multiple repetitions and projects, the system improves its strategy, corresponding to a variety of construction environments and barriers. RL also facilitates multi-agent coordination, where many agents manage various project aspects-that cooperate to optimizing a fleet of subcontractors, suppliers and equipment-united projects. This decentralized approach is very relevant for large, complex projects with many stakeholders.

One of RL's most promising programs is to manage uncertainties and disturbances. Construction projects are often affected by unexpected factors such as weather, regulatory changes or lack of supply. R.L. Agents can imitate different views, learn how to actively rearrange workflow to reduce the risks. This adaptability leads to more elastic project projects that maintain performance despite external shock. Moreover, the RL can integrate with IoT and real-time monitoring systems to create a closed-loop optimization framework. The data collected from the sensor feeds to the RL agent, which constantly updates its decisions depending on the status of the current site, ensures the best use of resources and maintains the speed of the project.

The AI and ML-powered process optimization provides a revolutionary approach to the management of construction projects by transforming steady plans into dynamic, intelligent systems. Through dynamic workflow modeling, construction management receives relief of timetable and resources in real time, increases response and reducing inaccuracy. The time provides strong tools to enable more knowledgeable and effective project decisions, to balance competitive objectives such as time, cost and durability. Reinforcement education further enhances these capabilities by enabling adaptive planning in a complex and uncertain environment, promoting elasticity and constant improvement.

Thus, AI -powered techniques contribute to significant benefits to productivity, cost savings and risk reducing, addressing long -lasting challenges in construction management. As digital transformation accelerates, adoption of AI and ML for process optimization will be a major competitive differential, driving smarter, safer and more durable construction methods worldwide. However, successful implementation requires not only technical investments, but also organizational readiness, data quality management and workforce upselling to fully use the possibility of these advanced equipment.

STRATEGIC FRAMEWORK FOR IMPLEMENTATION

Figure 3. Strategic framework for AI/ML adoption in construction management

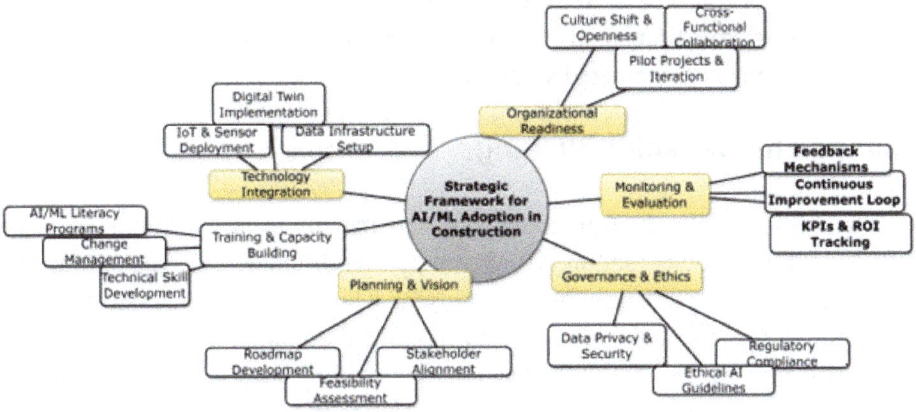

Roadmap for AI/ML Integration

The strategic framework for AI/ML adoption in construction management is depicted in Figure 3. A structured roadmap is required to guide the phased implementation of AI and ML technologies in construction management. This roadmap begins with the assessment and goal setting, where organizations evaluate their current digital maturity, identify the inaccuracies of the process, and define clear objectives arranged with business priority, such as reducing project delays, improving safety or increasing cost control. The next stage includes data infrastructure development - an important foundation for AI/ML success. Construction companies must invest in strong data collection systems such as IoT sensors, BIM platforms and cloud-based data storage to ensure the availability, quality and inter-efficiency of project data. A data governance policy should be established to manage data protection, privacy and compliance, with concerns that otherwise hinder the AI adoption. Subsequently, organizations should pilot AI/ML solutions on selected projects or procedures, gaining insights to improve algorithms and workflows from this trial. This pilot allows the stage to reduce the risk of reducing the risk, feedback from the stakeholder and demonstrating the tangible benefits, increasing confidence in the teams (Sharma et al., 2022b; Woschank et al., 2020).

Following the successful pilots, the enhanced Operational Parliamentary framework in scaling and integration includes embedding AI/ML tools, joining enterprise Resource Planning (ERP), project management and procurement systems. Continuous monitoring and evaluation of performance ensures that the AI application develops with organizational needs and technological advancement. Finally, the roadmap emphasizes cooperation and innovation by promoting partnerships with technical providers, research institutes and industry consortia to continue with emerging trends and co-development customized solutions.

Training and Capacity Building

AI and ML techniques require new skill sets that are clearly different from the roles of traditional construction. To overcome this distance, a dedicated training and capacity-construction program is important. This program should include both technical training and strategic education. Technical training focuses on equipping employees from AI fundamentals, data analytics, along with the interpretation of AI-powered insights and AI-operated insights. Hands-Works No Workshop, Courses and certification applications can increase the use of AI-powered platforms, understand machine learning output, and troubleshoot common issues (Waqar, 2024).

The purpose of strategic education is to cultivate AI-awareness culture that promotes innovation, experiments and critical ideas about AI application. Leadership development programs should empower officials and decisions to understand AI's potential, limitations, and moral considerations, which enables knowledgeable strategic planning. Capacity construction must also address interdisciplinary cooperation by promoting team work between construction professionals, data scales. Mano and IT experts. This integration ensures that AI solutions are technically appropriate and practically relevant. Considering the rapid evolution of AI technologies, it is necessary to include continuous learning framework.

Organizational Readiness and Change Management

The leadership of AI adoption plays a key role by setting clear expectations, allocation of resources, allocation of resources and accepting experiments and promoting failure. Leaders must adjust the enthusiasm for innovation to ensure responsible AI use. The regeneration of the process often requires the AI output to be integrated into workflow, which requires collaboration in sections and potentially redesigned job roles. This simplifies ownership and adoption by engaging in redesigning employees. In the end, organizations should establish a regime structure to monitor AI Ethics, compliance and risk management. These include data privacy, algorithmic transparency, and policies on the responsibility of increasing confidence in stakeholders.

The strategic structure for the implementation of AI and ML in construction management is multipurpose, which combines clear, phased roadmap with extensive training and active organizational change management. By developing methodological data infrastructure, piloting and scaling solutions, investing in human capital and promoting the culture of innovation and readiness, construction companies can unlike the full probability of AI-operated optimization. This holistic approach ensures that AI integrates technical adopts, constantly becoming the main driver of competitive advantage and industry change.

CHALLENGES AND LIMITATIONS

While AI and Machine Learning (ML) provide a transformative probability for construction management, their adoption is with significant challenges and limitations. These obstacles expand data-related issues, technology and skills deficiencies, and complex moral, legal and regime concerns. The AI-operated process in construction requires addressing these challenges to understand the full benefits of optimization (Soori et al., 2023; Tapeh & Naser, 2023; Waqar, 2024).

Figure 4. Challenges and limitations of AI and ML adoption in construction management

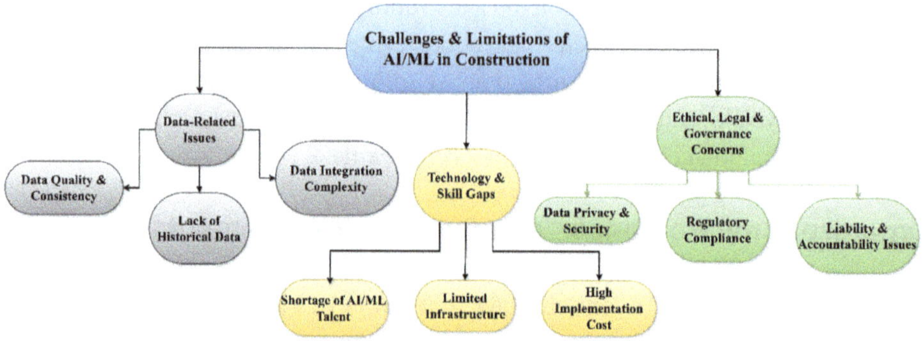

Data Availability and Quality Issues

Although data forms the back of AI and ML applications, the construction industry often struggles with the availability of data, access to sexuality and quality. Construction projects produce large amounts of data from multiple sources such as IoT sensors, BIM systems, drones and manual reports. However, this data is often fragmented, unstructured or incomplete, limiting its utility to train accurate AI models. Incompatible data standards in different contractors and projects further complicate data integration, leading to information that restricts comprehensive analysis. Additionally, there may be errors due to data manual entry or sensor defects, degrading model performance and reliability. The adoption of AI and ML in construction management faces both challenges and limitations as shown in Figure 4.

Limited history challenges the development of AI models, especially in new or unique projects, especially in new or unique projects. Moreover, confidentiality concerns and ownership data restrictions can hinder data sharing and collaborative AI progress in industry stakeholders. Strong data governance framework, standard data collection protocol and data cleansing and investment in IoT notation are important for eliminating these data challenges. Without high quality, wide datasets, AI-powered process optimization suboptimal or risks delivering misleading results.

Technological and Skill Gaps

The construction industry faces significant technical obstacles in adopting AI and ML. Many companies still rely on inheritance systems that lack intercourse with a modern AI platform. Digital infrastructure such as cloud computing, IoT devices and secure data storage requires significant investment to integrate AI solutions, which can be cost-demonstrators for small and medium-sized industries. Moreover, the complexity of AI models demands special skills for development, deployment and maintenance. There is a pronounced shortage of skilled employees that have a domain J Knowledge in both the construction and mastery in Data Vision, Machine, Machine Learning and AI Software Fatware Engineering. These skills go beyond technical roles up to gap management, where leaders often lack adequate understanding of AI capabilities and limitations, which hinders decisions and strategic planning. Targeting training programs, recruitment of interdependent talents and partnerships with technical providers and educational institutions are required to remove these gaps. In addition, user-friendly AI tools with intuitive interfaces can facilitate widespread adoption by reducing technical barriers for non-educated users.

Ethical, Legal, and Governance Concerns

The deployment of AI in construction presents complex moral, legal and governance challenges that must be carefully navigated to ensure responsible use. Morally, AI systems can permanent the partisans present in the training data, leading to unfair or unsafe consequences. For example, the partisan algorithms may introduce some labor groups to enhance safety monitoring or resource allocation, inequality. Concern about privacy arises from extensive data collection, especially when working work monitoring or sensitive project information. It is supreme to maintain confidence and to avoid legal penalties to ensure confidentiality of data and rules such as GDPR.

A.I. The legal structure of responsibility, liability and intellectual property rights is undeveloped in many jurisdictions, when AI -operated decisions lead to errors, accidents, or disputes of contracts. Governance Structures must establish clear policies for AI Ethics, Data Management, Transparency and Audit Duality. Regular AI system can help identify and reduce the risks of impact assessments. Moreover, the engagement of a stakeholder, including workers, consumers, regulators and technical vendors, to develop consensus on the use of moral AI and establish industry standards.

The way to AI-based process in construction is full of data quality and availability, technology and skills deficiencies, and challenges related to moral, legal and regime complications. Consolidated efforts are needed on multiple parameters to overcome these obstacles: strengthening data infrastructure and standards, investing in employees' development and digital transformation, and implementing a strong governance framework. Activating these limits will enhance the effectiveness of adoption of AI, enhance the stakeholder's confidence, and enable a sustainable, responsible innovation in construction management.

Figure 5. Emerging trends in AI & construction

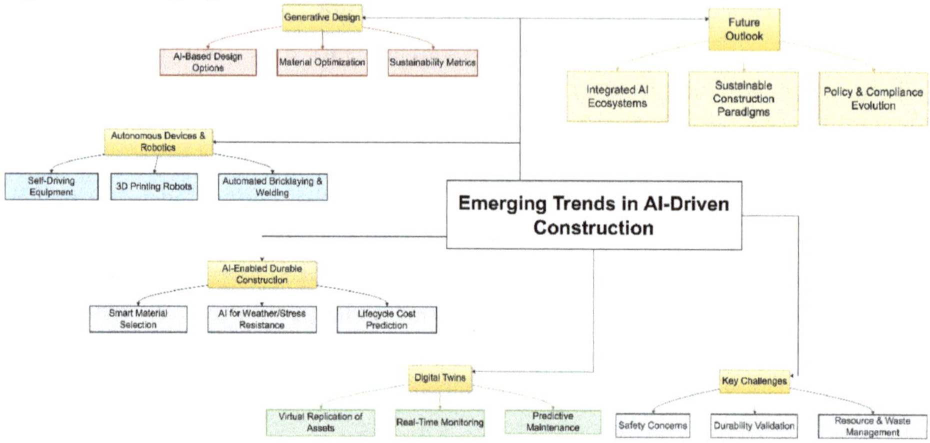

EMERGING TRENDS

The construction industry is going through a deeper change through the rapid progress in AI and related techniques. Emerging trends projects such as digital twins, generating designs, autonomous devices, robotics and AI-enabled durable construction shapes how to organize, implement and maintain. These innovations consider not only the benefits of efficiency, but also the serious challenges related to safety, durability and resource management, which determines the future view of construction management. Figure 5 reveals the emergence of new trends in AI and construction (Pan & Zhang, 2023a, 2023b; Waqar, 2024).

Digital Twins and Generative Design

Digital twins - virtual replicas of physical wealth or construction sites - Project monitoring and revolutionizing management. The digital twin project provides an accurate, dynamic simulation of the construction environment, integrating real-time data from IoT sensor, BIM (building information modeling) and AI analytics. This allows stakeholders to imagine the progress, find deviations, and predict the potential issues before. Supplementing digital twins, generating design AI algorithms uses to automatically produce optimize design options based on clear obstacles such as materials, costs, structural requirements and environmental influence. This approach enables architects and engineers to quickly explore enhanced design space and identify solutions to maximize efficiency, efficiency and aesthetics.

Together, these technologies promote more collaborative, data -based design and construction process. They reduce errors, accelerate decisions, and enable constant improvement through feedback loops between digital model and physical construction.

Autonomous Equipment and Robotics

Automation by autonomous devices and robotics is another main trend ready to reshape construction workflow. Autonomous machinery such as driverless meaning, drone and robotic brick leers can perform repetitive or risky tasks with higher precision and consistency than human operators. These technologies not only improve safety by reducing the exposure to workers in a dangerous environment, but also increase productivity and accuracy. Robotics associated with AI can adapt to complex site conditions, conduct quality observations, and assist in complex assembly functions. For example, drones equipped with AI-powered computer vision are used for site surveys, progress tracking and structural observations, capturing data that feeds back to project management systems for real-time analytics (Narra et al., 2024; Singh Madan et al., 2024).

AI for Sustainable Construction Practices

Durability is essential in modern construction, which is run by regulatory pressure, environmental concerns and corporate social responsibility. AI materials play a key role in expanding durable construction by the consumption of materials, energy consumption, debris reduction and carbon footprint. AI algorithms analyze large datasets to recommend environmentally-like materials, to pt over optima and reduce environmental disturbances. For example, AI-powered forecast models can predict the pattern of energy access, guide the renewable energy Sources and the establishment of efficient systems. The AI-enabled supply chain optimization increases waste reduction, ensures delivery of certain materials, and reduces over-ordering or deterioration. Additionally, AI-powered monitoring systems, emissions on the spot and track environmental parameters, enable compliance with durability standards and certificates.

Looking forward, the integration of AIs with durable construction methods, green materials, circular economy models and near-zero construction goals are expected to be operated by innovations in the regulatory framework. Industry analysts project that AI-enabled stability solutions will be the norm in construction in the next decade, significantly reducing the environmental influence of the sector.

Emerging trends such as digital twins, generating designs, autonomous devices, robotics and AI-powered stability are redefining the landscape of the construction. These technologies provide unprecedented opportunities to enhance efficiency, safety and environmental liability, addressing both current industry pain issues and future challenges. The ongoing evolution and integration of AI in construction will not only run the workflows, but also the more elastic infrastructure development worldwide. Strategic adoption and continuous innovation in these sectors is important for construction companies aiming to maintain competitive advantage and contribute to sustainable urban growth.

CASE STUDIES

Machine Learning (ML) and optimization technologies are increasingly influencing construction processes to improve productivity, safety, cost, and informed decision-making processes. Technologies provide predictive abilities, real time visibility and automated decision support. In this section, we present some examples of the application of ML and optimization methods to improve performance of operations in the construction sector(Soori et al., 2023; Tapeh & Naser, 2023; Waqar, 2024).

One of the most cited applications of ML in construction is predictive maintenance. Using historical sensor data from heavy equipment like excavators, cranes, and bulldozers, ML models can enable the prediction of failures to the equipment before they occur. For example, with supervised learning algorithms such as Random Forests and Support Vector Machines, operational data in the form of telemetry data can produce outlier anomalies in the minutiae of various forms of equipment behavioral patterns such as vibration, temperature, or hydraulic pressure. Predictive maintenance increases efficiencies from down time, extends equipment life and ultimately produces less unplanned costs. An additional example is safety management on construction sites. ML models that learn from images/video taken from cameras, drones or other type of sensors can identify unsafe activities, such as a worker removing a hard hat, or entering a restricted area. CNN or similar deep learning models are used to watch and analyze video streaming in real-time for safety violations. Safety management systems are valuable for ensuring compliance and consistency, while also reducing accident rates on a construction site (Kor et al., 2023; Pan & Zhang, 2023a).

Project schedule optimization is another area were ML compliments optimization. The typical planning tools such as Gantt charts and Critical Path Method (CPM), etc., don't really accommodate uncertainty. ML models can learn from past project execution data and find correlation to delays due to different risk factors, be it the weather, manpower availability or supply chain delays. Then various optimization algorithms such as Genetic Algorithms (GA) or Particle Swarm Optimization (PSO), could then suggest the best re-sequencing of project tasks that would have the least impact to project time and related costs. In the area of material input and procurement, predictive models can estimate material usage based on construction progress, seasonal environmental change and past usage patterns to mitigate the risk of under or over commitment of inventory. Optimization models can also refine procurement schedules based on supplier lead times, shipping costs, and storage constraints. For example, integer programming models can develop ideal delivery schedules for materials with regards to reducing waste and cost efficiency.

Another emerging practice is in energy efficiency, or lack thereof, for building operations. ML can be applied in the design phase of construction to analyze thousands of building designs and configurations and assess energy efficiency of designs. Reinforcement learning has also been applied to smart building systems and is used to optimize HVAC (heating, ventilation and air conditioning) controls, which differ based on occupancy patterns to minimize operating energy use. Finally, image-based ML systems are useful for assessing the quality of construction. Drones or robots take real-time images of construction sites, and deep learning models use those images to identify any structural defects, misalignment, or material inconsistency. This positively impacts quality control and allows for reductions in time and labor for manual inspections.

In conclusion, machine learning and optimization in organizing construction workflow is a strategic advantage for a complex project. Machine learning and optimization is a tool that can improve safety, efficiency, resource planning, and cost control and they are essential technological tools for modern, smart construction practices. The increasing prevalence and scale of data and computational power implies that ML and optimization will only become more ubiquitous and highly utilized in construction.

CONCLUSION

This chapter has invented the crucial role of artificial intelligence (AI) and machine learning (ML) in converting construction management through advanced algorithms, AI-powered approaches enhance enhanced decisions, cost efficiency, time savings and improved safety on construction sites, by benefiting from dynamic workflow modeling and real-time data analytics. AI/ML's foundational shows the possibility of revolutionizing traditional construction workflows, linked to practical applications such as quality control, safety inspection and IoT integration. However, the journey to adopt a complete AI is complex, challenged by data quality issues, technology and skills intervals, and moral and regime consideration. The strategic structure associated with clear roadmaps, training programs and transformation management is necessary to overcome these barriers and to realize sustainable benefits.

Further, emerging trends such as digital twins, generating designs, autonomous devices and AI-enabled durability promise to increase industry capabilities and elasticity. To earn these opportunities, construction companies must adopt a holistic, forward-thinking approach that balances innovation with responsibility. In essence, AI and ML are catalysts not only tools but to promote basic shifts, smarter, safe and more durable projects in construction management that meets the demand for a future built environment. The insights and framework presented in this chapter provide a solid foundation for stakeholders aiming to navigate and forward this transformative tour.

REFERENCES

Abioye, S. O., Oyedele, L. O., Akanbi, L., Ajayi, A., Delgado, J. M. D., Bilal, M., Akinade, O. O., & Ahmed, A. (2021). Artificial intelligence in the construction industry: A review of present status, opportunities and future challenges. *Journal of Building Engineering*, *44*, 103299.

Baduge, S. K., Thilakarathna, S., Perera, J. S., Arashpour, M., Sharafi, P., Teodosio, B., Shringi, A., & Mendis, P. (2022). Artificial intelligence and smart vision for building and construction 4.0: Machine and deep learning methods and applications. *Automation in Construction*, *141*, 104440.

Boopathi, S., Myilsamy, S., & Sukkasamy, S. (2021). *Experimental Investigation and Multi-Objective Optimization of Cryogenically Cooled Near-Dry Wire-Cut EDM Using TOPSIS Technique*. IJAMT PREPRINT.

Boopathi, S., & Sivakumar, K. (2013). Experimental investigation and parameter optimization of near-dry wire-cut electrical discharge machining using multi-objective evolutionary algorithm. *Springer: The International Journal of Advanced Manufacturing Technology*, *67*(9–12), 2639–2655.

Glady, J. B. P., D'Souza, S. M., Priya, A. P., Amuthachenthiru, K., Vikram, G., & Boopathi, S. (2024). A Study on AI-ML-Driven Optimizing Energy Distribution and Sustainable Agriculture for Environmental Conservation. In *Harnessing High-Performance Computing and AI for Environmental Sustainability* (pp. 1–27). IGI Global. DOI: 10.4018/979-8-3693-1794-5.ch001

Gunasekaran, K., Boopathi, S., Suresh, S., Suresh, P., & Gokulakannan, S. (2024). Sustainable Building Construction Practices for Transforming Waste Into Wealth Through Vernacular Architecture. In *Innovations in Energy Efficient Construction Through Sustainable Materials* (pp. 391–418). IGI Global. DOI: 10.4018/979-8-3693-3398-3.ch014

Kalaiselvi, T., Saravanan, G., Haritha, T., Babu, A. V. S., Sakthivel, M., & Boopathi, S. (2024). A Study on the Landscape of Serverless Computing: Technologies and Tools for Seamless Implementation. In *Serverless Computing Concepts, Technology and Architecture* (pp. 260–282). IGI Global. DOI: 10.4018/979-8-3693-1682-5.ch016

Kazeem, K. O., Olawumi, T. O., & Osunsanmi, T. (2023). Roles of artificial intelligence and machine learning in enhancing construction processes and sustainable communities. *Buildings (Basel, Switzerland)*, *13*(8), 2061.

Kor, M., Yitmen, I., & Alizadehsalehi, S. (2023). An investigation for integration of deep learning and digital twins towards Construction 4.0. *Smart and Sustainable Built Environment*, *12*(3), 461–487.

Madan, B. S., Viraj Dravid, S., Deenadayalan, G., Rathod, P., Arumugam, G., & Boopathi, S. (2024). Innovations and Applications of Engineered Cementitious Composites and Alternative Materials: Reimagining Construction. In *Advances in Chemical and Materials Engineering* (pp. 53–86). IGI Global. DOI: 10.4018/978-1-6684-8182-0.ch003

Mageswari, D. U., Kareemullah, H., Jithesh, K., Boopathi, S., Rachel, P. M. P. P., & Ramkumar, M. S. (2024). Experimental investigation of mechanical properties and multi-objective optimization of electronic, glass, and ceramic waste–mixed concrete. *Environmental Science and Pollution Research International.* Advance online publication. DOI: 10.1007/s11356-024-33751-7 PMID: 38806982

Manchana, R. (2022). Optimizing Real Estate Project Management through Machine Learning, Deep Learning, and AI. *Journal of Scientific and Engineering Research*, *9*(4), 192–208.

Muthukumar, R., Pratheep, G. V. G., S. J., S., Kunduru, K. R., Kumar, P., & Boopathi, S. (2024). Leveraging Fuel Cell Technology With AI and ML Integration for Next-Generation Vehicles: Empowering Electric Mobility. In *Advances in Mechatronics and Mechanical Engineering* (pp. 312–337). IGI Global. DOI: 10.4018/979-8-3693-5247-2.ch016

Myilsamy, S., & Sampath, B. (2017). Grey Relational Optimization of Powder Mixed Near-Dry Wire Cut Electrical Discharge Machining of Inconel 718 Alloy. *Asian Journal of Research in Social Sciences and Humanities*, *7*(3), 18–25.

Narra, M., Vij, M., Shreenidhi, K. S., Anandaram, H., Hari, B. S., & Boopathi, S. (2024). Redefining Sustainability in Building Practices in Circular Construction Process. In *Innovations in Energy Efficient Construction Through Sustainable Materials* (pp. 293–324). IGI Global., DOI: 10.4018/979-8-3693-3398-3.ch011

Pan, Y., & Zhang, L. (2023). Integrating BIM and AI for smart construction management: Current status and future directions. *Archives of Computational Methods in Engineering*, *30*(2), 1081–1110.

Rane, N. (2023). Integrating building information modelling (BIM) and artificial intelligence (AI) for smart construction schedule. *Cost, Quality, and Safety Management: Challenges and Opportunities.*

Reddy, P. Y., Madan, B. S., Anandaram, H., Rathod, P., Vasanthaseelan, S., & Boopathi, S. (2024). Analyzing Fuel Cell Vehicles Through Intelligent Battery Management Systems (BMS): AI and ML Technologies for E-Mobility. In *Advances in Web Technologies and Engineering* (pp. 335–355). IGI Global. DOI: 10.4018/979-8-3693-1487-6.ch016

Sharma, V., Zaki, M., Jha, K. N., & Krishnan, N. A. (2022). Machine learning-aided cost prediction and optimization in construction operations. *Engineering, Construction, and Architectural Management*, *29*(3), 1241–1257.

Singh Madan, B., Logeswaran, S., Veerapathran, M., Rai, P., Kumar Singh, S., & B., S. (2024). Engineered Cementitious Composites (ECC) Through Nano-Composite Integration: Enhancing Construction. In *Advances in Chemical and Materials Engineering* (pp. 210–243). IGI Global. DOI: 10.4018/978-1-6684-8182-0.ch008

Soori, M., Arezoo, B., & Dastres, R. (2023). Artificial intelligence, machine learning and deep learning in advanced robotics, a review. *Cognitive Robotics*, *3*, 54–70.

Tapeh, A. T. G., & Naser, M. (2023). Artificial intelligence, machine learning, and deep learning in structural engineering: A scientometrics review of trends and best practices. *Archives of Computational Methods in Engineering*, *30*(1), 115–159.

Tien, P. W., Wei, S., Darkwa, J., Wood, C., & Calautit, J. K. (2022). Machine learning and deep learning methods for enhancing building energy efficiency and indoor environmental quality–a review. *Energy and AI*, *10*, 100198.

Waqar, A. (2024). Intelligent decision support systems in construction engineering: An artificial intelligence and machine learning approaches. *Expert Systems with Applications*, *249*, 123503.

Woschank, M., Rauch, E., & Zsifkovits, H. (2020). A review of further directions for artificial intelligence, machine learning, and deep learning in smart logistics. *Sustainability*, *12*(9), 3760.

Zhang, W., Gu, X., Tang, L., Yin, Y., Liu, D., & Zhang, Y. (2022). Application of machine learning, deep learning and optimization algorithms in geoengineering and geoscience: Comprehensive review and future challenge. *Gondwana Research*, *109*, 1–17.

Chapter 7
CO2 Emission Calculations in Shipping for Sustainable Ship Design, Construction, and Operation:
Assessing Strategies and Regulatory Tools for Low-Carbon Maritime Innovation

Atanu Roy
https://orcid.org/0000-0001-5269-9122
The Neotia University, India

Sabyasachi Pramanik
https://orcid.org/0000-0002-9431-8751
Haldia Institute of Technology, India

Kalyan Mitra
Indian Maritime University, USA

Manashi Chakraborty
The Neotia University, India

ABSTRACT

Global trade considerably depends on maritime transportation, yet these activities generate substantial greenhouse gas (GHG) emissions, raising critical concerns

DOI: 10.4018/979-8-3373-2555-2.ch007

among policymakers about climate change. In response, the International Maritime Organization (IMO) has implemented stringent regulatory measures towards the construction of energy-efficient ships and operations to reduce carbon emissions from both new and existing vessels. This chapter explores key IMO approaches, including the Energy Efficiency Design Index (EEDI), the Energy Efficiency Existing Ship Index (EEXI), the Ship Energy Efficiency Management Plan (SEEMP), the Energy Efficiency Operational Index (EEOI), and the Carbon Intensity Indicator (CII). The study performs a quantitative analysis of GHG emissions for real ships to validate IMO's emission targets. This chapter highlights the importance of integrating sustainable design principles and operational strategies to mitigate unsustainable practices in shipbuilding and maritime operations, paving the way for environmentally responsible shipping.

INTRODUCTION

The maritime industry plays a critical role in global trade and commerce, with over 80% of goods being transported via ships (Roy & Chakraborty, 2025). However, despite its economic significance, the maritime sector is a major contributor to global greenhouse gas (GHG) emissions, particularly carbon dioxide (CO_2). Studies demonstrate that international maritime transport produces approximately 3% of global CO_2 emissions and predict this share will rise substantially because of a lack of effective reduction strategies (EEA, 2022). Ship engines burning fossil fuels produce emissions threatening climate systems and human health. The rising importance of environmental sustainability and decarbonization objectives makes shipping CO_2 emission reduction significant for stakeholders. Thus, the International Maritime Organization (IMO), the regulatory body of shipping, established stringent regulations to lower shipping GHG emissions by enhancing vessel design and operational efficiency measures for existing and new ships. IMO regulatory instruments such as the Energy Efficiency Design Index (EEDI), the Energy Efficiency Existing Ship Index (EEXI), the Ship Energy Efficiency Management Plan (SEEMP), the Energy Efficiency Operational Index (EEOI) and the Carbon Intensity Indicator (CII) serve as an essential mechanism to ensure compliance with global decarbonization targets (Alhasnawi et al., 2024).

The EEDI, introduced in 2013, mandates minimum energy efficiency construction standards for new ships. The EEDI requires ships to achieve specific energy performance levels based on their capacity and function, encouraging the adoption of fuel-efficient designs and technologies (Barreiro et al., 2022). IMO's short-term emission reduction strategy, the EEXI, serves as a requirement for existing vessel compliance. The EEXI system differs from EEDI because it assesses existing ships through technical modifications and operational improvements. Ships unable to meet EEXI requirements need to implement one of two solutions: engine power limitation or the integration of energy-saving technologies.

SEEMP becomes an organized framework for enhancing a ship's operational efficiency levels. SEEMP entered into effect in 2013 and underwent multiple changes before introducing SEEMP Part III in 2023. Ships must maintain ongoing energy efficiency assessments while meeting CII requirements under this newest version of SEEMP (SEEMP Part III., 2023). The CII system rates ship operational carbon intensity through annual assessments of shipping emissions relative to cargo volume and travelled mileage. Vessels that receive ratings of D or E in three consecutive years must develop action plans to boost their operational performance (Luo et al., 2024). Additionally, the EEOI provides performance metrics that allow tracking of a vessel's fuel efficiency and CO_2 emissions produced during transport operations. The EEOI functions independently from EEDI and EEXI since it is a non-obligatory performance assessment tool enabling operators to analyze and enhance actual voyage operations through live operational data. Shipowners who use SEEMP and CII together with EEOI can develop more efficient decarbonization approaches that enhance maritime transport sustainability.

The maritime sector shows increasing interest in using alternative fuels, including liquefied natural gas (LNG), biofuels and hydrogen (Roy et al., 2024). Such alternative fuels present reduced carbon emissions relative to marine diesel oil (MDO) and heavy fuel oil (HFO) conventional fuels. The combination of wind-assisted propulsion technologies with battery-electric systems and air lubrication technology serves to minimize fuel use and emissions. Operational strategies establish a central role in achieving emission reduction targets. Through speed optimization combinations with weather routing systems, along with proper hull maintenance practices and just-in-time (JIT) arrival approaches, ships can reduce both fuel waste and total emissions. Combining renewable energy with technological improvements and stringent regulatory oversight delivers an all-encompassing decrease in maritime industry carbon emissions (Mishra et al., 2024). Various obstacles continue to interfere with regulatory progress and technological development. Current global compliance costs, coupled with limited available alternative fuels and inconsistent regulatory enforcement measures, create barriers to effective emission control methods (Pal et al., 2024). Future investigations need to study how digitalization, along with artificial intelligence (AI), blockchain, and IoT frameworks (Pramanik et al., 2025), can improve transparent and efficient CO_2 monitoring and compliance practices.

The primary objective of this chapter is to analyze the regulatory framework governing CO_2 emissions in maritime transport and evaluate its effectiveness in achieving sustainability goals. The specific objectives include:

1. To examine the IMO's regulations on ship construction and operational strategies for emissions reduction.
2. To explore the detailed calculation methods of EEDI, EEXI, SEEMP, and CII in relation to sustainable ship design and operational efficiency.
3. To develop a quantitative approach for validating IMO's emission reduction targets using real-world ship data.
4. To identify key challenges in regulatory compliance and propose potential solutions for improved implementation.

This chapter analyses CO_2 emission mechanisms within the shipping industry to expand the sustainable discourse while providing systematic insights into IMO's emission strategies adaptation patterns. The study will aid maritime professionals and policymakers in making better decisions regarding CO_2 emission reduction along with operational efficiency maintenance and regulatory compliance achievements.

IMO STRATEGIES TOWARDS EMISSION MITIGATION

UNFCCC, in the Paris Climate Change Conference (adopted by 196 countries at COP 21[1]), agreed to limit global temperature change to less than 2°C (degrees Celsius), with a goal of 1.5°C contrasted with pre-industrial levels (Wang et al., 2020). Likewise, IMO adopted various measures in shipping to comply with the UNFCCC target of climate change requirements in MARPOL Annex-VI. The new green transport technologies were the global shipping theme for 2022 World Maritime Day, which specifies a motivation to focus on the worth of a greener recovery world by reduction of GHG. IMO actively supports the shipping sectors toward a greener transition to a sustainable future by deploying innovative technologies (W.M.T 2022, n.d.). IMO has adopted several initial resolutions to reduce the GHG effect from seaborne trading in April 2018.

Introduction to IMO GHG Strategy 2023

The IMO introduced the IMO GHG Strategy 2023 as a key initiative to address the growing concern of GHG emissions within the maritime sector. Considering the datum line to the 2008 emission level, the respective degrees of ambition driving the Initial Strategic Plan were mentioned as trails (IMO, 2021, 2023):

a) Short-Term Targets (up to 2030): Reducing Carbon footprints per transport work in trade compared to the 2008 emission level. The decline target is 40% by 2030, with a further target of 70% by 2050.

b) Mid-Term Targets (up to 2050): Reducing gross estimated GHG emissions produced by international shipping by at least 50% by 2050 compared to 2008 while considering strategies to phase them out to achieve CO_2 emission declines compatible with the Paris Agreement temperature objectives.
c) Carbon intensity (emissions per transport work) will reduce when further phases (Phase-I, II, and III) of the EEDI for newly constructed vessels are introduced as reduction factors.

A significant number of present literature have studied on ships' emissions and control methods, such as Review of Ship Energy Efficiency (Barreiro et al. 2022) for the marine division as well as commerce; Application of EEXI and costs of executing EEXI globally (Czermański et al., 2022); Fuel Consumption of vessel using speed loss parameters (MERSİN, 2020); Ship Speed Optimization using Ocean Current (Yang et al., 2020); Ship Loading effect on the bulk carrier by using marine diesel engine (Tran, 2020a); Ship Energy Efficiency operational measure by simulation method (Tran, 2020b). To diminish the CO_2 release from existing vessels, a method for assessing the energy-saving active measures is using the Carbon Intensity Index (CII) (Gianni et al., 2022). All these studies have underscored how harmful the Carbon footprint from ships is to the environment, individuals, and communities and how particular steps need to be taken to lessen those consequences.

IMO's strategy is critical to ensuring that international shipping contributes to global climate change mitigation efforts, aligning the sector with global sustainability goals, including the Paris Agreement and the United Nations Sustainable Development Goals (SDGs). The long-term goal is to achieve full ZERO Emission of the shipping industry, in line with the global ambition to limit the rise in global temperatures to below 1.5°C above pre-industrial levels.

IMO MEPC: Initial Strategies

In the 62nd session of the MEPC, the resolution MEPC.203 was adopted in MARPOL (Marine Pollution Protocol) Annex VI, dated 15th July 2011. In this resolution, MEPC.203 (62), IMO, targeted the minimum design standard requirement, EEDI, towards CO_2 emission reduction for the new and newly constructed vessels[2] of 400 GRT and above (IMO, 2023). The unit of EEDI is grams of CO_2 per ton-nautical mile (gCO_2/t-Nm). The lower the value of EEDI, the lesser the vessel's carbon footprint. Other than EEDI, in MEPC.203 (62), SEEMP is also adopted as an operational measure for the ships (new and existing) irrespective of their year of build. A SEEMP is a ship-specific measure applied in the company policy, SMS, in an effort by the ship's owner to lower operational expenses, eventually leading to lower fuel usage and greenhouse gas emissions.

On 13th April 2018, MEPC.304 (72) held that IMO set an initial objective to lessen GHG emissions, CO_2, from international shipping. By 2030, the target is set for reducing GHG emissions in marine transportation by at least 40%, following an attempt of lessening 70% by 2050 compared to the 2008 emission standard (V & L Ambition, n.d.). IMO, set a short-term target by 2023, a mid-term target by 2030 and long-term measures beyond 2030 as shown in Table.1.

Table 1. IMO initial objectives to lessen GHG emissions

Measure Category	Agreement Timeline	Measures
Short-term Measures	2018 - 2023	Improvements to ports, EEDI and SEEMP, national action plans, and IMO technical cooperation; optimization and decrease of speeds;
Mid-term Measures	2023 - 2030	Implementation plan for new emission reduction methods, market-based measures, and alternative low-carbon or zero-carbon fuels
Long-term Measures	Beyond 2030	The creation and distribution of fuels that do not contain any carbon or use any fossil fuels, as well as the promotion and support of new methods for reducing emissions.

EEDI: NEW VESSELS CONSTRUCTION CALCULATIONS TOWARDS EMISSION

The calculation for the EEDI determines legal control on carbon intensity measures for new building ships. The EEDI index calculations are getting more stringent bit by bit, phase-wise, as given in Tables 2 & 3. The shipping sector is free to adopt innovative technology when building commercial ships according to the EEDI, a non-prescriptive system that ensures these vessels fulfill specific energy efficiency standards.

a) EEDI Phase-0 required new building ship designs to meet specific energy standards for their ship type and carbon footprint reduction levels.
b) EEDI required ships to cut their carbon intensity from Phase I to Phase II.
c) Phase III requires all ships to reduce by at least 30%, with some ship types requiring a 50% reduction.

The projected amount is derived from the vessel's CO_2 emissions per unit of transportation work performed during operation, as determined by the EEDI formula. This EEDI consists of two parts:

a) EEDI Required, and
b) EEDI Attained

EEDI Measurement: Required

To comply with the EEDI emission conditions, the vessel's "Attained EEDI" value must be less than or equal to the vessel's "Required EEDI" value mentioned with a simple Eq.1 (IMO, 2023). Attained EEDI is the value the vessel achieved during the shop/sea trial, and "Required EEDI" is the maximum target value set by IMO.

$$Attained\ EEDI \leq Required\ EEDI = \left(1 - \frac{X}{100}\right) * Reference\ line\ value \quad (1)$$

Where,
Required EEDI = Target set value given by IMO.
Attained EEDI = Value achieved by vessel during sea trial; refer Eq.2 (IMO, 2023).
Reference line value = a x b $^{-c}$; Values of a, b, and c are specified in Table.3 (IMO, 2023).
X = EEDI reduction factor (RF) by vessel category and capacity in DWT (Dead Weight) are in Table.2 (IMO, 2023).

Table 2. Ship type-specific reference value-determination parameters

Vessel category	a	b	c
Bulk carrier ships	961.79	D.W.T of the vessel where D.W.T ≤ 279,000; & 279,000 where D.W.T > 279,000	0.477
Combination carrier ships	1,219.00	Ship D.W. T	0.488
Container ships	174.22	vessel D.W. T	0.201
Cruise passenger ships with non-traditional propulsion system	170.84	Ship GRT	0.214
Gas carrier ships	1,120.00	Ship D.W. T	0.456
General cargo ships	107.48	Ship D.W. T	0.216
LNG carrier ships	2,253.7	Ship D.W. T	0.474
Refrigerated cargo carrier ships	227.01	Ship D.W. T	0.244

continued on following page

Table 2. Continued

Vessel category	a	b	c
Ro-ro cargo ships	1405.15	Vessel D.W. T	0.498
	1686.17*	Vessel D.W.T where D.W.T ≤17,000* 17,000 where D.W.T > 17,000*	
Ro-ro cargo ships (vehicle carrier)	$(D.W.T/GRT)^{-0.7}$ 780.36 where D.W.T/GRT < 0.3; 1,812.63 where D.W.T/GRT ≥ 0.3	Ship D.W. T	0.471
Ro-ro passenger ships	752.16	Vessel D.W. T	0.381
	902.59*	Vessel D.W.T where D.W.T ≤10,000* 10,000 where D.W.T > 10,000*	
Tanker ships	1,218.80	Ship D.W. T	0.488

* To be used from Stage-II & thereafter.

The strictest (lowest) EEDI should be used for determining a ship's classification if its design permits it to fit into more than one of the categories provided in Table 2.

Table 3. Resolution MEPC.324(75): Revised EEDI Reduction Fac to r as a Percentage concerning the EEDI Reference Line Effective April 1, 2022

SHIP CATEGORY	SHIP SIZE	STAGE # 0 01.01.13 ~ 31.12.14	STAGE # I 01.01.15 ~ 31.12.19	STAGE # II 01.01.20 ~ 31.03.22	STAGE # II 01.01.20 ~ 31.12.24	STAGE # III 01.04.22 ~ onwards	STAGE # III 01.01.25 ~ onwards
Bulk carrier vessels	20k DWT & over	0	10		20		30
	10k & more but less than 20k DWT	N/R	0 to 10*		0 to 20*		0 to 30*
Gas carrier vessels	15k DWT& more	0	10	20		30	
	10k and more but less than 15k DWT						30
	2k &more but less than 10k DWT	N/R	0 to 10*		0 to 20*		0 to 30*

continued on following page

Table 3. Continued

SHIP CATEGORY	SHIP SIZE	STAGE # 0 01.01.13 ~ 31.12.14	STAGE # I 01.01.15 ~ 31.12.19	STAGE # II 01.01.20 ~ 31.03.22	STAGE # II 01.01.20 ~ 31.12.24	STAGE # III 01.04.22 ~ onwards	STAGE # III 01.01.25 ~ onwards
Tanker vessels	20k DWT& more	0	10		20		30
	4k and more but less than 20k DWT	N/R	0 to 10*		0 to 20*		0 to 30*
	200k DWT& more	0	10	20		50	
	120k and more but less than 200k DWT						
	80k &more but less than 120k DWT	0	10	20		40	
Container vessels	40k & more but less than 80k DWT	0	10	20		35	
	15k & more but less than 40k DWT	0	10	20		30	
	10k & more but less than 15k DWT	N/R	0 to 10*	0 to 20*		15 to 30*	
General Cargo vessels	15k DWT& more	0	10	15		30	
	3k & more but less than 15k DWT	N/R	0 to 10*	0 to 15*		0 to 30*	

continued on following page

Table 3. Continued

SHIP CATEGORY	SHIP SIZE	STAGE # 0 01.01.13 ~ 31.12.14	STAGE # I 01.01.15 ~ 31.12.19	STAGE # II 01.01.20 ~ 31.03.22	STAGE # II 01.01.20 ~ 31.12.24	STAGE # III 01.04.22 ~ onwards	STAGE # III 01.01.25 ~ onwards
Refrigerated cargo carrier vessels	5k DWT & more	0	10		15		30
	3k & more but less than 5k DWT	N/R	0 to 10*		0 to 15*		0 to 30*
Combination carrier vessels	20k DWT & more	0	10		20		30
	4k & more but less than 20k DWT	N/R	0 to 10*		0 to 20*		0 to 30*
LNG Carrier vessels ***	10k DWT & more	N/R	10**	20		30	
Ro-Ro cargo ship (vehicle carrier) ***	10k DWT & more	N/R	5**		15		30
Ro-Ro cargo ship***	2k DWT & more	N/R	5**		20		30
	1k & more but less than 2k DWT	N/R	0 to 5*, **		0 to 20*		0 to 30*
Ro-Ro passenger ship***	1k DWT & more	N/R	5**		20		30
	250 & more but less than 1k DWT	N/R	0 to 5*, **		0 to 20*		0 to 30*

continued on following page

Table 3. Continued

SHIP CATEGORY	SHIP SIZE	STAGE # 0 01.01.13 ~ 31.12.14	STAGE # I 01.01.15 ~ 31.12.19	STAGE # II 01.01.20 ~ 31.03.22	STAGE # II 01.01.20 ~ 31.12.24	STAGE # III 01.04.22 ~ onwards	STAGE # III 01.01.25 ~ onwards
Cruise Passenger Ship*** having non to traditional propulsion	85k GRT &more	N/R	5**	20		30	
	25k & more but less than 85k GRT	N/R	0 to 5*, **	0 to 20*		0 to 30*	

Note:

N/R: No Requirement of EEDI.

1k =1000

* Subject to the size of the vessel, the Reduction Factor (RF) will be linearly interpolated between the two numbers.

With the reduced ship size, the reduction factor (RF) should be applied at a lower value.

** On September 1, 2015, the vessels will begin Phase 1.

*** Vessels delivered on or after September 1, 2019, are subject to RF.

EEDI Measurement: Attained

The EEDI requirements apply to vessels like Containership, Gas carriers (LPG carriers), Bulk carriers, Refrigerated cargo vessels, Roll-on/roll-off, Tankers, LNG carriers, Combination carriers, General cargo ships and non-conventional propelled Cruise passenger ships, excluding ships with non-traditional propulsion such as turbine, diesel or hybrid thrust development systems, but excluding LNG ships and cruise ships in this context. The attained EEDI calculation follows the well-known calculation mentioned in Eq.2 (IMO, 2023). The conversion factor (CF) is mentioned in the Table 4 (Barreiro et al., 2022b), and parameter details of Attained EEDI are said in the Table 5 (Barreiro et al., 2022).

$$\textit{Attained EEDI} = \frac{\text{CO2 Emission (gms of CO2)}}{\text{Transport Work (ton − miles)}} = \frac{\text{Impact to the environment}}{\text{Benefit for the society}} \tag{2}$$

$$= \frac{EnginePower(P) \times SpecificFuelConsumption(SFC) \times CoversionFactorofCO_2(C_F)}{\text{CargoCarry} \times \text{DistanceTravel}}$$

$$= \frac{\text{(Main Engine Emission + Auxiliary Engine Emissions + Power Take In (PTI))} - \text{(Introduction of Energy Technology to Auxiliary Engine + Main Engine)}}{\text{Dead Weight x Distance Travel}}$$

$$
= \frac{\left(\prod_{j=1}^{n} f_j\right) x \left(\sum_{i=1}^{nME} P_{ME(i)} \cdot C_{FME(i)} \cdot SFC_{ME(i)}\right) + \left(P_{AE} \cdot C_{FAE} \cdot SFC_{AE}\right) + [(\prod_{j=1}^{n} f_j \sum_{i=1}^{nPTI} P_{PTI(i)} - \sum_{i=1}^{neff} f_{eff(i)} \cdot P_{AEeff(i)}) C_{FAE} \cdot SFC_{AE}] - (\sum_{i=1}^{neff} f_{eff(i)} \cdot P_{eff(i)} \cdot C_{FME} \cdot SFC_{ME})}{f_i . f_c . f_l . Capacity . f_w . V_{ref} . f_m}
$$

Table 4. Non-dimensional conversion factors (CF)

Fuel Types	Note	Carbon (C) matter x 10^{-3}	C_F (t-CO_2/t-Fuel)
LFO	ISO 8217 Categories: RMA ~ RMD	860	3.2060
Gasoil / Diesel	ISO 8217 Categories: DMX ~ DMC	875	3.2060
HFO	ISO 8217 Categories: RME ~ RMK	850	3.1144
LNG	---	750	2.7500
LPG	Propane Butane	819 827	3.0000 3.0300

Table 5. Parameter details of Attained EEDI equation

Parameter	Description	Source
C_F	Conversion Factor (Non-dimensional): fuel usage & CO_2 emissions. Table 4	MEPC245(66) "2014Guidelines on the calculation of the Attained EEDI for new ships"
V_{ref}	Ship speed in Knot	The speed-power curves were produced from the model testing during the design stage. Report on the Sea Trial at the Final Stage
Capacity	A function of deadweight, as stated in sections 2.3 and 2.4 of the 2014 MEPC 245(66) Guidelines; on the computation of the EEDI.	Stability Booklet
P_{ME}	75% MCR of the Main Engine's, measured in kW	NOx - Technical File
P_{AE}	Power from Auxiliary Engine	MEPC 245(66) "2014 Guidelines on the calculation of the Attained EEDI for new ships"
P_{PTI}	75% of shaft motor power consumption	
P_{eff}	Propulsion output using new mechanical energy-efficient technologies at 75% main engine power	
P_{AEeff}	Innovations in electrical energy efficiency have led to a decrease in auxiliary power.	
SFC	Certified SFOC (g/kWh)	NOx - Technical file

continued on following page

Table 5. Continued

Parameter	Description	Source
f_j	Correction factor to take into consideration things unique to the ship's design. (For instance, shuttle tankers and ships with an ice class)	MEPC 245(66) "2014 Guidelines on the calculation of the Attained EEDI for new ships"
f_w	The coefficient shows decrease of vessel speed in a typical sea situation concerning wind speed, frequency, and wave height.	
f_i	A capacity factor that accounts for any legislative or technological constraints on capacity	
f_c	Correction factor for cubic capacity (applicable to chemical and gas tankers)	
f_l	General cargo ships equipped by cranes and additional machinery linked to cargo have a deadweight compensation factor.	
f_{eff}	The accessibility of innovative energy-saving technologies	MEPC.1/Circ.815

EEXI: EXISTING VESSELS EMISSION CALCULATION

In 2021, the MEPC.328 (76) session was held to achieve the emission reduction target by IMO, and MARPOL Annex-VI was amended accordingly (MEPC.328(76)-2021, n.d.). In MARPOL Annex VI, amendments were adopted towards implementing a technical and operational approach, i.e., a hybrid approach towards emission control of GHG gases from the existing vessel, which was earlier for the new vessels only in MEPC.203 (62). IMO adopted the Energy Efficiency Existing Ship Index (EEXI) and Carbon Intensity Indicator (CII). These hybrid tools for emission control by vessels were adopted by 2023. IMO's mandatory EEXI measure sets new requirements to lessen the amount of CO_2 emissions on existing vessels. The steps to develop a practical analysis are shown in Figure 1.

Figure 1. EEXI compliance for a typical vessel requirement

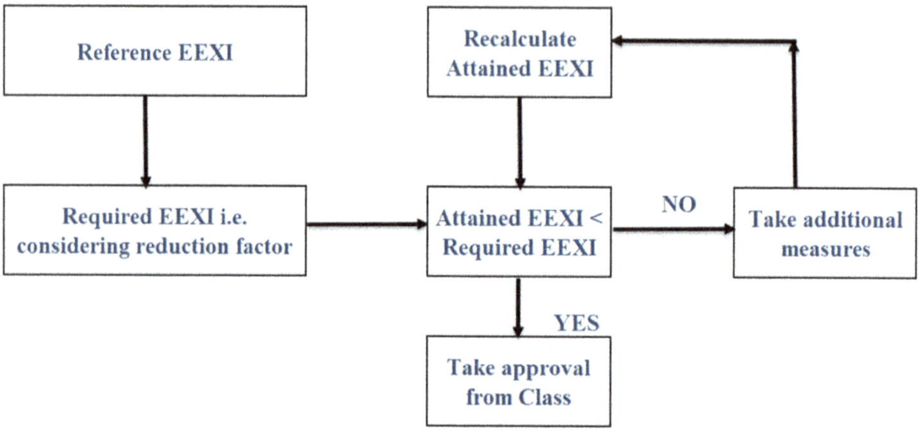

Methodology for IMO - EEXI implementation for Vessels

a) Initially, for the vessel, the Reference line / EEXI needs to be calculated based on the dead weight of the boat.
b) Considering the phase of reduction and ship types, the reduction factor needs to be determined.
c) Estimated by calculating the Required EEXI for that vessel. The Required EEXI is the maximum value an emission vessel can achieve.
d) The attained EEXI value must be identified from the vessel speed/power curve or sea trial data by calculation method.
e) If the boat Attained EEXI achieves lesser value than Required EEXI, the ship owner/manager can approach for certification to the classification society.
f) Failing to reach the target, the owner must take additional measures and recalculate the attained EEXI.
g) To comply with the IMO 2023 decarburization requirement, EEXI introduces the existing vessels of 400 GRT and above trading in international voyages.

EEXI Measurement: Required

The EEXI requirements apply to vessels like Bulk carriers, Gas carriers (LPG carriers), LNG carriers, Containerships, refrigerated cargo carriers, Tanker, Vehicle carriers, Cruise Vessel, Car carriers, Combination carriers, General cargo ships, non-conventional propelled Cruise vessels excluding vessels with hybridized transportation technologies or other nontraditional powering methods, but excluding LNG ships and cruise ships in this context.

The relations between both EEXIs are mentioned with a simple Eq.3 and 4, like the EEDI equation. Only here is the reduction factor considered Y, whereas in the case of EEDI, it was X, as mentioned in MEPC.328 (76) (MEPC.328(76)- 2021, n.d.).

$$Attained\ EEXI \leq Required\ EEXI \quad (3)$$

$$Required\ EEXI = \left(1 - \frac{Y}{100}\right) * Reference\ line\ value \quad (4)$$

Where,

- Attained EEXI= Vessel achieved value calculated during sea trial or, from vessel speed/power curve data, refer Eq.3.
- Required EEXI = Ship theoretical calculated target value to be calculated from above Eq.4.
- Reference Line (RL) Calculation = a x b^{-c}; Where values of a, b and c had expressed in IMO MEPC.328(76)
- Y = Reduction Factor (RF) of EEXI. The value of RF depends on ship type and capacity in DWT (Dead Weight) mentioned in IMO MEPC.328(76).

Required EEXI for Bulk Carrier

The Reference Lines for Bulk carrier Vessels of 400 and above GT are shown in Figure 2 concerning capacity (DWT). It can be identified that the bigger the vessel size, the more energy efficient the co-efficient will be (ClassNK, n.d.). The reduction factor in the Required EEXI estimation strengthens the IMO requirement, and vessels will be more energy efficient and have less carbon footprint.

Figure 2. Bulk carrier vessel required EEXI after considering reduction factor; Y=20

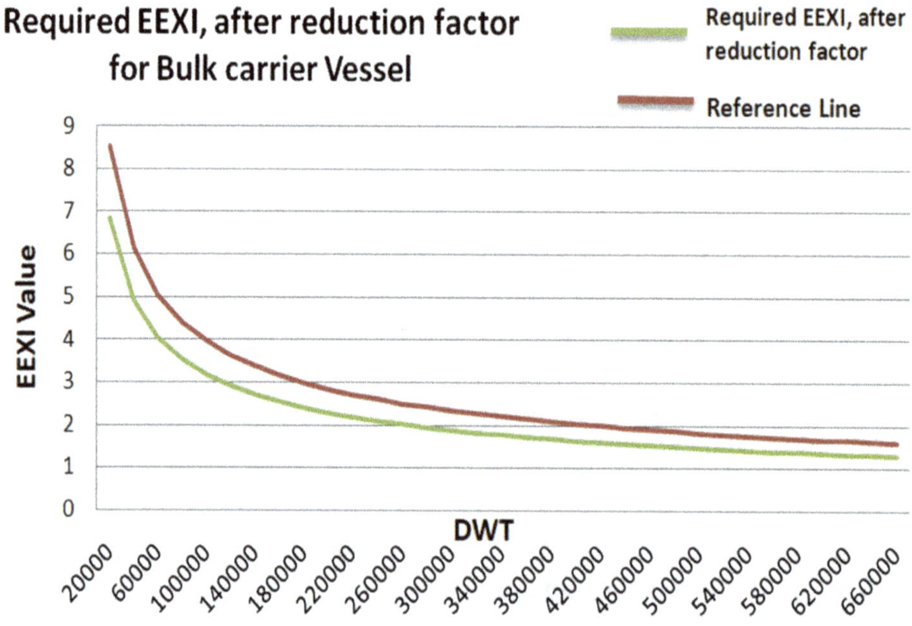

EEXI Measurement: Attained

The attained EEXI calculation follows the well-known attained EEDI calculation mentioned in Eq.5 (MEPC 78/17/Add.1 Annex 12, 2022, n.d.)

$$= \frac{\text{CO2 Emission (gms of CO2)}}{\text{Transport Work (ton – miles)}}$$

$$= \frac{\text{Impact to the environment}}{\text{Benefit for the society}}$$

$$= \frac{\textbf{Engine Power (P)x Specific Fuel Consumption (SFC)x Coversion factor of CO2 } (\mathbf{C_F})}{\textbf{Cargo carry x Distance travel}} \qquad (5)$$

Where.

1. CO_2 emission = Impact to the environment =

[(Auxiliary Engine (AE) Emissions + Main Engine (M/E) Emission + Power Take In (PTI)) –(Introduction of energy technology to main Engine + Auxiliary Engine)]

$$= [(P_{AE} \cdot C_{FAE} \cdot SFC_{AE})] + \left(\prod_{j=1}^{n} f_j\right)\left(\sum_{i=1}^{nME} P_{ME(i)} \cdot C_{FME(i)} \cdot SFC_{ME(i)}\right)$$
$$+[(\prod_{j=1}^{n} f_j \sum_{i=1}^{nPTI} P_{PTI(i)} - (\sum_{i=1}^{neff} f_{eff(i)} \cdot P_{eff(i)} \cdot C_{FME} \cdot SFC_{ME})$$
$$- \sum_{i=1}^{neff} f_{eff(i)} \cdot P_{AEeff(i)}) \, C_{FAE} \cdot SFC_{AE}]$$

2. Transport Work = Cargo carries x Distance travel = $f_l . f_c . f_i$. Capacity . f_m . f_w . V_{ref}

So; Attained EEXI (MEPC 78/17/Add.1 2022, n.d.

$$= \frac{[(P_{AE} \cdot C_{FAE} \cdot SFC_{AE})] + \left(\prod_{j=1}^{n} f_j\right)\left(\sum_{i=1}^{nME} P_{ME(i)} \cdot C_{FME(i)} \cdot SFC_{ME(i)}\right) +[(\prod_{j=1}^{n} f_j \sum_{i=1}^{nPTI} P_{PTI(i)} - (\sum_{i=1}^{neff} f_{eff(i)} \cdot P_{eff(i)} \cdot C_{FME} \cdot SFC_{ME}) - \sum_{i=1}^{neff} f_{eff(i)} \cdot P_{AEeff(i)}) \, C_{FAE} \cdot SFC_{AE}]}{f_l . f_c . f_i . \text{Capacity} . f_m . f_w . V_{ref}}$$

Note: Parameter details of Attained EEDI / EEXI equation are mentioned in Table.6 (Annex – 2014 EEDI, n.d.) and Table.7 (MEPC 78/17/Add.1 2022, n.d.)

Table 6. Non-dimensional conversion factors of CO_2 conversion's (ton of CO_2/ton of Fuel)

Fuel In Use	Standard Reference	Carbon (%C) in fuel	Conversion Factor; C_F
LPG: Liquefied Petroleum Gas	Butane (C_4H_{10})/ Propane (C_3H_8)	0.82640 0.81820	3.030 3.000
LFO: Light Fuel Oil	Grades RMA thru the RMD in ISO 8217	0.860	3.1510
LNG: Liquefied Natural Gas	-------	0.750	2.750
HFO: Heavy Fuel Oil	Grades RME thru the RMK in ISO 8217	0.850	3.1140
MGO: Gas Oil / MDO: Diesel Oil	Grades DMX thru the DMC in ISO 8217	0.8744	3.2060

Table 7. Attained EEDI/EEXI calculation; IMO.MEPC.245 (66), MEPC.350 (78)

Input-Data	Description	Standard Reference
Capacity	Deadweight (DWT)of the vessel; *IMO MEPC 245(66)- section 2.3 and 2.4*	Ship Stability Booklet
V_{ref}	Ship speed (nautical miles/ hour); *MEPC.350 (78)- section 2.2.3*	a. Speed-power curves. b. Sea Trial Report.
C_F	Correction Factors	IMO Standard
P_{AEi}	Power of auxiliary engines (A/E); *MEPC.350 (78)- section 2.2.2*	Ships Power data, Engine manual
P_{PTI}	Power consumption of shaft motor; *MEPC 245(66) section 2.5.3*	Ships Data
P_{MEi}	Power of main engines (M/E); *MEPC.350 (78)- section 2.2.1*	NOx Technical File
P_{AEeff}	Auxiliary energy saving by creative technology	*MEPC.350 (78)-section 2.5.5*
P_{eff}	Other Energy saving due to creative technology	*MEPC.350 (78)-section 2.5.4*
SFOC	Specific Fuel Oil Consumption (g/kWh); *MEPC.350 (78)-section 2.2.4*	NOx Technical file
f_w	Decrease in vessel speed due to wave height, wind speed and frequency;	Correction Factors: *IMO MEPC 245(66) Calculation Guidelines*
f_j	Vessel specific design, e.g., Ice classed vessel	
f_l	Vessel fitted with cargo gears and cranes	
f_c	Applicable togas carriers &chemical tankers	
f_i	Capacity Factor for any other requirements	
f_{eff}	Availability factor of creative energy equipment	MEPC.1/Circ.815

To fulfill the IMO 2023 decarburization requirement, EEXI introduces the existing vessels of 400 GRT and above trading in worldwide voyages. The relations between both EEXIs are mentioned with a simple Equation 2, like the EEDI equation. Only here is the reduction factor considered Y, whereas in the case of EEDI, it was X.

Y = EEXI reduction factor (RF) by vessel type and capacity in DWT (Dead Weight) are mentioned in Table.8 (MEPC 78/17/Add.1 EEXI, n.d.)

Table 8. EEXI reduction factor by vessel type and capacity

Vessel type	Size	Reduction factor
Bulk carrier	200, k D.W.T. and more	15
	20, k and more but less than 200, k D.W.T.	20
	10, k and more but less than 20, k D.W.T.	0-20*

continued on following page

Table 8. Continued

Vessel type	Size	Reduction factor
Gas carrier	15, k D.W.T. and more	30
	10, k and more but less than 15, k D.W.T.	20
	2, k and more but less than 10, k D.W.T.	0-20*
Tanker	200, k D.W.T. and more	15
	20, k and more but less than 200, k D.W.T.	20
	4, k and more but less than 20, k D.W.T.	0-20*
Container vessel	200, k D.W.T. and more	50
	120, k and more but less than 200, k D.W.T.	45
	80, k and more but less than 120, k D.W.T.	35
	40, k and more but less than 80, k D.W.T.	30
	15, k and more but less than 40, k D.W.T.	20
	10, k and more but less than 15, k D.W.T.	0-20*
General cargo vessel	15, k D.W.T. and more	30
	3, k and more but less than 15, k D.W.T.	0-30*
Refrigerated cargo carrier	5, k D.W.T. and more	15
	3, k and more but less than 5, k D.W.T.	0-15*
Combination carrier	20, k D.W.T. and more	20
	4, k and more but less than 20, k D.W.T.	0-20*
LNG carrier	10, k D.W.T. and more	30
Ro-ro cargo vessel (vehicle carrier)	10, k D.W.T. and more	15
Ro-ro cargo vessel	2, k D.W.T. and more	5
	1, k and more but less than 2, k D.W.T.	0-5*
Ro-ro passenger vessel	1, k D.W.T. and more	5
	250 and more but less than 1, k D.W.T.	0-5*
Cruise passenger vessel having non-conventional propulsion	85, k GRT and more	30
	25, k and more but less than 85, k GRT	0-30*

* The reduction factor should be calculated by linearly interpolating the two variables that depend on the size of the vessel.

Note: 1k = 1000

EXISTING VESSEL CONSTRUCTION EMISSION ANALYSIS: RTM PIRAMU

Verification of carbon footprint regulatory compliance as per Resolution MEPC.334 (76) for existing ship. The vessel considered in this work is the Bulk Carrier Ship. RTM PIIRAMU (IMO: 9341926) was built in 2008 and is sailing under the flag of Singapore. Figure 3

Vessel Particulars

- Length: 235.67 m
- Breadth: 43.00 m
- Draught: 12.837 m
- Summer Deadweight (DWT): 89861 tons
- Tonnage Gross: 53988
- Main engine type: B&W 6S60MC-C (Mark7) (2-stroke)
- Main engine power (MCR): (BHP) 13,500 kW x 105.0 rpm
- Service speed: 16.08 knots

Figure 3. Vessel view RTM PIIRAMU (IMO: 9341926), bulk carrier

Calculation of Required EEXI

RTM Piramu the summer deadweight (DWT): 89861 tons
Reference line value = a x b^{-c};

$= 961.79 \times (89861)^{-0.477} = 4.17$ (i)

Required EEXI $= \left(1 - \frac{X}{100}\right) *$ Reference line value

where the Reduction Factor is 20%,

$= \left(1 - \frac{20}{100}\right) * 4.17 = 3.336$ (ii)

Calculation of Attained EEXI

Basic data:	Main engine:	Auxiliary engines
Type of vessel: Bulk carrier Capacity D.W.T: 89861 Speed V_{ref} (knots): 16.08	MCR_{ME} (kW): 13,500 P_{ME} (kW): 11,475 Type of fuel: Diesel oil C_{FME}: 3.206 SFC_{ME} (g/kWh): 166.5, refer Fig.4	Type of fuel: Diesel oil P_{AE} (kW):625 SFC_{AE} (g/kWh):220.0 CF_{AE}:3.206

Electrical energy-efficient technology; N/A
Ice class: N/A
Cubic capacity correction factor: N/A
Mechanical energy-efficient technology: N/A

Figure 4. Engine SFOC chart

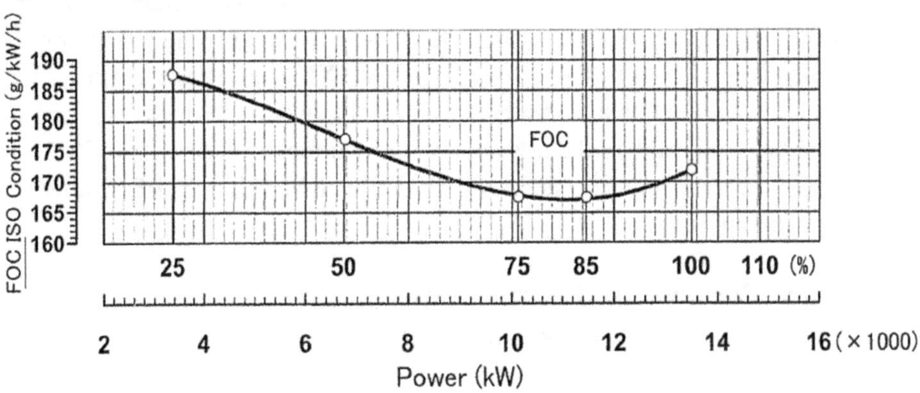

$$Attained\ EEXI = \frac{\text{CO2 Emission (gms of CO2)}}{\text{Transport Work (ton – miles)}}$$

$$= \frac{1x(11475x3.206x166.5) + (625x3.206x220) + 0 - 0}{1x1x1x89861x1x16.08x1} = 4.544 \text{ (iii)}$$

Figure 5. Comparison chart of EEXI, equation (ii) and (iii)

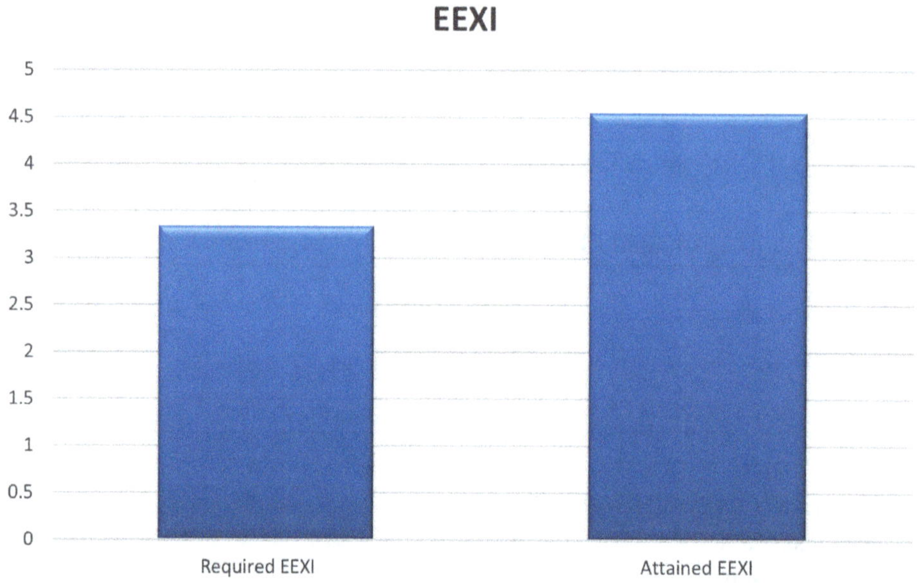

Comparing the analysis values in (ii) and (iii), the calculation shows in Fig.5 that the vessel attained EEXI, which is more than the Required EEXI. So, the vessel needs to take additional technological measures to achieve the target value. Suppose the ship's obtained EEXI does not meet the requisite EEXI. In that case, a verifier will request a survey application and an EEXI Technical File containing the appropriate background materials and all the information needed for the verification. The ship may undergo speed testing as is necessary to confirm the achieved EEXI after the main modification.

Calculation for Carbon Footprint in Actual Trading of the Vessel: RTM PIRAMU

For operational performance, IMO developed SEEMP and EEOI. The Energy Efficiency Operational Indicator (EEOI) is a voluntary tool to justify SEEMP—a simple calculation made in this section to identify the actual carbon footprint during trading by vessel. A quarterly fuel consumption data considers calculating the gms of CO_2 emission per cargo carrying. A calculation of the EEOI application (including one ballast voyage) is explained. The example illustrates based on the fuel consumption data taken from vessel flow meter readings. The average fuel consumption at sea is 19.0MT/day while loaded, 17.0 MT/Day on ballast voyage, and a Vessel Speed of 12.0 Knots. In port, the auxiliary machinery consumes diesel oil only, and at sea, it consumes HFO.

RTM PIIRAMU, BULK CARRIER				
Voyage	**Fuel consumption (FC) at sea and in port in tonnes**		**Voyage data**	
	Heavy Fuel Oil (HFO)	**Diesel/Gas Oil (MDO / MGO)**	**Cargo (MT)**	**Distance (NM)**
1	475	15	50000	300
2	425	16	0	300
3	1188	19	50000	750
4	237	13	25000	150

$$\text{Carbon foot print / EEOI value}^{3} = \frac{\text{CO2 Emission (gms of CO2)}}{\text{Transport Work (ton} - \text{miles)}}$$

$$= \frac{2325X3.114 + 63X3.206}{50000X300 + 0X300 + 50000X750 + 25000X150}$$

= 132.3 X 10^{-6} gms of CO_2 / (tons x nautical miles); the CO_2 calculation might be changed over from g/ton-mile to g/ton-km by multiplying 0.54.

Figure 6. Each voyage gms of CO_2 / tonne-miles

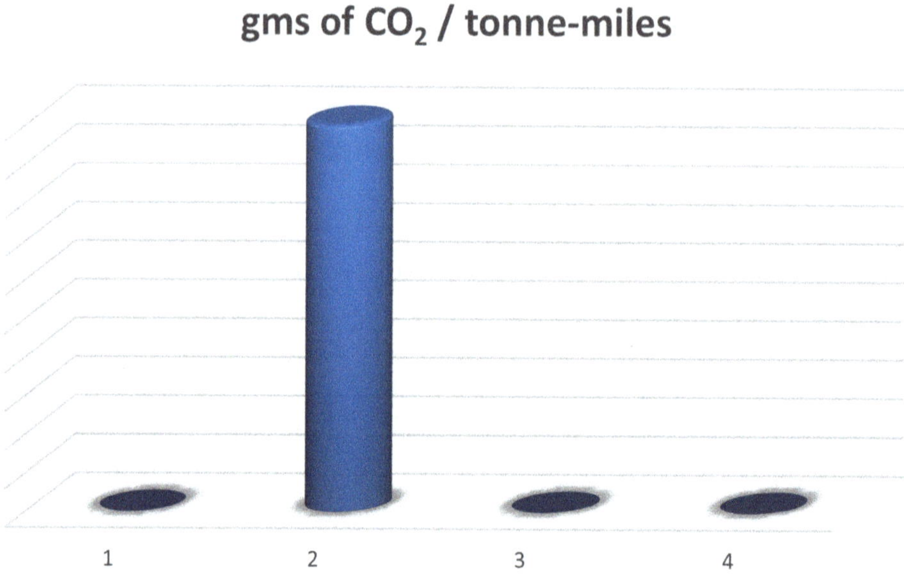

Therefore, Figure 6 shows that the effects on ballast voyage in ships are detrimental to the carbon efficiency indicator. As the denominator value in the calculation index goes to zero because of the ballast voyage, the value of the energy index goes to infinite. In the calculation, an attempt was made to identify the effect of the ballast voyage on the Energy Efficiency Index and the requirement of effective cargo trading for minimizing the ballast voyage. Carrying maximum cargo and minimum ballast is necessary to maintain a minimum CO_2 index for an operational vessel.

GRADING OF VESSEL

Carbon Intensity Indicator: CII

The global shipping sector aims to minimize greenhouse gas emissions from sea transport vessels—a resolution adopted by IMO in 2021 states that the organization will implement two strategies by January 2023. One is technical control, such as EEXI, and another is operational efficiency measures. This is known as the Carbon Intensity Indicator (CII) ship grading system. CII is used to gauge how well a vessel uses energy while trading from January 2023 for any vessels 5,000 GRT and above (MEPC.352(78)-2022 CII, G1), n.d.).

As discussed earlier in the EEXI calculation, Attained CII (Set target by IMO) and Required CII (Value achieved by vessel) to consider identifying sub-standard vessels. The measurement index is in grams of CO_2 released per nautical mile and cargo carrying capacity. The Attained CII score should be lower than the Required CII score.

Based on the CII calculation, the value vessel will be graded annually from A to E, where A is the highest rating, and E will be labeled inappropriate emission control measures. Figure 7 demonstrates that the diagram's acceptable grade and necessary minimum rating will be C (IMO Env.C, n.d.).

For calculation, the 2019 emission will be considered as a datum line. Ships with an E rating at any time or a D rating for 3 years in a row will need to create a remedial activity plan for SEEMP Part III CII, G2), n.d.). The rating thresholds will be lowered yearly along with the required CII considering the reduction factor, as mentioned in Table 9 for 2023 to 2026. The succeeding reduction factors will be determined at a review in 2025 (Bayraktar & Yuksel, 2023). The CII reference line will be calculated per IMO MEPC.337(76) guidelines (Carbon I.R, n.d.).

Figure 7. CII grading for the vessels

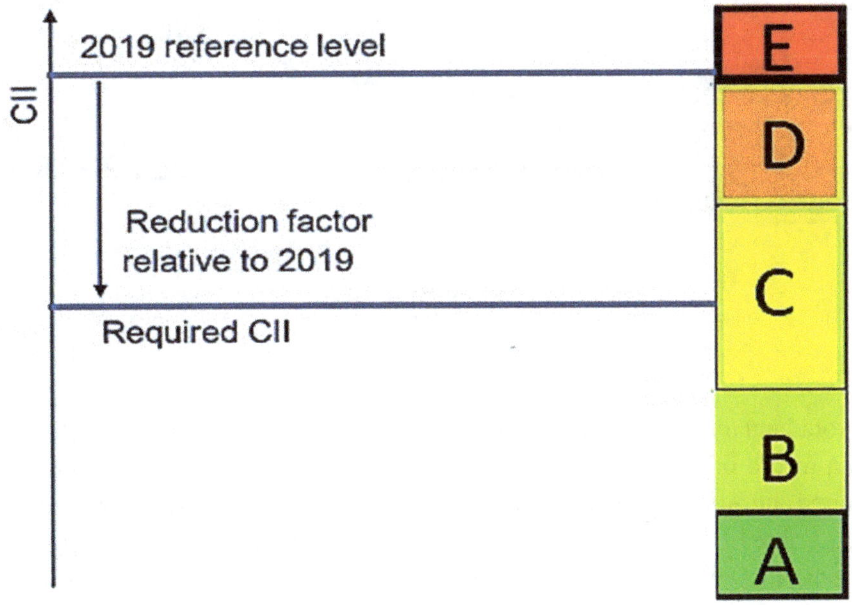

Table 9. CII reduction factor

YEAR	Decrease from 2019 baseline level to midpoint of C rating index in percentage
2023	5.00
2024	7.00
2025	9.00
2026	11.00
2027 - 2030	Not Yet Decided

OPERATIONAL MEASURES OF VESSEL

SEEMP Part III

MEPC.328 (76) highlighted that beginning in 2024, the CII should be computed and provided to the Data Collection System (DCS) verifier along with the total DCS data for the preceding year, as explained in Figure 8.

Figure 8. Relation between the DCS, SEEMP-III and CII

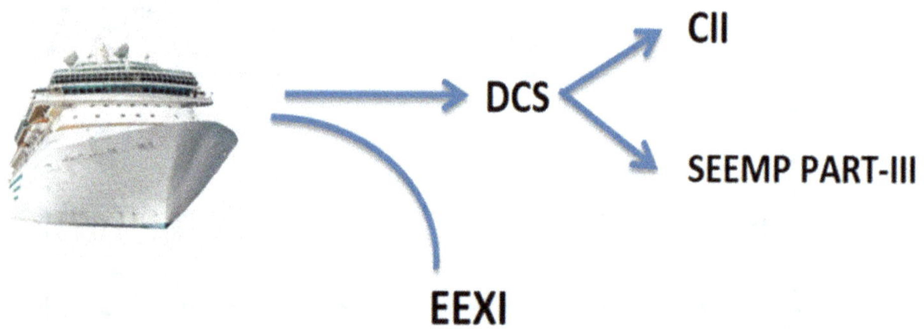

From January 1, 2023, SEEMP Part - III must be maintained onboard the vessel with a record of the vessel's CII goals and CII calculation technique to measure Required and Attained CII values (SEEMP Part III C.T., n.d.). IMO proposed EEXI in 2022, like EEDI, as a technological measure for running vessels. On the other hand, the IMO also introduced SEEMP in 2013 as an operational tool to reduce CO_2 emissions from vessels. Later, SEEMP Part III was incorporated in 2023 with Parts I and II of the current SEEMP to consider IMO Resolution MEPC.328 (76) (MEPC.328(76)- 2021, n.d.). The main goal of the EEXI and CII is to inculcate a mindset among those involved in the marine sector to continue to make efforts to reduce onboard greenhouse gas emissions.

Implementation of SEEMP-III onboard the vessel demanded all the parties to be involved. The responsibility lies with the Master and Crew onboard the ship to the top management of the vessel operators and managers. Ship staff will be responsible for -

a) Creating an environment of safety and environmental awareness on board to increase the ship's resource efficiency. Provide regular training and briefing on energy efficiency.
b) A well-developed work plan also effectively cuts carbon emissions during ship operations. Follow the vessel's planned maintenance system and soundly maintain the ship's machinery to prevent breakdowns and unplanned shipment delays while in port.
c) Employing accurate and contemporary navigational procedures for the most efficient voyage planning and weather routing.
d) Gather analyses and submit logs, performance measures, etc., to the company for review.
e) Tracking the effectiveness of SEEMP processes and how they impact vessel operations.

Similarly, the vessel management and operators should be familiar with the updated MARPOL Annex VI, and plans should be made for its effective implementation. Operators should upgrade their fleet's SEEMPs to incorporate SEEMP Part III plans. The company is responsible for setting up the technical file, assessments, and survey with the International Energy Efficiency (IEE) certification organization.

Company audits of SEEMP III shall be conducted by the standards established by the appropriate workshop.

The vessel manager should ensure that the SEEMP (Part III) for the ships subject to the CII rating must include the following in the manual onboard vessel:

a) CII computation approach
b) Processes for self-evaluation and enhancement.
c) Required CII values over the next three years.
d) Strategy and action plan for obtaining the Required CII.

CERTIFICATION OF ENERGY COMPLIANCE

a) An International Energy Efficiency Certificate (IEEC) certifying a ship's compliance with the SEEMP, EEDI, and EEXI standards is issued by the flag state defined in Annex-VI of MARPOL. This certificate must remain available for routine inspections and audits during the ship's lifetime (SEEMP, n.d.). Verification and certification will be conducted by the classification society acting on behalf of the flag state.
b) The vessel must develop a ship-specific EEXI Technical file for verification and certification. Ships that maintain EEDI technical files and comply with EEXI requirements will not need any EEXI Technical files to keep. In case the vessel does not comply with EEXI requirements, there will be a risk of detention by port authorities and risks of fines.
c) A statement of compliance (SoC) listing the CII and quality rating (A-E) will be retained onboard ship records for five years.
d) SEEMP III must be revised with a corrective action plan before the release of the SoC. It must also be verified if there have been three D grades in succession or one E grade. The corrective action plan should include a modified effective system, which should also review why the CII requirements were not satisfied.

ALIGNMENT OF IMO GHG STRATEGY WITH THE UN SDGS

The IMO GHG Strategy 2023 represents a significant milestone in defining the maritime industry's role in meeting global climate targets. The alignment of the strategy with the United Nations Sustainable Development Goals (SDGs) not only helps mitigate climate change but also promotes sustainable economic growth and innovation in the shipping sector, thanks to the IMO (Arora & Mishra, 2019).

SDG 13: Climate Action—Reducing Emissions in Maritime Shipping

The IMO GHG Strategy 2023 is a radical step towards climate change by ensuring that GHGs in the shipping industry are reduced to a considerable extent, as maritime activities contribute a notable number of emissions to a global level. This has been estimated to be realized by means of energy efficiency increases in current ships, the use of alternative fuels, and technological improvements (Gulev, 2023). The plan is encouraging the introduction of such technologies as wind-assisted propulsion, energy-efficient designs, and low-carbon fuels. The determinations that will assist in maintaining a continuous record of improved carbon performance within the sector include the EEXI and the CII.

SDG 9: Industry, Innovation, and Infrastructure in Shipping

Reduction in emissions is accompanied by the adoption of environmentally friendly materials, such as eco-friendly materials, the use of energy-saving engines, and green technology during the shipbuilding process (Arif et al., 2025). Another important area is ports, where the IMO has been promoting green initiatives that can help reduce pollution during loading and offloading processes, as well as encourage the use of shore-side electricity to lessen our dependency on fossil fuel.

SDG 7: Affordable and Clean Energy for Shipping

IMO GHG strategy focuses on alternative fuels that include H_2, ammonia, biofuels, and synthetic fuels to substitute conventional marine fuel. This change is necessary to reach carbon neutrality in transit in the middle of the century (Roy & Pramanik, 2024). Wind, solar, and wave energy are among the forms of renewable energy under consideration, as using them would decrease the dependency on fossil fuel in maritime operations. The strategy also recommends the need to increase the viability of renewable energy in large-scale shipping through research and development (R&D).

SDG 14: Life Below Water—Protecting Marine Ecosystems through Emission Reductions

The strategy by the IMO also eliminates the adverse impacts of air pollution on marine life since it can minimize the emission of GHGs by ships. The primary pollutants released through shipping include SOx and NOx, which cause ocean acidification and damage marine life (Akhai & Taneja, 2025). The decrease in carbon emissions by ships will also reduce the rate at which climate change affects the oceans and its problems, such as ocean acidification, ocean level rising, and coral bleaching. This aids in the overall objective of marine conservation.

SDG 8: Decent Work and Economic Growth

The IMO strategy provides not only prospects of green economy development in the marine sphere but also the possibility to create new workplaces in the domain of clean technology, the production of alternative fuels, and green shipbuilding. The IMO is also fostering sustainable operations, which promote the generation of long-term economic value by striking a balance between growth and environmental sustainability. Green shipping is also made to improve the competitiveness of the sector in the global context, given that sustainability is becoming the key priority in industries and countries (Roy & Chakraborty, 2024).

IMO COMPLIANCE CHALLENGES

The pushback to such challenges has created an acceptable level of digital innovations that are constantly being explored to increase the accuracy of monitoring, transparency of compliance, and adaptability of operations in maritime decarbonization, and they include blockchain, AI, and IoT (Pramanik et al., 2025, 2025). The decentralized systems can offer reliable and secure CO_2 reporting setups and support audits by classification societies through safe data channels. IoT-based systems allow ships to collect real-time information about their emissions and fuel use, which helps them follow SEEMP Part III rules and continuously check their efficiency.

In addition, the similarity with generative AI models is also finding application in the generation of adaptive SEEMP configurations optimized to current operating conditions and in the prediction of voyage profile emissions trajectories (Mamatha et al., 2025). At the same time, scalable database solutions, like the two-level sharding, enhance the processing capabilities of ship data, allowing the incorporation of high-frequency telemetry in the emissions calculation and decision support systems (Pramanik, Bag, Roy, & Rakshit, 2025). Also, data-balancing methods, such as SMOTE and ADASYN, have been used to improve the reliability of ML models in predicting vessel performance and emissions when the cargo load is variable and weather effects are present (Pramanik, Bag, Roy, Ghosh, et al., 2025). All these approaches help to supplement the regulatory frameworks like EEXI, EEDI, and CII by enhancing the use of data-driven decisions, real-time compliance monitoring, and dynamic emissions control, aligning with the advancing sustainability roadmap of IMO.

The maritime industry faces several challenges because of the critical need to reduce the carbon footprint of new and existing ships. The target is clear: zero emission from ships. IMO had set a course path for reducing GHG to achieve the target of decarburization. The IMO's Decarbonization initiative includes several operational and technological measures onboard vessels to cut emissions. "Operational measures" will push vessel managers to monitor any ship-based activity that lessens the effect of ship-generated pollutant emissions. Meanwhile, "Technical controls and measures" on equipment and systems installation result in owner budget constraints.

EEDI and EEXI are the IMO's baskets of measures towards energy efficiency control and reduction from shipping. In January 2023, EEXI came into force, which targets the existing vessel design efficiency. However, the carbon intensity indicator (CII) was adopted in the MEPC.328 (76) session to monitor and improve vessel operation to achieve IMO operational goals. If ships do not comply with IMO energy efficiency requirements, they may be detained by inspection authorities at various ports. Ship Operators must submit an annual fuel consumption report and annual distance travelled data, part of the Ship Fuel Consumption Data Collection System (DCS). Based on these data, the CII attained will be calculated and graded by the vessel. Each ship must indicate in the SEEMP (Part-III) the method of calculating the annual CII from the calendar year 2023 and the procedure for reporting the CII. For most vessels, the power limitation of the ship and alternative fuel choices are the easier ways to overcome the emission challenge.

By integrating CO_2 emission calculations, energy efficiency measures, and operational strategies, the IMO's framework is driving the maritime industry toward sustainable ship construction and operation. This chapter highlights the critical role of these regulations in promoting environmentally responsible shipping while addressing the challenges of regulatory compliance, economic feasibility, and technological adaptation in the transition toward a zero-emission future.

REFERENCES

Akhai, S., & Taneja, T. (2025). The Critical Role of Water Quality. In *Smart Water Technology for Sustainable Management in Modern Cities* (pp. 101–116). IGI Global. DOI: 10.4018/979-8-3693-8074-1.ch005

Alhasnawi, B. N., Almutoki, S. M. M., Hussain, F. F. K., Harrison, A., Bazooyar, B., Zanker, M., & Bureš, V. (2024). A new methodology for reducing carbon emissions using multi-renewable energy systems and artificial intelligence. *Sustainable Cities and Society*, *114*, 105721. DOI: 10.1016/j.scs.2024.105721

Annex – 2014 Guidelines on the Method of Calculation of the Attained Energy Efficiency Design Index (EEDI) for New Ships. (n.d.). https://imorules.com/GUID-507FC12C-452D-4925-8D8E-052333B40148.html

Arif, S. M., Akhai, S. S., & Dadhwal, R. (2025). High-Performance Aluminium Alloys for Shipbuilding. In *Innovative Materials for Next-Generation Defense Applications* (pp. 107–144). IGI Global. DOI: 10.4018/979-8-3373-0933-0.ch004

Arora, N. K., & Mishra, I. (2019). United Nations Sustainable Development Goals 2030 and environmental sustainability: Race against time. *Environmental Sustainability*, *2*(4), 339–342. DOI: 10.1007/s42398-019-00092-y

Barreiro, J., Zaragoza, S., & Diaz-Casas, V. (2022). Review of ship energy efficiency. *Ocean Engineering*, *257*, 111594. DOI: 10.1016/j.oceaneng.2022.111594

Bayraktar, M., & Yuksel, O. (2023). A scenario-based assessment of the energy efficiency existing ship index (EEXI) and carbon intensity indicator (CII) regulations. *Ocean Engineering*, *278*, 114295. DOI: 10.1016/j.oceaneng.2023.114295

ClassNK. (n.d.). https://www.classnk.com/hp/pdf/tech_info/tech_img/T1245e

Czermański, E., Oniszczuk-Jastrząbek, A., Spangenberg, E. F., Kozłowski, Ł., Adamowicz, M., Jankiewicz, J., & Cirella, G. T. (2022). Implementation of the Energy Efficiency Existing Ship Index: An important but costly step towards ocean protection. *Marine Policy*, *145*, 105259. DOI: 10.1016/j.marpol.2022.105259

Don't Wait. Act now. Manage your SEEMP Part III compliance today. (n.d.). https://www.lr.org/en/seemp-part-iii-ship-energy-efficiency

EEA. (2022). European Maritime Transport. *European Environment Agency*. https://www.eea.europa.eu/publications/maritime-transport/

Gianni, M., Pietra, A., Coraddu, A., & Taccani, R. (2022). Impact of SOFC Power Generation Plant on Carbon Intensity Index (CII) Calculation for Cruise Ships. *Journal of Marine Science and Engineering*, *10*(10), 1478. DOI: 10.3390/jmse10101478

Gulev, S. K. (2023). Global Climate Change and the Oceans. *Studies on Russian Economic Development*, *34*(6), 738–745. DOI: 10.1134/S1075700723060060

IMO. (2021). *Reducing greenhouse gas emissions from ships*. https://www.imo.org/en/MediaCentre/HotTopics/Pages/Reducing-greenhouse-gas-emissions-from-ships.aspx

IMO. (2023). *IMO's work to cut GHG emissions from ships*. https://www.imo.org/en/MediaCentre/HotTopics/Pages/Cutting-GHG-emissions.aspx

IMO Environment Committee approves amendments to cut ship emissions. (n.d.). https://www.imo.org/en/MediaCentre/PressBriefings/pages/42-MEPC-short-term-measure.aspx

Luo, X., Yan, R., & Wang, S. (2024). After five years' application of the European Union monitoring, reporting, and verification (MRV) mechanism: Review and prospectives. *Journal of Cleaner Production*, *434*, 140006. DOI: 10.1016/j.jclepro.2023.140006

Mamatha, G., Gupta, A., & Koujalagi, A. Pooja, Pramanik, S., & Roy, A. (2025). *Applying Generative AI in a Global Language School to Improve Curriculum Planning* (pp. 151–184). IGI Global Scientific Publishing. DOI: 10.4018/979-8-3693-8744-3.ch007

MEPC 78/17/Add.1 Annex 12, page 1 I:8 78-17-Add.1.docx ANNEX 12 RESOLUTION MEPC.350(78) (adopted on 10 June 2022) 2022 GUIDELINES ON THE METHOD OF CALCULATION OF THE ATTAINED ENERGY EFFICIENCY EXISTING SHIP INDEX (EEXI). (n.d.). https://www.register-iri.com/wp-content/uploads/MEPC.35078.pdf

MEPC.328(76)-AMENDMENTS TO THE ANNEX OF THE PROTOCOL OF 1997 TO AMEND THE INTERNATIONAL CONVENTION FOR THE PREVENTION OF POLLUTION FROM SHIPS, 1973, AS MODIFIED BY THE PROTOCOL OF 1978 RELATING THERETO 2021 Revised MARPOL Annex VI. (n.d.). https://www.ccs.org.cn/ccswzen//specialDetail?id=202206220255211398

MEPC.352(78)-2022 Guidelines on Operational Carbon Intensity Indicators and the Calculation Methods (CII Guidelines, G1). (n.d.). https://www.ccs.org.cn/ccswzen//articleDetail?id=202212210561846752

MEPC.353(78)-2022 Guidelines on the Reference Lines for Use with Operational Carbon Intensity Indicators (CII Reference Lines Guidelines, G2). (n.d.). https://www.ccs.org.cn/ccswzen/articleDetail?id=202212210517258440

Mersin, K. (2020). A New Method for Calculating Fuel Consumption by Using Speed Loss Function. *International Journal of Environment and Geoinformatics*, *7*(1), 64–67. DOI: 10.30897/ijegeo.700557

Mishra, B. K., Roy, A. I, V. V., Maity, A., Kumar, S., Sheet, A., & Dutta, D. (2024). A Review of India's Renewable Energy Transition: Policies, Achievements, and Integration Challenges. *2024 10th International Conference on Electrical Energy Systems (ICEES)*, 1–7. DOI: 10.1109/ICEES61253.2024.10776919

Pal, A., Maity, A., Roy, A., Goswami, S., Samanta, N., & Bera, S. (2024). A Review of Renewable Energy Systems and IMO's Advanced Strategies for Marine Pollution Mitigation. *2024 10th International Conference on Electrical Energy Systems (ICEES)*, 1–7. DOI: 10.1109/ICEES61253.2024.10776828

Pramanik, S., Bag, S., & Roy, A. (2025). *Grading the Blockchain Using Sharding*. IGI Global Scientific Publishing. DOI: 10.4018/979-8-3693-3956-5.ch009

Pramanik, S., Bag, S., Roy, A., Ghosh, R., & Rakshit, P. (2025). *SMOTE vs. ADASYN*. IGI Global Scientific Publishing. DOI: 10.4018/979-8-3693-8714-6.ch004

Pramanik, S., Bag, S., Roy, A., & Rakshit, P. (2025). *Comprehensive Modernity of Fish Smoking Techniques Using the IoT*. IGI Global Scientific Publishing. DOI: 10.4018/979-8-3373-4332-7.ch006

Pramanik, S., Roy, A., & Rakshit, P. (2025). Resuming Transfer Technique for Two-Level Sharding Point of Rupture for a Single Table. In *AI and the Revival of Big Data* (pp. 221–238). IGI Global Scientific Publishing. DOI: 10.4018/979-8-3693-8472-5.ch010

Pramanik, S., Roy, A., Rakshit, P., & Bag, S. (2025). *Blockchain-Based Decentralised Management of Data for the Cellular Internet of Things*. IGI Global Scientific Publishing. DOI: 10.4018/979-8-3693-8664-4.ch011

Roy, A., & Chakraborty, M. (2024). Effects of ship emissions on Asian haze pollution, health, and IMO strategies. *Societal Impacts*, *3*, 100055. DOI: 10.1016/j.socimp.2024.100055

Roy, A., & Chakraborty, M. (2025). A review of ship emissions impacts on environmental, health, societal impacts and IMO's mitigation policies. *Regional Studies in Marine Science*, *81*, 103964. DOI: 10.1016/j.rsma.2024.103964

Roy, A., & Pramanik, S. (2024). A review of the hydrogen fuel path to emission reduction in the surface transport industry. *International Journal of Hydrogen Energy, 49*, 792–821. DOI: 10.1016/j.ijhydene.2023.07.010

Roy, A., Pramanik, S., Mitra, K., & Chakraborty, M. (2024). *Application of Neural Network for Reducing Emission and Optimizing Performance of Hydrogen with Biofuel CI Engine*. DOI: 10.1007/978-981-97-7532-3_26

Rules on ship carbon intensity and rating system enter into force. (n.d.). https://www.imo.org/en/MediaCentre/PressBriefings/pages/CII-and-EEXI-entry-into-force.aspx

Ship Energy Efficiency Management Plan (SEEMP). (n.d.). https://www.amsa.gov.au/marine-environment/air-pollution/ship-energy-efficiency-management-plan-seemp

Tran, T. A. (2020a). Effect of ship loading on marine diesel engine fuel consumption for bulk carriers based on the fuzzy clustering method. *Ocean Engineering, 207*, 107383. DOI: 10.1016/j.oceaneng.2020.107383

Tran, T. A. (2020b). Simulation and analysis on the ship energy efficiency operational indicator for bulk carriers by Monte Carlo simulation method. *International Journal of Modeling, Simulation and Scientific Computing, 11*(04), 2050036. DOI: 10.1142/S1793962320500361

Vision and level of ambition of the Initial IMO Strategy. (n.d.). https://www.imo.org/en/OurWork/Environment/Pages/Vision-and-level-of-ambition-of-the-Initial-IMO-Strategy.aspx

Wang, X., Yu, Y., & Lin, L. (2020). Tweeting the United Nations Climate Change Conference in Paris (COP21): An analysis of a social network and factors determining the network influence. *Online Social Networks and Media, 15*, 100059. DOI: 10.1016/j.osnem.2019.100059

World Maritime Theme 2022. (n.d.). https://www.imo.org/en/About/Events/Pages/World-Maritime-Theme-2022.aspx

Yang, L., Chen, G., Zhao, J., & Rytter, N. G. M. (2020). Ship Speed Optimization Considering Ocean Currents to Enhance Environmental Sustainability in Maritime Shipping. *Sustainability, 12*(9), 3649. DOI: 10.3390/su12093649

KEY TERMS AND DEFINITIONS

CII: This is an efficiency indicator that tracks the carbon intensity of a vessel in real time. The CII controls the actual or operational CO_2 emissions emitted by vessels. From January 1, 2023, all ships must disclose their CII annually, depending on exact oil usage.

DCS: All ships with a gross tonnage of 5,000 or more are mandated by statute to utilize data collection systems, or DCS, to track their usage of several types of fuel oil.

EEDI: EEDI estimates the amount of CO_2 the ship releases per capacity mile (tonne-mile) or CO_2 emissions/Tonne-Mile. It is a technical measure that may be used throughout the design and building stages of the ship. i. Required EEDI: The reference / required value is calculated by IMO-based criteria. This value will be the limit or target value the vessel can produce. The value changes depending on the ship type. ii. Attained EEDI: The actual value achieved by the vessel after it has been constructed or has undergone a major conversion. The value will be calculated during the sea trial of the vessel. Attained EEDI should always be less than the Required EEDI.

EEOI: Energy Efficiency Operational Indicator (EEOI) enables vessels to track greenhouse gases, which may be calculated by the fuel consumed on a particular trip transporting a specific quantity of goods.

EEXI: Existing ships are establishing minimum requirements for technical efficiency to reduce CO_2 emissions as of January 1, 2023, for vessels 400GRT and above.

EPA: The Environmental Protection Agency oversees environmental protection issues for the US federal government.

IEA: The International Energy Agency (IEA) is a Paris-based autonomous intergovernmental organization that provides data, analysis, and legislative proposals for the worldwide energy sector, with a recent emphasis on reducing carbon emissions and achieving global climate targets, including the Paris Agreement.

IGF Code: The International Code of Safety for Ships Using Gases or Other Low-flashpoint Fuels.

IMO: The International Maritime Organization (IMO) is a specialized body of the United Nations that manages transportation and shipping-related issues.

MARPOL: One of the most significant international maritime environmental treaties is the International Convention for the Prevention of Pollution from Ships.

MEPC: IMO has five committees, and they oversee formulating, examining, revising, and adopting the Organization's rules and regulations. Any issue under the organization's purview and associated with avoiding and regulating ship-based pollution will be considered by the MEPC committee.

SEEMP: The SEEMP is an operational approach that generates a framework for increasing a ship's fuel economy in a logic-based way. The fundamentals of developing a SEEMP for a specific ship involve estimating and determining the ship's current energy usage and putting various strategies in place to lower it.

SOLAS: The most significant international treaty about the safety of commercial ships is the SOLAS Convention. The SOLAS Convention's primary goal is to lay down minimum requirements for the design, equipment, and supervision of vessels consistent with their safety.

ENDNOTES

1 Conference of Parties 21: COP 21

2 A newly constructed vessel means that it has undergone a significant conversion that the flag state will consider a newly built vessel.

3 The CO_2 emission factor of the heavy fuel oil ships use is 3,114 kg CO_2/tonne; MEPC.1/Circ.684

Chapter 8
Smart and Sustainable Strategies for Urban Energy Efficiency:
An Integrated Framework

Sitesh Kumar Singh
https://orcid.org/0000-0002-7108-0808
National University of Science and Technology, Oman

Jaishri Gothania
https://orcid.org/0000-0002-2656-642X
Lingaya's Vidyapeeth, India

Michael Toryila Tiza
https://orcid.org/0000-0003-3515-8951
Federal Polytechnic Wannune, Nigeria

ABSTRACT

Due to the overall growing trend of urbanisation, cities are primary areas of energy consumption and carbon emission and therefore highly need smart, efficient policies to sustain their existence. The chapter gives a framework consisting of five pillars: smart infrastructure and data, policy ecosystems, innovation platforms, governance coordination, and citizens' engagement. It discusses the change in the urban energy system through examples of Sweden, Germany, the UAE, Saudi Arabia, and the UK. Their effects on technologies, such as AI, digital twins, edge computing, and vehicle-to-grid systems, are explored. The important values of just energy transitions are identified as equity, digital inclusion, and resilience. The chapter provides policy, budgetary and planning tips and gives future research recommendations, in

DOI: 10.4018/979-8-3373-2555-2.ch008

behavioural science, multisector integration, and mutable governance, and they are meant to inform long-term sustainable city energy policies, particularly in rapidly expanding Global South cities.

INTRODUCTION

Background and Significance

It is in the urban areas where global challenges about energy and the environment stand out. Today, over 55 percent of the world population resides in cities, and this has been forecasted to increase to about 70 percent by 2050 (UN DESA, 2019). Such densely inhabited spaces account for more than two-thirds of worldwide energy consumption and carbon dioxide (CO2) emissions, mostly because of increasing demands within the fields of transportation, construction, and buildings. Urbanization proceeds at an increasing rate and therefore the need to adopt innovative strategies which facilitate energy efficiency, but which also guarantee environmental and social sustainability.

Traditional urban energy systems were mainly focused on reliability, growth and not on sustainability or flexibility. They are usually not integrated and hence bring energy wastefulness, high cost of operation and environmental strain. Moreover, they are poorly placed to address the dynamism of the changing climatic conditions, energy deficiency and urban sprawl. To eliminate these constraints, cities will be required to shift towards adaptive, intelligent, and sustainable energy models (IEA, 2021).

The connection between smart technologies and strategy that is focused on sustainability presents an opportunity to transform (Das et al., 2020). Artificial intelligence (AI) and the Internet of Things (IoT) along with data analytics are digital innovations that can enhance efficiency in energy usage, responsiveness in real-time, and the overall performance of a system. These tools can dramatically decrease urban energy footprints, and when sustainable practices other than renewable energy integration, green infrastructure, and principles of a circular economy are built in, these tools can increase urban resilience and livability (Batty et al., 2012).

Defining Smart and Sustainable Urban Energy Efficiency

In urban context, energy efficiency can be defined as efficient usage of energy to provide services like lighting, transport, heating and cooling, with little wastage. This definition, however, is expanded in smart and sustainable cities, as the emphasis is put on the environmental and systemic effects of energy usage. Smart energy systems dynamically manage energy using sensors, automation and prediction analytics. The long sustainable methods are focused on the feasibility of a long-term functioning of energy systems by reducing carbon dioxide emissions, evoking fair access and increasing the sustainability of ecological balance.

Collectively, the two views give rise to a more comprehensive view of energy efficiency. This involves the use not just the deployment of energy-efficient technologies but also the redesign of urban forms, a behavioral change process and the incorporation of energy-related planning into wider planning initiatives. Such multidimensional methods enable cities, as observed by Ji et al. (2018), to reach efficiency, equity, and environmental quality simultaneously.

The Need for an Integrated Framework

Smart or green initiatives have been introduced in many cities over the last few years, but little has been successfully achieved in integrating the two spheres. Smart city initiatives tend to focus on technological advances without any sustainability measures and green infrastructure initiatives do not look at the digital side of optimizing and tracking. Consequently, the urban energy plans are divided and not as effective as they should be.

To make a significant change, one must have an integrated framework in place a framework that will have an inter-sectoral coordination, integrate stakeholders, and infuse the advantages of smart and sustainable approaches. This chapter puts forward an integrative approach to this and identifies the connectedness among energy systems, governance mechanisms and urban socioeconomics. It is possible to pursue relevant synergies by integrating both sustainability with digital strategies and vice versa, to enhance the negative and positive knock-on effects on the environment and economy (GhaffarianHoseini et al., 2013).

Objectives and Scope of the Chapter

The chapter will serve to build a broad sense of intelligence and viable ways of making cities' energy efficient. It has a theoretical and applied aim:

- To assess the situation in contemporary urban energy consumption.
- To investigate the options of digital technologies that optimize energy in real-time.
- To explore urban sustainability practices that see to it that less energy is used and produce resilience.
- Introducing a unified model that helps to integrate smart technologies and planning to preserve the sustainability of the city.
- To study the case studies and the best practices of cities worldwide.

- To recommend practical ways of action to policymakers, planners and engineers.

The chapter is context sensitive as well as global. Its use is supposed to be in the cities of different capacities, starting with the technologically advanced urban centers and ending up with the resource-limited municipalities. The framework is flexible, operational and has the capacity to facilitate incremental changes and radical energy transitions in urban environments.

Relevance to Global Development Agendas

One of the primary topics of urban energy efficiency is in the core of various global frameworks with the most well-known being the United Nations Sustainable Development Goals (SDGs). SDG 7 suggests that there is need to have affordable, reliable, sustainable, and modern energy for all and, on the other hand, SDG 11 proposes sustainable cities and communities. Urban strategies on energy are also important in meeting national agreements under Paris Agreement and other climate agreements (IEA, 2021).

The ever-growing popularity of initiatives like the European Green Deal, the U.S. Infrastructure Investment and Jobs Act, or national smart city projects in countries like India and China can be seen globally as a demonstration of the new need to integrate digital transformation and sustainable energy policies. To obtain energy justice, cities can harmonize such agendas in efforts geared towards improving energy rich growth, carbon minimization, and inclusive development.

This chapter has a benefit to a variety of sectors of knowledge: improving the course of knowledge as well as supporting the active promotion of sustainable urban energy transitions.

Methodology and Approach

The chapter adopts an interdisciplinary approach that is a multi-methodology that originates in the fields of urban planning, environmental engineering and information system and must be created through a public policy approach. It involves the following methodology:

- Systematic literature review of scholarly articles, technical reports and policy documents.
- Comparative case study approach of other cities across the world with best integrated strategies.
- The synthesis of conceptual framework with cross-sector knowledge.

- Determination of implementations barriers and enablers in various urban environments.

This way guarantees high-level evidence-based discussion which is theoretically correct and practically implementable.

URBAN ENERGY EFFICIENCY: CONCEPTS AND METRICS

Understanding Energy Efficiency in Urban Contexts

Urban energy efficiency can be defined as the best utilization of available energy resources in urban areas to provide services of heating, cooling, lighting and transport, but with low costs of energy wasted, economic expenditure and environmental degradation. Urban energy efficiency is more complex and involves interdependent systems of building, transport networks, city infrastructure, and utilities as compared to more general concepts of efficiency in industry or within households. The rising population density in urban areas, energy use, and ecological destruction necessitate a broader perception of efficiency inclusive of the resilience and sustainability concepts (Kennedy et al., 2010).

A significant part of consumed energy is spent in the modern urban environment, as cities make up around 75 percent of worldwide primary energy consumption and generate more than 70 percent of greenhouse gas emissions (IEA, 2021). A very wide range of variables needs to be addressed in any urban energy efficiency work, such as demand side management, technological upgrade, behavior changes and policy responses. Efficiency is not technically a metric; it is a social-technical issue on how cities are planned assembled and functioning.

Dimensions of Urban Energy Efficiency

Considerations of energy efficiency in urban areas can be dealt with in four main dimensions:

i. **Technological efficiency:** Increasing the energy performance of systems e.g., HVAC systems, LED lighting, electric vehicles, and energy-efficient appliances. It is all about reducing input by the same output.
ii. **Spatial Efficiency:** This refers to the impact of the land use, the urban form as well as the direction of the buildings in changing the flow of energy. Space-saving and multi-purpose construction is likely to minimize travel distances and heat loss, and urban sprawl increases energy consumption.

iii. **Operational Efficiency:** It means that there is optimization of energy usage in real-time mode, which is performed with the help of ICT, automation, and predictive analysis. Examples of such systems are smart grids, traffic management systems and building energy management systems.
iv. **Behavioral Efficiency:** It considers the effects on energy use based on the behavior of the people, the lifestyle and level of awareness. The most advanced technology may fail to perform well compared to where the behavior of the user is not geared towards efficiency objectives.

These dimensions signify systems thinking in relationship to energy consumption in urban areas in that interventions cannot be solved but should be carried out comprehensively.

Key Metrics and Indicators

To assess the energy efficiency of cities, there is a need to have strong indicators around which planning, tracking, and policy making can be addressed. The commonly used metrics that are depicted in Fig ure 1.

Figure 1. Key energy efficiency metrics and indicators

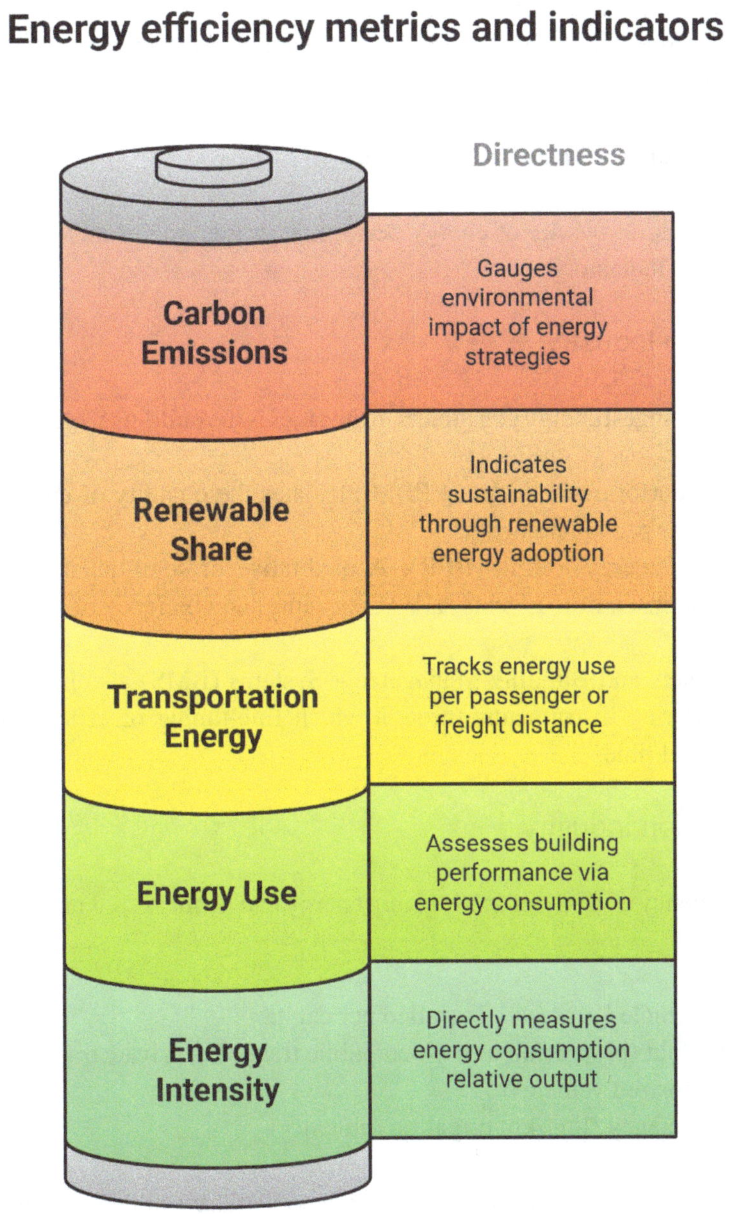

The data availability issue should not be the only challenge but also the standardization of data. Cities can also carry out reporting using different means making it hard to cross compare and benchmark cities. In its turn, different global organizations, like the International Organization for Standardization (ISO) and the World Bank, have introduced the idea of the standardization of urban energy indicators to introduce consistency and comparability (OECD, 2018).

Sector-Specific Efficiency Metrics

To have a closer picture of energy performance, it is important to focus on the sector-specific indicators:

Construction Industry

One of the biggest energy consumers in the cities is the buildings. Metrics include:

- Energy Performance Index (EPI): Calculates the quantity of energy needed every year per square meter.
- Building Energy Rating (BER): A qualitative or semi-quantitative rating, based on insulation, glazing, HVAC systems and efficiency of lighting.

Smart meters and Building Management Systems (BMS) also include performance monitoring instruments, through which fine-tuning of efficiency can be achieved in real time.

Transportation Industry

The efficiency of a city in transporting energy can be measured in the following ways:

- VKT: Vehicle Kilometer Travelled per capita
- Modal Split (proportion of trips on public transport, private transportation or non-motorized vehicles)
- Fuel Efficiency (liter/km or gallon /miles)

Electric mobility and the connected vehicle system have added new dimensions of efficiency into the mix; these involve the number of charging stations, the vehicle-to-grid (V2G) interactions.

Services and Infrastructure Urban

One of the usually neglected elements of energy audit is the public lighting, water distribution, and waste management systems. Such metrics as:

- Energy Consumption per liter of the water being pumped
- kWh/ Ton waste processed
- Public Lighting Lumens per Watt

can give information on the efficiency of municipal services.

Integrated Efficiency Appraisal Models

Several structures exist to help cities to approximate and improve energy effectiveness. They include some of the following:

- **ISO 37120:** Sustainable Cities and Communities indicators: Includes such indicators as energy problems, environmental issues, transportation and infrastructures.
- **LEED Cities:** A modification of the Leadership in Energy and Environmental Design (LEED) rating system scale to provide a measuring system of city performance.
- **Smart Urban Maturity Model:** It targets the determination of how well a city is integrating ICT as a way of making energy and other city operations leaner.

The measures are multidimensional in nature of a breadth of measure to evaluate the urban energy systems in a balanced image of economic, environmental and social performance.

Measurement and Evaluation Problems

Determining the level of urban energy efficiency is not an easy task:

- **Storage of Data Gaps:** There are also several cities whose data lack precision, disaggregation, or real-time view, particularly in Third World countries.
- **Silos:** Institutional Silos, more specifically, can hinder end-to-end collection of data by departments.
- **Temporal Mismatch:** Efficiency often takes place in long parts but political or investment situations are temporary.

- **Equity Blind Spots:** Metrics often focus on aggregate efficiency and may overlook distributional impacts, e.g., how benefits and burdens are shared among different communities

To resolve them, cities are growingly resorting to online systems, citizen science programs, and open data policies as the means of making the processes more transparent and involving stakeholders (Kennedy et al., 2010).

Future Directions for Metrics and Benchmarking

Newer practices point to the development of more dynamic, adaptive and real-time measures. Some of the significant developments are presented in Figure 2.

Figure 2. Future developments in energy metrics and benchmarking

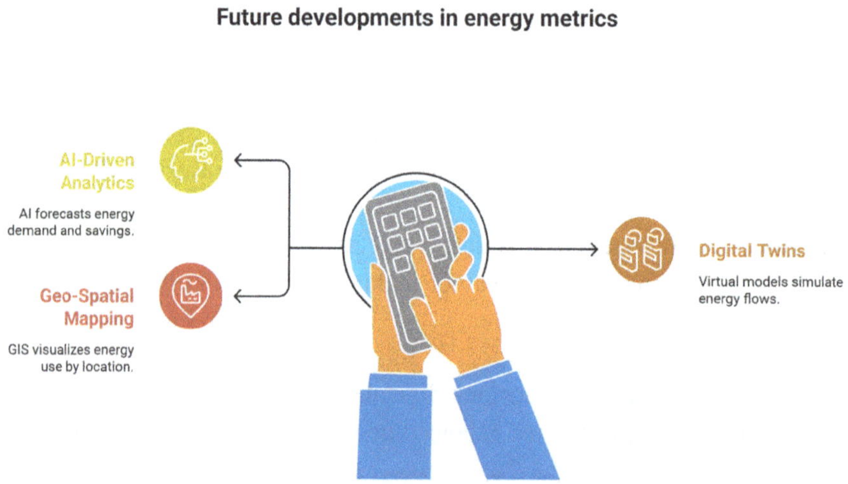

Besides, ethical and inclusive methods of efficiency evaluation are becoming popular. This means creating indicators which show not just technical performance but also access, affordability and social equity.

Conclusion

The energy efficiency of cities is a multidimensional notion as it concerns not only energy savings. It entails integrating technological, spatial, operational and behavioral variables into a complicated city system in a tactical way. Strong, sector-related and standard performance measures are paramount in measuring, determining goals and intervention strategies. Technical sophistication, however, needs to be complemented by institutional ability and cooperation of stakeholders to work. The contribution of dynamic, real time and inclusive energy metrics will increase as cities become smarter and more sustainable, and these metrics become vital in the process of their transformation.

SMART TECHNOLOGIES FOR URBAN ENERGY OPTIMIZATION

Introduction

The field of urban energy systems is becoming very keen on adopting smart technology to meet the current need of sustainability, efficiency, and resilience. Literature distinguishes traditional energy infrastructures that feature unidirectional flows and reactive control mechanisms with intelligent systems that can perform real time monitoring, adapting to the circumstances and predicting the future. By smart here means the utilization of digital technologies including the Internet of Things (IoT), Artificial Intelligence (AI), and Big Data analytics, as well as cyber-physical systems to enhance how, when, and why the urban energy flows operate, get controlled and planned. Specifically, this section delves into the main Smart technologies that can help in the process of optimization of energy use and their relation to larger objectives of smart and sustainable cities.

Internet of Things (IoT) and Sensor Networks

Smart energy systems used in cities are based on the Internet of Things (IoT). Cities can now capture data on energy use, temperature, humidity, occupancy, light levels and air quality continuously by placing sensors in buildings, transportation systems, grids and in public infrastructures. These data facilitate energy managers to take informed decisions, optimize the consumption pattern and even detect the wastage of energy.

IoT-controlled smart thermostats, occupancy sensors and lighting systems used in buildings can cut down wastage in energy as they dynamically adjust settings and remind the user to turn off equipment or adjust settings to address external factors.

Smart meters allow two-way communication between the consumers and the providers of the utility network and allow demand side management, which enables time-of-use pricing (Madakam et al., 2015).

Smart Grids and Microgrids

The development of smart grids is a once-in-a-lifetime changing paradigm of urban electricity grids that shift them to a decentral (interactive) system. The use of digital communication, automation and control technologies makes up a smart grid, aiming at improved sustainability, reliability and efficiency of electricity distribution systems.

Important aspects of smart grids as shown in Figure 3 and key IoT components and functions in Urban Energy Systems are presented in Table 1.

Figure 3. Key features of smart grid

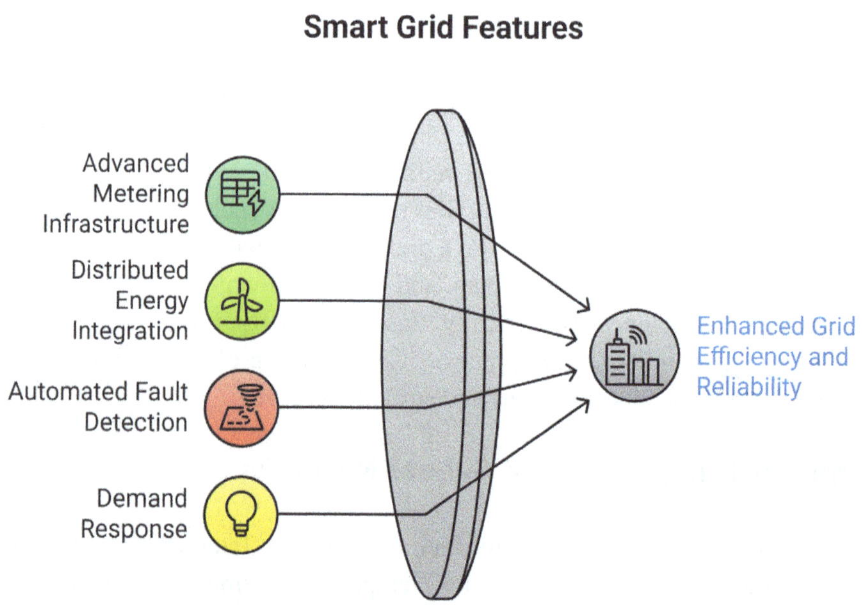

Table 1. Key IoT components and functions in urban energy systems

Component	Function	Example Use Case
Smart Meters	Real-time usage monitoring	Demand-side management
Occupancy Sensors	Detect presence for automation	HVAC & lighting control
Environmental Sensors	Monitor temp, humidity, air quality	Adaptive building control
Grid Sensors	Monitor voltage/load flow	Smart grid fault detection

Microgrids are local smart energy networks, which supplement wider smart grid infrastructure. They either work independently, or as part of the main grid, and this makes them especially useful when it comes to energy resiliency of critical infrastructure (e.g., hospitals, data centers) or underserved communities. Widely used elements in microgrids are renewable energy, storage and AI-based control algorithms that are used to optimize the energy delivery at the local level (Almihat & Munda, 2025).

Building Energy Management Systems (BEMS)

Building Energy Management systems (BEMS) presented in Figure 4 are hardware and software combinations that allow the monitoring of buildings, control of energy consumption, and optimization of the consumption. They form the core of smart building activity and are an important part of urban energy efficiency plan.

Figure 4. Components of building energy management systems

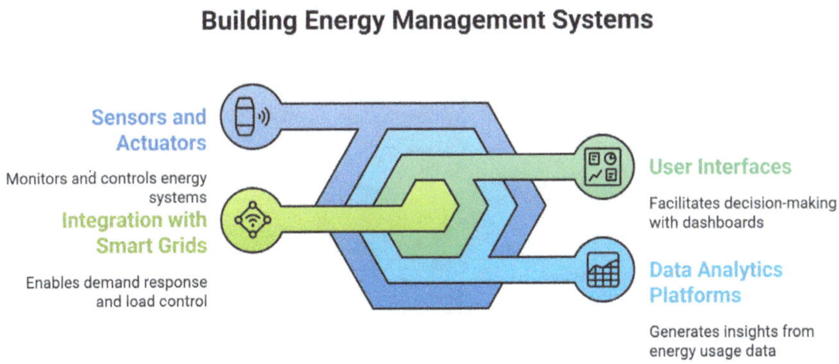

Advanced building energy management systems use AI and cloud to offer predictive control to adjust an energy system depending on the predicted weather, occupancy, or tariff variations. Such systems may lead to 15-30 percent energy savings in office buildings and to a substantial drop of peak demand loads (Wang & Xu, 2020).

Moreover, BEMS can help improve occupant comfort, and indoor environmental quality (IEQ) to provide co-benefits health and productivity. BEMS scalability across individual buildings to whole campuses or even districts ensures that they offer a flexible tool in terms of urban energy optimization.

Smart Mobility and Electrification

Cities are the largest consumers of energy and cause of emissions, mostly through transport systems. Not only is the industry being influenced by smart technologies, but this is encouraging electrification, automation, and coordination of intermodal transport (Singh & Jain, 2025). Smart charging infrastructure helps electric vehicles (EVs) to become independent of fossil fuels and generate flexibility in energy. Important technological advances are presented in Figure 5.

Figure 5. Key technological for development of smart mobility and electrification

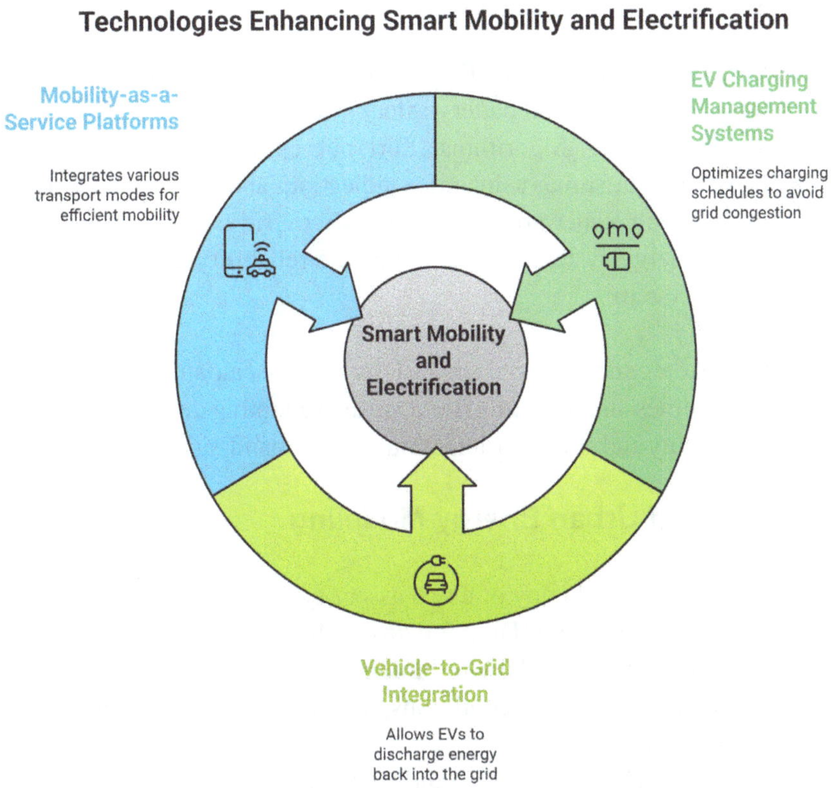

Artificial intelligence and IoT-based traffic management frameworks assist in utilizing the traffic networks, minimizing idling, and increasing fuel efficiency. Such systems watch traffic as it moves and either toggle signals or reroute or even buff tolls to ease traffic. All these technologies have worldly benefits in terms of energy preservation as well as an enhancement in air quality and decrease in the urban heat island effect (Elassy et al., 2024).

Energy Storage and Optimization Technologies

Storage plays a very crucial role in balancing the supply of invariable renewable energy and linking supply to demand. The technologies to consider most related to the city energy resilience systems are lithium-ion batteries, pumped hydro, compressed air, and new emergent solid or flow batteries.

The smart energy storage systems include:

- Battery Management Systems (BMS): Maximize the charging and discharging made to battery life and battery safety.
- AI-Based Scheduling Algorithms: Sort out optimal and least expensive charging/discharging times using price indications and projections of demand.
- Combination with Renewable Power and Microgrids: Incorporates the linking of local energy systems that can increase reliability and logistics of the local energy system.

Urban energy storage does not only mean large utility-scale storage. Increasingly, buildings, communities and transport fleets are also adopting decentralized storage to maximize flexibility and allow participation at the demand-side of energy markets.

Digital Twins and Urban Energy Modeling

A digital twin is a virtual copy of a real asset / system that uses real time measurements to model its behavior. Digital twins in the urban energy domain would help planners and operations model how buildings, energy networks, and transport systems could perform in various conditions, and thus are an essential element in planning and operations.

Combined data of IoT devices, GIS, and BEMS allow digital twins to support:

- **Scenario Analysis:** It tests policy intervention or deployment of technology.
- **Predictive Maintenance:** Inspection of faults at the point of failure prevention.
- **Urban Planning:** It encourages efficient use of land and design of infrastructure to be energy efficient.

Examples of cities which experiment with digital twin technologies include Singapore and Helsinki, using them to manage energy use and model the carbon footprint and infrastructure investments. The digital twin will play a more important role in smart urban energy management as computing capabilities and the availability of data increase (Batty, 2018).

Challenges and Considerations

Although the potential of smart technologies is massive, the implementation of these technologies is not free of problems:

- **Interoperability:** An interoperability issue is a concept that allows diverse systems and devices to interact and become optimized in an integrated way.
- **Cybersecurity Risks:** More connections will make any system vulnerable to cyber-attacks.
- Heavy upfront investment- Some of the smart systems have high original costs that can engage users.
- **Digital Divide:** Social inequalities can be increased where different people have unequal access to the digital infrastructure.
- **Data Governance:** the topics of data ownership, privacy, and ethical use are becoming vocal.

To deal with such issues, strong regulatory measures, private-public collaborations and planning involving everyone in a way that considers equity and resilience of sustainability in the long run are a must.

Cyber security plays a key issue in the ever-digitized urban energy systems. The vulnerability gives rise to malicious attacks on smart grids and devices, IoT, and AI-enabled energy platforms. The U.S. National Institute of Standards and technology (NIST, 2014) has produced the Smart Grid Cybersecurity Framework to assist cities in determining and addressing these hazards. Among the areas mentioned, decryption of data, real-time intrusion detection, and decentralized fail-safe protocols are the most important ones. With an awareness that new and innovative conditions must be tested in a controlled manner, some cities such as Singapore and Amsterdam have been experimenting with regulatory sandboxes, controlled environments where experimental new digital energy technologies can be given a test run with reduced compliance requirements, with close oversight and cybersecurity due diligence.

Conclusion

Energy use in cities is changing with the smart technologies that are transforming this process. By combining IoT, AI, smart grids, BEMS, and digital twins, cities may make the shift to more efficient, sustainable, and resilient energy systems. The benefits of these aspects are, however, dependent on considerate design, governance and implementation of these aspects bearing in mind the technical, economic and social realities. To continue the trend, cities need to adopt their thinking into a systems approach whereby technology-based solutions provide positive inputs towards energy optimization and the global aspirations of sustainability.

URBAN STRATEGIES FOR SUSTAINABLE ENERGY

Introduction

The urban energy systems need a radical overhaul that would help bring city energy systems in line with climate goals even as energy demands and the quality of life in cities continue to rise. The most common centralized infrastructures based on the use of fossil fuels rapidly become insufficient to provide cities with energy security and environmental performance. The movement towards sustainable energy models will have to entail both a critical systemic reconsideration of the supply dimension of energy, and the demand side of energy anchored in the integration of renewable energy, demand-based actions, the deployment of high-performance building strategies, as well as green urban planning. This part talks about these four pillars and their impact on the future of promoting sustainable energy in the urban environment.

Integration of Renewable Energy in Urban Areas

The built environment opens special possibilities regarding renewable energy installation even in the face of both spatial and infrastructure constraints. Solar photovoltaics (PV) is a large-scale solution, especially rooftop and building-integrated system, that has become prominent in cities because it is modular, increasingly cost-effective, and has low operational demands. More limited in saturated urban places, wind power can also be adopted by using micro-turbines and wind turbines with a vertical axis, which are more adapted to turbulent winds. Solar and wind hybrid systems have the advantage that they work well in cities systems by ensuring energy stability partially due to resource complementarity.

In one of the studies, it is stated that the integration of building-integrated PV and vertical-axis wind turbines (BIPvWt) would greatly enhance energy production in global cities and urban regions, particularly when supplemented with high-rise buildings (Lee et al., 2017). Besides, the biomass-to-energy plants based on organic household waste, fine-scaled geothermal installations contribute to the city renewable bucket, particularly within the context of district heating. Although this is the case, regulatory and financial aspects of renewable deployment, including the limitation of grid interconnections, start-up cost of capital, and acceptance of renewable by people, must be tackled by cities with special incentives and policy leeway's.

Demand-Side Management and Behavioral Interventions

Demand-side management (DSM) plays a significant role in the need to have not only low-carbon urban energy systems but also responsive and efficient systems. DSM describes a range of actions whose purpose is to change the consumer demand patterns, also by shifting loads, shaving peaks, and energy conservation or behavioral interventions usually through smart measures. Intelligent thermostats, time-of-use utilization fees as well as energy-efficient home appliance incentives are becoming increasingly common to lessen peak loads and enhance grid stability.

Behavioral change programs, such as consumption in real-time, gamification, and energy-saving competitions have been quite promising, as well. In the developing countries' situation, the DSM strategies can be very effective in areas of limited centralized supply. Research on DSM program in Bangladesh reveals that incentivized and education outreach buildings in Bangladesh can help to reduce significant peak demand, particularly with the assistance of the smart-metering infrastructure (Khan, 2019).

Technological aids present in the form of smart grids and building energy management systems (BEMS) provide an opportunity to automate the strategy of DSM, balancing the loads of HVAC according to occupancy or the prices. Nonetheless, successful execution remains in engaging the people, coordination within the institutions, and definite regulatory injunctions.

Net-Zero and Energy-Positive Buildings

The idea of net-zero energy buildings (NZEBs) and energy-positive buildings (EPBs) gains its primary role in the notion of sustainable urban development. NZEBs are such that they only use the amount of energy which they produce on-site using renewable sources per annum. EPBs even take one step forward and produce excess energy that can be sold to the grid or be used in local microgrids.

According to Singh, Toryila, and Alasali (2025), the energy performance of urban buildings may be improved by integrating smart digital technologies with the principles of sustainable design and material selection. Their model proves that the idea of implementing renewable materials and AI-based control mechanisms results in considerable changes in terms of energy efficiency and carbon reduction. To become more widespread, municipalities may assist NZEBs and EPBs with performance-oriented building codes, financial incentives, and public-private partnerships in invention.

Passive Design, Green Infrastructure, and Urban Form

Non-technological approaches (passive design and urban morphology) are also essential in the sustainability of urban energy. Urban buildings and districts impact greatly on heating, cooling and lighting loads by means of their orientation, shape and layout. Energy intensity can be quickly decreased using passive solar concepts (including maximization of window to wall ratios, shading devices, and employing the thermal mass).

Green infrastructure in the city environment such as green roofs, planted facades, city forests, and permeable street surfaces have two functions: alleviating heat island effects, and ecosystem services. Such interventions not only decrease cooling energy requirements but also boost stormwater management and city biodiversity.

Recent research indicates that the ability to connect renewable technologies and the distributed energy system is very much dependent on the urban form. To illustrate the point, the photovoltaic potential in compact urban designs being well-thought-out in terms of solar orientation might be twice that of the sprawling designs (Perera et al., 2019). The tools of spatial planning, which include the use of GIS, simulation models of solar radiation, and climate-sensitive urban planning can play an essential role in aligning physical development with objectives of energy efficiency.

There is a tight interdependence between urban energy efficiency and water systems that has been frequently referred to as energy-water nexus. The pumping, treatment and distribution of water demands a lot of power in urban areas, and water is an essential component in generating energy and cooling (Namany et al., 2021). The field that is becoming synergistic dissemination is the recovery of low-grade heat by wastewater that can be used in district heating. Such cities as Vancouver and Oslo have tested wastewater energy transfer systems which employ heat exchangers and heat pumps in the recovery of energy within the sewage to be utilised in the heating of homes showing potential reductions of up to 30% in heating energy requirements. The incorporation of these systems into the urban planning also helps diminish the carbon footprints, not forgetting to promote infrastructure circularity.

Besides green infrastructure, other passive cooling measures in mitigating Urban heat island (UHI) are cool roofs, i.e., the roof covering with high albedo materials reflecting the solar radiation as well as facade retrofits, e.g. consider dynamic shading structures or ventilated walls. Pervious pavements that render water percolate, can cool down surface temperatures alleviating stormwater management. Mediterranean and tropical cities have more successfully used strategic exploitation of urban canyons to maximize airflow and thermal comfort. When applied together, such techniques and those that rely on the use of vegetation provide a multi-layered shield against the impact of UHI and overall energy economy of cooling systems.

Barriers and Enablers

There are some impediments towards the implementation of urban sustainable energy policy. They are old and out of date building requirements, uncoordinated rules, expensive installed costs of renewable technology, and the opposition of people towards changes in behaviors. What is more, data interoperability and the digital divide can limit the success of smart systems, at least in low-income communities.

However, the transition can be fast tracked with the help of several enablers. Some of them are revised performance-based codes, expedited permitting of renewable installations, incentive-driven DSM policies, and planner-engineer capacity building. Novel modelling technologies such as Building Information Modeling (BIM) and digital twins will provide the cities with the ability of designing, simulate, and optimizing complex energy systems prior to their physical construction. According to a survey by Jurasz et al. (2019), it is necessary to combine different yet complementary sources of energy, energy planning tools, and stakeholder views because, in this way, the dynamics of the urban energy transition are characterized by the issues of adequate resilience and efficiency.

Conclusion

Sustainable energy urban policies should be inclusive, meaning that technological development should be combined with city planning, urban decision-making and community participation. Each of the strategies are renewable energy penetration, demand-side management, smart and energy-positive structures, and passive design must provide meaningful options toward minimizing energy intensity and environmental effect in urban regions.

These solutions, however, work best when they are implemented in concert in line with a vision for the whole and reinforced by suitable policy and when they are locally rooted. The following modification will introduce a synthetic structure that can help cities to coordinate these tools to a unified and flexible energy transition.

BUILT-IN STRUCTURE OF CITY ENERGY EFFICIENCY: AN INTEGRATED FRAMEWORK

Introduction

Incorporation of energy efficiency into urban systems involves phasing out the implementation of fragmented actions on energy efficiency to an integrated, dynamic framework that aligns infrastructure, governance, technologies, and social behavior. Individual efforts fail to take synergies into consideration, which are essential in scaling. With the input of both urban planning and energy governance, a systems-oriented framework will enable piecemeal actions to turn into large-scale resilient impact.

Theory: Systems Thinking and Integrated Planning

This framework is founded on systems that view cities as a changing interconnected web of systems that have feedback loops and emergent properties. To keep both the energy policies and land use, mobility, infrastructure, and social development in harmony and sustainable in the long-term, integrated planning would be maintained. Such globalized thinking will avert the occurrence of unintended consequences and drive resource efficiency (Singh et al., 2025).

An effective perspective to frame this combined framework is the Multi-Level Perspective (MLP) on socio-technical transitions since change, caused by the interactions among niche innovations, socio-technical regimes and landscape pressures in the society (Geels, 2002). Although MLP concentrates a great deal on path-dependencies and structural inertia, the framework that is proposed accentuates modular flexibility and co-creation among stakeholders (at the level of infrastructure intelligence and all the way to negotiation of citizen behavior). In contrast to conventional transition theory, it involves real time data, feedback systems, institutional interoperability to intervene, to achieve incremental and radical energy transitions in the urban ecosystems.

Essential Components of the Unified framework

Figure 6 shows the key components of the integrated structure of city energy efficiency built in framework and their descriptions as follows:

Figure 6. Core elements of integrated framework for built-in structure of city energy efficiency

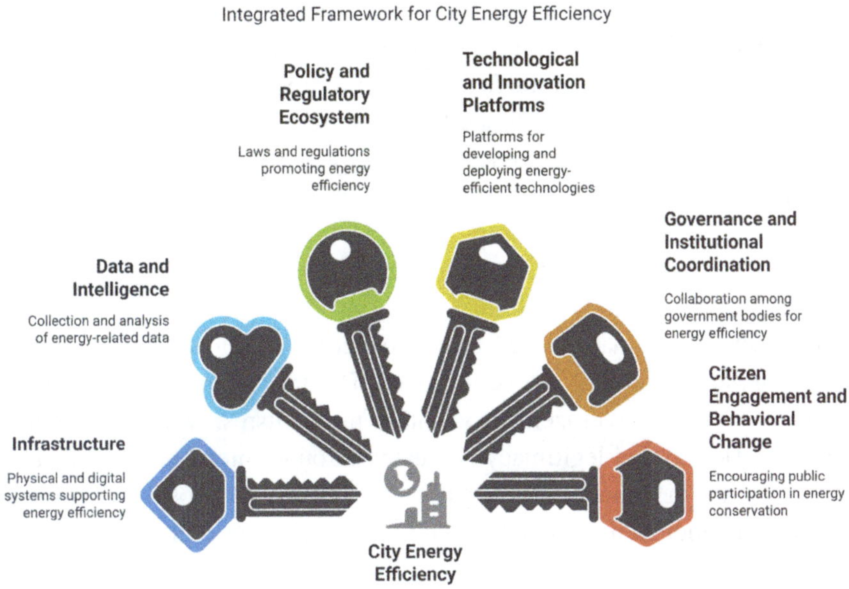

Infrastructure, Data, and Intelligence

An intelligent infrastructure is the core, with an inclusion of sensors, smart meters and digital twins as well as real time data analytics. Simulation of energy scenarios, anomaly detection and predictive optimization, a feature offered by digital twins now used in hundreds of cities, allows optimization of resilience and better decision making (Reuters, 2024). Interoperability and privacy needs should have robust data governance.

Ecosystem of Policy and Regulations

The regulatory scaffold is composed of performance-based building codes, energy disclosure requirements, green purchasing regulations and zoning aligned to climate change. The key enablers include streamlining the process of permitting distributed energy, harmonizing municipal plans with international climate targets (e.g. SDG7, SDG 11), and facilitating climate finance.

Italian Technological and Innovation Platforms

Important innovative platforms critical to piloting solutions, such as modular renewables, and predictive lighting, are public private testbeds. As an example, the urban twin of Sydney uses several data layers in a manner that models energy consumption, emissions, and traffic, contributing to more strategic sustainability planning (Sohail et al., 2024). Digital twin-powered energy centers will allow the organization management of multisector energy resources at the neighborhood level (Dulaimi et al., 2022).

Institutional, Institutional Coordination Governance

Cross-sector coordination, e.g., energy offices working with utility, transport, housing, and civil societies are very crucial. Digital forums and co-design platforms are ways to make citizen participation mechanisms, where the community values are reflected and legitimacy is created through interventions. The reforms of eco-efficiency in China have concentrated massively on boosting environmental co-production among citizens and the government (Zhang et al., 2025).

Citizen Engagement and Change of Behavior

Active citizen participation is necessary in sustainable impact. The involvement of the community feedback in digital twins (e.g., Bologna Civic Digital Twins) raises the level of transparency and encourages the shared decision-making (Luca et al., 2024). Examples of engagement tools include local energy dashboards, dynamic pricing and energy challenges to promote behavior change and collective action.

Adaptability and Scalability

The framework is scalable and fits within the capacity of the cities:

- Advanced cities have an opportunity to implement AI-twined cities and grids.
- Passive design and simple DSM can be started by resource-constrained cities.

The modular structure also enables an iterative development process, so with the introduction of updated policy, to pilots and potential scaling of activities according to the data findings and the alignment of the activities with the climate goals through performance indicators.

Case Illustration: A Mid-Sized City Scenario

A city that wants to reduce its emissions by 40 percent over a decade may implement the framework as follows:

- **Infrastructure:** school and whole city halls sensor networks.
- **Policy:** updated green roofs construction codes and solar-ready construction code.
- **Technology:** digital twins showing the energy flows in the neighborhoods.
- **Governance:** an interagency task force that also manages energy activities.
- **Citizens:** competitions, education campaigns and creating energy dashboards that are open to the public.

With the difficulties it has experienced, adopting an integrated approach, avoiding the pairing of elements, the city establishes resilience, investment, and becomes a role model in implementing solutions applicable elsewhere.

Implementation Challenges and Mitigation

Barriers:

- There are silos in the institution that limit coordination.
- Investment obstacles are caused by funding constraints.
- Digital divides can leave out the lower-income bracket.
- Behavioral resistance is a hindering force to culture adoption.

Enablers:

- Multi departmental task teams or innovation labs.
- Blended financing e.g. green bonds or energy-as-a service model.
- Integration into digital platforms and capacity building.
- Evidence-based communication to create trust among the masses.

Toolkits for Launching a District Microgrid and Performance-based Codes

Toolkit: Steps to Launch a District Microgrid

a. Conduct feasibility study & stakeholder mapping

b. Identify anchor loads (e.g., hospitals, schools)
c. Secure regulatory permissions and utility coordination
d. Choose DERs (solar, storage, etc.) and control platform
e. Define business model (PPP, ESCO, cooperative)
f. Pilot, monitor KPIs, scale iteratively

Toolkit: Drafting Performance-Based Energy Codes

a. Set outcome metrics (e.g., $kWh/m^2/year$)
b. Align with climate targets and building typologies
c. Include compliance tools (e.g., energy modeling)
d. Ensure update cycles and enforcement protocols

Conclusion

The integration of energy efficiency into urban systems will necessitate coordination between physical infrastructure, policies tools, innovative platforms, coordination of the institutions, as well as the actions of the people. This comprehensive system outlines a direct plan to help cities to evolve out of piece-meal initiatives to make more systematic, flexible, and resilient change.The implementation of this framework enables cities to provide consistent, just, and sustainable energy systems. The second section will demonstrate how it is used, presenting the world case studies and discussing successes and troubles with practice.

CASE STUDIES AND GOOD PRACTICES

Introduction

To illustrate the practicality of the Integrated Framework (Section 5) it is necessary to investigate cities and neighborhoods in which the sustainable energy strategies have been successfully operationalized. This part contains four examples of case studies covering ambitious urban districts to low-resource city environments which illustrate the different enablers, innovations, and difficulties. All the cases illustrate the effectiveness of the pillars of the framework that are implemented in a precise way to make a difference.

Hyllie, Malmö, Sweden

Hyllie is a representative of a district-level deployment that is based on the idea of public-private partnership. The 25,000-populated development, initiated in 2010, is based on the integration of energy systems (solar, wind, biogas, waste heat, and industrial thermal) and matching them through the system created by E.ON, ectogrid, and ectocloud (digital) (Sandström et al., 2025). Smart control makes excess heat redirect, storage to be optimized and thermal response to be demand responsive. This coincides with our framework pillar of infrastructure and intelligence and is a pilot innovation platform (Sohail et al., 2024).

The sustainable components of infrastructure and governance that Hyllie integrated in their early stages portrays how integrating energy systems into the urban planning at the beginning stages is worth consideration. The district also demonstrates interaction with the citizens, where the microgrids and education programs are clearly seen, providing support and confidence in the community.

Bahnstadt, Heidelberg, Germany

The Bahnstadt is the largest neighborhood in the world that is carbon-free, which is a layering of energy efficiency guidelines, clever meters, green roofs, and low consumption building designs (Time, 2024). Buildings can reduce heating energy by up to 80% compared to conventional districts in the cities and smart meter information provides real time feedback and community driven optimization.

These three pillars of the framework (AML-enabled infrastructure, policy and regulation (building standards, permitting), and citizen engagement especially community access to energy usage data) fit this case. Bahnstadt was designed in such a way that it could replicate the same idea in other neighborhoods since it is modular and scalable.

Abu Dhabi (Masdar City and Broader Reforms)

The Masdar City project of Abu Dhabi portrays the weaknesses and advantages of prototype eco-cities. The district includes passive cooling measures, Solar PV, and smart sensors in an urban testbed. Nevertheless, limited funds led to delay and a limitation in scale, which was the case and entailed the need to have a solid innovation platform and a grown governance model (Ajaj, 2024). Critiques aside, it is on Masdar where relevant pilot projects and green building innovation can still be found. Such initiatives, as Singh et al. (2025) point out, reveal both the potential and the constraints of an urban energy innovation driven by prototypes, yet it is not thoroughly incorporated into the citywide municipal governance and funding systems.

The wider program of One Less Nuclear Power Plant in Seoul can provide contrast here: it shows a possible impact of integrated policies, citizen participation, and DSM programs on an entire city, making it energy independent and thus less dependent on centralized power generation, allowing citizens to share energy with others (Seoul Metropolitan Government, 2023).

The cases present the need to interlink pilot creativity and broader policy landscapes and institution integration to achieve widespread accomplishment at the city level.

Low-Income & Emerging City Contexts – Jizan, Saudi Arabia

The pilot project of Al-Khaledia in the city of Jizan adopted urban computing technology to enable a Smart and Sustainable Urban Neighborhood Framework (SSUCF), by combining data monitoring, municipal indexes, and SDG-based indicators (Aina, 2023). It is possible to present a scalable model of data-driven decision-making, despite the obstacles to data privacy, capacity, and interoperability.

This shows that the aspects of the framework of infrastructure and intelligence and coordination of policy can work even when the conditions are low capacity, in the event of institutional commitment and adaptive planning.

Greenwich, London – District Energy Hub

Greenwich can be considered as an example of multi-vector assets such as intelligent control systems. Communal heating and EV charging were also controlled in terms of a digital twin, streamlining operating processes across technical boundaries (O Dwyer et al., 2020). The case promotes both technology platform and governance coordinates pillars in terms of open-source architecture to embrace energy system interoperability and local control.

Synthesis of Key Lessons

Mapping the cases onto the integrated framework, there are common success enablers as reflected in Table 2:

Table 2. Mapping of integrated framework pillar with various case studies

Framework Pillar	Malmö	Bahnstadt	Masdar	Jazan	Greenwich
Infrastructure & Intelligence	✓	✓	✓	✓	✓
Policy & Regulatory System	✓	✓		✓	✓
Technology Platform	✓	✓	✓		✓
Governance & Coordination	✓	✓		✓	✓
Citizen Engagement	✓	✓			✓

- Adaptive and scalable energy systems are made possible by innovative platforms and digital infrastructure, i.e., district digital twins (London), solar-thermal hub (Malm), and smart-energy neighborhood.
- Uptake based on policy instruments such as building codes and permitting reforms sets the basis on a system wide uptake.
- Mechanisms of governance such as interagency coordination are put in place and such mechanisms make sure that different stakeholders converge towards a common energy agenda.
- Adoption and long-term success depend on community engagement which must be aided by feedback technologies and all-encompassing planning.

The following two examples offer short vignettes showing how cities overcame institutional inertia to adopt one or two key smart-energy measures.

- Vignette: Citizen-level energy audit in Seoul

Seoul was able to minimize household energy consumption by 12 percent in pilot neighborhoods through neighborhood energy councils and digital dashboards to circumvent resistance through gamified savings campaigns.

- Vignette: The Digital Twin deployment in London

Simulating EV charging and heating conditions to power homes and businesses, the cross-departmental energy task force deployed the city applied an open-source twin of long-established silos between transport and energy departments.

Barriers Encountered

The international cases share such common challenges as:

- **Financial sustainability:** The fact that Masdar took so long to deploy demonstrates that the prototype projects require their own financial backing and monetization.
- **Technical interoperability:** Information and systems must exchange information; open standards and APIs are essential.
- **Change with time passing:** Bahnstadt and Hyllie illustrate the value of delayed planning and living labs to adapt a strategy.
- **Equity issues:** Equity issues necessitate governance considerations on fairness and accessibility, as inclusion of underserved communities as Jizan has shown.
- **Community engagement failures case:** In Chicago, a smart street lighting modernization project was not a welcome idea among the people there since they failed to communicate to the community well- things were misunderstood and they had concerns of being spied when using the bulbs, and their employment would be rendered redundant. Its implementation was put on hold and eventually reworked including town halls and privacy assurances.

Conclusion

The cases certify that integrated city energy efficiency frameworks can be effective in various geographic, economic, and governance environments. Successful implementations use a combination of top-tech solutions, compliance, harmonization of stakeholders, and citizen participation.

In the following section, lessons will be extracted based on these case studies to clearly describe the implementation guidelines as well as policy recommendations, depending on typologies of cities.

POLICY AND FINANCING

Introduction

To achieve its goals regarding efforts to transform to a world of sustainable energy within urban settings, there is the need to have powerful policy frameworks as well as innovative ways of financing. Although technical capabilities have become very important, they must be supported by logical systems of governance, fiscal incentives, and market channels to allow for sustainable implementation. This part describes major policy tools, economic frameworks, and the governance means that can fit into the proposed integrated framework, so that the sustainability energy programs could be quite scalable, resilient, and cost-efficient.

Policy Instruments for Sustainable Urban Energy

Smart integration of energy on a large scale can be propelled by effective policy mechanisms. The important tools are:

- **Code construed on performance:** Performance-based building codes, i.e., the code that does not specify measures but sets energy performance or renewables installation requirements, tends to ensure performance achievement (Meacham et al., 2005).
- **Energy disclosure and benchmarking:** There is a legal requirement that the use of energy in buildings should be reported, building disclosure and benchmarking do this as evident in success of policies in New York and London (Sunikka-Blank & Galvin, 2012).
- **Grid-interactive tariff reform:** Avoiding a system made vulnerable by utilization of distributed generation and DSM, the time-of-use tariffs and peak demand charges speed up the minimization of system vulnerability, and cost recovery (A, 2024).
- **Zoning and land-use incentive:** Green density incentive, solar easement and co-location bonus promote the inclusion of renewables in high-density urban settings (Merili, 2025).

These policy coincidences facilitate the integration of renewable, the demand response, and distributed energy to complement, instead of working the fields in silos.

Financing Models and Tools

Initial costs in sustainable energy projects are usually high. Funding gap can be filled using the following models (shown in figure 7):

- **Green bonds:** Municipal green bonds allow cities to develop low-carbon infrastructure capital with low interest rate (Flammer, 2020).
- **Energy Service Company (ESCO) programs:** Energy efficiency projects in the public sector can also be executed with no need to use capital because the cost is paid through Performance Contracts in which the amount saved is used to finance the energy improvement project (Heffernan et al., 2020).
- **Blended finance:** It is a combination of several different forms of financing, namely, pooling public grants, private funds, and concessional loans that interest investors in sustainability projects that are neither too risky nor too safe (Oppermann et al., 2020).

- **Community investment and crowdfunding:** The ownership of solar co-operatives and local energy are democratic, and the crowds can take part in these local energy schemes (Menyeh & Acheampong, 2024).
- **Water smart city:** Cities meeting SDG 7 and NDC aims and objectives can access climate funding such as the Green Climate Fund or GEF to capacity building and projects funding.

Figure 7. Funding models for sustainable energy initiatives

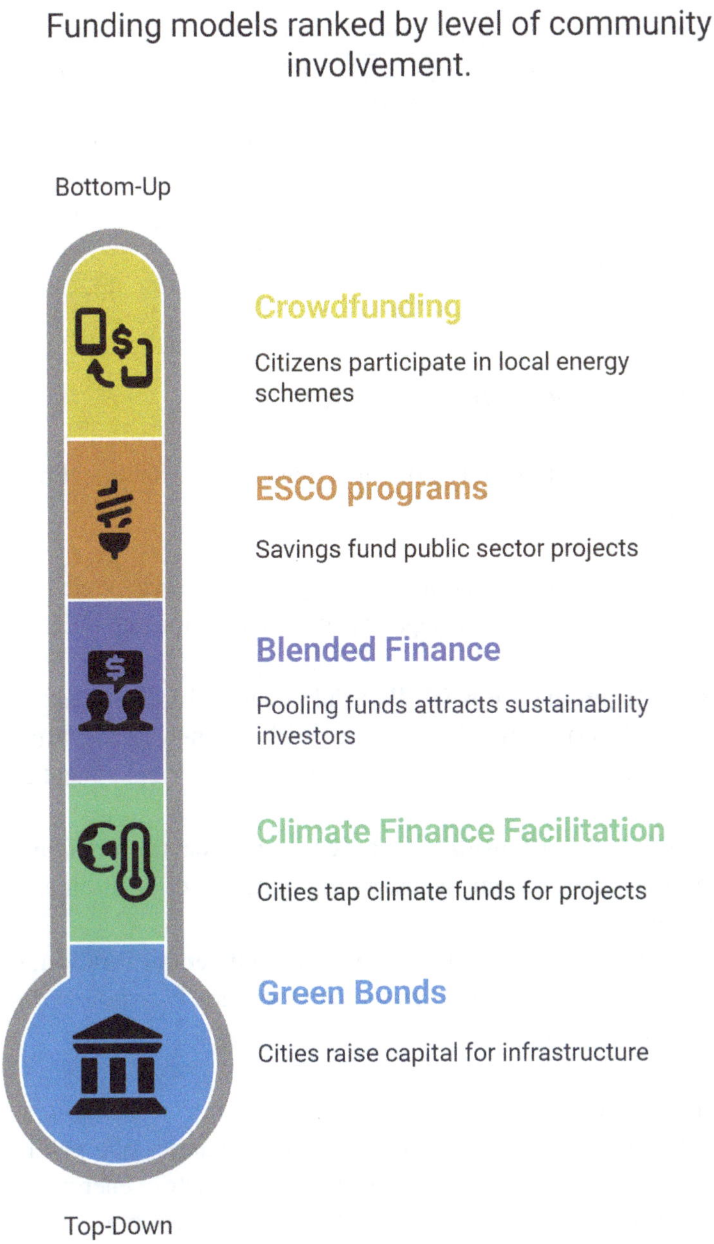

To strengthen the discussion on financial models, Table 3 discusses the Cost-Benefit overview of key smart urban energy interventions.

Table 3. Cost-benefit overview of key smart urban energy interventions

Intervention	CapEx ($/m²)	Payback Period	Energy Savings	Co-Benefits
BIPV (Building PV)	$200–300	7–10 years	20–30%	Carbon reduction, grid relief
Smart Lighting (LED + sensors)	$25–40	3–5 years	40–60%	Crime reduction, maintenance cut
BEMS in public buildings	$50–75	4–6 years	20–35%	Occupant comfort, productivity

Institutional and Governance Enablers

Institutional capacity makes the difference, regardless of laws and money. Implementation should be:

- **Central energy coordination offices:** They lie within the municipal governments, and they organize inter-departmental cooperation and implementation policy.
- **Public private partnerships (PPPs):** Public and private parties collaborate and, as a result, bear some of the risk and develop new smart infrastructure (i.e., smart grid or metering systems) to enable faster deployment and innovation (Jain et al., 2025).
- **Clear; transparent procurement framework:** Competitive bidding by clear and criteria-based procurement fosters accountability on sustainable energy technologies.
- **Hard-nosed Performance monitoring and intelligent evaluation:** Frequent updates in performance, as well as financial rewards based on performance, enhance responsibility.

Meaningful stakeholder management is needed in urban energy strategy. To define roles and responsibilities that cut across the boundaries of a public-private partnership (especially in complicated projects, such as district energy systems or digital twin platforms), it is possible to use RACI matrices (Responsible, Accountable, Consulted, Informed). The use of this tool guarantees transparency as well as balanced participation by the stakeholders and accountability.

Moreover, the usage of lifecycle cost-benefit frameworks (LCBF) must be implemented while planning the infrastructure. These models do not only analyze capital expenditure (CapEx) but also the operational expenditure (OpEx), the costs of maintenance, environmental externalities, and the social consequences of the asset over their lifetime. As an example, a cool roof retrofit can be viewed as expensive in the short term but its savings in energy, minimized UHI effects, and foregone health expenses can tend to offset the initial cost over the long term.

Barriers to Policy and Financing

Important issues are:

- **Short political horizons:** Local politicians tend not to engage in long term investment that brings returns in the distant future.
- **Financial fragmentation:** There are many stakeholders who lack defined jurisdiction in slowing the movement of capital.
- **Reactive regulation:** archaic codes limit innovation and regulation compliance burden.
- **Equity implications:** In absence of inclusive models, the gains of efficiency can skip at-risk households.

Mitigation Measures and Best Practices

To address such impediments, cities could:

- Pass project-level pilot policies based on election cycles that create early successes and prove viability.
- Create municipal green banks to mobilize the private funds with governmental seed capital (Flammer, 2020).
- Train the municipal officials on capacity-building programs about handling of energy contracts and use of climate finance tools.
- Establish inclusive finance regimes like subsidized on-bill-repayment or slinky payment schedules to help allow access among low-income populations.
- To lower the cost of transactions, promote standardized templates of clean-energy products and DSM technology procurement.

Conclusion

Policy and financing are the enablers that support the sustainability energy strategies in the administration of the cities. Incentive-based building codes, blended finance and PPP models are all options to cities looking to leverage resources and delivery. The integration of the framework into the practices of the institutions and making it adaptable, inclusive, and result-driven is the critical aspect of the actualization of the integrated framework.

Case Example

New York has issued a total of $1B of municipal green bonds to retrofit more than 200 public buildings, make energy savings of 25%, and create employment (Flammer, 2020).

ESCO Success

Energy Efficiency Services Limited (EESL) in India installed LED streetlights through ESCOs saving the municipal 50 percent on its energy bill that was paid back in savings.

FUTURE PROSPECT

Introduction

The urban energy systems are experiencing transition because of climate emergency, population growth, digital revolution, and the new socio-economic dynamics. Although the integrated framework suggested in this chapter is an efficient framework for organizing the present energy strategies, the future of urban energy efficiency lies in the ability to pursue adaptability to uncertainty, realize innovation use, and institutionalize resilience at various scales. The cities of tomorrow will not be more intelligent or environmentally friendly, but they should be dynamic, equity-based, decentralized and highly inter-operatory.

In this section, the path of the future urban energy strategies is studied in multidimensional directions. It puts an emphasis on technological and integrative forms of governance, systems operated based on behavior, convergence across sectors, and equity-oriented energy transitions along with the changing contribution of emerging cities in global energy decarbonization processes.

Technological Frontiers in Urban Energy Efficiency

AI and Machine Learning for Predictive Optimization

Most studies have arrived at the point that AI-supported HVAC systems are effective in minimizing energy consumption in cities. According to Lee and Lee

(2023), AI-optimized HVAC with dynamic thermal models result in 30 percent energy saving. On the same note, a study of Jain et al. (2025) and Batty (2018) depicts synergy in the application of AI, along with occupancy sensors, and adaptive control algorithms throughout real-time energy management. All in all, these studies indicate that a combination of predictive analytics and environmental sensing can flatten the energy peaks substantially and improve the comfort in a minimal manual interference manner.

Digital Infrastructure: Edge Computing and 5G

With the decentralization of energy systems, computation must be brought near to the source of data. 5G connectivity with Edge computing allows fast decisions to be made at the building or neighborhood level and limits latency and offloading of the load to centralized cloud systems (Alnoman, 2021). An example is edge-enabled smart meters and appliances are capable of autonomously reacting to tariff signal, demand response, and optimizing operations without the need of human input.

Such integration of distributed intelligence is very beneficial in disaster resilience cases. When the central servers go awry, Edge devices can maintain their operations an added benefit to cities struggling with climate risks.

Urban Energy Storage and Vehicle-to-Grid Systems

Storage is the key to switching to renewable energy systems with high penetration. The pillars of higher and more advanced battery storage, thermal storage and hydrogen-based solutions are being piloted to offset variability and offer services, including ancillary services. In addition, electric vehicles can be mobile storage facilities because of Vehicle-to-Grid (V2G) technology. In Singapore and Tokyo, the community level integration of V2G into pilot programs allows EV fleet to discharge power to their peak load (Kumar et al., 2024).

With the falling costs of mobility electrifications and battery prices today, cities can also look to neighborhood-scale storage nodes which can enable grid stability, peak reduction, and the ability to operate within the framework of rooftop solar panels and local wind installations.

Digital Twins and the Orchestration of Urban Systems

Smart dynamic, virtual, digital replicas of physical urban systems digital twins are developing to be something between models of static simulation and operational tools in real time. Digital twins are being deployed in energy planning to model grid loads, test energy retrofit efficiency alternatives and track the operation of buildings.

The next generation of urban twins will be multi-layered, adopting energy, water, waste, transporting and emergency services into a single platform.

An analysis on European cities indicates that the combination of digital twins and AI-enhanced simulations will enhance predictive capabilities and policy setting to fine tune what interventions cities choose to focus on to provide the most climate and social bang-per-buck (Luca et al., 2024). In addition, digital twins are also promoting transparency and civic engagement by incorporating citizens feedback systems.

In their article introducing AI in a sustainable urban system, Singh and Alasali (2025) point to digital twins as their most significant contribution to the system of decentralized but aligned infrastructure, allowing energy systems to compensate based on environmental and usage demands.

Behavioral and Social Innovation

Net-zero targets cannot be achieved by technological effectiveness alone. Human behavior is at the middle of the process of shaping energy use. The future city will continue evolving to rely more on behaviorally informed systems participating in the system in an active proactive manner rather than passive consumers.

Behavioral Feedback and Gamification

Behavioral economics shows signs that regular feedback on a comparative basis will result in a steady decline in the use of energy in households. Research indicates that peer-to-peer dashboards and to some extent gamification challenges (such as competitions to reduce energy consumption) can reduce consumption between 5-15 (Peschiera & Taylor, 2012). When (these interventions) are incorporated into apps or interfaces of buildings, they create motivation, reinforcement, and social recognition.

In the future, smart cities will be equipped with dynamic nudges systems real-time messages helping people understand how to behave with energy given the current state of the grid or that of the community.

Citizen Co-Creation

Neighborhood energy councils, participatory planning platforms, and Civic innovation labs, are taking the role of essential governance structures. The co-designed solutions do not only increase the level of legitimacy but also effectiveness because they take into consideration the local knowledge and social phenomena. In Bologna,

Civic Digital Twin introduces residents in the construction of local patterns of energy and mobility (Luca et al., 2024).

Such activities represent a trend towards democratized energy governance whereby residents have a collective ownership of the process (and in some cases the infrastructure e.g., energy co-ops).

Cross-Sector and Cross-Scale Integration

Other systems are those systems related to the energy transition. The next generation cities will need systems integration strategies where the energy can be aligned with water, food, mobility, housing and health to optimize intersection and minimize wastage of resources.

The Industrial-Urban Symbiosis

Industrial cities such as in the Kalundborg, Denmark are becoming frontiers of industrial urban symbiosis where waste heat and carbon emissions as well as industrial by-products can be reused in district heating, agriculture and construction (D. Li et al., 2020). These types of models minimize overall energy demand and involve the establishment of circular economies that will concurrently resolve emissions, resource security and jobs.

Urban Metabolisms within Urban Metabolisms

Metropolitan poles will have more and more nested infrastructures concepts, with building system level joining block-level microgrids, district-level, and city-level platforms. Local resilience and optimization within these nested systems are facilitated but so are the larger-scale climate and energy objectives.

Equity-Oriented Energy Transitions

Successful interventions require multi-level coordination amongst households, developers, municipalities, utilities and national agencies. Shared data platforms, standard energy metrics, cross-jurisdiction planning regimes will form some important tools.

Normative dimensions of urban energy systems can also be unpacked with the help of energy justice frameworks. According to Sovacool et al. (2016), three tenets of energy justice offer an important step towards the examination of urban energy interventions: distributional justice, procedural justice, and recognition justice. An example is that even though decentralized solar can seem to be equitable, its advantages tend to fall to the wealthier populations unless policy protection mechanisms such as tiered subsidy systems or universal financing schemes are in place. Specifying the principles of justice within energy justice will make the process of urban energy transitions sensitive to not only cutting down emissions but also doing so in a way that makes energy systems more equitable in terms of access, power, and representation of marginalized groups.

The shift towards energy must not shove up inequality. Being efficient, smart systems have the risk of enhancing the digital divides unless they are built inclusively.

Energy Poverty and Accessibility

The inability of an individual or household to access proper energy services at reasonable cost has been termed as urban energy poverty, which is also common even among the developed economies. Policies that will be developed in future must entrench the principle of energy justice; the upgrading of infrastructure should not leave out the low-income or marginalized neighborhood.

On-bill financing, tiered subsidies, community solar, and shared storage agreements are some inclusionary policies that allow the vulnerable populations to enter the clean energy economy (Ravikumar et al., 2022). Some of these measures ought to be incorporated even at the very onset of planning of smart cities.

Gender and Generational Equity

Energy systems can significantly differ between impoverished populations of women, young and old individuals. It will require planning in terms of their unique needs and will require them to take part in decision-making. Intersectional equity in energy projects can be enhanced through participatory budget, citizen jury and gender audit.

Resilience and Climate Adaptation

Given that events of climate shocks are becoming frequent, it is paramount that urban energy planning includes a resilient nature.

Microgrids and Disaster Preparedness

Durable microgrids, particularly those in hospitals, schools and emergency facilities can provide life-saving continuity of energy during outages. These systems can disconnect with the grid and work on renewable energy, and in the case of storage.

Solar-powered microgrids have been invested in cities, such as San Juan, Puerto Rico, after blackouts imposed by disasters.

Scenario-Based Urban Energy Planning

The cities should be able to embrace anticipatory governance that entails anticipating the future using scenario planning and digital simulations to review how the city could respond to a shock, like a flood, heatwave or cyberattack. Once integrated digital twins, mentioned above, can have a critical role to play in resilience foresight and response orchestration.

The Role of Emerging Cities

In the coming 30 years, most of the urban growth will be witnessed in the cities of the Global South. The way they treat energy will determine the climatic state of the world.

Leapfrogging Potential

New cities have an opportunity to bypass the old con-infrastructure by simply adopting distributed renewables, smart microgrids, and digital systems of governance. Solar kiosks, pay-as-you-go energy systems, and local community energy systems are scalable locally adapted solutions that can achieve high-impact with an increased level of flexibility and reduced costs of delivery (Singh and Alasali, 2025).

Capacity-Building and South-South Knowledge Sharing

International cooperation should focus on capacity-building that assists cities to build technical, financial and planning capacities. Innovation can be enhanced through south-South Collaborations (e.g., between India, Kenya and Brazil) which entail the exchange of contextually relevant experience.

Conclusion

The upcoming state of urban energy efficiency does not switch back and forth. It is dynamic, multi-dimensional, and greatly interconnected with more extensive urban systems. The way it will be influenced will be through the way cities:

- Incorporate intelligent technologies as AI, edge computing, and digital twins
- Transform the citizens to co-create and design behaviorally

- Optimize energy in association with the water, waste, transport and housing industries
- Institutionalize equity, inclusion and digital access
- Develop system resilience to the risks of climate
- Encourage cross-border partnership and mutual learning particularly among future cities

The task is enormous, however, so is the potential. The urban energy systems of the future will be smarter, fairer and truly sustainable through intentional design, cooperative governance, and inspired leadership.

CONCLUSION AND RECOMMENDATIONS

Conclusion

In this chapter a holistic approach and integrated framework in the development of urban energy efficiency adopting smart and sustainable strategies has been postulated. It offers an opportunity to reassess energy systems in cities by considering the synergy between emerging technologies, urban governance, behavioral science and systemic resilience and that is both theoretical and practical. Postulated framework is aligned by five pillars of foundation namely infrastructure and intelligence, policy and regulatory ecosystems, technology innovation platforms, institutional coordination, and citizen engagement.

Case studies involving practical application of these pillars in the real sense of Hyllie to Bahnstadt, Masdar to Jazan have proved day-to-day applications of these pillars with different levels of accomplishments relying on situational circumstances, scope and the ability to do so. The major themes arising in discussion of these examples are that co-designed solutions are crucial, that integrating across sectors is key, digital readiness, as well as inclusive planning. In addition, the future perspectives name revolutionizing trends, including AI-empowered energy systems, urban digital twins, conservation through behavior, and fair and decentralized energy regulations.

What comes out strongly is that urban energy efficiency is no longer a silo goal; it is a converging goal, which means that the convergence of climate policy, urban design, economic inclusion and social justice are essential. It is not a unidimensional problem that requires some cross-disciplinary cooperation and systemic change.

The chapter concludes with Urban management challenges that are mapped with smart technologies in Table 4 as trend challenge matrix.

Table 4. Mapping smart technologies to urban management challenges

Technology	Managerial Challenge Addressed	Challenge Exacerbated (if any)
Digital Twins	Siloed data; delayed planning decisions	High initial cost: skills gap
V2G	Grid stress; peak demand costs	Interoperability, regulatory lag
AI-based HVAC	Energy inefficiency; user comfort	Privacy; reliance on data integrity
Smart Meters	Billing inefficiency; leak detection	Data governance; consumer trust

Key Takeaways

Key takeaways are represented in Figure 8 in the form of urban energy transition pyramid and described below:

Figure 8. Key takeaways

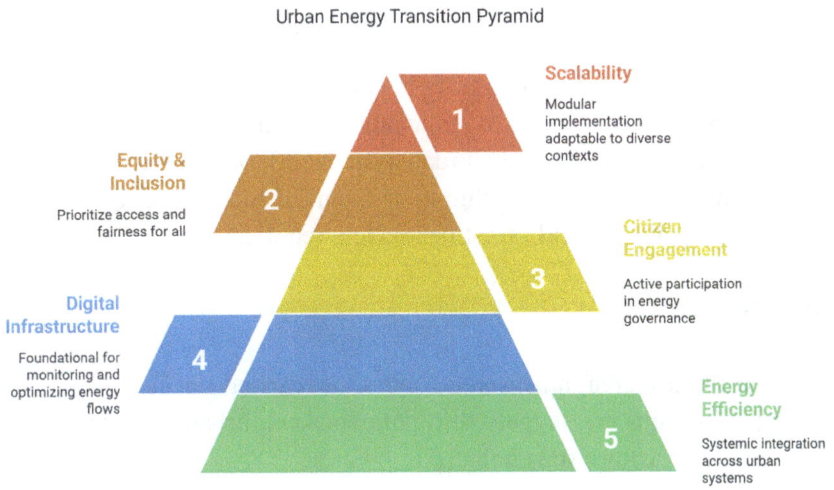

a) **Energy Efficiency as Systemic Imperative:** Energy efficiency should be incorporated not only in buildings and infrastructure but also in transport, water, waste and mobility systems in the city. Systemic thinking helps cities coordinate the objectives and actions in different areas (Geels et al., 2017).

b) **The Centrality of Digital Infrastructure:** Digital twin, smart meters, and Internet of Things have ceased to be optional but essential components of monitoring and controlling energy flows in an urban environment as well as the optimization of those flows. Predictive planning, resilience and real-time decision-making are favored by data-driven systems (Singh et al., 2025).

c) **Citizen-Centric Energy Governance:** Co-governance platforms, behavioral interventions and gamified platforms are promising in involving the citizens as active stakeholders. The legitimacy of the strategy and long-term compliance of energy conservation strategy is promoted by participatory design (Bulkeley & CastAN Broto, 2013).

d) **Equity and Inclusion are Non-Negotiable:** Energy transitions need to be based on accessibility, affordability, and equity made to underserved people, particularly the vulnerable. Community solar, on-bill financing, and inclusive procurement systems are some of the programs that prevent the digital divide and prioritize justice (Ravikumar et al., 2022).

e) **Modular Frameworks:** The integrated model recommended in the chapter can be realized in phases in modular form, applicable to resource-rich and resource-scarce environments in the urban setting. This is flexible, which follows greater applicability in a variety of geographies.

New technologies, such as digital twins or V2G are not the out-of-this-world inventions because they respond to real challenges which managers face in urban management. Digital twins address silo data systems, offering an end-to-end platform, and V2G allows cities to work within their financial limits because EVs become energy resources, and the grid investment is less required.

Recommendations

Based on the research, analysis and synthesis carried out in this chapter, the following policy steps are suggested to policymakers, urban planners, municipal leaders, and infrastructure stakeholders:

Recommendations on Policy and Governance

- Develop performance-based codes and standards which will reward energy-positive construction and retrofit.
- Implementation of cross-sector coordination should be institutionalized by creating energy transition offices in municipal governments.
- Match urban energy plans with the national and global goals on climate change taking inspiration through the SDG 7 and Paris agreements.
- Degrade the administration, allow neighborhood councils and citizens panels to provide their input into the energy decision-making process.

Recommendations on Finance and Economic

- Municipal green banks: develop municipal green banks to spur public-private investment in energy efficiency and renewable.
- Install participatory financing models like tiered subsidies, pay-as-you-save (PAYS) systems and performance contracts.
- Reward energy social clubs and local possession of energy, especially those located in low-income or unserved regions (Menyeh & Acheampong, 2024).

Recommendations on Infrastructure and Technology

- Require commercial and residential buildings to be smart meter-ed and have real-time data visibility.
- Invest in digital twins of an urban scale with a special focus on integration among sectors (e.g. water, waste, and transport).
- Normalize open data and interoperability standards, and allow vendor-independent, scalable smart city platforms.

Behavioral and Social Recommendations

- Create community-based literacy programs on energy, including the youth, women and the marginalized groups.
- Integrate behavioral science in policy development, providing feedback, gamification and norm messages to establish lasting behavior (Peschiera & Taylor, 2012).
- Signatures on civic tech programs like digital twins, dashboards and open platforms by which people in the community co-design and track energy strategies.
- The cities will have to invest in upscaling their workforce and implementing technology. Smart systems cannot perform sufficiently unless there is training of building operators, city planners, and technicians. Internationally, such programs as Singapore Smart Nation Academy can be used as a model of municipal digital capacity-building.

Future Research Directions

To promulgate further in this domain, the following knowledge areas prove by future research the necessity to be studied the best, both at the scholastic academic level, and at the policy review level:

- Comparative analysis of the integrated systems in various socio-political reality, particularly developing regions.
- Equity and cultural aspects of behavioral interventions affect, as well as longitudinal research concerning the effectiveness of behavioral interventions.
- Resilience and adaptive energy systems and simulation modeling, a convergent AI digital twin, and participatory data testing.
- The studies of governance innovation examine multi-level, multi-colored and polycentric structures, as well as the interaction between the public, the private, and civil society.
- Dependency, Privacy and AI ethics and AI governance in urban energy systems.

Final Reflection

These are not technological advances of the city to energy-efficient cities so much as it is the transformation of the way we design and govern the cities we inhabit. With urban expansion and an increase in climate pressures, such correlated, comprehensive, inclusive, and clever structures will be mandatory. The chapter does not only offer a roadmap to be implemented urgently but also lays the foundation of long-term structural change founded on collaboration, equity, and evidence.

Urban energy efficiency is no longer a mere technical issue that is the sphere of engineers or architects; it is one of the keystones of the urban sustainable civilization. The principles are similar in Malm, in Masdar, in Mumbai: think smart, act inclusively and govern adaptively.

REFERENCES

A, B. M. (2024, February 18). *Unlocking Solar Power: A Comprehensive guide to solarpermits*. https://www.linkedin.com/pulse/unlocking-solar-power-comprehensive-guide-permits-brij-mohan-arya-yastc/

Aina, L. B. &. H. M. a. &. Y. A. (2023). Developing an urban computing framework for smart and sustainable neighborhoods: a case study of Alkhaledia in Jizan City, Saudi Arabia. ideas.repec.org. http://ideas.repec.org/a/gam/jsusta/v15y2023i5p4057-d1077795.html

Ajaj, R. (2024). A case study on the barriers towards achieving sustainable smart city for Abu Dhabi. *IET Smart Cities*, *6*(2), 112–128. DOI: 10.1049/smc2.12077

Almihat, M. G. M., & Munda, J. L. (2025). The role of smart grid technologies in urban and sustainable energy planning. *Energies*, *18*(7), 1618. DOI: 10.3390/en18071618

Alnoman, A. (2021). Edge computing services for smart Cities: A review and case study. *2022 International Symposium on Networks, Computers and Communications (ISNCC)*, 1–6. DOI: 10.1109/isncc52172.2021.9615785

Batty, M. (2018). Digital twins. *Environment and Planning. B, Urban Analytics and City Science*, *45*(5), 817–820. DOI: 10.1177/2399808318796416

Batty, M., Axhausen, K. W., Giannotti, F., Pozdnoukhov, A., Bazzani, A., Wachowicz, M., & Portugali, Y. (2012). Smart cities of the future. *The European Physical Journal. Special Topics*, *214*(1), 481–518. DOI: 10.1140/epjst/e2012-01703-3

Bulkeley, H., & Castán Broto, V. (2013). Government by experiment? Global cities and the governing of climate change. *Transactions of the Institute of British Geographers*, *38*(3), 361–375. DOI: 10.1111/j.1475-5661.2012.00535.x

Das, S., Rawat, A., & Singh, S. K. (2020). Impacts of temperature, precipitation and population growth on energy demands of Delhi. *International Journal of Design & Nature and Ecodynamics*, *15*(4), 539–544. DOI: 10.18280/ijdne.150411

Dulaimi, A., Hamida, R., Naser, M., & Mawed, M. (2022). Digital twin solution implemented on energy hub to foster sustainable smart energy city: Case study of sustainable smart energy hub. *ISPRS Annals of the Photogrammetry, Remote Sensing and Spatial Information Sciences*, *X-4*(W3), 41–48. DOI: 10.5194/isprs-annals-X-4-W3-2022-41-2022

Elassy, M., Al-Hattab, M., Takruri, M., & Badawi, S. (2024). Intelligent transportation systems for sustainable smart cities. *Transportation Engineering*, *16*, 100252. DOI: 10.1016/j.treng.2024.100252

Flammer, C. (2020). Green Bonds: Effectiveness and implications for public policy. *Environmental and Energy Policy and the Economy*, *1*, 95–128. DOI: 10.1086/706794

Geels, F. W., Sovacool, B. K., Schwanen, T., & Sorrell, S. (2017). Sociotechnical transitions for deep decarbonization. *Science*, *357*(6357), 1242–1244. DOI: 10.1126/science.aao3760

GhaffarianHoseini, A., Dahlan, N. D., Berardi, U., GhaffarianHoseini, A., Makaremi, N., & GhaffarianHoseini, M. (2013). Sustainable energy performances of green buildings: A review of current theories, implementations, and challenges. *Renewable & Sustainable Energy Reviews*, *25*, 1–17. DOI: 10.1016/j.rser.2013.01.010

IEA. (2021). *Empowering Cities for a Net Zero Future: Unlocking Resilient, Smart, Sustainable Urban Energy Systems.* International Energy Agency. https://www.iea.org/reports/empowering-cities-for-a-net-zero-future

Jain, R., Singh, S. K., Palaniappan, D., & Associates. (2025). Integrating smart cyber-physical systems into urban planning: The future of smart cities. In *Smart Cyber Physical Systems: Innovations and Real-World Implications* (Ch. 5). CRC Press.

Jain, R., Singh, S. K., Palaniappan, D., & Parmar, K., & T, P. (2025). Data-Driven Civil Engineering: Applications of artificial intelligence, machine learning, and deep learning. *Turkish Journal of Engineering*, *9*(2), 354–377. DOI: 10.31127/tuje.1581564

Ji, X., Ren, J., & Ulgiati, S. (2018). Towards urban-rural sustainable cooperation: Models and policy implication. *Journal of Cleaner Production*, *213*, 892–898. DOI: 10.1016/j.jclepro.2018.12.097

Jurasz, J., Canales, F., Kies, A., Guezgouz, M., & Beluco, A. (2019). A review on the complementarity of renewable energy sources: Concept, metrics, application and future research directions. *Solar Energy*, *195*, 703–724. DOI: 10.1016/j.solener.2019.11.087

Kennedy, C., Pincetl, S., & Bunje, P. (2010). The study of urban metabolism and its applications to urban planning and design. *Environmental Pollution*, *159*(8–9), 1965–1973. DOI: 10.1016/j.envpol.2010.10.022

Khan, I. (2019). Energy-saving behaviour as a demand-side management strategy in the developing world: The case of Bangladesh. *International Journal of Energy and Environmental Engineering*, *10*(4), 493–510. DOI: 10.1007/s40095-019-0302-3

Kumar, P., Channi, H. K., Kumar, R., Rajiv, A., Kumari, B., Singh, G., Singh, S., Dyab, I. F., & Lozanović, J. (2024). A comprehensive review of vehicle-to-grid integration in electric vehicles: Powering the future. *Energy Conversion and Management, X*, 100864. DOI: 10.1016/j.ecmx.2024.100864

Lee, D., & Lee, S. (2023). Artificial intelligence enabled energy-efficient heating, ventilation and air conditioning system: Design, analysis and necessary hardware upgrades. *Applied Thermal Engineering, 235*, 121253. DOI: 10.1016/j.applthermaleng.2023.121253

Lee, J., Park, J., Jung, H., & Park, J. (2017). Renewable energy potential by the application of a building integrated photovoltaic and wind turbine system in global urban areas. *Energies, 10*(12), 2158. DOI: 10.3390/en10122158

Li, D., Heimeriks, G., & Alkemade, F. (2020). The emergence of renewable energy technologies at country level: Relatedness, international knowledge spillovers and domestic energy markets. *Industry and Innovation, 27*(9), 991–1013. DOI: 10.1080/13662716.2020.1713734

Luca, M. L., Lepri, B., Gallotti, R., Paolazzi, S., Bigi, M., & Pistore, M. (2024). Towards civic digital twins: Co-design the citizen-centric future of Bologna. *arXiv*.

Madakam, S., Ramaswamy, R., & Tripathi, S. (2015). Internet of Things (IoT): A literature review. *Journal of Computer and Communications, 3*(5), 164–173. DOI: 10.4236/jcc.2015.35021

Meacham, B., Bowen, R., Traw, J., & Moore, A. (2005). Performance-based building regulation: Current situation and future needs. *Building Research and Information, 33*(2), 91–106. DOI: 10.1080/0961321042000322780

Menyeh, B. O., & Acheampong, T. (2024). Crowdfunding renewable energy investments: Investor perceptions and decision-making factors in an emerging market. *Energy Research & Social Science, 114*, 103602. DOI: 10.1016/j.erss.2024.103602

Merili. (2025, January 15). *Solar Energy in Urban Design: Powering Sustainable Cities*. Impact Day. https://impactday.eu/blog/solar-energy-in-urban-design-powering-sustainable-cities/

Namany, S., Govindan, R., Di Martino, M., Pistikopoulos, E. N., Linke, P., Avraamidou, S., & Al-Ansari, T. (2021). An Energy-Water-Food nexus-based decision-making framework to guide national priorities in Qatar. *Sustainable Cities and Society, 75*, 103342. DOI: 10.1016/j.scs.2021.103342

NIST. (2014). NIST Framework and Roadmap for Smart Grid Interoperability Standards, Release 3.0.

OECD. (2018). *Building Resilient Cities: An Assessment of Disaster Risk Management Policies in Southeast Asia, OECD Green Growth Studies*. OECD Publishing. DOI: 10.1787/9789264305397-

Perera, A. T. D., Coccolo, S., & Scartezzini, J. (2019). The influence of urban form on the grid integration of renewable energy technologies and distributed energy systems. *Scientific Reports*, *9*(1). Advance online publication. DOI: 10.1038/s41598-019-53653-w

Peschiera, G., & Taylor, J. E. (2012). The impact of peer network position on electricity consumption in building occupant networks utilizing energy feedback systems. *Energy and Building*, *49*, 584–590. DOI: 10.1016/j.enbuild.2012.03.011

Ravikumar, A. P., Baker, E., Bates, A., Nock, D., Venkataraman, D., Johnson, T., Ash, M., Attari, S. Z., Bowie, K., Carley, S., Castellanos, S., Cha, M., Clark, D. L., Deane-Ryan, D., Djokic, D., Ford, J. C., Goldstein, A., Grubert, E., Hu, L., & Tuominen, M. (2022). Enabling an equitable energy transition through inclusive research. *Nature Energy*, *8*(1), 1–4. DOI: 10.1038/s41560-022-01145-z

Reuters. (2024, May 23). *How AI is arming cities in the battle for climate resilience*. Reuters.

Sandström, P. (2025, March 18). *Sweden's city of tomorrow: How Scandi innovation is leading climate-smart urban development.* The Guardian.

Singh, S. K., & Alasali, B. (2025). *Urban Mobility and Smart Energy: Pathways for Global South Cities*. IGI Global.

Singh, S. K., & Jain, R. (2025). Future of mobility: potential challenges and ethical considerations. In *AIoT for Smart Transportation: Transforming the future of mobility* (pp. 235–253). IET., DOI: 10.1049/pbtr047e_ch12

Singh, S. K., Toryila, M., & Alasali, B. (2025). Utilization of sustainable resources for energy efficiency using smart technologies. In *Artificial Intelligence for Environmental Sustainability*. Apple Academic Press – CRC Press.

Sohail, A., Shen, B., Cheema, M. A., Ali, M. E., Ulhaq, A., Babar, M. A., & Qureshi, A. (2024). Beyond data, towards sustainability: A Sydney case study on urban digital twins. *arXiv*.

Sovacool, B., Heffron, R., & McCauley, D. (2016). Energy decisions reframed as justice and ethical concerns. *Nature Energy*, *1*, 16024. DOI: 10.1038/nenergy.2016.24

Sunikka-Blank, M., & Galvin, R. (2012). Introducing the prebound effect: The gap between performance and actual energy consumption. *Building Research and Information*, *40*(3), 260–273. DOI: 10.1080/09613218.2012.690952

UN DESA. (2019). *World Urbanization Prospects: The 2018 Revision.* United Nations Department of Economic and Social Affairs. https://population.un.org/wup/

UN-Habitat. (2020). *World Cities Report 2020: The Value of Sustainable Urbanization.* United Nations Human Settlements Programme. https://unhabitat.org/world-cities-report-2020-the-value-of-sustainable-urbanization

Yigitcanlar, T., Kamruzzaman, M., & Teriman, S. (2015). Neighborhood sustainability assessment: Evaluating residential development sustainability in a developing country context. *Sustainability*, *7*(3), 2570–2602. DOI: 10.3390/su7032570

Zhang, R., Nie, L., Zhao, C., & Zhao, X. (2025). Eco efficiency as a catalyst for citizen co production: Evidence from Chinese cities. *arXiv*.

Chapter 9
Advancing Smart and Sustainable Construction:
Integrating Circular Economy, Energy Efficiency, and Economic Viability in the Built Environment

Anandharajah Sagadhevan
https://orcid.org/0009-0005-4354-2680
Marwadi University, India

Atheek Ahamed
https://orcid.org/0009-0008-5831-0067
Marwadi University, India

Faruk Abdulla
https://orcid.org/0009-0007-1478-7376
Parul University, India

Amit Kumar Yadav
https://orcid.org/0009-0000-7886-0615
Marwadi University, India

ABSTRACT

Nearly 39% of all global CO_2 emissions and about 30% of natural resources are used in the construction industry, which makes it very important for sustainability. Although more businesses choose renovations, pay attention to energy saving, and focus on CSR, sustainable changes move at a slow pace due to lacking economic and environmental integration. Progress in sustainable products, circular economy models, energy-saving building techniques, and digital system applications are reviewed

DOI: 10.4018/979-8-3373-2555-2.ch009

in this chapter. The use of low-carbon cement, recycled composites, passive design, prefabrication, and AI is changing the construction field. At the same time, the high expenses, strict rules, and reserved attitude among investors delay the industry's progress. It points out that policy incentives, financial instruments, and effective regulations are the most important for a change to green and efficient construction.

INTRODUCTION

Improvements in construction are being pushed to the forefront of the world's sustainability issues; with the end product producing almost 39% of total carbon dioxide emissions (CO_2) and almost 30% of natural resources. The mandate for renovation, rehabilitation, and energy integration in the built environment has led to a shift to intelligent and sustainable construction practices. Adoption of the mentioned CSR initiatives, however, continues to be slow, and it is primarily because of the absence of integrated approaches that consider the environmental as well as the economic feasibility factors. This review examines state-of-the-art progress in sustainable materials, circular economy approaches, energy-saving methods of construction, and digital construction technologies. The study identifies such low carbon cement substitutes, recycled composites, passive design concepts, modular prefabrications, and AI-powered optimization which seem to be changing the face of construction. In addition, such things as the role of carbon capture, life cycle assessment and smart urban energy systems are analyzed to determine the impact of modern construction to the environment. In spite of such innovations, the high prices, regulatory hurdles and reticence of the market continue to stymie a wide spread use. This paper aims at offering an integral assessment of incentives through the government, financial models and regulatory strategies that could boost the transition to net zero construction. The construction industry can realize a resilient, low carbon, and resource-efficient future by combining advances in technology with circular economy principles and the economics of feasibility.

The Need for Smart and Sustainable Construction

The construction industry is among the most extreme contributors to global carbon emissions, exhaustion of natural resources and energy usage. With the acceleration of urbanization, there is increased demand for infrastructure and housing, which has meant that the use of energy intensive materials like concrete, steel and bricks as become too prevalent. The built environment is responsible for about 39% of CO2 emissions globally with about 28% derived from operational energy use and 11% due to embodied carbon in buildings (Terzis, 2025). In addition, construction and demolition waste (CDW) accounts for approximately 25-30% at the total global waste generation and thus necessitate the application of circular economy principles in the industry (Soares et al., 2025).

Even though there have been substantial developments in building materials that take sustainability, energy-efficient designs and digital construction methods seriously, the process of adoption lags and is patchy. One of the major barriers is the absence of an integrated framework by which environmental sustainability could be merged with economic viability thereby; allowing stakeholders to transition into smart resilient construction practice (Zhang et al., 2024). This paper discusses how circular economy strategy, low carbon materials and digital technologies can coordinate to advance urban energy efficiency while achieving cost effectiveness and long term functioning.

Challenges in Renovation, Rehabilitation, and Energy Integration

The undertakings involve a combination of two problems—an acute need for new infrastructure, eased by the rehabilitation and up-gradation of older building stock. In developing regions around the world, the demand for new housing, commercial buildings and infrastructure have encouraged resource intensive growth, which has increased environmental degradation. In contrast; the developed countries; focus has turned to renovation and retrofitting of old buildings for energy efficiency and carbon neutrality (Wang et al.; 2024). Nevertheless, these are both challenged by high costs in the initial capital outlay, absence of policy incentives, and slow uptake of digital construction technologies (Santos et al., 2024).

Energy inefficiency is one of the major aspects, which account for unsustainability in construction. Many existing buildings have inadequate insulation, antiquated HVAC systems and use of fossil fuels resulting in high operational energy consumption and emission of greenhouse gases. Various studies have indicated that passive design strategies, modular building, and smart energy systems can drastically minimize footprints of energy in the built environment without negatively affecting occupant comfort (Hernando Castro et al., 2024). However, it has been a slow process towards the low energy, decarbonized built environments for economic factors and regulatory factors (Zhang et al. 2024).

The Role of Circular Economy and Digital Transformation

In order to overcome the environmental and economic aspects of construction, the industry must move towards circular economy models centered on resource efficiency, material recycling and waste minimization. Research shows that recycled aggregates, low-carbon cement and bio-based materials can act as substitutes to the conventional ones while decreasing embodied carbon emissions (Yang et al., 2024). In addition, the use of various advanced digital tools including Artificial Intelligence (AI), Building Information Modeling (BIM), 3D printing etc., has a potential to change the course of design optimization, material selection, and waste reduction strategy (Li et al., 2024).

One promising innovation is a material life cycle, carbon footprint and energy efficiency analysis using AI-driven sustainability analytics on an ongoing basis. With AI-driven predictive models and thus with automated design tools, it will be possible to enhance energy management and optimize material usage thus causing up to 30% less construction waste (Meng et al., 2024). The emergence of 3D printing in construction is also beneficial because the technology allows the production of customized, low-cost, low-carbon structures with reduced waste of materials and corresponds to the framework of the circular economy (Khan et al., 2024).

SUSTAINABLE CONSTRUCTION MATERIALS AND CIRCULAR ECONOMY STRATEGIES

High-Performance and Low-Carbon Materials

The transition out of the traditional construction materials towards low carbon, high performance alternatives has been a major focus in the area of sustainable construction. Traditional materials like Portland cement and steel also play a major role massively in the depletion of carbon emissions and resources. Consequently, scientists have introduced alternate cementitious material, bio-based composites and recycled materials to increase the sustainability.

Low-carbon cement and geopolymer binders are of the most promising solutions to decrease the environmental footprint of concrete production. Research reports that the manufacturing of geopolymer-based concrete (manufactured from waste materials like slag, fly ash) can reduce CO_2 emissions by up to 80, compared to Portland cement (Yang et al. 2024). Likewise, alkali-activated cementitious systems use waste-derived materials giving them superior chemical resistance and durability while reducing carbon intensity (Meng et al., 2024).

The other sustainable option is the use of recycled PET-based composite panels. Studies show that recycled PET panels have superior thermal and acoustic insulation properties that help minimize energy requirements in buildings (Cavagnoli et al. 2024). In addition, fiber-reinforced composites that include basalt, glass or agricultural waste fibers have shown high structural integrity and long term durability thus are potential alternatives for existing reinforcement materials (Poopalam et al, 2024).

Circular Economy in Construction: Recycling and Waste Utilization

Circular economy thinking advocates for waste reduction, material reuse and resource efficiency that develops a closed-loop construction system. Reuse of construction and demolition waste (CDW), industrial waste products, and organic waste is essential towards realizing sustainability goals.

The recycled concrete aggregates (RCA) have received much attention as material in place of natural aggregate production in construction. Research has shown that the use of RCA-enhanced concrete/eccentric concrete can retain the same level of mechanical performance as virgin aggregate concrete, and reduce the emissions of its environmental footprint (Yu et al., 2024). In addition, the use of mining waste as part of construction materials such as iron ore tailing and copper slag has been found to improve strength and durability and decrease the use of natural resources (Lai & Chen, 2024).

Another avenue to material circularity is through the use of agri-food waste to make construction. Based on research there is tremendous promise (regarding the use of rice husk ash, sugarcane bagasse ash, and oil palm biomass as pozzolanic additives and bio-based reinforcements) in sustainable composites (Soares et al. 2025). These materials provide good level of embodied carbon, airy feel, improved insulation performance enabling the design of energy-efficient building.

The recycling of glass waste in soil stabilization and cementitious products has also been investigated. Research tells that glass micro-powder improve mechanical properties of stabilized soil and minimizes the environmental burden of waste disposal (Lai & Chen, 2024). In addition, incorporation of recycled gypsum in the drywall manufacturing practice reduces landfill accumulation and sulfur dioxide emissions hence, making interior construction sustainable.

Challenges and Future Prospects

Although low-carbon materials and circular construction have considerable environmental advantages, their issue is rather scalability, cost and standardization. In most cases high production costs and challenges associated to lack of regulatory frameworks ensure mass use of geopolymer concrete and bio-based composites is prevented (Zhang et al., 2024). In addition, performance variability of recycled materials is yet to be optimized as well as quality control measures (Santos et al., 2024).

Further research should dig into material processing techniques improvement, development of certification guidelines, and application of AI-based material selection tools for enhancing the marketability of the technology, and economic viability (Li et al., 2024). With circular economy policies being adopted by governments and industries, there has been an increased adoption of recycled, low carbon, and high performance materials that will contribute to more resource efficient and environment friendly sector of construction.

Figure 1. Diagram

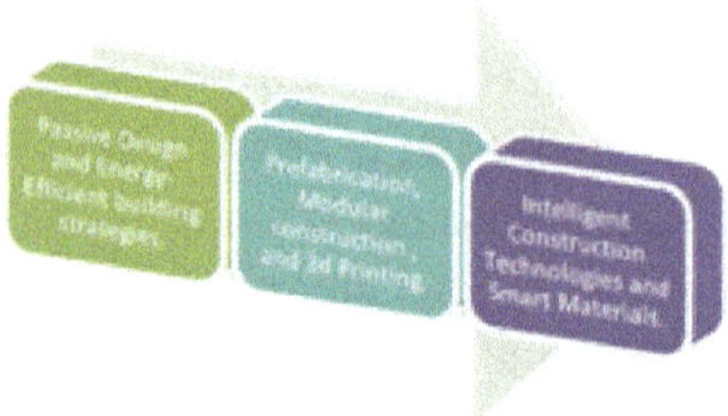

ENERGY-EFFICIENT AND SMART CONSTRUCTION TECHNIQUES

Passive Design and Energy-Efficient Building Strategies

Operational carbon emissions, and long-term sustainability, will be relatively dependent on energy efficiency in buildings. In building passive design strategies that involve maximum use of natural ventilation, insulation, and that of solar energy, have proved very useful in the reduction of heating and cooling energy consumption.

Studies in thermal insulation materials have revealed that bio based insulators, aerogels and phase change materials (PCMs) help increase energy efficiency considerably. Research shows that insulation materials like volcanic rock based composites and hemp based panels minimize thermal conductivity whilst maintaining mechanical strength, thus lowering the need for mechanical heating and cooling systems (Wang et al., 2024). Electro chromic smart glass has also been incorporated to building envelopes for controlling incidental lighting and thermal indoor comfort hence reduced consumption of energy in commercial and residential structures (Santos et al., 2024). response Likewise electro chromic smart glass has been integrated into building envelopes to govern indoor lighting and thermal comfort thus reducing commercial and residential buildings energy consumption (Santos et al., 2024).

Another programming of the visual appearance regarding energy efficient is the application of reflective and self-cooling coats on building facades. These coatings use nanoparticles and infrared-reflective pigments to lessen heat absorption and keep low indoor temperatures and reduce dependence on air conditioning systems (Firoozi et al., 2024).

Prefabrication, Modular Construction, and 3D Printing

Prefabrication and Modular construction methods are being adopted more and more to enhance energy savings, promote waste reduction and increase construction speed. Prefabricated components such as lightweight fiber reinforced panels and 3D printed geo polymer have better thermal insulation and lower embodied carbon than conventional components (Cavagnoli et al., 2024).

Research of the 3D-printed concrete and prefabricated modular units shows that the result is 60% material waste reduction and shorter construction periods (Khan et al., 2024). Moreover, the use of geopolymer based 3D printing offers a low-carbon alternative to conventional cement since investigations has proven that the CO_2 emissions from 3D-printed geopolymeric structures can be reduced up to 80% over the conventional concrete (Meng et al., 2024).

The recycled PET-based modular panels also improve energy performance by eliminating thermal bridging and air leakage. Sufficient research has pointed out that modular and prefabricated components that come with integrated insulation are capable of substantially reducing heating and cooling loads, meaning they are an integral part of the energy efficient building solutions (Cavagnoli et al., 2024).

Intelligent Construction Technologies and Smart Materials

The combination of technology with artificial intelligence (AI), Building information modeling (BIM), and smart sensors consequently driving energy optimization and resource management in construction. AI-powered predictive modeling facilitates immediate analysis of energy consumption patterns for automatic HVAC settings, lighting control and even material selection for the greatest efficiency (Li et al., 2024).

Self –healing concrete and nanomaterial- enhanced composites are also contributing significantly to building longevity and energy efficiency. Studies have shown how bacteria-based self-healing concrete can repair micro-cracks as a self-repairing process and reduce maintenance costs and enhance durability (Yang et al., 2024). In addition, Carbon nanotube-reinforced composites increase thermal conductance and strength thus making them suitable for high performance structural and insulation applications (Firoozi et al., 2024).

Smart materials including phase change materials (PCMs) and nanoparticle enhanced coatings are being incorporated into the building envelopes to give automatic response to any ambient temperature variations to thermal conditions. Some research shows that PCMs can slow down fluctuations in indoor temperatures by up to 30% which means considerable energy savings in hot and cold climates (Wang et al., 2024).

Challenges and Future Prospects

Although there are abundant advantages of implementing the energy-efficient constructions techniques, there are several barriers to the broad adoption. Before seizing such an opportunity as smart materials, prefabrication technologies, and AI-based monitoring systems, developers have to account for high costs, which impede the initial stages of promoted development (Zhang et al., 2024). Moreover, the absence of standardized regulations for 3D-printed buildings and smart energy systems makes it difficult for these systems to gain acceptance on large scales (Santos et al., 2024).

Future development should be devoted to strategies of cost reduction, incentives in policies, and digital integration in order to induce the use of energy saving and smart forms of construction. The integrations between energy management, modular prefabrication, and low-carbon materials enabled by AI are easily seen to lead to a net zero sustainable future industry (Li et al. 2024).

CARBON CAPTURE, LIFE CYCLE ASSESSMENT (LCA), AND ENVIRONMENTAL IMPACT REDUCTION

Carbon-Neutral and Carbon-Negative Materials

Construction is accountable for a large part of the world's CO_2 emissions, largely because of the use of cement-based materials, and steel that is generated, as well as high-energy-demand building techniques. To minimize this impact, PhD students and industry stakeholders are concentrating on carbon capture technologies and negative carbon materials that not only replace but actively remove CO_2 emissions throughout the lifecycle of buildings, from manufacturing through to end-of-life.

One of the most promising innovations in the making is a carbon-sequestering concrete, which combines CO_2 mineralization methods to capture carbon dioxide during the hardening process. Research indicates that up to 50% of emissions can be switched to carbon-mineralized concrete compared to conventional Portland cement-based concrete (Yang et al., 2024). In a similar way, geo polymer binders that use industrial waste products like fly ash and blast furnace slag obviate the necessity for clinker production that consumes vast amounts of energy and decrease CO_2 emissions by up to 80 per cent (Meng et al., 2024).

Apart from cementitious solutions, bio-based materials like hemp Crete, mycelium composites, bamboo are being paid a lot of attention because of their carbon sequestration properties. For instance, hemp Crete can absorb more CO_2 during its growth process than it emits during its manufacturing process making it a net carbon-negative material (Firoozi et al., 2024). Similarly, mycelium-based insulation panels are highly thermally effective and biodegradable and carbon neutral. Another dilution strategy that is emerging includes the use of biochar soaked-type construction material as it has the potential to lock in carbon forever with improved performance on soil and concrete (Poopalam et al., 2024).

Carbon Capture and Utilization (CCU) in Construction

In order to reduce construction-related emissions even more, carbon capture and utilization (CCU) technologies are incorporated in building materials and construction process. CCU refers to capturing the CO_2 from industrial emissions or even from the air and incorporating it for useful building components such as aggregates, fillers and alternative binders.

The built environment is more and more interested in exploring the opportunities of the DAC (direct air capture) technology, with the buildings being capable of functioning as long-term carbon sinks. From research, it is evident that DAC-integrated building facades and coatings have the ability to continuously absorb atmospheric CO_2 hence support carbon-neutral building designs (Santos et al. 2024). In addition, self-carbonating cementitious systems like magnesium-based cements, are capable of reacting with the CO_2 in the air over time resulting in an improvement in durability and the structural integrity as well as ongoing reduction of its carbon footprint (Yu et al, 2024).

Studies have also examined this concern of the use of aggregates that absorb CO_2 that reacts with atmospheric CO_2 and forms mineralized carbonates, an idea that can lead to strengthening of the concrete and increase chemical stability (Lai & Chen, 2024). These innovations have long term carbon storage solutions and thus buildings contribute to carbon neutrality for the lifetime of the buildings.

Life Cycle Assessment (LCA) in Sustainable Construction

One of the central challenges in sustainable construction is the ability to adequately measure the environmental footprint of materials and the construction process itself. Life cycle assessment (LCA) is a widely used method to investigate embodied carbon, energy consumption and environmental impacts within the building life cycle from raw material extraction to end of life disposal.

LCA studies found that the use of geo polymer concrete, recycled PET panels, and prefabricated modular components can decrease, substantially, their embodied carbon compared with the conventional material (Hernando Castro et al., 2024). Also, digitalized tools like AI-driven sustainability analytics and BIM-integrated LCA simulations have enhanced carbon construction and tracking and development of environmental performance models (Li et al., 2024). According to research, LCA-optimized material selection can achieve a 30–40% span of total carbon footprint of a building and therefore confirms the imperative of data-driven decision making in construction sustainability (Santos et al., 2024).

In addition, comparative LCAs between traditional and sustainable materials demonstrate the advantages of recycled and bio-based alternatives. For instance, LCA studies on insulation materials show that bio-based insulators like hemp fiber, sheep wool and mycelium based panels, are far superior (in terms of GWP) to synthetics (Hernando Castro et al. 2024). Similarly, modular and prefabricated construction approaches have been discovered to have reduced embodied energy by up to 50% as compared to traditional onsite construction approaches (Meng et al., 2024).

Challenges and Future Directions

Despite all the successful emerging innovations in carbon capture, sustainable materials and LCA methods, a number of challenges remain. Economic viability, regulatory regulations, and large-scale scalability still prevent the implementation of carbon-neutral and carbon-negative construction practices by a large scale number of people. The saccharification of deacrylic cellulose acetate and of CCU-integrated materials is much too expensive to be applied broadly (Zhang et al., 2024).

In addition, there is a lack of standardized certification frameworks for up-and-coming materials; for instance, CO_2 absorbing aggregates and geopolymer concrete (Santos et al., 2024), which is why they are not widely accepted in commercial building. Policymakers and industry leaders have an important task of jointly creating clear guidelines, performance metrics, and financial incentives in order to ensure that low-carbon/carbon-negative materials are adopted (Zhang et al., 2024).

Further studies should be aimed at increasing the efficiency and affordability of carbon capture technologies, LCA assessment accuracy, and construction hybrid methods which combine the carbon sequestration, digital modeling and material recycling (Li et al. 2024). With application of net-zero policies by governments and organizations, the integration of carbon capture, sustainable material selection, and LCA-led decision making will assume a critical role in determining construction sustainability (Santos et al., 2024).

Figure 2. Diagram

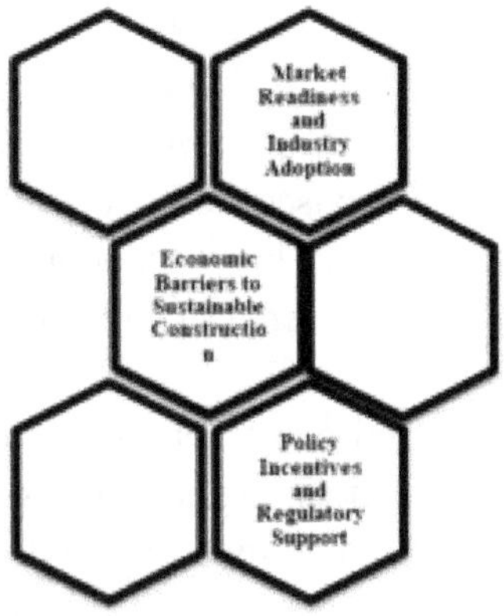

ECONOMIC FEASIBILITY AND MARKET ADOPTION OF SUSTAINABLE PRACTICES

Economic Barriers to Sustainable Construction

There are relatively high economic restrictions, high initial investments, and insufficient financial incentives in making the switch to sustainable construction materials, circular economy approaches, and energy saving technologies. Even though the environmental and long/run cost efficiency is gaining increasing recognition, the front capital cost needs to install green construction materials, prefabrication methods, and smart building schemes continue to be a barrier (Zhang et al., 2024).

The higher price of production of sustainable materials is one of the major economic issues. Traditional building materials such as Portland cement, steel, and synthetic insulation are common and mass-produced enjoying their well-developed supply chains, economies of scale. However, low-carbon alternatives are geopolymer concrete, bio-based and recycled PET-based insulation panels, which necessitate specialized manufacturing processes and inputs of advanced technology making them expensive for production (Yang et al., 2024).

According to studies, Low-carbon cement-based alternatives can be up to 20-50% more expensive than the traditional cement, largely because of increased cost of raw materials and beneficiation requirements and absence of large-scale production facilities (Hernando Castro et al., 2024). In a parallel manner, although geo polymer concrete has environmental values, it is currently not maximally used because of high production costs and the heterogeneous quality control standards (Meng et al., 2024).

The long period for recovery of investment in energy-efficient construction solutions is another economic barrier. Although modern insulation materials, smart control of energy consumption, and passive strategies can substantially reduce heating, cooling and operation expenses, and pay back their costs in over years, or, even a century (Santos et al., 2024). This discourages developers/others to pursue the green building initiatives unless there is a high incentive/finance or government subsidies to follow.

Also, high prices of the digital construction technologies (AI-driven material selection, BIM-integrated energy modeling, automated waste management systems) prevent their wide spread use among small and mid-sized construction firms (Li et al. 2024). These technologies demand specialized skills, needs of infrastructure improvements and costly software licenses, which are not very feasible for the smaller players in the industry to incorporate into their work flows.

Policy Incentives and Regulatory Support

With these economic limitations, the government policies, financial incentives and regulatory frameworks are key drivers in instilling rapid adoption of sustainable construction practices. Geopolitical climates of many governments hence have adopted carbon pricing mechanisms, green building certification programs and tax incentives to encourage utilization of environment friendly materials and energy efficient technologies (Terzis, 2025).

The major policy strategy in this regard is the tax implementation for carbon and emission reduction mandates. Studies indicate that nations with high carbon pricing policies (EU's Emissions Trading System, ETS) have recorded a significant increase in sustainable construction investments (Santos et al., 2024). Carbon pricing puts a financial penalty on high-carbon materials, pushing developers to use low-carbon alternatives and, in return, making sustainable materials more viable in the market, (Zhang et al, 2024).

In addition to carbon pricing, green building certifications and sustainability assessment frameworks have turned into the indispensable regulatory instruments for encouraging energy efficient and low-carbon building. Inducement programs like LEED (Leadership in Energy and Environmental Design), BREEAM (Building Research Establishment Environmental Assessment Method) and DGNB (German Sustainable Building Council) attracts incentives including, rebate on tax, expeditious process of getting breeds, and monetary grants for undertaking the projects that lie within the parameter of sustainability (Santos et al. 2024). It is shown that, on average, certificated LEED buildings consume 25% fewer energy while featuring lower cost of operation, which makes them worth financial investment in the long term (Li et al., 2024).

The other significant policy intervention is the inclusion of Life Cycle Assessment (LCA) into codes and regulations for buildings. Several countries are beginning to require LCA-based environmental impact assessments to be carried out during the construction of large scale buildings, obliging developers to assess the carbon footprint of their materials and construction processes, before issuance of building permit (Hernando Castro et al. 2024). Such extension of LCA requirements worldwide can drastically accelerate the transition towards the circular economy approach and material choice that is sustainable (Soares et al., 2025).

In addition, public-private partnerships (PPPs) and sustainability-linked financing models have momentum as solutions to close the difference of finances in green construction. Many governments are partnering with banks and private investors to implement low-interest green loans, tax credits and grant programs that promote infrastructural changes to the circular economy principles on a large scale (Yu et al., 2024). Besides, green bonds and ESG (Environmental, Social, and Governance) investments are gaining a shared popularity enabling companies to gain financing for sustainable buildings' construction (Santos et al., 2024).

Market Readiness and Industry Adoption

However, the level of market readiness toward sustainable construction solutions differs quite considerably between regions and industry sectors despite the growth in policy support and financial instruments (Zhang et al., 2024).

A major challenge is the absence of education and training programs for architects, engineers and construction professionals on sustainable materials, digital construction tools and circular economy strategies (Li et al. 2024). Sufficient training is essential to overcome the hesitation to embrace innovations in the industry because of the reliability of performances, regulatory approvals, and cost consequences (Soares et al. 2025).

Another important barrier is the fractured supply chain of sustainable materials. Although recycled aggregates, and bio-based insulation and geo polymer binders are propitious to the environment, their availability is not uniform especially among the developing regions where access to bidding manufacturing facilities as well as green infrastructure is limited (Yu et al., 2024). To overcome these supply chain constraints, investment in local production hubs and supply chain digitalization, real-time logistics, must be made to increase the global availability of sustainable materials (Santos et al., 2024).

Moreover, the requirement for sustainable construction practices is more and more determined by the interests of investors. Businesses adopting ESG oriented strategies and adopting net-zero initiatives are more likely to receive green investments as well as green financing avenues and benefit from a shift in market trends towards greener building solutions (Terzis, 2025).

Challenges and Future Directions

Figure 3.

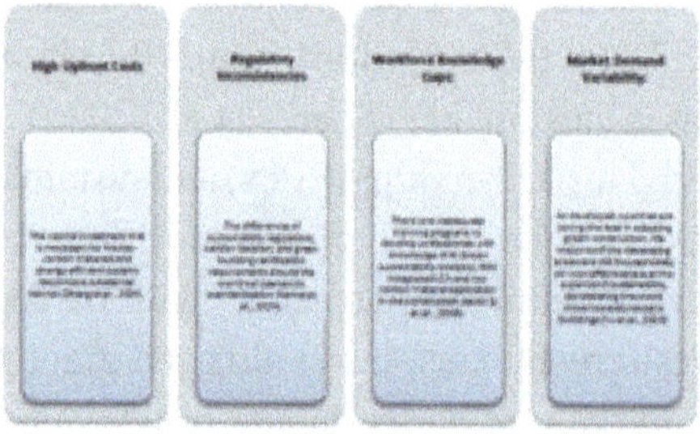

Despite the encouraging improvement in economic and policy arrangements for sustainable construction, several critical problems exist:

- High Upfront Costs: The capital investment that is necessary for the low-carbon materials and energy-efficient systems becomes a substantial barrier (Zhang et al., 2024).
- Regulatory Inconsistencies: The differences of sustainability regulations, carbon taxation, and green building certification requirements around the world set barriers to standardization (Santos et al., 2024).
- Workforce Knowledge Gaps: There are inadequate training programs to develop professionals with knowledge of AI driven sustainability analytics, BIM integrated LCA and low carbon material application in the construction sector (Li et al., 2024).
- Market Demand Variability: As developed countries are taking the lead in adopting green construction, the majority of the developing areas are still focused solely on cost-effectiveness at the expense of sustainability, decelerating the world move towards net zero buildings (Yu et al., 2024).

Future research should target costs reduction in sustainable materials, and beefing up use of financing, and sales of digital solutions for cost efficiency. With more of governments, and industries stepping up their efforts towards net zero goals, there will be increased growth in the market for sustainable construction solutions, thus making green construction the norm in the decades to come (Santos et al 2024).

FUTURE TRENDS AND CHALLENGES IN SMART AND SUSTAINABLE CONSTRUCTION

Emerging Trends in Sustainable Construction

The building construction industry is dramatically changing the technologies with the recent innovations being material sustainable, energy efficient, and digital construction methodologies. These developments are motivated by the demand to minimize carbon emissions, economize on use of resources, and strengthen the built environment. Some of the major trends driving the future of smart and sustainable construction in the next generation include next generation material, AI driven optimization, and circular economy frameworks.

One of the most exciting world-changing innovations is a next generation of sustainable materials that outperform traditional concrete and steel in terms of strength, thermal performance and environmental impact. Studies show that geopolymer concrete, bio-based composites, and self-healing materials will be common substitutes of high-carbon construction materials (Yang et al. 2024). Likewise, carbon-absorbent aggregates and cement filled with biochar receive attention with respect to their capacity to absorb the CO_2 from the atmosphere, providing the carbon-negative solutions for the future buildings (Santos et al., 2024).

Another important trend is the combination of 3D printing and prefabrication processes to support waste minimization, cost efficiency and energy efficient construction. Research demonstrates that the use of 3D-printed concrete decreases material waste by up to 60%, and permits the production of custom complex and high-performance structures (Khan et al., 2024). In addition, prefabricated modular construction is being integrated with recycled and bio-based material in order to lower embodied carbon and increase the efficiency of the construction process (Cavagnoli et al., 2024).

Increasing number of the phase change materials (PCMs) and aerogels used in building insulation is also expected to increase the energy efficiency by controlling indoor temperatures and minimizing the load of both heating and cooling (Wang et al., 2024). Although these materials, along with self-cleaning coatings and dynamic facades, will allow buildings to regulate conditions and live accordingly and save energy (Firoozi et al. 2024).

Digital Transformation and Artificial Intelligence in Construction

Digital construction technologies are to be combined with the optimizing abilities of AI and the automated monitoring systems, which will change sustainable building practices. With the help of AI, various tools are being created to examine matter features, streamline the processes of construction, and estimate building performance from real-time data and climate conditions (Li et al. 2024).

AI application that has had a big impact on construction industry is AI-inspired sustainability analytics, which allows real-time carbon footprint tracking, LCA calculations, and predictive maintenance. Studies indicate that AI-based material selection algorithms can lower the embodied carbon by up to 30%, through suggesting low impact alternative based on structural and environmental needs (Santos et al., 2024).

In addition, Building Information Modeling (BIM) and digital twins are used to simulate energy performance, optimize material efficiency, and optimize waste management during the entire lifecycle of a building. with digital twin technology they are able to forecast maintenance as well as energy modeling thereby ensuring that the buildings function optimally and at the same time reduce their environmental footprint (zhang et al., 2024).

Another major innovation is the Internet of Things (IoT) in smart buildings through sensor-based monitoring systems providing live-time reading of energy consumption, indoor air quality and seeking environmental impacts. These sensors can automatically regulate lighting, HVAC systems, and renewable energy sources reducing the operational demand for energy (Santos et al., 2024).

Circular Economy and Sustainable Urban Development

The circular economy model is expected to take a leading role of curtailing construction waste, encouraging reuse and maximizing resource efficiency. Governing bodies and industry leaders are investing in closed-loop material cycles of which construction and demolition waste (CDW), industrial by-products and bio-based residues are reinvented to new high performance materials (Soares et al., 2025).

One area of attention is the reuse of building components through design for disassembly (DfD), and adaptable architecture. Research makes it known that with modular and prefabricated construction systems that are easily disassemble-able can help with prolonging the lifespan of building materials by up to 50% which will reduce waste generation and also lead to resource depletion (Yu et al., 2024).

Furthermore, urban sustainability planning is changing to adopt net-zero energy buildings, green infrastructure and climate responsive design to future cityscapes. Research shows that cities engaging in low-carbon building policies and decentralized renewable energy systems can have up to 70% reduction in urban energy related emissions (Santos et al., 2024).

Challenges in Implementing Smart and Sustainable Construction

Figure 4. Challenges in implementing smart and sustainable construction

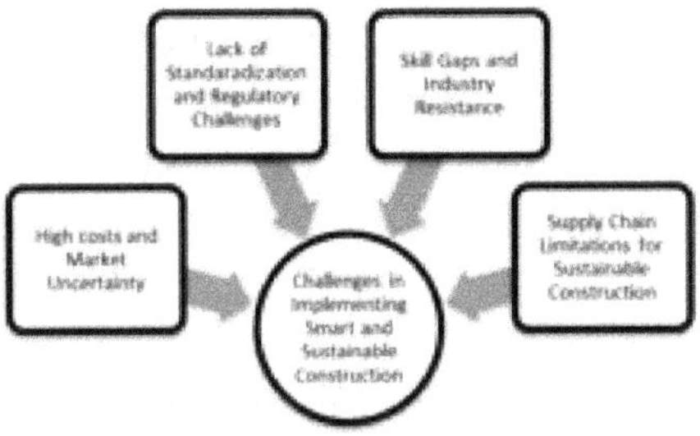

Although there have been significant advances in material science, digital construction, and circular economy techniques, it is increasingly important to develop workable solutions for further scaling up the smart and sustainable construction—several persistent challenges still stand:

High Costs and Market Uncertainty

Although sustainable materials and AI-powered construction tools offer perpetual benefits, the astronomical costs at implementation keep on holding them back from being used on a large scale. According to recent research, there is significant need for R&D financing to develop and implement geo polymer concrete, self-healing materials, and innovative digital buildings technologies (Zhang et al., 2024). These solutions become inaccessible to small and medium-size companies due to high costs.

Lack of Standardization and Regulatory Challenges

The absence of consistent global standards for sustainability and certification of systems has discouraged adoption of low-carbon materials and digital building technology. Leed and Breeam have remained a sound green building certification while their lack in the regulatory approval of emerging buildings materials and techniques is a barrier to market entry (Santos et al., 2024).

Skill Gaps and Industry Resistance

The construction industry is confronted with a clear lack of knowledge that can be used to rollout AI-enhanced sustainable practices, circular economy approaches, and technology innovations to building efficiency. There is a deficit in training widely among architects, engineers and construction managers (declined ability to apply advanced digital technologies and low-carbon materials in project design) (Li et al., 2024). In an effort to get over this barrier, such strategies as strengthening of education, promotion of collaborative research and stimulating technology transfer should be implemented (Soares et al., 2025).

Supply Chain Limitations for Sustainable Materials

The availability and price of sustainable materials range widely, and developing countries are particularly restricted in terms of access to recycled aggregates, bio-based composites, and low-carbon cement. To improve green material access globally, the policymakers have to put extra effort into developing the regional manufacturing networks; negotiating for more favorable trade deals, as well as, building the reinforcing infrastructure (Yu et al., 2024).

The Road Ahead: Future Research and Development Priorities

As for succeeding in tackling these problems, further research should involve:

- Increasing the supply of the following next generation of materials including self-healing concrete, carbon-negative cement, and bio-based insulation (Yang et al., 2024).
- The development of zero spend strategy for AI driven sustainability analytics, BIM integrated LCA tools, digital twin cost-reduction strategies (Li et al., 2024).
- Expanding government incentives and financial models to facilitate investment in circular economy construction and low carbon infrastructure (Santos et al., 2024).
- Standardizing global sustainable material to guarantee unified regulatory approval and adoption in the market (Zhang et al., 2024).

With continuous evolution of clever and sustainable construction, integration of advanced materials, AI-driven optimization and those applying the principles of circular economy will play an important role in the development of the built environment in the future. With further research and intervention by the policy bodies, and technological advancement the industry is projected to shift towards a fully net-zero, resource efficient and intelligent construction ecosystem (Soares et al., 2025).

CONCLUSION

The construction industry is on a significant crossroad because of increased pressure from all quarters for change in line with sustainable, energy efficient and environmentally friendly practices. As one of the biggest global polluters of carbon emission and depletion of resources and a waste producer, the sector must change to smart and sustainable construction methodologies that will enable it to conform to global climate goals and circular economy models. This review has invited the discussion on the role played by innovative materials and energy-efficient strategies, digital construction technologies as well as policy frameworks which have driven this transformation.

One of the most promising sustainable construction is emergence of low carbon and high performance materials. Traditional building materials including Portland cement and steel have been deemed critical sources of CO_2 emissions and embodied carbon footprints in the world. Nevertheless, the uptake of geo polymer concrete, bio-based composites, and carbon capturing materials has demonstrated significant ability to reduce the greenhouse gases and enhance the environmental efficiency of the buildings (Yang et al. 2024). Recycled aggregates, industrial by-products and agri-waste-based materials have also been successfully integrated in recent construction activities and can be seen to reflect the viability of the circular economy approach in reducing consumption of materials and landfill waste (Soares et al, 2025).

Advancements in energy efficient approaches to construction have also become a basic element of sustainable construction practices alongside the material innovation. The use of passive, improved insulation system and smart building material is making structures able to decrease operational energy demand, improve indoor comfort, and reduce dependence on mechanical heating and cooling systems (Wang et al ., 2024). The introduction of modular construction, prefabrication use of building components, and 3D printing has paved the way for further optimization of resource efficiency, reduced onsite waste, and accelerated project delivery times as cost effective and environmentally friendly solution as opposed to conventional construction methods (Khan et al., 2024).

Apart from material and construction approaches, building artificial intelligence-based sustainability analytics, BIM, and digital twins have transformed the construction industry. These tools have been used to improve material selection, maximize energy performance, and assist in minimizing wastage in construction planning and construction execution, hence a data-driven and resource-intensive strategy (Li et al

2024). AI-driven carbon footprint bloodstream and energy monitoring algorithms have been reliable in the reduction of emissions, knowledge of inefficiencies and improved overall performance of buildings in term of long-term (Santos et al., 2024).

However, in spite of such developments, movement towards smart and sustainable building is hampered by economic, regulatory, and market influences, respectively. The high first cost related to sustainable materials, digital technologies, and energy-efficient layouts is one of the major challenges. Most construction firms and developers are reluctant to engage in green building initiatives because their payback periods are long and there is uncertainty in the market and non availability of financial incentives (Zhang et al., 2024). Moreover, the discrepancies in regulations, as well as the absence of standardization in terms of upcoming materials, have slowed these new materials' spread, bringing skepticism among the industry professionals and investors (Santos et al., 2024).

A key challenge, in this regard, remains supply chain and workforce knowledge gap. The lack of coherent deployment of sustainable materials, scarce infrastructure for massive recycling, and scarce work-round training programs have created logistical and technological impediments to the implementation of circular economy construction (Yu et al., 2024). Lacking better industry synergy, research spending's, and educational initiatives, numerous sustainable solutions will become niche applications instead of mainstream industry approaches (Soares et al., 2025).

As we project into the future, technological innovation, policy interventions and changing market forces will shape the future of smart and sustainable construction. Governments and industry leaders have to work together to scale-up low carbon materials, broaden financial incentives for building energy efficiency and enforce regulatory framework for sustainability. The use of carbon pricing, tax incentives for green construction projects and worldwide standardization for sustainability assessment will be necessary in breaking down economic and regulatory barriers (Terzis, 2025).

Furthermore, an increasing importance and influence of corporate sustainability commitments and investor initiated ESG (Environmental, Social, and Governance) initiatives will only enhance the need for net-zero resource –efficient construction solutions (Santos et al. 2024). As significant real estate developers and construction firms focus on carbon reduction targets, the market of sustainable building materials as well as AI-powered energy management systems and modular prefabrication techniques is likely to grow substantially (Zhang et al., 2024).

There are several critical actions that must be taken for the industry to enter a net-zero, resilient, and resource efficient future fully.

1. Expand Research & Development: Saving large sums of money on materials like carbon-negative cement, bio char composites, and self-healing concrete thereby

increasing material durability, carbon sequestering potentials and overall high performance (Yang et al., 2024).

2. Improve Financial Accessibility: Promoting green financing models, sustainability linked-loans, and government assisted incentives in order to increase financial feasibility of sustainable construction (Santos et al., 2024).
3. Enhance Digital Integration: Utilizing AI driven material selection, digital twin simulations and IoT based smart energy monitoring to optimize the performance and efficiency of building (Li et al. 2024).
4. Strengthen Global Regulatory Standards: Forming reliable sustainability assessment frameworks and obligatory carbon footprint reporting to help build green practices across the market (Zhang et al., 2024).

5.Expand Circular Economy Infrastructure: Establishing regional recycling hubs, upcycling facilities, and circular supply chain systems in as far as possible maximizing waste reductions and improving the efficiency of resources (Soares et al., 2025).

6. Take up Investments in Workforce Training & Education. Building training frameworks for architects, engineers, and construction professionals that will mainstream sustainable practice in building design (Yu et al., 2024).

With climate change issues on the increase, there is a unique opportunity for the construction industry to become a leader for world sustainability initiatives. Drawing on next-generation materials, AI-driven optimization and values, the industry can create a purposeful movement towards net zero carbon targets. With further innovation, collaboration, and regulatory support, smart and sustainable construction will transform the way in which built environments are delivered creating long term economic, environmental, as well as social benefit for generations to come.

REFERENCES

Cavagnoli, S., Fabiani, C., de Albuquerque Landi, F. F., & Pisello, A. L. (2024). Advancing sustainable construction through comprehensive analysis of thermal, acoustic, and environmental properties in prefabricated panels with recycled PET materials. *Energy and Building*, *312*, 114218. Advance online publication. DOI: 10.1016/j.enbuild.2024.114218

Chen, L., Yang, M., Chen, Z., Xie, Z., Huang, L., Osman, A. I., Farghali, M., Sandanayake, M., Liu, E., Ahn, Y. H., Al-Muhtaseb, A. H., Rooney, D. W., & Yap, P. S. (2024). Conversion of waste into sustainable construction materials: A review of recent developments and prospects. *Materials Today Sustainability*, *27*, 100930. DOI: 10.1016/j.mtsust.2024.100930

Chen, L., Yang, Y., Liu, J., & Xie, P. (2024). Advancing eco-friendly cementitious composites: Integrating carbon capture and sustainable resource utilization. *Cement and Concrete Research*, *170*, 106302. DOI: 10.1016/j.cemconres.2024.106302

Firoozi, A. A., Firoozi, A. A., Oyejobi, D. O., Avudaiappan, S., & Saavedra Flores, E. (2024). Emerging trends in sustainable building materials: Technological innovations, enhanced performance, and future directions. *Results in Engineering*, *24*, 103521. DOI: 10.1016/j.rineng.2024.103521

Hernando Castro, S. M., Rodríguez, R. C., López Arquillo, J. D., & Álvarez de Eulate, M. P. (2025). Comparative evaluation of construction insulation materials: Environmental performance across production, end-of-life, and beyond system boundaries. *Journal of Building Engineering*, *99*, 111570. DOI: 10.1016/j.jobe.2024.111570

Huang, H., Lu, J., Jin, L., & Ren, H. (2024). The future of environmental engineering technology: A disruptive innovation perspective. *Engineering*, *41*(10), 153–160. DOI: 10.1016/j.eng.2024.06.009

Khan, S. A., Jassim, M., Ilcan, H., Sahin, Ö., Bayer, İ. R., Sahmaran, M., & Koç, M. (2023). 3D printing of circular materials: Comparative environmental analysis of materials and construction techniques. *Case Studies in Construction Materials*, *18*, e02059. DOI: 10.1016/j.cscm.2023.e02059

Lai, Z., & Chen, Y. (2024). Enhancing the mechanical and environmental performance of solidified soil using construction waste and glass micro-powder. *Heliyon*, *10*(22), e40187. DOI: 10.1016/j.heliyon.2024.e40187 PMID: 39584091

Li, L., Yi, Z., Jiang, F., Zhang, S., & Zhou, J. (2023). Exploring the mechanism of digital transformation empowering green innovation in construction enterprises. *Developments in the Built Environment*, *15*, 100199. DOI: 10.1016/j.dibe.2023.100199

Meng, Q., Hu, L., Li, M., & Qi, X. (2024). Assessing the environmental impact of building life cycle: A carbon reduction strategy through innovative design, intelligent construction, and secondary utilization. *Environmental Science & Technology*, *58*(4), 765–778. DOI: 10.1021/acs.est.3c07876

Poopalam, K. D., Ismail, T. N. M. T., Hanzah, N. A., Alias, A. H., Wahab, N. A., Ibrahim, Z., Ibrahim, Z., Subramaniam, V., Armylisas, A. H. N., & Idris, Z. (2024). Utilization of oil palm biomass and polyurethanes as sustainable construction materials: A review. *Developments in the Built Environment*, *17*, 100380. DOI: 10.1016/j.dibe.2024.100380

Santos, A. K., Ferreira, V. M., & Dias, A. C. (2025). Promoting decarbonisation in the construction of new buildings: A strategy to calculate the embodied carbon footprint. *Journal of Building Engineering*, *103*, 112037. DOI: 10.1016/j.jobe.2025.112037

Soares, M., Faria, L., Miranda, T., Pereira, E., Vilarinho, C., & Carvalho, J. (2025). The potential of agri-food waste to solve construction's environmental problems: A review. *Cleaner and Circular Bioeconomy*, *10*, 100138. DOI: 10.1016/j.clcb.2025.100138

Terzis, D. (2022). Monitoring innovation metrics in construction and civil engineering: Trends, drivers and laggards. *Developments in the Built Environment*, *9*, 100064. DOI: 10.1016/j.dibe.2021.100064

Wang, X., Zhang, Y., Hu, X., Jia, X., Li, K., Wang, C., & Wang, Y. (2024). Sustainability-oriented construction materials for traditional residential buildings: From material characteristics to environmental suitability. *Building and Environment*, *234*, 109944. DOI: 10.1016/j.buildenv.2024.109944

Yang, M., Lin, C., Lai, J., Osman, A. I., Farghali, M., Rooney, D. W., & Yap, P. S. (2024). Advancing environmental sustainability in construction through innovative low-carbon, high-performance cement-based composites: A review. *Materials Today Sustainability*, *26*, 100712. DOI: 10.1016/j.mtsust.2024.100712

Yu, H., Zahidi, I., Chow, M. F., Liang, D., & Madsen, D. Ø. (2024). Reimagining resources policy: Synergizing mining waste utilization for sustainable construction practices. *Journal of Cleaner Production*, *464*, 142795. Advance online publication. DOI: 10.1016/j.jclepro.2024.142795

Yu, Z., Nurdiawati, A., Kanwal, Q., Al-Humaiqani, M. M., & Al-Ghamdi, S. G. (2024). Assessing and mitigating environmental impacts of construction materials: Insights from environmental product declarations. *Journal of Building Engineering*, *98*, 110929. DOI: 10.1016/j.jobe.2024.110929

Zhang, R., Tang, Y., Liu, G., Wang, Z., & Zhang, Y. (2024). Unveiling the impact of innovation on competitiveness among construction projects: Moderating and mediating role of environmental regulation and sustainability. *Ain Shams Engineering Journal, 15*(4), 102558. DOI: 10.1016/j.asej.2023.102558

Zhu, J., Zhang, C., Zhang, M., & Yuan, J. (2024). Modeling and simulation of technological innovation diffusion of high-performance construction materials in transportation engineering. *Transportation Research Part D, Transport and Environment, 120*, 103838. Advance online publication. DOI: 10.1016/j.trd.2024.103838

Chapter 10
Sustainability in Practice: The Intersection of Nature, Technology, and Collaborative Design

Anosh Nadeem Butt
https://orcid.org/0000-0002-9662-9473
University of Glasgow, UK

ABSTRACT

This chapter will explore the dynamic intersection of nature-based design, cutting-edge technological solutions, and interdisciplinary collaboration in the pursuit of sustainable architecture. In an era marked by increasing environmental challenges, the integration of natural systems with advanced technologies offers a transformative approach to building design that can significantly reduce resource consumption, enhance resilience, and promote ecological stewardship. The chapter will provide a comprehensive overview of how these elements—nature, technology, and collaboration—interact in practice, offering case studies, theoretical insights, and practical recommendations for architects, engineers, and other design professionals working together to achieve sustainability goals in the built environment.

INTRODUCTION

The Need for Sustainable Architecture

The built environment is a significant contributor to global environmental challenges, including climate change, resource depletion, and biodiversity loss. Buildings account for nearly 40% of global energy consumption and 33% of greenhouse gas

DOI: 10.4018/979-8-3373-2555-2.ch010

emissions, making sustainable architecture a critical area for intervention (IEA, 2021). As urban populations grow, the demand for new construction continues to rise, further amplifying environmental impacts. Consequently, the architecture, engineering, and construction (AEC) industry must adopt innovative strategies that reduce carbon footprints, enhance energy efficiency, and promote ecological balance (Kibert, 2016).

Sustainability in architecture goes beyond energy efficiency; it requires a holistic approach that integrates environmental, social, and economic considerations. The shift toward sustainable buildings is driven by global policies, such as the United Nations' Sustainable Development Goals (SDGs) and international frameworks like BREEAM (Building Research Establishment Environmental Assessment Method) and LEED (Leadership in Energy and Environmental Design) (Canton, 2021). While regulatory frameworks provide benchmarks for sustainability, achieving meaningful environmental impact demands more than compliance; it necessitates innovative, interdisciplinary, and context-sensitive approaches (Guy & Farmer, 2001; Carbajal-Pina & Acur, 2024).

Limitations of Conventional Design and Construction Methods

To fully appreciate the transformative potential of technological innovations in sustainable architecture, it is crucial to first acknowledge the limitations of conventional design and construction practices. Traditional methods have historically prioritized short-term cost savings, speed, and standardized processes over environmental performance and long-term resilience. These shortcomings include: (1) high carbon footprint: conventional construction materials such as concrete and steel are energy-intensive to produce and contribute significantly to greenhouse gas emissions (Fenner et al., 2018; Sizirici et al., 2021); (2) resource inefficiency: linear construction models result in substantial waste generation, with minimal reuse or recycling of materials (Janda et al., 2015; Kamali & Hewage, 2017); (3) lack of environmental integration: traditional buildings often overlook site-specific ecological conditions, leading to poor energy performance, excessive water use, and habitat disruption (Handfield et al., 2002; Leoto & Lizarralde, 2019); (4) limited adaptability: rigid design and construction techniques often result in buildings that are less adaptable to changing environmental conditions or user needs (Heidrich et al., 2017; Mlote et al., 2024); and (5) fragmented workflows: siloed professional roles and limited cross-disciplinary coordination contribute to inefficiencies and missed opportunities for sustainability integration (Kim, 2012; Wortmann & Tunçer, 2017).

In contrast, sustainable approaches integrate emerging technologies, ecological principles, and collaborative processes to deliver buildings that are not only more efficient but also regenerative and future-ready. The following sections explore key technological innovations that address these limitations and support the transition to sustainable architecture.

The Trialogue for Sustainable Building Design

To address contemporary environmental challenges, sustainable building design must integrate three critical dimensions:

Nature-Based Solutions (NbS)

Nature-Based Solutions refer to strategies that draw from the natural environment to help buildings perform better while enhancing ecological resilience. Instead of relying solely on mechanical or technological systems, these solutions tap into natural processes to achieve sustainability goals. For example, green roofs, rain gardens, and urban forests can manage stormwater, reduce urban heat islands, and improve air quality. NbS focuses on using ecosystem services to benefit both the built environment and the surrounding ecosystem. Additionally, incorporating elements such as passive solar design and natural ventilation can reduce a building's reliance on energy-consuming systems. NbS help foster biodiversity and provide co-benefits like improved aesthetic value and public health.

Technological Innovations

This dimension encompasses the use of advanced technologies to reduce the environmental footprint of buildings. These innovations range from energy-efficient building materials, such as insulation made from recycled materials or low-emission concrete to smart technologies like sensor-based energy management systems. Smart systems allow for dynamic adjustments to heating, cooling, and lighting based on occupancy and environmental conditions, leading to more efficient energy use. Other technological tools include building information modelling (BIM) and simulation software, which enable architects and engineers to design buildings that optimize energy flow, reduce waste, and simulate the building's performance over time. Innovations like these minimize energy consumption and help buildings adapt to climate change impacts.

Interdisciplinary Collaboration

Creating sustainable buildings requires the collective expertise of various professionals. Architects, engineers, urban planners, and environmental scientists must collaborate closely from the outset of a project to ensure that sustainability is embedded throughout the design, construction, and operational phases of a building. This interdisciplinary approach encourages the integration of diverse perspectives, ensuring that environmental, social, and economic sustainability goals are met. For instance, an architect's vision can be aligned with an engineer's technical solutions, while urban planners ensure that the building fits well within its urban context, considering factors such as transportation, green spaces, and infrastructure. Collaboration also extends beyond just professionals; it involves engaging local communities and stakeholders to ensure the project meets the needs of both people and the planet.

Nature-based Solutions: Learning from Ecosystems

Nature-Based Solutions (NbS) have gained prominence as a means to enhance urban resilience, improve biodiversity, and mitigate climate change (Seddon et al., 2021). Inspired by ecological systems, NbS encompasses design strategies such as biophilic design, green infrastructure, and regenerative architecture (Kellert, 2011; Macaione et al., 2024). Green roofs, living walls, and urban tree canopies are examples of how NbS can reduce urban heat islands, manage stormwater, and improve air quality (Hayes et al., 2022).

A growing body of research supports the idea that integrating NbS into the built environment yields economic and social benefits in addition to environmental gains. For example, buildings designed with natural ventilation and daylighting strategies have been shown to reduce energy consumption by up to 30% while enhancing occupant well-being and productivity (Altomonte, 2008). Such findings highlight the importance of reconnecting architecture with nature as part of a sustainable development strategy.

Technological Innovations for Sustainable Architecture

Advancements in digital technology have revolutionized how architects and engineers design, simulate, and construct sustainable buildings. Among the most influential technologies are:

- Building Information Modelling (BIM) – A digital tool that allows for precise planning and resource optimization, reducing material waste and improving building lifecycle performance (Najjar et al., 2019; Xu et al., 2021).
- Smart Building Systems – The use of Internet of Things (IoT) sensors, automation, and artificial intelligence (AI) to monitor and optimize energy use in real-time (Shah et al., 2022; Alijoyo, 2024).

- Renewable Energy Integration – Solar photovoltaics (PV), wind energy, and geothermal systems that allow buildings to generate clean energy and achieve net-zero emissions (Ahmed et al., 2022; Reddy et al., 2024).
- Material Innovations – The development of low-carbon concrete, bio-based insulation, and self-healing materials that reduce the environmental impact of construction (Di Sarno et al., 2024).

While technology provides powerful tools for sustainability, its full potential can only be realized when paired with thoughtful design and interdisciplinary collaboration.

The Role of Interdisciplinary Collaboration in Sustainable Design

Sustainable architecture is inherently interdisciplinary, requiring architects, engineers, urban planners, environmental scientists, and policymakers to work together toward shared objectives (Williams, 2007). However, collaboration is often hindered by disciplinary silos, lack of shared knowledge, and conflicting project priorities (Nwulu et al., 2023). Successful sustainable projects demonstrate the value of integrated design teams, where knowledge-sharing across disciplines leads to innovative and high-performance solutions (Oxenswärdh, 2019). The success of projects such as The Edge in Amsterdam, one of the world's most energy-efficient buildings illustrates how combining nature-based strategies, smart technologies, and interdisciplinary expertise can result in exceptional sustainability outcomes (Lawrence, 2020).

This chapter explores the intersection of nature-based solutions, technological innovation, and interdisciplinary collaboration in achieving sustainability in architectural design. It will:

- Examine the principles and benefits of nature-based design in the built environment.
- Investigate technological advancements that support energy efficiency and low-carbon construction.
- Highlight the importance of interdisciplinary collaboration in achieving sustainable outcomes.
- Present case studies showcasing successful integration of these principles.
- Discuss challenges and barriers, offering strategies to overcome them.

By bridging the gap between nature, technology, and people, this chapter aims to provide a comprehensive framework for architects and designers seeking to create buildings that are resilient, regenerative, and future-ready.

NATURE-BASED SOLUTIONS (NBS) IN SUSTAINABLE ARCHITECTURE

Introduction

The increasing global demand for sustainable urban development has led to a growing emphasis on Nature-Based Solutions (NbS) in architecture. As cities expand and environmental challenges such as climate change, biodiversity loss, and resource depletion intensify, architects and urban planners are increasingly turning to ecological principles to enhance sustainability in the built environment. NbS refers to design approaches that integrate natural processes, mimic ecological systems, and harness the benefits of biodiversity to create energy-efficient, resilient, and environmentally responsive buildings (Davies et al., 2021). Unlike conventional construction techniques that often prioritize short-term efficiency over long-term sustainability, NbS seeks to harmonize the built environment with natural systems, fostering climate resilience, energy efficiency, and improved urban livability (Anguelovski & Frumkin, 2025).

The application of NbS in architecture extends beyond aesthetics and energy conservation; it plays a pivotal role in addressing urban heat islands, air and water pollution, carbon sequestration, and mental well-being (Anguelovski & Frumkin, 2025). Strategies such as biophilic design, green roofs, passive cooling, and regenerative urban planning contribute to more sustainable urban landscapes, benefiting both ecosystems and communities. This section explores the core principles of NbS, its key design strategies, environmental and social benefits, case studies of successful implementation, and challenges associated with its integration into architectural practice.

Principles of Nature-based Solutions in Architecture

Nature-Based Solutions are guided by several fundamental principles that ensure their effectiveness in promoting sustainability in the built environment. First, ecosystem integration is essential, as architectural designs must work in harmony with natural systems rather than against them (Hegazy et al., 2025). Buildings designed with NbS principles leverage ecosystem services such as passive temperature regulation, natural ventilation, and water filtration, reducing their environmental footprint while enhancing efficiency. Second, climate adaptation is a key principle, ensuring that buildings are designed to withstand climate-related challenges such as heatwaves, flooding, and extreme weather events (Wang et al., 2025). This requires the integration of green infrastructure, blue-green corridors, and water-sensitive urban design to enhance resilience.

A third guiding principle is energy efficiency, which entails reducing reliance on mechanical heating, cooling, and artificial lighting through passive design strategies and natural material selection (Gijón Mancheño et al., 2025). Additionally, biodiversity conservation is integral to NbS, as urban developments must foster ecological diversity by incorporating native vegetation, pollinator habitats, and regenerative landscapes (La Notte & Zulian, 2021). Finally, NbS must be designed to generate social and economic co-benefits, including improved air quality, community engagement, public health, and long-term cost savings (Anguelovski & Frumkin, 2025). By adhering to these principles, NbS facilitates a shift toward architecture that is environmentally regenerative, resource-efficient, and socially inclusive.

Nature-based strategies in Architecture

A range of architectural strategies has been developed to integrate NbS into contemporary design. Among the most widely adopted approaches is biophilic design, which incorporates natural light, ventilation, vegetation, and water elements into built environments to enhance human well-being and productivity (Kellert, 2011). Studies indicate that biophilic spaces can significantly reduce stress, increase cognitive performance, and improve indoor air quality, leading to healthier and more efficient work and living environments (Wang et al., 2025). One notable example is Amazon's Spheres in Seattle, where over 40,000 plants create a climate-controlled workspace that enhances both air purification and psychological well-being (Jamawat, 2019).

Another critical NbS strategy is green roofs and living walls, which mitigate urban heat islands, improve insulation, and support biodiversity (Wang et al., 2025). These features provide temperature regulation, carbon sequestration, and stormwater management, making them particularly effective in densely populated cities. The Parkroyal Collection Hotel in Singapore demonstrates the effectiveness of this approach, where extensive green facades reduce ambient temperatures by up to 4°C, significantly lowering energy demands for cooling (Rowe et al., 2019).

Passive cooling and heating systems further enhance NbS by utilizing natural ventilation, solar shading, and thermal mass materials to optimize indoor comfort without excessive energy consumption. These strategies include cross-ventilation, solar chimneys, tree canopies, and shading devices, reducing reliance on artificial climate control systems (Lawrence, 2020). The Edge in Amsterdam, one of the most energy-efficient buildings in the world, integrates passive cooling and dynamic shading to achieve 70% lower energy demand compared to conventional office buildings.

Additionally, water-sensitive urban design (WSUD) focuses on incorporating rainwater harvesting, permeable pavements, and bioswales into urban environments to manage flooding, improve water quality, and enhance groundwater recharge (Su et al., 2025). The Copenhagen Cloudburst Management Plan showcases how cities can leverage green and blue infrastructure to mitigate flood risks while simultaneously enhancing urban aesthetics and biodiversity (Mudunuri & Attaluri, 2025).

Environmental and Social Benefits of NbS

Nature-Based Solutions offer significant environmental, economic, and social benefits. One of the primary advantages is carbon sequestration, as NbS strategies like green roofs, afforestation, and soil regeneration contribute to reducing atmospheric CO_2 levels (Wang et al., 2025). Additionally, NbS improves air quality by filtering pollutants, reducing particulate matter, and increasing oxygen production through vegetation. Another notable benefit is heat mitigation and microclimate regulation, where NbS strategies lower urban temperatures and prevent heat stress-related illnesses (Hayes et al., 2022).

Beyond environmental advantages, NbS contributes to public health and well-being. Access to green spaces has been shown to reduce stress, anxiety, and depression, while also fostering community interaction and social cohesion (Yevenes et al., 2025). NbS also promotes biodiversity conservation, restoring ecosystems that are often disrupted by urbanization (Sowińska-Świerkosz & García, 2022).

Summary

Section 2 examines the growing role of Nature-Based Solutions (NbS) in architecture as a critical response to the escalating challenges posed by climate change, biodiversity loss, and urban resource depletion. As cities expand and environmental pressures intensify, NbS has emerged as a design philosophy that integrates ecological processes into the built environment. This section outlines how NbS can contribute to the creation of buildings and urban spaces that are more energy-efficient, resilient, and socially inclusive.

The discussion begins with an overview of the foundational principles that underpin NbS. These include the integration of ecosystem services into design, climate adaptation through green and blue infrastructure, and the enhancement of energy efficiency through passive strategies. NbS also promotes biodiversity conservation by embedding natural habitats within urban contexts and seeks to generate co-benefits such as improved air quality, public health, and long-term economic savings.

The section then explores a range of architectural strategies that operationalise NbS in practice. Biophilic design, for example, incorporates elements such as natural light, ventilation, and vegetation to improve human well-being and indoor environmental quality. Green roofs and living walls serve to mitigate the urban heat island effect, enhance insulation, and support ecological diversity. Passive cooling and heating techniques make use of natural ventilation and shading to regulate indoor temperatures while reducing energy demand. Water-sensitive urban design further complements these approaches by integrating stormwater management features that improve flood resilience and water quality.

In addition to outlining these strategies, the section highlights the environmental and social benefits of NbS. These include carbon sequestration, microclimate regulation, air purification, and ecosystem restoration. On a social level, NbS contributes to mental well-being, fosters community cohesion, and supports equitable access to healthier living environments. The section concludes by emphasising that NbS not only addresses pressing ecological issues but also offers a transformative approach to architecture that aligns with broader goals of sustainability, resilience, and human-centred design.

TECHNOLOGICAL INNOVATIONS IN SUSTAINABLE ARCHITECTURE

The integration of advanced technologies into sustainable architecture is transforming the built environment, leading to enhanced energy efficiency, reduced environmental impact, and improved occupant comfort. This section explores key innovations, including smart building technologies, sustainable materials and construction techniques, and examines case studies of high-tech green buildings.

Smart Building Technologies

Smart building technologies leverage artificial intelligence (AI), the Internet of Things (IoT), and automation to optimize building performance and sustainability.

Role of AI, IoT, and Automation in Sustainable Design

Artificial intelligence, the Internet of Things, and automation work synergistically to create intelligent building systems that can monitor, control, and adjust various building functions in real time (Rane, 2023). These technologies collect and analyze vast amounts of data through embedded sensors placed throughout the building. Sensors can measure factors such as temperature, humidity, occupancy levels, and energy consumption, providing real-time insights into how a building is performing (Saha et al., 2019).

For instance, in smart buildings, AI-driven systems can adjust the heating, ventilation, and air conditioning (HVAC) systems based on occupancy and external weather conditions, ensuring that energy is only used when needed (Lee & Lee, 2023). This allows buildings to dynamically manage their internal environments, optimizing energy consumption while maintaining occupant comfort. By analyzing data from these sensors, AI algorithms can predict future energy demands, enabling the system to proactively adjust and minimize energy waste (Ukoba et al., 2024). For example, an AI-powered HVAC system can predict cooling and heating needs based on weather forecasts and the building's occupancy patterns, leading to significant reductions in energy usage. Studies have shown that these intelligent systems can reduce energy consumption by up to 30%, significantly lowering operating costs and minimizing the building's environmental footprint (Nguyen & Aiello, 2013; Shaikh et al., 2014).

Furthermore, IoT-enabled devices in smart buildings can integrate with smart grids to manage energy more efficiently (Goudarzi et al., 2022). A smart grid is a modernized electrical grid that uses digital communication and IoT devices to manage the distribution of electricity (Hossain et al., 2013). It can balance energy supply and demand by incorporating renewable energy sources such as solar panels and wind turbines. This enables buildings to use energy more efficiently to generate and store their own energy, supporting the transition towards zero-carbon buildings. By connecting building systems with renewable energy sources, smart buildings can optimize energy distribution, reduce reliance on fossil fuels, and contribute to the overall decarbonization of the built environment.

Energy-Efficient Smart Buildings: Passive Cooling, Dynamic Facades, Adaptive Shading

Smart buildings are also designed to incorporate energy-efficient features that reduce the reliance on mechanical systems for heating and cooling. Some of these innovations include passive cooling techniques, dynamic facades, and adaptive shading systems.

Passive Cooling refers to using architectural design elements and materials to naturally cool a building (Kamal, 2012). This technique capitalizes on natural ventilation, the use of thermal mass, and shading strategies to regulate internal temperatures without the need for air conditioning. For example, buildings can be designed to encourage airflow through strategic window placement, and materials with high thermal mass can absorb heat during the day and release it at night, maintaining a comfortable indoor temperature (Zhang et al., 2016).

Dynamic Facades are facades that can change their properties in response to environmental conditions (Shafaghat & Keyvanfar, 2022). These facades use sensors connected to IoT systems to automatically adjust based on factors such as temperature, humidity, and sunlight intensity (Chiesa et al., 2020). For example, building facades can be designed to open or close windows or alter the building's insulation properties based on weather conditions, helping regulate indoor temperature and light levels. This reduces the need for artificial heating and cooling and enhances the building's aesthetic appeal, creating a building that is both functional and visually engaging.

Adaptive Shading Systems work in a similar way by adjusting their position based on the sun's angle and intensity (Konstantoglou & Tsangrassoulis, 2016). These shading devices optimize the use of natural light, minimizing glare and reducing heat gain (De Luca et al., 2022). For instance, a smart shading system can automatically adjust window coverings to block direct sunlight during peak hours, while still allowing diffused natural light to enter, thus reducing the need for artificial lighting and mechanical cooling. These technologies contribute to improved occupant comfort and substantial energy savings.

Sustainable Materials and Construction Techniques

The adoption of sustainable materials and innovative construction methods is crucial for reducing the environmental footprint of buildings.

Use of Bio-Based Materials, Nanomaterials, and Circular Design

Sustainable materials are central to reducing the carbon footprint of buildings. Bio-based materials, derived from renewable sources such as wood, hemp, and agricultural residues, offer a sustainable alternative to traditional construction materials like concrete and steel (Yadav & Agarwal, 2021). These materials reduce reliance on finite natural resources and also sequester carbon, contributing to climate change mitigation (Sutkowska et al., 2024). For example, wood is a carbon-neutral material that, when responsibly sourced, can store carbon dioxide for the life of the building, offsetting emissions from other materials.

Nanomaterials are another innovative development in sustainable construction. Nanotechnology involves the manipulation of materials at the molecular or atomic level, enhancing properties such as strength, durability, and thermal performance (Muhammed & Tsakalakos, 2003). When incorporated into construction materials, nanomaterials can result in lighter, stronger, and more energy-efficient buildings (Hanus & Harris, 2013). For example, nanomaterials can be used in insulation to improve heat retention, or in coatings that reduce the need for frequent maintenance and repairs. These advancements lead to longer-lasting, more resilient buildings with reduced energy consumption over time.

Circular Design is a growing trend in sustainable construction, focusing on creating closed-loop systems where building materials are reused or recycled at the end of their life cycle (Charef et al., 2022). This concept aims to minimize waste and reduce the consumption of raw materials. By prioritizing the reuse and recycling of materials, circular design helps reduce landfill waste and lowers the environmental impact of construction (Rahla et al., 2021). This approach aligns with the principles of a circular economy, which seeks to preserve natural capital, optimize resource yields, and minimize risks associated with resource depletion.

Prefabrication and Modular Construction for Sustainability

Prefabrication and modular construction are innovative construction methods that contribute to sustainability by reducing waste, improving quality control, and minimizing environmental disruption (Jiang et al., 2019; Adeyemi et al., 2024). Prefabrication involves manufacturing building components off-site in controlled factory settings, where they are assembled before being transported to the construction site. This approach offers several advantages:

- Reduced Waste: By using precise manufacturing techniques, prefabrication minimizes material waste. Materials are cut and assembled with high accuracy, reducing excess materials and improving the efficiency of resource use (Tam et al., 2005; Ferreira Silva et al., 2020).
- Improved Quality Control: Manufacturing components in a controlled factory environment ensures higher standards of quality and durability (Boafo et al., 2016). Buildings constructed with prefabricated components are typically stronger and more resilient, requiring fewer repairs and less maintenance over time (Fleischman & Seeber, 2016).
- Accelerated Construction Timelines: Prefabrication allows for concurrent site preparation and the fabrication of building components, shortening construction timelines (Smith, 2010). This reduces environmental disruption at construction sites and minimizes the overall carbon footprint of the project.
- Enhanced Energy Efficiency: Buildings constructed using prefabricated components are typically more energy-efficient due to their tight construction and better insulation (Sadineni et al., 2011).

These advantages make prefabrication and modular construction ideal methods for creating energy-efficient, sustainable buildings. Recent advancements in the use of bio-based materials for prefabrication have further enhanced the sustainability of these techniques, enabling the creation of energy-efficient, environmentally friendly buildings at scale.

Summary

Chapter 3 examines the transformative role of technological innovation in advancing sustainable architecture. As environmental pressures intensify and the demand for efficient buildings grows, the integration of advanced technologies has become critical to reducing energy consumption, improving indoor environmental quality, and achieving long-term environmental resilience. This chapter explores developments in smart building systems, sustainable construction materials, and innovative methods such as prefabrication, supported by real-world case studies.

The chapter begins by analysing smart building technologies, which utilize artificial intelligence, the Internet of Things, and automation to monitor and optimise building performance. These technologies enable intelligent systems to regulate temperature, lighting, and energy use based on real-time data, thereby improving both efficiency and occupant comfort. The implementation of dynamic facades, adaptive shading, and passive cooling systems illustrates how buildings can respond to environmental conditions with minimal reliance on mechanical systems. Through automated adjustments informed by occupancy and weather data, smart buildings contribute significantly to energy savings and the reduction of carbon emissions.

The discussion then turns to sustainable materials and construction techniques. The adoption of bio-based materials, including timber, hemp, and agricultural by-products, reflects a shift towards renewable and carbon-sequestering alternatives to concrete and steel. The chapter also highlights the application of nanomaterials, which enhance the structural and thermal properties of construction components at the molecular level. These innovations contribute to improved durability, thermal regulation, and overall building performance.

Circular design principles are presented as a foundational approach to reducing construction waste and promoting material reuse. This concept supports a closed-loop system in which building components can be disassembled, recycled, or repurposed at the end of their life cycle. By reducing landfill waste and reliance on virgin resources, circular design aligns with broader efforts to transition to a circular economy.

Finally, the chapter explores prefabrication and modular construction as forward-looking methods that enhance both sustainability and efficiency. These approaches allow for precise manufacturing in controlled environments, resulting in reduced material waste, improved quality control, and shorter construction timelines. Buildings constructed through prefabrication also benefit from enhanced energy performance due to improved insulation and airtightness. The integration of bio-based materials into prefabricated systems further elevates the environmental benefits of this technique, making it suitable for scalable, low-impact development.

Overall, Chapter 3 underscores how the convergence of digital technologies, material science, and innovative construction practices is reshaping the architectural landscape. By embedding sustainability into each stage of the building lifecycle, these technological advancements support the transition toward environmentally responsible and future-ready built environments.

INTERDISCIPLINARY COLLABORATION IN SUSTAINABLE DESIGN

Sustainable design is no longer the sole domain of architects and engineers; it requires the concerted efforts of a wide range of professionals, including urban planners, ecologists, sustainability experts, and other stakeholders (Murtagh et al., 2016). This shift toward interdisciplinary collaboration is reshaping how buildings and urban spaces are planned, constructed, and maintained (McClure & Bartuska, 2011). The emphasis on collaboration across different domains of expertise helps ensure that sustainability goals are achieved and foster innovation, reduces inefficiencies, and ensures a holistic approach to project development (Adams et al., 2016). This section delves into the necessity of interdisciplinary collaboration in sustainable design, examining the role of frameworks like the RIBA Plan of Work, sustainable certification systems, and highlighting successful case studies that demonstrate the power of such collaborative efforts.

The Need for Interdisciplinary Collaboration

The increasing complexity of sustainable design requires a collaborative approach that integrates diverse skill sets, knowledge, and expertise (Mauser et al., 2013). Traditional architectural and engineering practices often operate in silos, where architects focus on design and aesthetics, while engineers handle the structural and systems aspects of the building (Hansen, 2014). However, this separation can lead to inefficiencies, conflicts, and missed opportunities for optimizing sustainability. By contrast, an interdisciplinary approach encourages professionals from various fields to work together from the outset, ensuring that sustainability is integrated into every aspect of the project (Annan-Diab & Molinari, 2017).

Challenges of Siloed Approaches in Architecture and Engineering

One of the key challenges of siloed approaches in architectural and engineering practices is the poor integration of sustainability measures (Ikudayisi et al., 2023). For example, passive design strategies, such as the use of natural ventilation and passive solar heating, may not be properly integrated with mechanical systems like heating, ventilation, and air conditioning (HVAC). As a result, a building might require additional energy inputs to compensate for inefficiencies that could have been avoided through early collaboration. Similarly, renewable energy systems such as solar panels or wind turbines might not be optimally positioned or integrated with the building's energy needs if not considered from the start (Omer, 2008).

These types of disconnects lead to wasted energy and higher operational costs, but they can also result in higher project costs due to inefficiencies, redesigns, and delays. Communication breakdowns between different stakeholders can exacerbate these issues, making it more difficult to solve problems and find solutions that meet sustainability objectives (Hirsch et al., 2001). Moreover, a siloed approach makes it harder to adapt to emerging environmental challenges and new regulatory requirements. As environmental and climate challenges become more urgent, the ability to adapt and respond quickly is crucial, and interdisciplinary collaboration facilitates this process.

Benefits of a Collaborative, Integrated Process

An interdisciplinary, integrated approach to sustainable design addresses these challenges by bringing together experts from different fields early in the design process. This type of collaboration fosters a more holistic understanding of sustainability, enabling a range of environmental, social, and economic considerations to be factored into the design from the beginning.

Key benefits of interdisciplinary collaboration include:

- Holistic Sustainability Strategies: By involving a range of professionals, sustainability can be embedded into every stage of the project, from site selection and energy efficiency to material choices and waste management (Sorensen et al., 2020). For example, a collaboration between architects and urban planners might lead to a building design that reduces energy consumption and simultaneously minimizes its impact on the surrounding ecosystem, promoting biodiversity and ecological resilience.
- Innovative Problem-Solving: A diverse group of professionals brings different perspectives and expertise, encouraging creative solutions to complex environmental and technical challenges (von Davier et al., 2017). This can

result in more effective strategies for addressing issues like water conservation, renewable energy generation, and waste management.

- Improved Project Efficiency: Early-stage collaboration helps avoid costly redesigns and reduces delays (Nwulu et al., 2023). By identifying potential issues and solutions early, stakeholders can allocate resources more efficiently, leading to a more streamlined design and construction process.
- Enhanced Compliance and Certification Achievement: Sustainable certification systems, such as BREEAM and LEED, require collaboration across all design phases to ensure compliance with environmental standards (Johnson et al., 2019). These systems provide valuable benchmarks and ensure that projects are meeting key sustainability goals. By integrating expertise from various disciplines, projects can more effectively achieve these certifications, demonstrating their commitment to sustainability.

In summary, the benefits of interdisciplinary collaboration in sustainable design are manifold, including better sustainability outcomes, more efficient processes, and the ability to meet emerging environmental standards.

The Role of the RIBA Plan of Work

The Royal Institute of British Architects (RIBA) Plan of Work is a structured framework that guides the design and construction process, ensuring that sustainability considerations are incorporated into every phase of a project (Segara et al., 2024). By providing clear stages for the development of a building or urban space, the RIBA Plan of Work can help ensure interdisciplinary coordination and alignment of sustainability goals across all stakeholders.

How the RIBA Plan Supports Sustainability at Each Design Stage

The RIBA Plan of Work is divided into several stages, each of which involves a different phase of the project. Sustainability goals are addressed at each stage, ensuring that they are integrated into the design process from start to finish (Halliday & Atkins, 2019). The following outlines how the Plan of Work supports sustainability:

- Stage 0-1 (Strategic Definition & Preparation and Brief): At the beginning of the project, sustainability objectives are established, and input is gathered from environmental consultants, engineers, and other key stakeholders. This stage lays the foundation for the sustainable design by considering environmental impacts, energy requirements, and material selection.

- Stage 2 (Concept Design): During the concept design stage, architects, engineers, and ecologists collaborate to refine strategies for energy efficiency, sustainable material selection, and site planning. This is where the first holistic integration of passive and active sustainable strategies, such as renewable energy systems or green roofs, takes place.
- Stage 3-4 (Spatial Coordination & Technical Design): Interdisciplinary coordination continues as the team works together to refine the technical aspects of the building. At this stage, renewable energy systems, water conservation strategies, and waste management systems are integrated into the design, ensuring that all systems work together efficiently. The technical design phase also includes making decisions about how the building will be constructed and how to minimize environmental impacts during construction.
- Stage 5-7 (Construction, Handover & Use): The construction phase ensures that sustainability goals are adhered to, and post-occupancy evaluations assess how well the building performs environmentally. During this phase, sustainability experts monitor energy consumption, indoor air quality, and other factors to ensure that the building continues to meet its environmental objectives after it is completed.

Aligning NbS with the Work Stages of RIBA

Nature-based solutions (NbS) are increasingly recognized as essential components of sustainable urban design (Dorst et al., 2019). NbS include strategies that integrate ecological and natural processes into urban planning, such as green roofs, permeable surfaces, and urban forests (Biswal et al., 2022). Aligning these solutions with the RIBA Plan of Work may ensure that ecological considerations are embedded in every stage of the project:

- Early-Stage Ecological Assessments: Ecological experts can perform site assessments early in the project to guide decisions about site selection, biodiversity conservation, and the preservation of natural habitats (Alshehri et al., 2024).
- Sustainable Urban Drainage Systems (SuDS): In the spatial coordination phase, SuDS and other green infrastructure strategies are incorporated into the design to manage stormwater, enhance biodiversity, and improve resilience to flooding (Sagala et al., 2022).
- Post-Occupancy Monitoring: Post-occupancy evaluations help assess the long-term environmental and social benefits of NbS, ensuring that the ecological interventions are working as intended and providing ongoing benefits

to the building's occupants and the surrounding community (Mahmoud et al., 2024).

Aligning NbS with the RIBA framework may ensure ecological considerations are prioritized throughout the project lifecycle, contributing to both environmental sustainability and social well-being.

BREEAM and Sustainable Certification Systems

Sustainable certification systems play a pivotal role in formalizing and standardizing sustainability efforts in architecture and engineering (Herazo & Lizarralde, 2015). These frameworks provide quantifiable benchmarks, ensuring that projects meet rigorous environmental and performance criteria. Certification systems such as BREEAM (Building Research Establishment Environmental Assessment Method), LEED (Leadership in Energy and Environmental Design), and the WELL Building Standard encourage interdisciplinary collaboration by necessitating input from architects, engineers, environmental scientists, and urban planners throughout the design and construction process (Di Gaetano et al., 2023).

The Function and Comparative Analysis of BREEAM, LEED, and WELL

Each of these certification systems has unique assessment criteria, but they share the common objective of enhancing sustainability, optimizing resource efficiency, and promoting occupant well-being.

BREEAM: European Leadership in Environmental Performance. Developed by the Building Research Establishment (BRE) in the United Kingdom, BREEAM is one of the most established sustainability assessment methods globally (Polli, 2020). It evaluates building performance across a wide spectrum of sustainability factors, including energy use, materials, waste management, and ecological impact. A BREEAM certification is awarded on a scale from Pass to Outstanding, with the latter denoting exceptional sustainability performance (Shibani et al., 2021).

BREEAM necessitates the involvement of structural engineers, energy consultants, and environmental planners to optimize aspects such as thermal efficiency, passive solar design, and low-carbon construction materials. Collaboration between mechanical engineers and architects is particularly crucial in maximizing the benefits of renewable energy technologies and water management systems.

LEED: A Performance-Oriented Certification System. LEED, administered by the U.S. Green Building Council (USGBC), places a strong emphasis on performance metrics (Cidell, 2009). The rating system evaluates buildings across categories such as energy efficiency, water conservation, and sustainable site development, with certification levels ranging from Certified to Platinum.

Unlike BREEAM, which employs a prescriptive approach, LEED follows a performance-based methodology that encourages data-driven optimization (Dos Santos, 2020). Engineers and sustainability consultants play a fundamental role in the certification process, ensuring that mechanical, electrical, and structural systems align with LEED's rigorous performance thresholds.

WELL: The Intersection of Sustainability and Human Well-Being. The WELL Building Standard diverges from traditional sustainability frameworks by prioritizing human health and well-being (Well, 2014). Developed by the International WELL Building Institute (IWBI), WELL evaluates buildings based on air quality, thermal comfort, lighting design, and biophilic integration. This approach necessitates interdisciplinary collaboration between environmental psychologists, medical professionals, interior designers, and mechanical engineers to create spaces that promote cognitive function, respiratory health, and overall well-being.

Interdisciplinary Success Stories in Sustainable Architecture

Interdisciplinary collaboration is the foundation of contemporary sustainable architecture, enabling designers, engineers, ecologists, and urban planners to coalesce their expertise into transformative projects (Qian et al., 2023). The integration of knowledge across multiple disciplines is not merely an enhancement of design but a necessity in addressing climate resilience, urban liveability, and ecological conservation. Several exemplary projects worldwide highlight the remarkable results that can be achieved when diverse fields work in unison to reimagine the built environment.

Transforming a City's Waterways: Bishan-Ang Mo Kio Park, Singapore

In the heart of Singapore, an urban oasis thrives where once a sterile, concrete drainage canal dictated the landscape. The Bishan-Ang Mo Kio Park project redefines the relationship between urban infrastructure and nature by transforming a 2.7-kilometer-long concrete storm drain into a meandering, naturalized river system (An et al., 2020). This transformation exemplifies an interdisciplinary synthesis of hydrology, ecology, landscape architecture, and urban planning, creating a flood-resilient solution and a biodiverse and socially engaging public space (Jagadisan & Sen, 2024).

Before the redesign, the Kallang River was encased in a rigid concrete channel, engineered for flood management but devoid of ecological or social value (Triman, 2014). Recognizing the limitations of this outdated approach, Singapore's Public

Utilities Board (PUB) partnered with the National Parks Board (NParks), landscape architects, hydrologists, and ecologists to devise a more resilient and regenerative solution (Sini & Sini, 2020).

- Hydrological and Engineering Innovation: Civil engineers and hydrologists reimagined the river using the Active, Beautiful, Clean (ABC) Waters Programme, replacing the straight-lined canal with a widened, meandering river (Dreiseitl, 2020). The reconfigured waterway can now retain and absorb excess rainwater, naturally expanding during heavy downpours and contracting during dry spells, mitigating urban flood risks.
- Ecological and Landscape Architecture Contributions: A multidisciplinary team of ecologists and landscape designers meticulously restored wetland ecosystems and riparian habitats, incorporating native vegetation that stabilizes riverbanks, improves water quality, and supports local wildlife, including otters, herons, and dragonflies (Wang & Li, 2024).
- Urban Planning and Community Integration: Urban planners and social researchers ensured that the park serves as an interactive public space, seamlessly integrating cycling paths, boardwalks, and recreational areas. The design prioritizes community engagement, offering opportunities for education, leisure, and environmental stewardship (Lim & Xenarios, 2021).

The project's success shows the importance of nature-based solutions (NbS) in urban planning, proving that sustainable infrastructure can be both ecologically restorative and socially enriching. It is a powerful testament to how engineering, architecture, and ecology can work harmoniously to reshape urban landscapes.

Revolutionizing Urban Skyscrapers: Bosco Verticale, Milan

At first glance, Milan's Bosco Verticale ("Vertical Forest") appears almost otherworldly; two residential skyscrapers rising above the city, their facades bursting with the lush greenery of over 20,000 trees, shrubs, and plants (Giacomello, 2015). This visionary architectural achievement, completed in 2014, redefines the traditional high-rise by integrating urban forestry and ecological resilience into its very structure. The project represents a groundbreaking collaboration between architects, botanists, structural engineers, and climate scientists, demonstrating the profound impact of interdisciplinary thinking in urban sustainability.

Designed by architect Stefano Boeri, the Bosco Verticale seeks to combat air pollution, mitigate urban heat islands, and enhance biodiversity within dense cityscapes (Castaldo, 2022). The innovative concept of vertical afforestation required extensive cooperation across disciplines:

- Structural and Environmental Engineering Challenges: Unlike conventional facades, Bosco Verticale's balconies needed to support the weight of full-grown trees, requiring innovative load-bearing calculations. Structural engineers developed a reinforced concrete framework with deep cantilevered balconies, carefully balancing wind resistance, seismic stability, and soil retention (Tokuç & İnan, 2017).
- Botanical and Climate Adaptation Expertise: Botanists played a crucial role in species selection, ensuring that trees and plants could withstand urban microclimates, varying wind conditions, and seasonal shifts (Wang et al., 2020). The vegetation functions as a natural air purifier, absorbing CO_2 and fine particulate matter while increasing urban biodiversity.
- Energy and Climate Mitigation Benefits: The green facade significantly reduces the buildings' energy demands, providing natural shading, improved insulation, and humidity regulation (Di Paola, 2021). This passive climate control minimizes the need for mechanical cooling, making the towers energy-efficient despite their height.

Bosco Verticale has become an internationally recognized model for sustainable high-rise living, inspiring similar projects in cities like Nanjing, Utrecht, and Paris. The interdisciplinary nature of its development highlights how architecture, botany, and engineering can unite to create a new paradigm for urban dwelling; one that merges human habitation with nature.

Setting a New Standard for Net-Positive Buildings: The Bullitt Center, Seattle

Nestled in Seattle's Capitol Hill neighborhood, the Bullitt Center stands as one of the most sustainable commercial buildings in the world (Homchick Crowe, 2020). Often referred to as the "greenest office building", this six-story structure generates more energy than it consumes, serving as a model for net-positive energy design. The project's pioneering approach involved a diverse team of architects, renewable energy specialists, materials scientists, and sustainability engineers, proving that interdisciplinary collaboration is fundamental to creating truly self-sufficient buildings.

- Innovative Structural and Material Design: The building is constructed from locally sourced, FSC-certified mass timber, showcasing a low-carbon alternative to steel and concrete (Doub et al., 2019). Structural engineers designed the timber frame for both seismic resilience and carbon sequestration, reinforcing its long-term environmental benefits.

- Net-Positive Energy Generation: A rooftop solar array covering over 14,000 square feet produces 105% of the building's annual energy needs (Murphy, 2016). Electrical engineers and computational modelers optimized the photovoltaic system's orientation and efficiency, ensuring maximum solar gain even in Seattle's cloudy climate.
- Water Self-Sufficiency and Closed-Loop Systems: The Bullitt Center operates a rainwater collection and purification system, supplying 100% of its potable water needs (Fatima, 2024). Civil engineers and hydrologists worked together to develop a comprehensive filtration and storage mechanism, aligning the system with Seattle's stringent water regulations.
- Adaptive and Smart Building Technologies: Sensors throughout the building monitor air quality, temperature, and occupancy, allowing automated adjustments that minimize energy waste (Eley & Eley, 2016). The naturally ventilated workspace enhances indoor air quality, reducing reliance on mechanical HVAC systems.

The Bullitt Centre represents the culmination of interdisciplinary expertise, proving that commercial buildings can function as self-sustaining ecosystems, minimizing resource dependency while maximizing environmental benefits.

Retrofitting for Sustainability and Social Equity

While new green developments often capture attention, the reality in many urban environments is that the majority of building stock already exists and much of it is outdated, inefficient, and vulnerable to climate risks. Retrofitting, or the process of updating existing buildings with sustainable technologies and practices, offers an essential path to reducing emissions, conserving resources, and promoting environmental justice.

Sustainable Retrofitting: Global Examples and Recommendations

Retrofitting delivers immediate benefits by minimizing demolition waste, preserving embodied carbon, and extending a building's lifecycle (Luo & Oyedele, 2021; Iwuanyanwu et al., 2024). Cities such as Copenhagen, Toronto, and Tokyo have embraced this strategy by mandating energy retrofits in public housing and civic buildings.

- In Toronto, the Tower Renewal Program retrofits post-war high-rises with better insulation, triple-glazed windows, and solar water systems, reducing energy use by 30-40% while improving tenant comfort (Valzania, 2022).

- Berlin's Energiesprong-inspired initiatives focus on prefabricated external cladding and rooftop solar, dramatically reducing heating demand in ageing housing blocks (Loorbach, 2017; van der Schoor, 2022).
- The Barbican Centre in London demonstrates that even heritage sites can achieve improved environmental performance through LED lighting retrofits, HVAC upgrades, and smart energy management systems (Harwood, 2011; Behar, 2014).

Common retrofit strategies include:

- Building envelope improvements: enhanced insulation, airtightness, and glazing.
- On-site renewables: solar panels, heat pumps, and green roofs.
- Water efficiency: greywater systems, rainwater harvesting, and low-flow fixtures.
- Digital controls: IoT-based sensors and smart meters to optimize energy use.

These upgrades reduce emissions and make buildings healthier and more resilient, particularly important as extreme climate events increase in frequency.

Addressing Social Equity Through Sustainable Design

Sustainability is not only about the environment; it must also address social and economic disparities. Vulnerable communities often located in poorly maintained housing with high energy bills and minimal access to green spaces stand to gain the most from sustainable retrofitting.

- The Via Verde development in the Bronx integrates affordable housing with rooftop gardens, solar panels, and passive ventilation, offering residents a healthier and lower-cost living environment (Perini & Sabbion, 2016).
- In Cape Town, retrofitting informal housing with modular insulation and community-managed solar microgrids has increased safety and energy access while reducing household expenses (Shortt & Hammett, 2013).
- BedZED (Beddington Zero Energy Development) in London offers another model for low-income, low-carbon housing, combining super-insulated structures, energy-efficient appliances, and pedestrian-friendly design (Andrews, 2008).

By adopting inclusive design principles such as barrier-free access, culturally appropriate community spaces, and participatory planning, retrofitting projects can enhance not only sustainability metrics but also equity, health, and well-being.

Interdisciplinary Support for Equitable Retrofitting

Successful retrofit projects rely on collaboration between policymakers, engineers, social scientists, and community groups. Architects and designers must work hand-in-hand with residents to tailor interventions that meet both technical goals and lived realities. Programs like RE:FIT London and PACE financing in the USA show that financial models and policy incentives are equally critical to widespread adoption (Partridge, 2019).

Consequences of Unsustainable Construction Practices

While sustainable architecture presents promising pathways, it is important to recognize the severe consequences of conventional construction approaches that dominate many global contexts. Business-as-usual development often leads to significant environmental degradation and social inequities: (1) environmental degradation: traditional construction accounts for nearly 40% of global carbon dioxide emissions, largely due to the intensive use of concrete and steel. Land clearing for development contributes to biodiversity loss, while impermeable surfaces increase urban flooding and disrupt hydrological cycles (Omopariola et al., 2024; Satyam & Sharma, 2024); (2) resource depletion and waste: linear construction models consume vast amounts of virgin resources and generate over one-third of global solid waste. In many regions, demolition waste is not reused or recycled, further exacerbating landfill overflow and pollution (Spence & Mulligan, 1995; Al-Numan, 2024); (3) heat island effects and poor air quality: dense urban development lacking green infrastructure contributes to elevated temperatures and deteriorating air quality. Cities built with minimal vegetation suffer from extreme heat events and smog, endangering public health (Singh et al., 2020; Piracha & Chaudhary, 2022); and (4) social inequities and displacement: business-as-usual development often prioritizes economic return over community needs, resulting in gentrification, displacement of vulnerable populations, and lack of access to green spaces in underserved neighborhoods (Ijabadeniyi & Vanclay, 2020; Mosneaga, 2022).

Real-world examples highlight these costs: Jakarta's subsidence crisis, driven in part by unregulated urban growth and hardscaping, is displacing thousands (Abidin et al., 2015; Octavianti & Charles, 2018). In contrast, cities like Copenhagen, Denmark and Melbourne, Australia that have invested in green infrastructure demonstrate that alternative, nature-integrated approaches yield more resilient and equitable outcomes (Liu & Jensen, 2018).

Summary

Chapter 4 explores the essential role of interdisciplinary collaboration in advancing sustainable design. As environmental, social, and technical challenges become more complex, effective architectural and urban planning solutions increasingly depend on cooperation among a diverse range of professionals. The chapter begins by identifying the limitations of traditional siloed practices in architecture and engineering, which often result in fragmented decision-making, inefficiencies, and suboptimal integration of sustainability principles. In contrast, collaborative approaches enable professionals from different disciplines to contribute their expertise early in the design process, ensuring a more cohesive and comprehensive response to sustainability objectives.

The RIBA Plan of Work is examined as a framework that supports interdisciplinary coordination across all project phases. By embedding sustainability considerations from strategic planning to post-occupancy evaluation, the Plan of Work facilitates consistent engagement among stakeholders. The chapter also discusses how Nature-Based Solutions can be aligned with this framework to enhance ecological performance, climate resilience, and urban liveability.

Sustainable certification systems such as BREEAM, LEED, and WELL are presented as mechanisms that formalize sustainability goals and encourage multidisciplinary involvement. Each system focuses on specific priorities, including energy efficiency, environmental performance, and occupant well-being. Achieving these certifications typically requires collaborative input from architects, engineers, ecologists, public health experts, and other professionals.

A series of case studies demonstrates the real-world benefits of interdisciplinary design. Projects such as Bishan-Ang Mo Kio Park in Singapore, Bosco Verticale in Milan, and the Bullitt Centre in Seattle illustrate how collaboration across disciplines can result in innovative, high-performance buildings and landscapes that are environmentally restorative and socially inclusive. These examples show how teams combining ecological knowledge, engineering skills, and community planning can deliver sustainable outcomes at both building and urban scales.

The chapter also addresses the underrepresentation of retrofitting and social equity in sustainable design discourse. Retrofitting existing buildings is presented as a critical strategy for reducing emissions and conserving resources, particularly in cities where much of the built environment is aging or inefficient. Global examples from Toronto, Berlin, and London show how retrofits can improve energy performance and occupant comfort while preserving embodied carbon. Furthermore, the chapter emphasizes that sustainability must address social as well as environmental goals. Projects in the Bronx, Cape Town, and London illustrate how design can serve vulnerable communities through affordable housing, energy access, and participatory planning.

The final section outlines the environmental and social consequences of conventional construction methods. High carbon emissions, biodiversity loss, poor air quality, and community displacement are cited as common outcomes of unsustainable development. The chapter concludes by reinforcing the need for interdisciplinary, inclusive, and forward-looking approaches to transform the built environment into a regenerative and equitable system.

CONCLUSION AND FUTURE DIRECTIONS

The intersection of sustainability, technology, and interdisciplinary collaboration has reshaped contemporary architectural discourse, driving the built environment toward greater resilience, efficiency, and ecological harmony. The adoption of nature-based solutions (NbS), advanced digital technologies, and integrative design processes has demonstrated immense potential in addressing pressing environmental and socio-economic challenges. However, while significant strides have been made, barriers to implementation, knowledge gaps, and systemic constraints remain. This chapter has examined the role of interdisciplinary collaboration in achieving sustainable architecture and urban development. Here, we summarize key insights, identify critical challenges, and propose directions for future research and practice.

Critical Insights

A comprehensive and integrative approach is imperative in sustainable architecture. The following insights specify the central themes explored in this discussion:

Nature-based Solutions (NbS) and Urban Resilience

Nature-based solutions (NbS) have emerged as a fundamental strategy for enhancing urban resilience, promoting biodiversity, and mitigating climate change impacts. As cities become increasingly vulnerable to environmental challenges ranging from rising temperatures and air pollution to flooding and biodiversity loss, NbS offers a means of reconciling urban development with ecological conservation.

- The integration of green roofs, vertical gardens, permeable pavements, and urban forests helps regulate microclimates, reduce heat island effects, and improve air quality.
- Sustainable Urban Drainage Systems (SuDS) and ecological restoration initiatives such as wetland preservation and river naturalization projects enhance flood resilience and water retention while fostering habitat creation.
- NbS addresses environmental challenges and provides psychological and social benefits, improving public well-being through access to green spaces.

Despite the demonstrated benefits, the adoption of NbS remains inconsistent due to challenges related to land availability, economic feasibility, and institutional resistance. Expanding their implementation requires a shift in urban planning paradigms, stronger policy frameworks, and increased public-private partnerships.

The Role of Smart Technologies in Sustainable Architecture

Recent advancements in digital technologies, artificial intelligence (AI), and smart building systems have redefined energy efficiency and sustainability in architecture. The integration of AI, Internet of Things (IoT), and data analytics allows for real-time energy monitoring, predictive maintenance, and adaptive environmental controls that optimize building performance.

- Smart sensors and automation systems enable buildings to self-regulate energy use, reducing waste and operational costs.
- Digital twin technology, which creates virtual models of buildings and urban environments, aids in simulating environmental impacts, energy flows, and material efficiencies, allowing for more precise decision-making in design and maintenance.
- AI-driven predictive climate modelling helps architects and urban planners design spaces that respond dynamically to changing weather patterns and energy demands.
- Blockchain-based carbon tracking and smart contracts have begun revolutionizing supply chain transparency and sustainability metrics in construction projects.

As these technologies mature and become more accessible, they will fundamentally alter the way buildings are designed, constructed, and maintained. However, their widespread application is hindered by high initial costs, cybersecurity concerns, and the need for specialized technical expertise.

The Critical Role of Interdisciplinary Collaboration

No single discipline can achieve sustainability in isolation. The success of sustainable architecture and urban design hinges on effective collaboration between architects, engineers, ecologists, urban planners, policymakers, and even social scientists and economists.

- Architects and engineers must integrate environmental considerations from the earliest stages of project development.
- Urban planners and policy experts must ensure that sustainability principles are codified within zoning laws, urban policies, and economic incentives.
- Material scientists and environmental researchers contribute to advancing low-carbon construction materials and circular economy solutions.
- Social scientists and behavioural researchers provide insight into how users interact with built environments, ensuring that sustainability measures align with human well-being and equity.

A more cohesive and multidisciplinary approach can bridge the gap between innovative theory and real-world implementation, leading to buildings and cities that are not just sustainable in design but functional and equitable in practice.

Policy Incentives, Barriers, and Regional Examples

Policies and regulations play a crucial role in shaping the adoption of sustainable design and construction practices. In many regions, policy frameworks serve as both enablers and barriers. For example, the European Union's Green Deal and Energy Performance of Buildings Directive (EPBD) actively mandate higher energy efficiency standards, encouraging innovations such as net-zero buildings and retrofitting initiatives (Congedo et al., 2024; Dulian, 2024). Similarly, countries like Singapore and the United Arab Emirates have introduced green building codes that promote technologies like smart grids, passive design, and rainwater harvesting (Najini et al., 2020; Rodriguez-Ubinas et al., 2020). In contrast, regions with outdated building codes or limited enforcement mechanisms often hinder sustainable development.

Financial incentives, including tax rebates, grants, and green certification subsidies, are critical in offsetting the higher upfront costs associated with sustainable technologies (Krass et al., 2013). For instance, the United States offers federal tax credits for solar installations and energy-efficient home improvements (Dixon et al., 2010). In Germany, the KfW Development Bank supports sustainable retrofitting with low-interest loans (Galvin & März, 2025). Conversely, in many developing nations, the absence of financing mechanisms remains a barrier to the adoption of sustainable architecture.

A strong policy environment that aligns with long-term climate goals and is supported by practical economic tools can accelerate the widespread adoption of sustainable architecture. Public-private partnerships, improved code enforcement, and international knowledge-sharing will be essential to achieving this transformation.

Challenges and Areas for Future Research

Implementing sustainable design strategies is not without its obstacles. Despite the growing awareness and technological advancements, several persistent barriers slow the adoption of these approaches:

1. Industry Resistance and Conservatism: Many stakeholders in the construction sector remain hesitant to change well-established practices due to perceived risks, lack of familiarity, or inertia. This cultural resistance can impede innovation and delay the uptake of sustainable methods (Van Tonder, 2017).
2. High Upfront Costs: While sustainable buildings often offer long-term cost savings, the initial investment for green technologies, renewable systems, or sustainable materials can be significantly higher. These costs can deter developers, particularly in markets where short-term profit margins are prioritized over lifecycle performance (Agwu et al., 2024).
3. Skills and Knowledge Gaps: The construction and design industries are facing a skills deficit in green building techniques, digital tools, and interdisciplinary collaboration. Without adequate education and training, implementing complex sustainability measures remains a challenge (Ipsen et al., 2021).
4. Regulatory Misalignment: Inconsistent or outdated building codes in many regions fail to accommodate or encourage innovative sustainable practices, further complicating project approvals (Crilly & Lemon, 2018).

Numerous challenges persist in achieving truly sustainable and regenerative architecture. The following areas require further research, policy development, and industry-wide transformation:

Stronger Policies and Incentives for Sustainable Construction

Governments and regulatory bodies play a crucial role in shaping sustainable building practices. However, in many regions, sustainability policies remain fragmented or inadequately enforced.

- There is a need for more stringent carbon reduction mandates for new and existing buildings.
- Financial incentives, such as tax breaks for energy-efficient retrofits or subsidies for green building materials, could accelerate the transition toward net-zero and carbon-negative buildings.
- Stricter building codes that require the integration of NbS, renewable energy systems, and life-cycle carbon analysis could standardize sustainability as a non-negotiable requirement rather than an optional feature.

Education and Training in Sustainable Design

One of the fundamental barriers to widespread adoption of sustainable architecture is the lack of comprehensive sustainability training in architectural and engineering education.

- University curricula must integrate systems thinking, regenerative design principles, and interdisciplinary project-based learning to equip future professionals with the skills needed to tackle sustainability challenges.
- Greater emphasis on climate literacy and ecological understanding is needed in architectural education, ensuring that sustainability is treated as a core competency rather than a niche specialization.
- Professional development programs and continuing education for industry practitioners should be expanded to include emerging fields such as AI-driven energy modelling, bio-integrated design, and circular economy construction practices.

Advancing Material Innovation and Circular Economy Practices

Building materials remain one of the largest contributors to global carbon emissions. Research into low-impact, bio-based, and recyclable materials must be accelerated.

- The development of biodegradable, carbon-sequestering materials such as mycelium-based composites, algae-based insulation, and hempcrete offers promising avenues for reducing the carbon footprint of buildings.
- Greater adoption of modular construction and prefabrication can minimize material waste and improve resource efficiency.
- Circular economy principles, such as adaptive reuse, material reclamation, and cradle-to-cradle design, must be mainstreamed into construction practices.

Enhanced Performance Monitoring and Long-Term Sustainability Assessment

Post-occupancy evaluations and continuous monitoring of sustainability metrics remain underutilized in the architectural field. Many buildings that are designed with sustainability in mind fail to meet their targets due to insufficient follow-up assessments.

- AI-powered building management systems should be employed to monitor real-time energy consumption, indoor air quality, and thermal efficiency.
- Long-term studies assessing the life-cycle performance of sustainable buildings will be critical in refining future design methodologies.

Final Thoughts

The next frontier in sustainable architecture lies in moving beyond harm reduction toward regenerative design, where buildings actively contribute to ecological restoration, carbon sequestration, and social well-being. The built environment must be seen as a dynamic, adaptive system that supports, rather than disrupts, natural ecosystems.

A Vision for the Future

- Adopting regenerative principles that restore biodiversity, enhance ecological resilience, and create self-sustaining urban ecosystems.
- Leveraging cutting-edge digital technologies to design buildings that are energy-positive, resource-efficient, and responsive to environmental changes.
- Expanding interdisciplinary collaboration to ensure that sustainability efforts are technically sound, socially just, and economically viable.

The architecture of the future must be not just sustainable but life-enhancing, a force that heals, rather than exploits, the planet. Through bold innovation, systemic policy reform, and a commitment to interdisciplinary cooperation, we can shape a built environment that thrives in harmony with nature and future generations alike.

REFERENCES

Abidin, H., Andreas, H., Gumilar, I., & Brinkman, J. (2015). Study on the risk and impacts of land subsidence in Jakarta. *Proceedings of the International Association of Hydrological Sciences*, *372*(372), 115–120.

Adams, R., Jeanrenaud, S., Bessant, J., Denyer, D., & Overy, P. (2016). Sustainability-oriented innovation: A systematic review. *International Journal of Management Reviews*, *18*(2), 180–205.

Adeyemi, A. B., Ohakawa, T. C., Okwandu, A. C., Iwuanyanwu, O., & Ifechukwu, G.-O. (2024). Integrating modular and prefabricated construction techniques in affordable housing: Architectural design considerations and benefits. *Journal of Construction Innovation*, *18*(2), 67–82.

Agwu, K., Oraefo, O., Jude, B., Onuorah, I., Onyemaechi, C., & Chinenye, C. (2024). Exploring the impacts of sustainable design practices on construction cost. *International Journal of Progressive Research in Engineering Management and Science*, *4*, 1531–1537.

Ahmed, A., Ge, T., Peng, J., Yan, W.-C., Tee, B. T., & You, S. (2022). Assessment of the renewable energy generation towards net-zero energy buildings: A review. *Energy and Building*, *256*, 111755.

Al-Numan, B. S. O. (2024). Construction Industry Role in Natural Resources Depletion and How to Reduce It. In *Natural Resources Deterioration in MENA Region: Land Degradation, Soil Erosion, and Desertification* (pp. 93-109). Springer.

Alijoyo, F. A. (2024). AI-powered deep learning for sustainable industry 4.0 and internet of things: Enhancing energy management in smart buildings. *Alexandria Engineering Journal*, *104*, 409–422.

Alshehri, K., Chen, I.-C., Rugani, B., Sapsford, D., Harbottle, M., & Cleall, P. (2024). A novel uncertainty assessment protocol for integrated ecosystem services-life cycle assessments: A comparative case of nature-based solutions. *Sustainable Production and Consumption*, *47*, 499–515.

Altomonte, S. (2008). Daylight for energy savings and psycho-physiological well-being in sustainable built environments. *Journal of Sustainable Development*, *1*(3), 3–16.

An, Z., Chen, Q., & Li, J. (2020). Ecological strategies of urban ecological parks–a case of Bishan Ang Mo Kio park and Kallang river in Singapore. E3S Web of Conferences.

Anguelovski, I., & Frumkin, H. (2025). Safeguarding nature and cities in a burning country. *Nature Cities*, 1-2.

Annan-Diab, F., & Molinari, C. (2017). Interdisciplinarity: Practical approach to advancing education for sustainability and for the Sustainable Development Goals. *International Journal of Management Education*, *15*(2), 73–83.

Behar, C. (2014). Utilising resident feedback to inform energy-saving interventions at the Barbican. *Local Environment*, *19*(5), 539–559.

Biswal, B. K., Bolan, N., Zhu, Y.-G., & Balasubramanian, R. (2022). Nature-based Systems (NbS) for mitigation of stormwater and air pollution in urban areas: A review. *Resources, Conservation and Recycling*, *186*, 106578.

Boafo, F. E., Kim, J.-H., & Kim, J.-T. (2016). Performance of modular prefabricated architecture: Case study-based review and future pathways. *Sustainability*, *8*(6), 558.

Canton, H. (2021). United Nations Environment Programme—UNEP. In *The Europa directory of international organizations 2021* (pp. 188–214). Routledge.

Carbajal-Pina, C., & Acur, N. (2024). From principles to practice: Responsible implementation of emerging technologies through innovation dilemmas. *IEEE Transactions on Engineering Management*.

Castaldo, N. F. (2022). *Buildings That Breathe: Greening the World's Cities*. Lerner+ ORM.

Charef, R., Lu, W., & Hall, D. (2022). The transition to the circular economy of the construction industry: Insights into sustainable approaches to improve the understanding. *Journal of Cleaner Production*, *364*, 132421.

Chiesa, G., Di Vita, D., Ghadirzadeh, A., Herrera, A. H. M., & Rodriguez, J. C. L. (2020). A fuzzy-logic IoT lighting and shading control system for smart buildings. *Automation in Construction*, *120*, 103397.

Cidell, J. (2009). Building green: The emerging geography of LEED-certified buildings and professionals. *The Professional Geographer*, *61*(2), 200–215.

Congedo, P. M., Baglivo, C., D'Agostino, D., & Albanese, P. M. (2024). Overview of EU building envelope energy requirement for climate neutrality. *Renewable & Sustainable Energy Reviews*, *202*, 114712.

Crilly, M., & Lemon, M. (2018). The misalignment of policy and practice in sustainable urban design. *Smart Futures, Challenges of Urbanisation, and Social Sustainability*, 69-87.

Davies, C., Chen, W. Y., Sanesi, G., & Lafortezza, R. (2021). The European Union roadmap for implementing nature-based solutions: A review. *Environmental Science & Policy, 121*, 49–67.

De Luca, F., Sepúlveda, A., & Varjas, T. (2022). Multi-performance optimization of static shading devices for glare, daylight, view and energy consideration. *Building and Environment, 217*, 109110.

Di Gaetano, F., Cascone, S., & Caponetto, R. (2023). Integrating BIM processes with LEED certification: A comprehensive framework for sustainable building design. *Buildings (Basel, Switzerland), 13*(10), 2642.

Di Paola, L. (2021). Milan's private Vertical Forests vs. horizontal urban greening. In *The Green City and Social Injustice* (pp. 25–34). Routledge.

Di Sarno, L., Albuhairi, D., & Medeiros, J. M. P. (2024). Exploring innovative resilient and sustainable bio-materials for structural applications: Hemp-fibre concrete. *Energy Policy, 38*(11), 6398–6408.

Dorst, H., Van der Jagt, A., Raven, R., & Runhaar, H. (2019). Urban greening through nature-based solutions–Key characteristics of an emerging concept. *Sustainable Cities and Society, 49*, 101620.

Dos Santos, L. F. B. S. (2020). *Efficient modeling strategies for performance-based building design supported by daylight and building energy simulations*. University of California.

Doub, S., Hanford, J., Sprug, M., Hellstern, C., & Misel, K. (2019). Living Buildings: The Bullitt Center. In *Architecture and Health* (pp. 260-279). Routledge.

Dreiseitl, H. (2020). Water and sustainable design. In *Sustainable Built Environments* (pp. 187–205). Springer.

Dulian, M. (2024). Revision of the Energy Performance of Buildings Directive–Fit for 55 package. *EU Legislation, 19*(1).

Eley, C., & Eley, C. (2016). Smart Building Design: Contextual Design, Energy Efficiency, and Curtailment. *Design Professional's Guide to Zero Net Energy Buildings*, 9-60.

Fatima, K. (2024). Building Better from LEED to Living: An Approach to Net Zero Water Management.

Fenner, A. E., Kibert, C. J., Woo, J., Morque, S., Razkenari, M., Hakim, H., & Lu, X. (2018). The carbon footprint of buildings: A review of methodologies and applications. *Renewable & Sustainable Energy Reviews, 94*, 1142–1152.

Ferreira Silva, M., Jayasinghe, L. B., Waldmann, D., & Hertweck, F. (2020). Recyclable architecture: Prefabricated and recyclable typologies. *Sustainability, 12*(4), 1342.

Fleischman, R. B., & Seeber, K. (2016). New construction for resilient cities: The argument for sustainable low damage precast/prestressed concrete building structures in the 21st century. *Scientia Iranica, 23*(4), 1578–1593.

Galvin, R., & März, S. (2025). A finance scheme to help Germany's small private landlords sharply increase their buildings' energy performance: Tapping into the banking system. *Energy Research & Social Science, 120*, 103929.

Giacomello, E. (2015). Case study: Bosco verticale, milan: A new urban forest rises in Milan. *CTBUH Journal*, (1), 12–18.

Gijón Mancheño, A., Jafino, B. A., Hofland, B., van Wesenbeeck, B. K., Kazi, S., & Urrutia, I. (2025). Nature Meets Infrastructure: The Role of Mangroves in Strengthening Bangladesh's Coastal Flood Defenses. *Sustainability, 17*(4), 1567.

Goudarzi, A., Ghayoor, F., Waseem, M., Fahad, S., & Traore, I. (2022). A survey on IoT-enabled smart grids: Emerging, applications, challenges, and outlook. *Energies, 15*(19), 6984.

Guy, S., & Farmer, G. (2001). Reinterpreting sustainable architecture: The place of technology. *Journal of Architectural Education, 54*(3), 140–148.

Halliday, S., & Atkins, R. (2019). *Sustainability: RIBA Plan of Work 2013 Guide.* RIBA Publishing.

Handfield, R. B., Melnyk, S. A., Calantone, R. J., & Curkovic, S. (2002). Integrating environmental concerns into the design process: The gap between theory and practice. *IEEE Transactions on Engineering Management, 48*(2), 189–208.

Hanus, M. J., & Harris, A. T. (2013). Nanotechnology innovations for the construction industry. *Progress in Materials Science, 58*(7), 1056–1102.

Harwood, E. (2011). The Barbican, City of London. *DASH| Delft Architectural Studies on Housing,* (4), 22-33.

Hayes, A. T., Jandaghian, Z., Lacasse, M. A., Gaur, A., Lu, H., Laouadi, A., Ge, H., & Wang, L. (2022). Nature-based solutions (nbss) to mitigate urban heat island (UHI) effects in Canadian cities. *Buildings (Basel, Switzerland), 12*(7), 925.

Hegazy, B., Khodeir, L., & Fathy, F. (2025). Achieving Socio-economic Resilience in Neighborhood Through Nature-based Solutions: A Systematic Review. *Results in Engineering*, 104266.

Heidrich, O., Kamara, J., Maltese, S., Re Cecconi, F., & Dejaco, M. C. (2017). A critical review of the developments in building adaptability. *International Journal of Building Pathology and Adaptation*, *35*(4), 284–303.

Herazo, B., & Lizarralde, G. (2015). The influence of green building certifications in collaboration and innovation processes. *Construction Management and Economics*, *33*(4), 279–298.

Hirsch, P. L., Shwom, B. L., Yarnoff, C., Anderson, J. C., Kelso, D. M., Olson, G. B., & Colgate, J. E. (2001). Engineering design and communication: The case for interdisciplinary collaboration. *International Journal of Engineering Education*, *17*(4/5), 343–348.

Homchick Crowe, J. (2020). Architectural advocacy: The Bullitt Center and environmental design. *Environmental Communication*, *14*(2), 236–254.

Hossain, M. R., Oo, A. M., & Ali, A. S. (2013). Smart grid. *Smart grids: opportunities, developments, and trends*, 23-44.

IEA. (2021). Buildings: a source of enormous untapped efficiency potential. International Energy Agency (IEA).

Ijabadeniyi, A., & Vanclay, F. (2020). Socially-tolerated practices in environmental and social impact assessment reporting: Discourses, displacement, and impoverishment. *Land (Basel)*, *9*(2), 33.

Ikudayisi, A. E., Chan, A. P., Darko, A., & Adedeji, Y. M. (2023). Integrated practices in the Architecture, Engineering, and Construction industry: Current scope and pathway towards Industry 5.0. *Journal of Building Engineering*, *73*, 106788.

Ipsen, K. L., Pizzol, M., Birkved, M., & Amor, B. (2021). How lack of knowledge and tools hinders the eco-design of buildings—A systematic review. *Urban Science (Basel, Switzerland)*, *5*(1), 20.

Iwuanyanwu, O., Gil-Ozoudeh, I., Okwandu, A. C., & Ike, C. S. (2024). Retrofitting existing buildings for sustainability: Challenges and innovations. *Engineering Science & Technology Journal*, *5*(8), 2616–2631.

Jagadisan, S., & Sen, J. (2024). The Role of Community Engagement in Building Sustainable Blue-Green Infrastructure–Best Practices and Case Studies. In *Blue-Green Infrastructure for Sustainable Urban Settlements: Implications for Developing Countries Under Climate Change* (pp. 193-215). Springer.

Jamawat, J. (2019). *Redesign, redeploy, and re-envision urban corporate headquarters: Amazon's Seattle campus case study*. Massachusetts Institute of Technology.

Janda, K. B., Wilson, C., Bartiaux, F., & Moezzi, M. (2015). Improving efficiency in buildings: conventional and alternative approaches. *Global Energy: Issues, Potentials, and Policy Implications*, 163-188.

Jiang, Y., Zhao, D., Wang, D., & Xing, Y. (2019). Sustainable performance of buildings through modular prefabrication in the construction phase: A comparative study. *Sustainability, 11*(20), 5658.

Johnson, C., Tilt, J. H., Ries, P. D., & Shindler, B. (2019). Continuing professional education for green infrastructure: Fostering collaboration through interdisciplinary trainings. *Urban Forestry & Urban Greening, 41*, 283–291.

Kamal, M. A. (2012). An overview of passive cooling techniques in buildings: Design concepts and architectural interventions. *Acta Technica Napocensis: Civil Engineering & Architecture, 55*(1), 84–97.

Kamali, M., & Hewage, K. (2017). Development of performance criteria for sustainability evaluation of modular versus conventional construction methods. *Journal of Cleaner Production, 142*, 3592–3606.

Kellert, S. (2011). *Biophilic design: The theory, science and practice of bringing buildings to life*. John Wiley & Sons.

Kibert, C. J. (2016). *Sustainable construction: green building design and delivery*. John Wiley & Sons.

Kim, K. (2012). A model-driven workflow fragmentation framework for collaborative workflow architectures and systems. *Journal of Network and Computer Applications, 35*(1), 97–110.

Konstantoglou, M., & Tsangrassoulis, A. (2016). Dynamic operation of daylighting and shading systems: A literature review. *Renewable & Sustainable Energy Reviews, 60*, 268–283.

Krass, D., Nedorezov, T., & Ovchinnikov, A. (2013). Environmental taxes and the choice of green technology. *Production and Operations Management, 22*(5), 1035–1055.

La Notte, A., & Zulian, G. (2021). An ecosystem services-based approach to frame NBS in Urban Context. In *Nature-Based Solutions for More Sustainable Cities–A Framework Approach for Planning and Evaluation* (pp. 47–65). Emerald Publishing Limited.

Lawrence, R. (2020). *Creating built environments: Bridging knowledge and practice divides*. Routledge.

Lee, D., & Lee, S.-T. (2023). Artificial intelligence enabled energy-efficient heating, ventilation and air conditioning system: Design, analysis and necessary hardware upgrades. *Applied Thermal Engineering, 235*, 121253.

Leoto, R., & Lizarralde, G. (2019). Challenges in evaluating strategies for reducing a building's environmental impact through Integrated Design. *Building and Environment, 155*, 34–46.

Lim, M., & Xenarios, S. (2021). Economic assessment of urban space and blue-green infrastructure in Singapore. *Journal of Urban Ecology, 7*(1), juab020.

Liu, L., & Jensen, M. B. (2018). Green infrastructure for sustainable urban water management: Practices of five forerunner cities. *Cities (London, England), 74*, 126–133.

Loorbach, D. (2017). *Urban Sustainability Transition: Retrofitting the City. Retrofitting Cities for Tomorrow's World.* John Wiley & Sons Ltd.

Luo, X., & Oyedele, L. O. (2021). Assessment and optimisation of life cycle environment, economy and energy for building retrofitting. *Energy for Sustainable Development, 65*, 77–100.

Macaione, I., Raffa, A., & Andaloro, B. (2024). Climate-Adaptive Nature-Based Regenerative Urban Green Streetscapes: Design Exploration from the City of Matera. *Sustainability, 16*(16), 6811.

Mahmoud, I., Morello, E., Bisello, A., & Kolokotsa, D. (2024). *Embedding technologies for improving Nature-Based Solutions performance and fostering social inclusion in urban greening strategies: Augmented NBS for cities* (Vol. 93). Elsevier.

Mauser, W., Klepper, G., Rice, M., Schmalzbauer, B. S., Hackmann, H., Leemans, R., & Moore, H. (2013). Transdisciplinary global change research: The co-creation of knowledge for sustainability. *Current Opinion in Environmental Sustainability, 5*(3-4), 420–431.

McClure, W. R., & Bartuska, T. J. (2011). *The built environment: a collaborative inquiry into design and planning*. John Wiley & Sons.

Mlote, D. S., Budig, M., & Cheah, L. (2024). Adaptability of buildings: A systematic review of current research. *Frontiers in Built Environment, 10*, 1376759.

Mosneaga, A. (2022). Unsustainable development, disasters and displacement: Revisiting the governance challenge. *International Journal of Disaster Risk Reduction, 79*, 103172.

Mudunuri, L. N. R., & Attaluri, V. (2025). Urban Development Challenges and the Role of Cloud AI-Powered Blue-Green Solutions. In *Integrating Blue-Green Infrastructure Into Urban Development* (pp. 507–522). IGI Global Scientific Publishing.

Muhammed, M., & Tsakalakos, T. (2003). Nanostructured materials and nanotechnology: Overview. *Journal of the Korean Ceramic Society*, *40*(11), 1027–1046.

Murphy, K. M. (2016). *Sustainable and energy-efficient development interventions and their application toward net-zero or net-positive energy and water building development.* Massachusetts Institute of Technology.

Murtagh, N., Roberts, A., & Hind, R. (2016). The relationship between motivations of architectural designers and environmentally sustainable construction design. *Construction Management and Economics*, *34*(1), 61–75.

Najini, H., Nour, M., Al-Zuhair, S., & Ghaith, F. (2020). Techno-economic analysis of green building codes in United Arab Emirates based on a case study office building. *Sustainability*, *12*(21), 8773.

Najjar, M., Figueiredo, K., Hammad, A. W., & Haddad, A. (2019). Integrated optimization with building information modeling and life cycle assessment for generating energy efficient buildings. *Applied Energy*, *250*, 1366–1382.

Nguyen, T. A., & Aiello, M. (2013). Energy intelligent buildings based on user activity: A survey. *Energy and Building*, *56*, 244–257.

Nwulu, E. O., Elete, T. Y., Omomo, K. O., Akano, O., & Erhueh, O. (2023). The importance of interdisciplinary collaboration for successful engineering project completions: A strategic framework. *World Journal of Engineering and Technology Research*, *2*(3), 48–56.

Octavianti, T., & Charles, K. (2018). Disaster capitalism? Examining the politicisation of land subsidence crisis in pushing Jakarta's seawall megaproject. *Water Alternatives*, *11*(2).

Omer, A. M. (2008). Renewable building energy systems and passive human comfort solutions. *Renewable & Sustainable Energy Reviews*, *12*(6), 1562–1587.

Omopariola, E. D., Olanrewaju, O. I., Albert, I., Oke, A. E., & Ibiyemi, S. B. (2024). Sustainable construction in the Nigerian construction industry: Unsustainable practices, barriers and strategies. *Journal of Engineering. Defence Technology*, *22*(4), 1158–1184.

Oxenswärdh, A. (2019). Knowledge sharing and sustainable development. In *Encyclopedia of sustainability in higher education* (pp. 1057–1065). Springer.

Partridge, C. C. (2019). *Green Municipal Bonds and the Financing of Green Infrastructure in the United States UCL.* University College London.

Perini, K., & Sabbion, P. (2016). Green-blue infrastructure in urban areas, the case of the Bronx River (NYC) and Paillon (Nice). *Techne : Journal of Technology for Architecture and Environment*, 97–103.

Piracha, A., & Chaudhary, M. T. (2022). Urban air pollution, urban heat island and human health: A review of the literature. *Sustainability*, *14*(15), 9234.

Polli, G. H. B. (2020). A comparison about european environmental sustainability rating systems: Breeam UK, DGNB, LiderA, ITACA and HQE. *U. Porto Journal of Engineering*, *6*(2), 46–58.

Qian, H., Zhang, F., & Qiu, B. (2023). Deciphering the evolution, frontier, and knowledge clustering in sustainable city planning: A 60-year interdisciplinary review. *Sustainability*, *15*(24), 16854.

Rahla, K. M., Mateus, R., & Bragança, L. (2021). Selection criteria for building materials and components in line with the circular economy principles in the built environment—A review of current trends. *Infrastructures*, *6*(4), 49.

Rane, N. (2023). Integrating leading-edge artificial intelligence (AI), internet of things (IOT), and big data technologies for smart and sustainable architecture. *Engineering and Construction (AEC) Industry: Challenges and Future Directions.*

Reddy, V. J., Hariram, N., Ghazali, M. F., & Kumarasamy, S. (2024). Pathway to sustainability: An overview of renewable energy integration in building systems. *Sustainability*, *16*(2), 638.

Rodriguez-Ubinas, E., Alzarouni, S., Alhammadi, N., Alantali, M., Behzad, F., & Sgouridis, S. (2020). Sustainability through energy conservation building codes: Comparative analysis of green building regulations in the middle east. *WIT Transactions on Ecology and the Environment*, *249*, 85–100.

Rowe, P. G., Hee, L., Rowe, P. G., & Hee, L. (2019). Gardens, parks and green reserves. *A City in Blue and Green: The Singapore Story*, 82-113.

Sadineni, S. B., Madala, S., & Boehm, R. F. (2011). Passive building energy savings: A review of building envelope components. *Renewable & Sustainable Energy Reviews*, *15*(8), 3617–3631.

Sagala, S., Murwindarti, A., Avila, B. E., Rosyidie, A., & Azhari, D. (2022). Sustainable urban drainage system (SUDS) as nature based solutions approach for flood risk management in high-density urban settlement. IOP Conference Series: earth and environmental science. *Energy and Building*, *188*, 278–285.

Satyam, V., & Sharma, P. (2024). Unsustainable construction material and process leading to severe environmental degradation. *Global Change Biology*, *27*(8), 1518–1546.

Segara, S., Li, Q. J., Gallotta, A., Wang, Y., Gosling, J., & Rezgui, Y. (2024). Taxonomy of circularity indicators for the built environment: Integrating circularity through the Royal Institute of British architects (RIBA) plan of work. *Journal of Cleaner Production*, *446*, 141429.

Shafaghat, A., & Keyvanfar, A. (2022). Dynamic façades design typologies, technologies, measurement techniques, and physical performances across thermal, optical, ventilation, and electricity generation outlooks. *Renewable & Sustainable Energy Reviews*, *167*, 112647.

Shah, S. F. A., Iqbal, M., Aziz, Z., Rana, T. A., Khalid, A., Cheah, Y.-N., & Arif, M. (2022). The role of machine learning and the internet of things in smart buildings for energy efficiency. *Applied Sciences (Basel, Switzerland)*, *12*(15), 7882.

Shaikh, P. H., Nor, N. B. M., Nallagownden, P., Elamvazuthi, I., & Ibrahim, T. (2014). A review on optimized control systems for building energy and comfort management of smart sustainable buildings. *Renewable & Sustainable Energy Reviews*, *34*, 409–429.

Shibani, A., Agha, A., Hassan, D., & Naomi, F. (2021). The impact of green certification BREEAM on occupancy rates of commercial buildings in the UK. *International Journal of Innovations in Engineering Research and Technology, 8*(9), 153-167.

Shortt, N. K., & Hammett, D. (2013). Housing and health in an informal settlement upgrade in Cape Town, South Africa. *Journal of Housing and the Built Environment*, *28*(4), 615–627.

Singh, N., Singh, S., & Mall, R. (2020). Urban ecology and human health: implications of urban heat island, air pollution and climate change nexus. In *Urban ecology* (pp. 317–334). Elsevier.

Sini, R., & Sini, R. (2020). The Singapore Playground: System of Themed Public Parks that Addresses Environmental, Social and Cultural Sustainability. *Singapore's Park System Master Planning: A Nation Building Tool to Construct Narratives in Post-Colonial Countries*, 253-328.

Sizirici, B., Fseha, Y., Cho, C.-S., Yildiz, I., & Byon, Y.-J. (2021). A review of carbon footprint reduction in construction industry, from design to operation. *Materials (Basel)*, *14*(20), 6094.

Smith, R. E. (2010). *Prefab architecture: A guide to modular design and construction*. John Wiley & Sons.

Sorensen, N. L., Rasmussen, F. N., Oien, T. B., & Frandsen, A. K. (2020). Holistic sustainability: Advancing interdisciplinary building design through tools and data in Denmark. *Construction Economics and Building*, *20*(2), 25–44.

Sowińska-Świerkosz, B., & García, J. (2022). What are Nature-based solutions (NBS)? Setting core ideas for concept clarification. *Nature-Based Solutions*, *2*, 100009.

Spence, R., & Mulligan, H. (1995). Sustainable development and the construction industry. *Habitat International*, *19*(3), 279–292.

Su, W., Zhang, W., & Chang, Q. (2025). Tailored green and blue infrastructure for heat mitigation under renewal planning of urban blocks in Beijing. *The Science of the Total Environment*, *967*, 178759.

Sutkowska, M., Stefańska, A., Vaverkova, M. D., Dixit, S., & Thakur, A. (2024). Recent advances in prefabrication techniques for biobased materials towards a low-carbon future: From modules to sustainability. *Journal of Building Engineering*, 109558.

Tam, C., Tam, V. W., Chan, J. K., & Ng, W. C. (2005). Use of prefabrication to minimize construction waste-a case study approach. *International Journal of Construction Management*, *5*(1), 91–101.

Tokuç, A., & İnan, T. (2017). A Green Outlook to Tall Building Facades via Milan's "Vertical Forest". Proceedings of the International Advanced Technologies Symposium (IATS).

Ukoba, K., Olatunji, K. O., Adeoye, E., Jen, T.-C., & Madyira, D. M. (2024). Optimizing renewable energy systems through artificial intelligence: Review and future prospects. *Energy & Environment*, *35*(7), 3833–3879.

Valzania, G. (2022). Towers once in the park: Uprooting Toronto's welfare landscapes. *Geografiska Annaler. Series B, Human Geography*, *104*(3), 227–249.

van der Schoor, T. (2022). Lessons from EU-projects for energy renovation. *Urban Planning*, *7*(2), 123–130.

Van Tonder, E. (2017). Passive innovation resistance–a conservative consumer perspective. *European Business Review*, *29*(6), 642–663.

von Davier, A. A., Hao, J., Liu, L., & Kyllonen, P. (2017). Interdisciplinary research agenda in support of assessment of collaborative problem solving: Lessons learned from developing a collaborative science assessment prototype. *Computers in Human Behavior*, *76*, 631–640.

Wang, M., Xu, H., Zhao, J., Sun, C., Liu, Y., & Li, J. (2025). The Carbon Sequestration Potential of Skyscraper Greenery: A Bibliometric Review (2003–2023). *Sustainability*, *17*(5), 1774.

Wang, X., Gard, W., Borska, H., Ursem, B., & Van De Kuilen, J. (2020). Vertical greenery systems: From plants to trees with self-growing interconnections. *European Journal of Wood and Wood Products*, *78*, 1031–1043.

Wang, X., & Li, H. (2024). Enhancing Public Education of Water Resource Conservation Through Urban Landscape: A Case Study of Bishan-Ang Mo Kio Park in Singapore. *Communications in Humanities Research*, *45*, 25–31.

Well, W. (2014). *WELL building standard.* International Well Building Institute.

Williams, D. E. (2007). *Sustainable design: ecology, architecture, and planning.* John Wiley & Sons.

Wortmann, T., & Tunçer, B. (2017). Differentiating parametric design: Digital workflows in contemporary architecture and construction. *Design Studies*, *52*, 173–197.

Xu, X., Mumford, T., & Zou, P. X. (2021). Life-cycle building information modelling (BIM) engaged framework for improving building energy performance. *Energy and Building*, *231*, 110496.

Yadav, M., & Agarwal, M. (2021). Biobased building materials for sustainable future: An overview. *Materials Today: Proceedings*, *43*, 2895–2902.

Yevenes, M. A., Kolok, A. S., & Araneda, A. (2025). Citizen science to promote a nature-based solution: Barriers and lessons learned from a vegetated vermifilter implementation experience in a Chilean community. *Environmental Monitoring and Assessment*, *197*(2), 1–12.

Zhang, T., Tan, Y., Yang, H., & Zhang, X. (2016). The application of air layers in building envelopes: A review. *Applied Energy*, *165*, 707–734.

Chapter 11
Sustainable Construction Strategies that Guarantee Humanistic, Inclusive, and Technologically Innovative Urban Design

Lara Vélez-González
https://orcid.org/0009-0002-5145-3600
CEU San Pablo University, Spain

ABSTRACT

Cities are undergoing a profound transformation driven by technological advances such as artificial intelligence (AI), the Internet of Things (IoT), and digitalisation, which are redefining how people interact with their urban environment. These changes underscore the need to design more sustainable, inclusive, and resilient cities, capable of balancing technological innovation with social well-being and environmental protection. In this context, the chapter offers a comprehensive analysis of emerging urban design paradigms within the framework of smart and sustainable cities. Based on a systematic literature review and a multidisciplinary perspective, it explores the intersections between technology, sustainability and urbanism, with special attention to key issues such as cybersecurity, resilience, and the circular economy. It also identifies innovative strategies, including green and blue infrastructure, humanistic and inclusive urbanism, and hybrid and modular housing. Finally, it proposes actions to foster urban spaces that are resilient, accessible, and aligned with contemporary challenges.

DOI: 10.4018/979-8-3373-2555-2.ch011

INTRODUCTION

Technological progress has never been as rapid or pervasive as it is today. From robotics to artificial intelligence, innovation is not only transforming industry and communication, but also reshaping the way we inhabit and design our cities. These advances not only have an irreversible impact on social, economic and territorial organisation, but are also redefining the way in which people interact with their immediate environment. This is forcing cities, buildings and therefore housing to adapt to new demands. In this context, cities, buildings and housing must become key factors in responding to global changes, where the need for indoor comfort has become an essential requirement to ensure the quality of life of citizens.

The necessity for a healthier, more accessible and efficient environment has brought the importance of urban design to the forefront. Smart cities have emerged as a clear example of how technologies can enhance citizens' life quality, integrating advanced infrastructures and providing an interconnected, inclusive, healthy and sustainable urban system. This concept goes beyond urban spaces digitisation; it is about creating ecosystems that harmonise social well-being, economic development and environmental sustainability, (Anthopoulos, 2015).

However, widespread and abusive use of such technologies, for example, those connected to smart grids, transport systems, buildings, or health services, introduces risks that may become serious if not properly addressed. Cybersecurity therefore becomes a cornerstone in cities design and management, ensuring sensitive data protection and urban services reliability,m, (Ma, 2021).

This technological approach must be accompanied by a humanistic conception of the city. In an increasingly digitalised world, cities must be conceived as spaces that promote social inclusion, active citizen participation and respect for diversity. Urban design that takes into account the needs of all people, including the most vulnerable, is essential to ensure that progress does not leave any sector of society behind. This involves creating accessible and adapted spaces, integrating innovative practices and solutions that foster social justice, equity and cohesión, (Costa, 2021).

Sustainable city concept, based on energy efficiency, reduction of polluting emissions and optimisation of natural resource consumption, has also become a central component of contemporary urban design. It aims to create more environmentally friendly places and resilient communities capable of addressing climate change challenges, mass urbanisation and social inequalities. Renewable energy integration, efficient water resources management, sustainable transport promotion and green spaces creation are among key elements of this approach, (Pérez-Cornejo et al., 2023).

Cities future will be defined not only by their capacity to integrate technologies, but also by their ability to balance economic development, environmental sustainability, and human well-being. In this regard, smart cities should be seen as reference models for technological innovation and for their capacity to promote inclusive, equitable, and healthy urban environments. The transformation towards a smart, inclusive and sustainable city represents multidimensional challenge that requires collaboration between governments, businesses, communities, and citizens to create urban spaces that serve as true resilience models, equity, and life quality for all, (Anthopoulos, 2015; Toli & Murtagh, 2020).

This chapter aims to provide comprehensive and up-to-date analysis of new urban design paradigms in smart and sustainable cities context. The specific objectives guiding this work are the following ones:

- To analyse the smart city concept from a multidisciplinary perspective, addressing both its potentials and limitations, particularly in relation to sustainability and social welfare.
- To explore the intersections between technology, sustainability and urban design, with particular attention to the role of AI, IoT, and other digital tools in the transformation of urban environments.
- To investigate how the Sustainable Development Goals (SDGs) are being localised in contemporary cities, with an emphasis on SDG 11 and its impact on urban planning and management.
- To identify sustainable urban strategies, such as green and blue infrastructure, humanistic and inclusive urbanism, the circular economy, and resilience to certain natural phenomena, such as climate change.
- To reflect on the main technological and ethical challenges, such as the digital divide, citizen participation, and cybersecurity, which directly affect the development of truly inclusive smart cities.
- To study new housing typologies and multifunctional buildings, assessing how they respond to the current challenges of flexibility, energy efficiency, accessibility, and social cohesion.

In summary, these objectives allow for a critical examination of the challenges facing cities in the twenty-first century, integrating technological, social and environmental approaches. The chapter offers a diagnosis of the current state of smart and sustainable urban design and provides proposals that contribute to the creation of more resilient and inclusive environments, adapted to contemporary transformations.

METHODOLOGY

For this study, a literature review was conducted across different databases, such as Scopus or Web of Science (WOS), which offer a wide range of research on smart cities, sustainable urban design and new multifunctional living spaces. Several keywords related to the above topics were selected in order to cover different disciplines such as architecture and health.

Selection criteria included thematic relevance, multidisciplinary approach (architecture, urban planning, engineering, sociology, technology and environment), as well as the geographical origin and institutional affiliation of the authors. To this end, indicators comprising the number of publications per year and the distribution of authors by country and university were analysed.

Specifically, the searches focused on five broad thematic areas: smart city concept, sustainability in urban design, humanistic approach to urbanism, development and integration of sustainable technologies in the city, and urban regeneration processes linked to circular economy. These categories guided both the selection of keywords and the organisation of the analysis, allowing the review to be structured from a comprehensive, contemporary and cross-disciplinary perspective.

Figure 1 illustrates the evolution of academic interest in the concepts through the number of annual publications associated with the terms 'smart city' and 'innovative urban design'. The analysis shows a growth in these areas over recent years, reflecting a growing concern to rethink urban design through more sustainable, technological and inclusive approaches.

Figure 1. Documents published by year on 'innovative urban design' (Scopus)

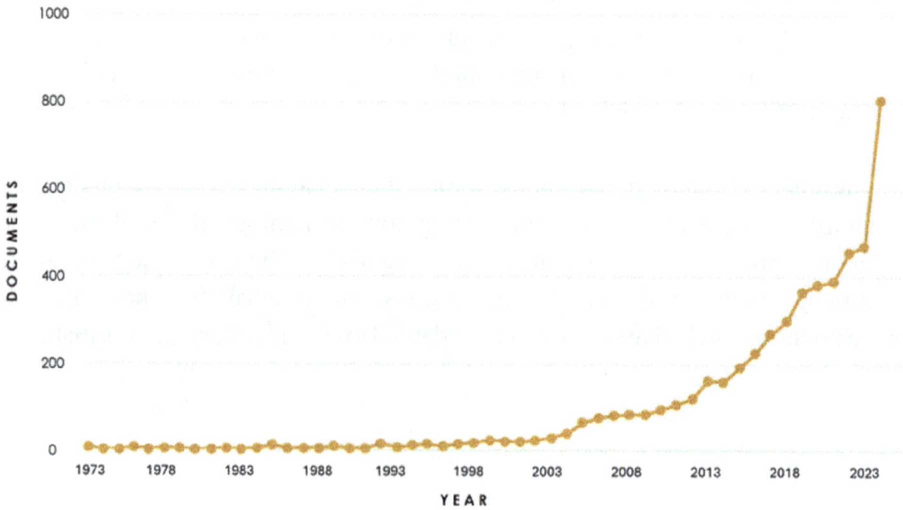

Figure 2 shows a graph representing the connections between authors, countries and institutional affiliations associated with publications on 'smart city', 'humanistic urbanism', 'inclusive urbanism', 'technologies in urban design' and 'urban regeneration'. This visualisation enables the identification of the universities and research centres most active in these areas, as well as the existing links of international collaboration. The analysis reveals which countries lead scientific production and how authorship is distributed among the different institutions, highlighting the main hubs of knowledge generation that are shaping a more integrated and global approach to contemporary urban design.

Figure 2. Graph of relationships between authors, countries and affiliations in urban design studies

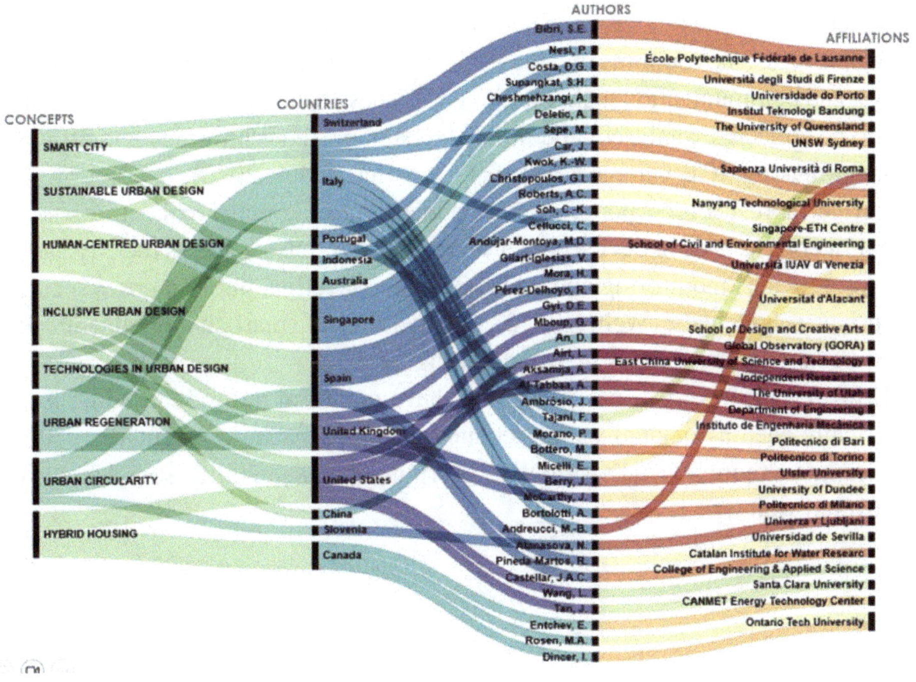

The results show a clear trend towards the consolidation of the smart city concept as a central axis in the academic debate on urban design. The publications growth from 2015 onwards indicates a growing concern with incorporating more sustainable and inclusive approaches into urban design. Furthermore, the analysis of the distribution of publications by country and institutional affiliation reveals a strong concentration of scientific production in universities and research centres in Europe and Asia, with highly active research hubs in countries such as the Netherlands,

South Korea, Germany, and Singapore. These findings allow the identification of the priority topics, as well as the main actors and regions leading the transformation of contemporary urban thinking from a global, technological, and humanistic perspective.

'SMART CITY' CONCEPT

The term 'smart city' has become the main objective of current urban strategies, integrating technology with a sustainable and welfare-oriented approach. This model seeks to optimise resources, using improvement of citizens' life quality and the efficient, sustainable management of city itself as drivers, (Benites & Simões, 2021).

A smart city is more than an urban environment filled with technologies, screens and electronics. It is an interconnected ecosystem, where infrastructure, public and private services, the economy and the environment interact through the use of information and communication technologies (ICT These technologies enable real-time data to be collected, analysed, and utilised to support more informed and efficient decision-making, (Gil-Garcia, Chen, & Casco-Hernandez, 2023).

Furthermore, IoT incorporation is embedded as an emerging architecture base on global Internet, enabling devices interconnection and sensors across the city, facilitating continuous monitoring and real-time data collection, but also poses security and privacy challenges, (Weber, 2010). An overview of smart city architecture and its interconnected systems is presented in Figure 3.

However, smart city concept does not have universal definition. While it has historically been associated with a technological approach, in recent years it has evolved into a more holistic vision that incorporates sustainability, social inclusion and overall well-being.

Figure 3. Smart city architecture

Smart Cities and Sustainability

Although smart and sustainable cities share objectives related to efficiency and resource optimisation, each prioritises different approaches. First of all, sustainable cities focus on balancing urban demands with environmental protection, aiming to minimise ecological impact through practices such as emissions reduction, resource efficiency, and the implementation of renewable energy. On the other hand, smart cities address urban challenges by using technology to improve citizens' life quality, although their ability to effectively and concretely promote sustainability remains a subject of debate, (Kennedy, Pincetl, & Bunje, 2011).

`For smart cities to truly contribute to sustainability, it is crucial to move beyond the mere implementation of innovative technologies. Smart cities have significant potential to integrate technology and sustainability, which not only helps reduce pollutant emissions but also contributes to urban health. An example is the implementation of smart energy grids or the use of sensors to monitor air quality and traffic, which not only helps to reduce pollutant emissions, but also contribute to urban health, (Bibri & Krogstie, 2017).

For effective integration of technology and sustainability, is essential to understand the concept of 'urban metabolism', which describes how cities consume resources, produce energy and generate waste. By optimising these flows, cities can achieve a more efficient balance between human activity and natural ecosystems. This requires not only the integration of more technologies, but also rethinking urban design to reduce resource consumption, minimise waste generation and limit environmental impact, given that we live on a finite planet. This holistic approach enables cities to become not only technological hubs and models of sustainability and resilience, (Barles, 2010).

In this context, the concept of the 'Sustainable Smart City' (SSC) emerges, combining advanced technology with a clear focus on environmental protection, responsible urbanisation and citizen well-being, (Lement, Ruysschaert, & Crutzen, 2023).

In Asia, Seoul and Singapore stand out as benchmark models, each offering distinct approaches to smart and sustainable urban development. Seoul combines technological innovation with strong social policies, promoting governmental transparency and citizen participation through platforms like the Seoul Smart City initiative. The city integrates smart grids, high-efficiency public transport, and digital inclusion to enhance both quality of life and environmental resilience, (Joo, 2023). Singapore, on the other hand, adopts a centralised, state-led planning model through its flagship Smart Nation programme. By integrating AI, sensor networks and digital governance tools in areas such as water management and mobility, it aims to maximise efficiency while maintaining a high standard of living and sustainability performance, (Mahizhnan, 1999).

Localising the Sustainable Development Goals (SDGs).

A key challenge for smart and sustainable cities is tha ability to immplement the Sustainable Development Goals (SDGs) at the local level, adapting them to the specific needs of each city and its inhabitants. The SDGs established by the United Nations constitute a global framework ranging from poverty eradication to environmental protection, require contextualisation to be effective, (United Nations, n.d.).

This process is particularly relevant in case of SDG 11, which promotes 'Sustainable Cities and Communities' by integrating sustainability into all levels of urban planning. This includes improving infrastructure, enhancing accessibility, and reducing ecological footprints. However, the localisation of these goals faces several barriers, including limited funding, challenges in gathering reliable data to monitor progress, and a lack of awareness among local authorities and citizens, all of which hinder the effective implementation of sustainable policies, (Alshuwaikhat, Adenle, & Almuhaidib, 2022).

Figure 4. Targets of SDG 11

1. Access for all to adequate, safe and affordable housing and basic services and upgrade slums.
2. Access to safe, affordable, accesible, and sustainable transport systems for all.
3. Enhance inclusive and sustainable urbanization and capacity for participatory, integrated and sustainable human settlement planning.
4. Strengthen efforts to protect and safeguard the world's cultural and natural heritage.
5. Significantly reduce disaster-related deaths, minimize economic losses, and enhance the protection of vulnerable populations, particularly in water-related disasters.
6. Reduce the adverse per capita environmental impact of cities.
7. Provide universal Access to safe, inclusive and accesible, green and public spaces.

a. Support positive economic, social and environmental links between urban, per- urban and rural areas.

b. Promote the adoption of integrated urban policies that enhance inclusion, resource efficiency, and resilience to climate change and disasters, in alignment with the Sendai Framework for Disaster Risk Reduction

c. Provide financial and technical support to least developed countries for the construction of sustainable and resilient buildings using locally sourced materials.

Figure 4 outlines the key targets established by SDG 11 in the context of urban sustainability. Figure 5 presents the evolution in the number of scientific publications related to various SDG 11 targets over the last 25 years, enabling the identification of key research trends and emerging areas of interest within this field.

Figure 5. Documents per year for each SDG 11 target (Scopus)

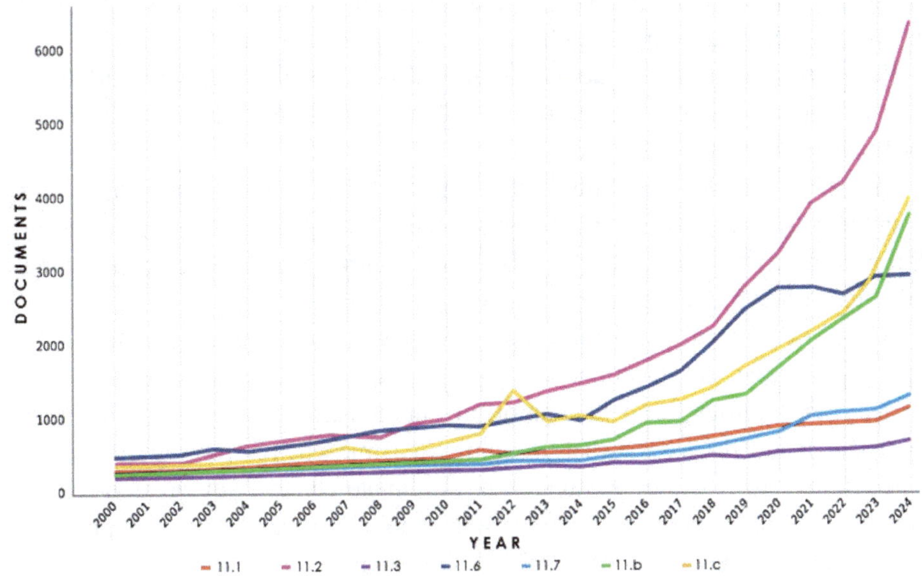

This bibliometric analysis of SDG 11 targets over the last 25 years reveals exponential growth in academic interest in topics related to urban sustainability.

Particularly are publications related to sustainable transport (11.2), sustainable building materials (11.c) and urban air quality (11.6), which a sharp increase since 2020. This pattern reflects a growing concern with climate change mitigation and the transition towards healthier and more resilient cities. In contrast, concepts such as participatory urban planning (11.3) or public green spaces (11.7) present a comparatively lower volume of research, which may indicate areas with potential for future development. This type of visualisation helps to identify thematic imbalances within urban sustainability research and highlights the need for a more integrated and balanced approach to addressing all SDG 11 targets comprehensively.

In this context, smart city strategies play a key role. Through the use of digital technologies such as big data, citizen participation platforms, and urban monitoring systems, it becomes possible to identify which targets are being least addressed, adapt local policies more rapidly, and improve the tracking of progress towards the SDGs. This enables cities to operate more efficiently and answer more effectively to the real challenges faced by their inhabitants, (Giles-Corti et al., 2016).

Sustainable Planning Strategies in Cities

Priority should be given to the efficient use of natural resources and the integration of nature-based solutions, such as green and blue infrastructure, which combine natural elements with technological approaches to promote sustainability and enhance the quality of urban life.

Green infrastructure includes the creation and maintenance of parks, urban gardens and green corridors that improve well-being, offset carbon emissions, promote biodiversity and improve air quality.

In contrast, blue infrastructure refers to urban water systems that manage water sustainably, such as urban wetlands, green roof, and sustainable drainage systems (SuDS). These types of solutions enable flood control, aquifer recharge and runoff reduction, while preserving the city's natural environment.

It is also essential to consider the orography and morphology of the terrain. Appropriate urban site selection, taking into account natural slopes, soil infiltration capacity, and hydrological behaviour of surrounding areas, can significantly reduce the risk of natural disasters such as flooding. At a time when extreme weather events are becoming increasingly frequent due to climate change, anticipation through strategic orographic planning becomes a key tool for climate change adaptation.

This enhances the resilience of cities to climate change and optimises both energy and water consumption. It also contributes social cohesion by facilitating access to high-quality public spaces for all citizens and reducing urban inequalities, (Jim, 2013).

Quantitative Impact of Smart Energy System in Selected Cities

To complement the theoretical discussion, this section presents several real-world cases in which smart city strategies have led to measurable improvements in energy efficiency and sustainability.

Barcelona (Spain)

As one of the pioneering European cities in adopting the smart city model, Barcelona has implemented a broad range of technological innovations through its Urban Platform and sensor network. The deployment of smart lighting systems in public spaces—capable of adjusting intensity based on movement and daylight—has achieved electricity savings of approximately 30% in several districts. Additionally, smart water irrigation systems installed in parks and green areas, which monitor humidity and forecast weather, have reduced water consumption by 25%, (Bakici, Almirall, & Wareham, 2013). These measures are part of a broader energy optimisation strategy that also includes public transport electrification and smart mobility platforms.

From an economic perspective, the city's digital infrastructure strategy has enabled budgetary savings in public services and attracted international investment, consolidating Barcelona's role as a hub for urban innovation and green technology. The Barcelona Activa programme and the 22@ Innovation District have generated thousands of jobs in ICT and energy sectors, while promoting a knowledge-based economy closely linked to sustainability and digital transformation.

While the city's strategy has received international praise, it has also been subject to critical review for privileging techno-centric solutions driven by private actors. Critics argue that the Barcelona model must move further toward democratising urban governance, ensuring that smart infrastructures serve inclusive and equitable urban development, (Charnock, March, & Ribera-Fumaz, 2021).

Amsterdam (Netherlands)

The city has implemented smart grid systems that allow real-time energy monitoring and dynamic demand management in neighbourhoods like Nieuw-West and IJburg, contributing to a 15–20% reduction in residential energy consumption, (Noori, Hoppe, & de Jong, 2020). In partnership with energy providers and citizens, projects such as *City-zen* and *Smart Solar Charging* have promoted the integration of renewable energy, electric vehicle infrastructure, and building energy retrofits.

From an economic perspective, Amsterdam's smart city agenda has fostered an ecosystem of public-private-civil collaboration that attracts startups and green tech companies, generating local employment and enhancing the city's international competitiveness. Through initiatives like StartupAmsterdam, the city has supported over 750 startups in sectors including clean energy, mobility, and circular economy, reinforcing its position as a digital innovation hub in Europe, (Yigitcanlar et al., 2019).

However, the Amsterdam model has also raised critical reflections. While civil society actors are formally included in many smart city projects, their influence is often limited to consultation rather than co-decision-making. This creates a dynamic of selective inclusion, where certain voices—particularly from grassroots or marginalised communities—remain underrepresented in decision-making processes, (Mello Rose, Thiel, & Grabher, 2022). As such, while Amsterdam demonstrates strong performance in technological and environmental innovation, ongoing attention is needed to ensure that social inclusion and equity are fully embedded in the smart transition.

Dubai (United Arab Emirates)

Dubai represents a top-down, state-led model of smart city development, characterised by centralised governance, strong public investment, and rapid implementation of large-scale digital infrastructure. The city exemplifies a "technology-push" approach, with innovation driven primarily by government-led strategies and top-level planning, (Noori, Hoppe, & de Jong, 2020).

A cornerstone of its transformation is the Smart Dubai 2021 initiative, which integrates more than 100 digital services to improve efficiency, mobility and sustainability. In areas like Dubai Silicon Oasis, the adoption of smart lighting, IoT-integrated energy systems and real-time monitoring has led to energy savings of up to 20%. Moreover, the digitisation of administrative procedures has contributed to the elimination of over 336 million printed documents per year, reinforcing Dubai's goal of becoming a paperless government.

From an economic perspective, Dubai's smart city strategy supports its global positioning as a hub for smart tourism and digital entrepreneurship, especially through AI-based urban services in the hospitality, mobility and planning sectors. These innovations have improved the operational efficiency of urban systems while contributing to sustainable economic growth, (Khan et al., 2017).

Additionally, the city's high level of technological readiness—both among its institutions and citizens—has been fundamental to the scalability of smart services. This includes the successful deployment of advanced ICT infrastructure, energy optimisation technologies and e-governance platforms, (Shwedeh et al., 2022). However, Dubai's model also raises concerns about the lack of participatory mechanisms, as top-down governance may limit civic engagement and inclusivity in decision-making processes, (Noori, Hoppe, & de Jong, 2020).

TECHNOLOGICAL CHALLENGES, AI AND CYBERSECURITY

Despite the many benefits that smart cities offer, they also face a number of significant challenges that must be addressed to ensure fair and equitable development.

One of the main challenges is unequal access to technology. Digital divides can exclude certain population groups from the benefits of these innovations, potentially deepening existing social inequalities, (Gil-Garcia, Chen, & Gasco-Hernandez, 2023).

Citizen participation, which is fundamental to inclusive cities design, faces notable barriers related to technological accessibility. Although digital platforms have the potential to mobilise a greater number of citizens, they tend to attract specific user profiles. Young people, despite being digitally adept, often participate less, while older citizens —typically more politically engaged— encounter technological obstacles. Designing platforms that foster inclusive participation requires targeted strategies, such as messaging that reinforces inclusive social norms, intuitive interfaces adapted to varying levels of digital literacy, and tools accessible to individuals with disabilities, (van den Berg et al., 2020).

However, the most critical challenge for cities lies in cybersecurity. Highly interconnected infrastructures that support these cities —such as electricity grids, transport systems and communication services— are vulnerable to cyberattacks. These threats can compromise the operation of essential services, leading to widespread disruptions and significantly impacting the daily lives of citizens.

An example of the severity of these attacks is a case study involving the Azure HDInsight platform, which demonstrated how cyberattacks are systematic and difficult to detect due to high volume and complexity of network data. By applying Big Data and unsupervised learning techniques, anomalies were identified with 96% accuracy, (Terzi, Terzi, & Sagiroglu, 2017). This highlights advanced technologies importance, such as AI algorithms, in detecting threats in real time and protecting critical infrastructure.

Moreover, the collection and management of large volumes of sensitive data —including individuals' locations, behaviours and personal preferences — pose significant privacy risks. Without appropriate safeguards, such as data encryption, minimisation of collected information, and strict access control, personal data may become vulnerable to unauthorised access and misuse, (Almeida, 2023).

In IoT ecosystems, challenges arise due to devices diversity and communication technologies range they employ. Many devices have limited capacity to implement advanced security measures, such as strong passwords or data encryption. In addition, the AI used to manage these networks can also be manipulated by attackers to trigger faulty decision-making. A common example is the use of devices in large-scale attacks, such as DDoS (Distributed Denial of Service) attacks, in which attackers attempt to overwhelm the target with excessive data requests, rendering the system inoperable. To enhance security, it is essential to design devices with built-in protection from the outset and to promote global collaboration in establishing standards to safeguard these systems, (Andrade et al., 2020).

AI is playing an increasingly important role in cities development. Its application in urban planning enables land use optimisation, energy efficiency improvements, environmental impact reductions, and citizen well-being promotion. AI, through Big Data and machine learning, can process real-time urban data to support strategic decision-making in areas such as mobility, waste management, and urban security, (Tomorrow.City, n.d.).

These algorithms can anticipate traffic patterns, optimise public resources allocation and reduce energy waste. In addition, they automate administrative processes, streamlining the management of urban planning permissions, environmental monitoring and the planning of new urban developments, (Phillips, Jiao, & Clubb, 2024).

In relation to urban design, AI enables the simulation and modelling of urban scenarios, allowing for the assessment of changes in the distribution of public space, pedestrian flows and green areas. Cities such as Singapore and Amsterdam have integrated these technologies to enhance mobility and sustainability, (Burry, 2022).

However, the implementation of AI in cities presents significant challenges. One of the main risks concerns privacy and ethics in urban data use, as large-scale collection and data analysis can lead to the infringement of individual rights if appropriate regulatory frameworks are not in place. Public perceptions of AI are often mixed —marked by acceptance on the one hand, and concerns on the other about the lack of transparency in its use and the accessibility of personal data, (Lehtiö et al., 2023).

The interconnectivity of urban systems also increases their vulnerability to cyberattacks, placing critical infrastructure —such as electricity grids, transport systems and emergency services —at risk. Moreover, the lack of transparency in AI models could lead to algorithmic bias, favouring certain sectors of the population over others and undermining equity in urban decision-making. In the absence of clear regulation, the automation of urban processes may reinforce existing inequalities rather than mitigate them, ultimately resulting in less inclusive cities, (Kamrowska-Zaluska, 2021).

In this context, Explainable Artificial Intelligence (XAI) emerges as an essential solution for improving the transparency, reliability and ethical accountability of AI systems applied to urban planning. Unlike conventional AI systems, which operate as 'black boxes', XAI enables a clearer understanding of how and why certain automated decisions are made, thereby facilitating evaluation and oversight by citizens, urban planners, and public authorities, (Yigitcanlar et al., 2020).

This is particularly relevant in areas such as traffic management, where AI models may adjust traffic light timings or propose modifications to transport routes without the public understanding the criteria behind these changes. In infrastructure planning, XAI supports resources allocation justification and new urban spaces design, ensuring that decisions are equitable and based on verifiable data. Furthermore, in urban safety field, XAI helps to explain how patterns of risk are identified within the urban environment, ensuring that crime prevention measures are applied fairly and without algorithmic discrimination (Javed et al., 2023), as illustrated in Figure 6, which contrasts the functioning of traditional AI systems with the transparency offered by XAI models.

Figure 6. Difference between AI and XAI working methodology

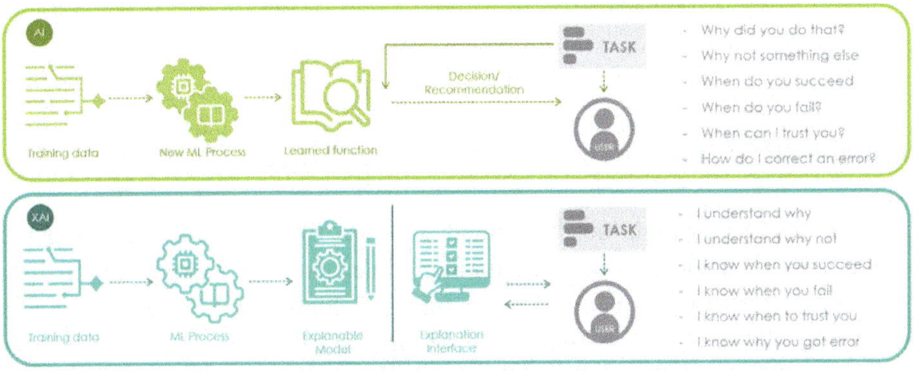

SUSTAINABILITY IN URBAN DESIGN

Including sustainability in cities design is no longer optional. Currently, more than 50% of the global population resides in urban areas, and this figure is projected to reach 66% by 2050. This accelerated growth of cities has led to unsustainable development patterns, where increasing demands on infrastructure, natural resources and basic services such as water, sanitation and transport exert considerable pressure on both urban and natural ecosystems, (Gil & Duarte, 2013; Puchol-Salort et al., 2021).

Sustainability in urban design involves a broad spectrum of interrelated challenges and approaches that go beyond environmental considerations alone. To provide a structured overview, Table 1 summarises the key dimensions of sustainability addressed in this section.

Table 1. Summary of advantages and limitations of smart cities

Dimension	Focus Area	Key Challenge	Strategy
Perception Well-being	5.1. Humanistic urbanism	Discomfort in public spaces; lack of emotional connection to place	Lighting design, noise mitigation, human-centred planning
Equity Accessibility	5.2. Inclusive urbanism	Spatial and social exclusion; barriers for people with disabilities	Universal design, participatory planning, inclusive public space
Environment Resources	5.3. Sustainable technologies	Pollution, climate risks, resource overuse	Green & blue infrastructure, smart tech integration, energy efficiency, resilience strategies
Circularity Resilience	5.4. Circular economy and regeneration	Waste, urban decay, material scarcity	Selective deconstruction, material reuse, adaptive reuse, regenerative urbanism

Humanistic Urbanism and its Impact on Urban Planning

Urban planning should not be limited solely to structural and environmental considerations, but must also account for intangible factors such as perception, lived experience, and the functionality of urban space. Lighting and the detailed design of public spaces are particularly relevant in shaping how citizens experience the city. Effective lighting in urban environments can enhance quality of life by providing safety, comfort and improved aesthetic integration with the built environment, (Chunmei, 2017).

In addition, urban noise management is another crucial element for well-being. Excessive noise—whether from traffic, commercial or industrial operations—directly impacts comfort, as well as the physical and mental health of citizens. A humanistic approach to urban planning should incorporate strategies to reduce noise pollution, such as the design of natural barriers, the use of sound-absorbing materials in infrastructure and land-use planning that effectively separates residential areas from high noise intensity zones. Creating pleasant acoustic environments, alongside adequate lighting and accessible public spaces, can contribute to making more liveable, healthy and people-centred cities, (Radicchi et al., 2021).

Inclusive Urbanism and Universal Accessibility

The concept of smart and inclusive citizenship involves placing people at the centre of urban design, ensuring equitable access to public spaces and encouraging active participation in planning and transformation of cities. In this regard, people with disabilities are especially valuable, as their perspectives and needs offer a vital perspective for creating environments that are accessible, comfortable and functional for all, (Shahraki, 2021).

A truly smart city adapts to the needs of all its citizens, not only from a physical perspective, but also from a cognitive, sensory, and emotional dimensions Inclusive urbanism, therefore, must incorporate universal accessibility principles, designing spaces that can be used by everyone, regardless of their abilities. This includes accessible routes, adapted street furniture, legible signage and integrated assistive technologies, (Insituto de Arquitectura Avanzada de Cataluna, n.d).

Furthermore, promoting the active involvement of groups traditionally excluded from urban design —such as people with reduced mobility or sensory disabilities —is not only social justice act but also a strategy that enriches the design process. By incorporating a wider range of perspectives, the creation of more empathetic, efficient, and sustainable environments is encouraged. Participatory planning initiatives in various cities have demonstrated that including these voices directly contributes to improving the quality and functionality of urban spaces, (Wu, Wang, & Yang, 2019).

Sustainable Technologies in Urban Design

Urban design faces challenges stemming from climate change, rapid urbanisation and increasing human influence on the environment. These factors place growing pressure on natural resources and pose considerable risks to urban ecosystems. Air and water pollution, the urban heat island effect and uncontrolled land sprawl are phenomena that not only affect environmental sustainability, but also alter citizens life quality and their psychological well-being.

Deteriorating air quality, driven by climate change, has a direct impact on public health, contributing to the rise in cardiovascular and respiratory diseases, (Short & Farmer, 2021). Similarly, urban sprawl and ecosystem degradation lead to biodiversity loss, habitat disruption, and, consequently, a decline in overall quality of life, (Johnson & Munshi-South, 2017). In response to these challenges, it is essential to adopt preventive strategies aimed at mitigating climate change effects and strengthening cities' capacity to adapt to and withstand its impacts, (Bradley & Altizer, 2007).

Green and blue infrastructures are increasingly recognised as effective solution for addressing urban environmental challenges. These include urban parks, gardens, integrated water management systems, green roofs and water bodies, offering both environmental and social benefits, (Donati et al., 2022). They contribute not only to thermal regulation and the enhancement of biodiversity, but also to social cohesion by providing spaces for citizen interaction and well-being. Green infrastructures improve air quality, restore biodiversity and helps reduce extreme temperatures, thereby supporting climate change mitigation efforts, (Barreira et al., 2023). Blue infrastructure plays a vital role in managing the urban water cycle, enabling rainwater harvesting and storage through sustainable drainage systems and urban wetlands, reducing flood risk and improving water quality, (Radinja et al., 2021).

As environmental pressures intensify, urban planning must prioritise the efficient use of resources by incorporating sustainable technologies in construction, such as photovoltaic systems and electric machinery that replaces combustion engines, thereby reducing polluting emissions, (Hiu et al., 2023).

In addition, smart technologies integration into urban resource management enables the optimisation of water and energy use, minimises waste and ensures the availability of these resources. The collection and analysis of real-time data improve system efficiency and support more informed decision-making processes, (Zucaro, Maselli, & Ulgiati, 2022).

Future urban design must also prioritise resilience to disasters and extreme weather events. This includes the integration of natural or artificial firebreaks in strategic locations, the use of fire-resistant construction materials, and the implementation of passive solutions to thermal phenomena such as heatwaves and cold spells. Examples include the incorporation of urban trees, sunshades, green roofs, and urban windbreaks, which help to regulate temperature and reduce direct exposure to sunlight or wind.

Furthermore, power outages or other energy-related risks, it is essential to design urban back-up systems based on lithium batteries, gravitational storage solutions, or emerging technologies such as thermal sand batteries. These innovations not only enhance energy security in emergency contexts, but also strengthen building autonomy through a decentralised, resilient, and sustainable approach.

Circular Economy and Urban Regeneration

In this context, it is essential to incorporate circular economy principles into cities design and development. This model proposes a paradigm shift in construction and urban planning, moving away from the linear 'use and dispose' approach toward a system based on reuse, recycling and resource maximisation. This approach not only focuses on reuse of materials from building demolition, but also promotes broader recovery strategies, emphasising rehabilitation and revaluation of existing infrastructures by assigning them new functions. It encourages more sustainable and resilient planning that supports urban environments longevity and flexibility, (Akhimien, Latif, & Hou, 2021).

One of the cornerstone of circular economy as applied to architecture is selective deconstruction, which allows existing buildings to be dismantled in a planned manner in order to recover materials with high reuse potential, such as timber, steel, concrete, bricks, and glass. This method, unlike conventional demolition, avoids large-scale waste generation and reduces virgin raw materials consumption, thereby decreasing non-renewable natural resources demand, (Tirado et al., 2022).

Beyond material recovery, circular economy promotes revaluation and adaptive reuse of pre-existing buildings, enabling a second life for urban infrastructure without demolition need. This strategy supports revitalisation of deteriorated neighbourhoods and urban spaces, while preserving cities identity. Former administrative buildings, factories or industrial complexes could be rehabilitated and transformed into cultural, residential or coworking spaces, adapting to new social and economic needs. This approach not only contributes to architectural heritage preservation and urban identity, but also strengthens social cohesion by providing functional and accessible spaces within the consolidated urban fabric, (Bilal et al., 2020).

By implementing these strategies, waste generation is reduced and environmental footprint is minimised, while available resources are optimised within a continuous cycle of innovation and reuse, (Bertino et al., 2021). This urban model development fosters city resilience by reducing dependence on newly produced materials and promoting a more efficient system of resource management. In this way, materials that would otherwise be discarded are repurposed, promoting a regenerative economy that turns waste into new opportunities for sustainable architecture and balanced urban growth, (Donati et al., 2022; Rahla, Mateus, & Bragança, 2021).

NEW DESIGN OF LIVING SPACES AND MULTIFUNCTIONAL BUILDINGS

Cities evolution and shifts in lifestyles driven by technology, sustainability and digitalisation have led a living spaces design transformation. Future housing must adapt to increasingly dynamic lifestyles and the demand for flexibility and resource efficiency. Spaces design can no longer follow the traditional model of static construction; instead, it must adapt to new ways of living and working, integrating concepts of multifunctionality, flexibility, and efficiency in both architectural and urban planning, (Zivkovic, Stamenkovic, & Petrovic, 2021).

One of the most innovative approaches is functions hybridisation within buildings, allowing same infrastructure to accommodate multiple uses without generating conflicts. Coworking spaces within residences, commercial and leisure areas integrated into residential buildings, and even temporary modular housing creation are some of the strategies that are being implemented in several cities around the world. These types of buildings can adapt to population changing needs, fostering social cohesion and maximising use of urban space, (Edge Workspaces, n.d.).

In addition, smart technologies housing integration enables energy, water, and material consumption optimisation. Through home automation systems and IoT sensors, buildings can automatically regulate lighting, temperature and waste management, thereby reducing their environmental impact and enhancing energy efficiency. Moreover, innovative materials incorporation—such as bio-based construction using laminated timber, advanced ceramics, and recycled components—supports more sustainable architecture approach aligned with circular economy principles, (Aldossari & Sidorova, 2020).

Meanwhile, modular and prefabricated housing is gaining prominence as an alternative to traditional construction model. These solutions involve manufacturing modules in specialised factories, followed by on-site assembly, thereby reducing construction time, waste, and costs. They also offer the possibility of reconfiguring spaces over time, adapting user needs and facilitate the relocation of units to different urban contexts, (Kamali, Hewage, & Sadiq, 1011; Meire, Liñares, & Hermo, 2024). A key çomponent of this model is the standardisation of building elements, enabling consistent design and integration of components such as doors, windows, staircases, kitchens, and sanitary fittings. This accelerates production, reduces on-site errors, and optimises logistics, enhancing efficiency and making construction faster, more adaptable, and affordable.

The transformation of housing models responds not only to sustainability and technological demands, but also to a growing need for adaptability and accessibility. Housing design must consider aspects such as accessibility for individuals with reduced mobility, adaptability to different types of family units, and the creation of spaces that encourage interaction and coexistence. This highlights the need to rethink residential urban planning to ensure it is more inclusive, resilient, and aligned with emerging social and labour dynamics.

Within this new paradigm, dwellings may dispense with conventional elements such as private kitchens in favour of shared facilities—such as communal industrial kitchens—which optimises available space, foster social interaction and reduce energy and material consumption. Additionally, mobile and modular furniture will facilitate fast spatial transformation, enabling a single area to serve multiple functions—such as remote work, rest or social activities—depending on changing needs, (Vestbro & Horelli, 2012).

DISCUSSION AND CONCLUSIONS

Contemporary cities face complex challenges that demand a fundamental revision of existing development models. Smart technologies, sustainability and social inclusion promotion are central priorities in this transformation. It is becoming increasingly evident that tools such as AI, green and blue infrastructures, and circular economy can contribute to the creation of more efficient and resilient urban environments.

However, technological progress also introduces a range of challenges that must not be overlooked. The absence of adequate regulation in areas such as emerging technologies, urban data governance, and cybersecurity generates uncertainty and risk. These concerns must be addressed through the development of robust, ethical, and transparent regulatory frameworks, (Yigitcanlar et al., 2020).

In terms of urban sustainability, the design of cities must go beyond energy efficiency. Humanistic and inclusive approaches that ensure citizens' holistic well-being need to be incorporated. Humanistic urbanism and universal accessibility are essential dimensions for creating emotionally liveable and functionally inclusive urban environments for all, (Insituto de Arquitectura Avanzada de Cataluna, n.d.; Shahraki, 2021).

Furthermore, sustainable smart city concept must also embrace a decentralised territorial vision, supported by digital tools that reduce population and service concentration in urban centres. Emerging technologies applied to housing, remote work, and mobility are reshaping the way we inhabit cities, (PwC Espana, 2024). This shift is essential to address the housing crisis affecting many European countries, where rising prices and land scarcity are making it increasingly difficult to access decent and affordable housing, (Lehtiö.

New living spaces design must prioritise multifunctionality, spatial flexibility, climatic comfort, and energy self-sufficiency. This requires sustainable technologies integrating and resilience strategies to events such as heat waves, fires and power outages—challenges that are becoming increasingly urgent in the context of climate change, (Gordon et al., 2022).

While there is broad consensus in the literature on the need to move towards more sustainable and technologically integrated urban models, divergent approaches remain regarding how to prioritise sustainability dimensions. Some studies focus primarily on environmental efficiency and technological innovation, often sidelining the social implications of smart urban development. Others, by contrast, emphasise participatory governance, equity, and inclusiveness as core to urban sustainability. This chapter positions itself within an integrative perspective that acknowledges the value of technological advancement, but insists on embedding it within human-centred, inclusive, and equitable urban frameworks. By linking sustainability not only to environmental performance but also to accessibility, social cohesion, and democratic participation, the chapter advocates for a multidimensional understanding of urban transformation.

While the concept of the "smart city" has gained wide traction in policy, academia, and planning practice, it remains a somewhat ambiguous and contested term. Table 2 summarises the main advantages and limitations commonly associated with this model, structured around three key dimensions: technology, governance, and inclusion.

Table 2. Summary of advantages and limitations of smart cities

Dimension	Advantages	Limitations
TECHNOLOGY	- Optimisation of urban services. - Real-time data analysis. - Efficient infrastructure management.	- Overreliance on digital systems. - Cybersecurity vulnerabilities. - Technocentric bias
GOVERNANCE	- Improved decision-making through data. - Greater transparency. - Smart platforms for participation.	- Lack of clear regulation. - Risk of surveillance or control. - Data ownership issues.
INCLUSION	- Potential to improve accessibility. - Customisable urban experiences. - Citizen engagement tools	- Digital divide excludes vulnerable groups. - Uneven access to benefits. - Risk of reinforcing inequalities

In addition to technological and social considerations, the economic dimension plays a decisive role in determining the feasibility, scalability, and timing of sustainable urban strategies. Although sustainability agendas are often driven by long-term goals—such as climate resilience, health improvement, or social inclusion—their implementation depends heavily on short-term budgetary realities. Many of the measures proposed in this chapter, such as modular housing, circular rehabilitation, and smart infrastructures, require substantial initial investment. This poses a significant barrier, especially in smaller municipalities or cities operating under austerity conditions, where limited access to funding can delay or block innovation, (ICATIECE, 2022).

For instance, applying circular economy principles to building rehabilitation can increase initial costs by 15–20% compared to conventional demolition and reconstruction methods. However, life-cycle analyses reveal that these interventions generate notable economic returns over a period of 10–15 years through reductions in energy consumption, material replacement, and waste management expenses, (Batty et al., 2012). Likewise, smart lighting systems—such as those deployed in Barcelona—required multimillion-euro investments, yet led to over 30% energy savings and significant reductions in municipal electricity bills, proving their cost-efficiency over time, (Chatzitheodoridis et al., 2023).

In the housing sector, prefabricated modular construction offers strong cost-saving potential: it typically reduces expenses by 10% to 25% per square metre, shortens construction timelines, and optimises labour requirements. Nevertheless, these advantages depend on up-front investment in manufacturing infrastructure, standardisation protocols, and regulatory adaptation. Similarly, green and blue infrastructures—while offering clear benefits in climate adaptation, public health, and social cohesion—incur expenses in terms of land acquisition, hydrological studies, ecological design, and maintenance. These costs are often underestimated or omitted in the early phases of planning (ICATIECE, 2022).

Therefore, ensuring the economic viability of sustainable urban transformation requires a multidimensional financial strategy. This includes incorporating return-on-investment models and life-cycle cost analyses into the planning process, as well as identifying funding sources such as EU structural funds, green bonds, climate resilience grants, or public-private partnerships, (Chatzitheodoridis et al., 2023). Cities that integrate these tools are not only more likely to implement their sustainability agendas effectively, but also to generate the political legitimacy and citizen support necessary for long-term urban transformation. Furthermore, research shows that citizens tend to perceive a higher quality of life when smart urban investments are implemented transparently and with clear social benefits, strengthening the long-term social and economic justification for these projects, (Wang & Zhou, 2023).

Limitations

This study has certain limitations that should be acknowledged. Cities evolution depends on multiple economic, political, social and technological factors, which may affect feasibility and effectiveness of the proposals outlined. Moreover, transition towards more sustainable, inclusive, and technologically advanced urban models requires long-term commitment, as well as coordinated cooperation between public authorities, private sector, scientific community, and citizens. Despite these limitations, this analysis offers a comprehensive and up-to-date perspective on the challenges and opportunities facing contemporary cities, emphasising the importance of advancing towards balanced urban development with a focus on equity, accessibility, and collective well-being.

Recommendations and Proposals

Based on the results and reflections, the following lines of action are proposed:

- Establish clear regulatory frameworks for urban technologies use, particularly those related to AI, big data and urban surveillance systems. Personal data protection must be guaranteed, and transparency must be promoted, enabling citizens to understand how automated decisions are made and ensuring equity in public services distribution.
- Promote inclusive urban planning by incorporating the active participation of people with disabilities and other vulnerable groups. This enriches the decision-making process with diverse perspectives and lived experiences. Embedding their vision into urban design is an essential step towards transforming cities into truly inclusive spaces that place people at the centre of urban development.

- Encourage territorial decentralisation as a response to rising rental costs and land scarcity, which affect both Spain and other European countries. In this context, it is essential to rethink the traditional urban model. Emerging technologies—such as remote working, telemedicine, and online education—offer the possibility of reduce population concentration from urban centres, favouring a more equitable and sustainable territorial distribution.
- Reinforce investment in green and blue infrastructure as a strategy for climate change adaptation and to promote ecological regeneration and social cohesion. These infrastructures support thermal regulation, improve air quality, enable sustainable water management, and contribute to the creation of accessible and healthy public spaces for all citizens.
- Promote urban rehabilitation based on circular economy principles, prioritising interventions that allow for material reuse, spatial flexibility, and waste and energy consumption minimisation. Rehabilitation, rather than new construction, is an effective strategy to reduce environmental impact of building sector and to revitalise established urban areas.
- Develop multifunctional housing solutions, including models promotion such as modular housing, shared residences, and buildings with communal spaces for work, leisure, or care. These typologies offer more flexible and affordable alternatives, adapted to contemporary dynamics of mobility, cohabitation, and sustainability.
- Incorporate economic feasibility and financial tools into sustainability planning, by applying return-on-investment analysis and life-cycle cost assessments, and leveraging instruments such as green bonds, EU funds, or public-private partnerships. This enhances implementation capacity, strengthens political legitimacy, and ensures that sustainability agendas are both socially equitable and economically viable.

REFERENCES

Akhimien, N. G., Latif, E., & Hou, S. S. (2021, June 1). Application of circular economy principles in buildings: A systematic review. *Journal of Building Engineering*. Elsevier Ltd. DOI: 10.1016/j.jobe.2020.102041

Aldossari, M. Q., & Sidorova, A. (2020). Consumer Acceptance of Internet of Things (IoT): Smart Home Context. *Journal of Computer Information Systems, 60*(6), 507–517. DOI: 10.1080/08874417.2018.1543000

Almeida, F. (2023). Prospects of Cybersecurity in Smart Cities. *Future Internet, 15*(9). Advance online publication. DOI: 10.3390/fi15090285

Alshuwaikhat, H. M., Adenle, Y. A., & Almuhaidib, T. (2022). A Lifecycle-Based Smart Sustainable City Strategic Framework for Realizing Smart and Sustainability Initiatives in Riyadh City. *Sustainability (Switzerland), 14*(14). Advance online publication. DOI: 10.3390/su14148240

Andrade, R. O., Yoo, S. G., Tello-Oquendo, L., & Ortiz-Garces, I. (2020). A Comprehensive Study of the IoT Cybersecurity in Smart Cities. *IEEE Access : Practical Innovations, Open Solutions, 8*, 228922–228941. DOI: 10.1109/ACCESS.2020.3046442

Anthopoulos, L. (2015). Defining Smart City Architecture for Sustainability. *Electronic Government and Electronic Participation: Joint Proceedings of Ongoing Research and Projects of IFIP WG 8.5 EGOV and EPart 2015*, 140–147.

Bakıcı, T., Almirall, E., & Wareham, J. (2013). A Smart City Initiative: The Case of Barcelona. *Journal of the Knowledge Economy, 4*(2), 135–148. DOI: 10.1007/s13132-012-0084-9

Barles, S. (2010). Society, energy and materials: The contribution of urban metabolism studies to sustainable urban development issues. *Journal of Environmental Planning and Management, 53*(4), 439–455. DOI: 10.1080/09640561003703772

Barreira, A. P., Andraz, J., Ferreira, V., & Panagopoulos, T. (2023). Perceptions and preferences of urban residents for green infrastructure to help cities adapt to climate change threats. *Cities (London, England), 141*. Advance online publication. DOI: 10.1016/j.cities.2023.104478

Batty, M., Axhausen, K. W., Giannotti, F., Pozdnoukhov, A., Bazzani, A., Wachowicz, M., & Portugali, Y. (2012). Smart cities of the future. *The European Physical Journal. Special Topics, 214*(1), 481–518. DOI: 10.1140/epjst/e2012-01703-3

Benites, A. J., & Simões, A. F. (2021). Assessing the urban sustainable development strategy: An application of a smart city services sustainability taxonomy. *Ecological Indicators, 127*. Advance online publication. DOI: 10.1016/j.ecolind.2021.107734

Bertino, G., Kisser, J., Zeilinger, J., Langergraber, G., Fischer, T., & Österreicher, D. (2021). Fundamentals of building deconstruction as a circular economy strategy for the reuse of construction materials. *Applied Sciences (Switzerland), 11*(3), 1–31. DOI: 10.3390/app11030939

Bibri, S. E., & Krogstie, J. (2017, May 1). Smart sustainable cities of the future: An extensive interdisciplinary literature review. *Sustainable Cities and Society*. Elsevier Ltd. DOI: 10.1016/j.scs.2017.02.016

Bilal, M., Khan, K. I. A., Thaheem, M. J., & Nasir, A. R. (2020). Current state and barriers to the circular economy in the building sector: Towards a mitigation framework. *Journal of Cleaner Production, 276*. Advance online publication. DOI: 10.1016/j.jclepro.2020.123250

Bradley, C. A., & Altizer, S. (2007, February). Urbanization and the ecology of wildlife diseases. *Trends in Ecology & Evolution*. Advance online publication. DOI: 10.1016/j.tree.2006.11.001 PMID: 17113678

Burry, M. (2022). A new agenda for AI-based urban design and planning. In *Artificial Intelligence in Urban Planning and Design: Technologies, Implementation, and Impacts* (pp. 3–20). Elsevier. DOI: 10.1016/B978-0-12-823941-4.00005-6

Charnock, G., March, H., & Ribera-Fumaz, R. (2021). From smart to rebel city? Worlding, provincialising and the Barcelona Model. *Urban Studies (Edinburgh, Scotland), 58*(3), 581–600. DOI: 10.1177/0042098019872119

Chatzitheodoridis, F., Melfou, K., Kontogeorgos, A., & Kalogiannidis, S. (2023). Exploring Key Aspects of an Integrated Sustainable Urban Development Strategy in Greece: The Case of Thessaloniki City. *Smart Cities, 6*(1), 19–39. DOI: 10.3390/smartcities6010002

Chunmei, W. (2017). Light art design of public architecture based on the urban humanistic environment. *Agro Food Industry Hi-Tech, 28*(1), 322–326. https://blog.iaac.net/course/moec-23-24-urbanismo-humanista/

Costa, E. M. (2021). From smart cities to more Humane and Sustainable Smart Cities: the world in transformation. In *Humane and Sustainable Smart Cities: A Personal Roadmap to Transform Your City after the Pandemic* (pp. 41–63). Elsevier. DOI: 10.1016/B978-0-12-819186-6.00005-1

Donati, G. F. A., Bolliger, J., Psomas, A., Maurer, M., & Bach, P. M. (2022). Reconciling cities with nature: Identifying local Blue-Green Infrastructure interventions for regional biodiversity enhancement. *Journal of Environmental Management, 316*. Advance online publication. DOI: 10.1016/j.jenvman.2022.115254 PMID: 35576714

Eberhardt, L. C. M., Birkved, M., & Birgisdottir, H. (2022). Building design and construction strategies for a circular economy. *Architectural Engineering and Design Management, 18*(2), 93–113. DOI: 10.1080/17452007.2020.1781588

Edge Workspaces. (n.d.). *Integrated co-living and co-working spaces: The future of urban living and working.* https://www.edgeworkspaces.com/integrated-co-living-and-co-working-spaces-the-future-of-urban-living-and-working/

Gil, J., & Duarte, J. P. (2013). Tools for evaluating the sustainability of urban design: A review. *Proceedings of the Institution of Civil Engineers: Urban Design and Planning*. Thomas Telford Services Ltd. DOI: 10.1680/udap.11.00048

Gil-Garcia, J. R., Chen, T., & Gasco-Hernandez, M. (2023). Smart City Results and Sustainability: Current Progress and Emergent Opportunities for Future Research. *Sustainability (Switzerland), 15*(10). Advance online publication. DOI: 10.3390/su15108082

Giles-Corti, B., Vernez-Moudon, A., Reis, R., Turrell, G., Dannenberg, A. L., Badland, H., … Owen, N. (2016, December 10). City planning and population health: a global challenge. *The Lancet*. Lancet Publishing Group. DOI: 10.1016/S0140-6736(16)30066-6

Gordon, K. S., Kyung, T., Perez, C. R., Holec, P. V., Ramos, A., Zhang, A. Q., & Birnbaum, M. E. (2022). Screening for CD19-specific chimaeric antigen receptors with enhanced signalling via a barcoded library of intracellular domains. *Nature Biomedical Engineering, 6*(7), 855–866. DOI: 10.1038/s41551-022-00896-0 PMID: 35710755

Hui, C. X., Dan, G., Alamri, S., & Toghraie, D. (2023). Greening smart cities: An investigation of the integration of urban natural resources and smart city technologies for promoting environmental sustainability. *Sustainable Cities and Society, 99*. Advance online publication. DOI: 10.1016/j.scs.2023.104985

Instituto de Arquitectura Avanzada de Cataluña (IAAC). (n.d.). *MOeC 23/24 Urbanismo Humanista*. https://blog.iaac.net/course/moec-23-24-urbanismo-humanista/

.Javed, A. R., Ahmed, W., Pandya, S., Maddikunta, P. K. R., Alazab, M., & Gadekallu, T. R. (2023, February 1). A Survey of Explainable Artificial Intelligence for Smart Cities. *Electronics (Switzerland)*. MDPI. DOI: 10.3390/electronics12041020

Jim, C. Y. (2013). Sustainable urban greening strategies for compact cities in developing and developed economies. *Urban Ecosystems*, *16*(4), 741–761. DOI: 10.1007/s11252-012-0268-x

Johnson, M. T. J., & Munshi-South, J. (2017, November 3). Evolution of life in urban environments. *Science*. American Association for the Advancement of Science. DOI: 10.1126/science.aam8327

Joo, Y. M. (2023). Developmentalist smart cities? the cases of Singapore and Seoul. *International Journal of Urban Sciences*, *27*(S1), 164–182. DOI: 10.1080/12265934.2021.1925143

Kamali, M., Hewage, K., & Sadiq, R. (2022). Economic sustainability benchmarking of modular homes: A life cycle thinking approach. *Journal of Cleaner Production*, *348*. Advance online publication. DOI: 10.1016/j.jclepro.2022.131290

Kamrowska-Załuska, D. (2021, November 1). Impact of ai-based tools and urban big data analytics on the design and planning of cities. *Land*. MDPI. DOI: 10.3390/land10111209

Kennedy, C., Pincetl, S., & Bunje, P. (2011, August). The study of urban metabolism and its applications to urban planning and design. *Environmental Pollution*. Advance online publication. DOI: 10.1016/j.envpol.2010.10.022 PMID: 21084139

Khan, , M. S., Woo, M., Nam, K., & Chathoth, P. K. (2017). Smart city and smart tourism: A case of Dubai. *Sustainability (Switzerland)*, *9*(12). Advance online publication. DOI: 10.3390/su9122279

Lehtiö, A., Hartikainen, M., Ala-Luopa, S., Olsson, T., & Väänänen, K. (2023). Understanding citizen perceptions of AI in the smart city. *AI & Society*, *38*(3), 1123–1134. DOI: 10.1007/s00146-022-01589-7

Lehtiö, A., Hartikainen, M., Ala-Luopa, S., Olsson, T., & Väänänen, K. (2023). Understanding citizen perceptions of AI in the smart city. *AI & Society*, *38*(3), 1123–1134. DOI: 10.1007/s00146-022-01589-7

.Lement, J., Ruysschaert, B., & Crutzen, N. (2023, November 1). Smart city strategies – A driver for the localization of the sustainable development goals? *Ecological Economics*. Elsevier B.V. DOI: 10.1016/j.ecolecon.2023.107941

Ma, C. (2021). Smart city and cyber-security; technologies used, leading challenges and future recommendations. *Energy Reports*, *7*, 7999–8012. DOI: 10.1016/j.egyr.2021.08.124

Mahizhnan, A. (1999). Smart cities: The Singapore case. *Cities (London, England), 16*(1), 13–18. DOI: 10.1016/s0264-2751(98)00050-x

Meire, C., Liñares, P., & Hermo, V. (2024). El Sistema constructivo modular Walluminium, análisis de la envolvente hermética y termoacústica y su sistema de producción. *Informes de la Construcción, 76*(573), 6451. DOI: 10.3989/ic.6451

Mello Rose, F., Thiel, J., & Grabher, G. (2022). Selective inclusion: Civil society involvement in the smart city ecology of Amsterdam. *European Urban and Regional Studies, 29*(3), 369–382. DOI: 10.1177/09697764221092587

Noori, N., Hoppe, T., & de Jong, M. (2020). Classifying pathways for smart city development: Comparing design, governance and implementation in Amsterdam, Barcelona, Dubai, and Abu Dhabi. *Sustainability (Switzerland), 12*(10). Advance online publication. DOI: 10.3390/SU12104030

Pérez-Cornejo, C., Rodríguez-Gutiérrez, P., & de Quevedo-Puente, E. (2023). City reputation and the role of sustainability in cities. *Sustainable Development, 31*(3), 1444–1455. DOI: 10.1002/sd.2459

Phillips, C., Jiao, J., & Clubb, E. (2024). Testing the Capability of AI Art Tools for Urban Design. *IEEE Computer Graphics and Applications, 44*(2), 37–45. DOI: 10.1109/MCG.2024.3356169 PMID: 38241102

Puchol-Salort, P., O'Keeffe, J., van Reeuwijk, M., & Mijic, A. (2021). An urban planning sustainability framework: Systems approach to blue green urban design. *Sustainable Cities and Society, 66*. Advance online publication. DOI: 10.1016/j.scs.2020.102677

PwC España. (2024, febrero 26). *Los expertos señalan la falta de oferta y una legislación inadecuada como las principales causas del problema de la vivienda en España*. PwC España. https://www.pwc.es/es/sala-prensa/notas-prensa/2024/expertos-oferta-legislacion-problema-vivienda-espana.html

Radicchi, A., Cevikayak Yelmi, P., Chung, A., Jordan, P., Stewart, S., Tsaligopoulos, A., & Grant, M. (2021). Sound and the healthy city. *Cities & Health, 5*(1–2), 1–13. DOI: 10.1080/23748834.2020.1821980

Radinja, M., Atanasova, N., & Zavodnik Lamovšek, A. (2021). The water-management aspect of blue-green infrastructure in cities. *Urbani Izziv, 32*(1), 98–110. DOI: 10.5379/URBANI-IZZIV-EN-2021-32-01-003

Rahla, K. M., Mateus, R., & Bragança, L. (2021). Implementing circular economy strategies in buildings—From theory to practice. *Applied System Innovation, 4*(2). Advance online publication. DOI: 10.3390/asi4020026

Shahraki, A. A. (2021). Urban planning for physically disabled people's needs with case studies. *Spatial Information Research, 29*(2), 173–184. DOI: 10.1007/s41324-020-00343-9

Short, J. R., & Farmer, A. (2021). Cities and Climate Change. *Earth (Switzerland), 2*(4), 1038–1045. DOI: 10.3390/earth2040061

Shwedeh, F., Hami, N., Bakar, S. Z. A., Yamin, F. M., & Anuar, A. (2022). The Relationship between Technology Readiness and Smart City Performance in Dubai. *Journal of Advanced Research in Applied Sciences and Engineering Technology, 29*(1), 1–12. DOI: 10.37934/araset.29.1.112

Terzi, D. S., Terzi, R., & Sagiroglu, S. (2017). Big data analytics for network anomaly detection from netflow data. In *2nd International Conference on Computer Science and Engineering, UBMK 2017* (pp. 592–597). Institute of Electrical and Electronics Engineers Inc. https://doi.org/DOI: 10.1109/UBMK.2017.8093473

Tirado, R., Aublet, A., Laurenceau, S., & Habert, G. (2022). Challenges and Opportunities for Circular Economy Promotion in the Building Sector. *Sustainability (Switzerland), 14*(3). Advance online publication. DOI: 10.3390/su14031569

Toli, A. M., & Murtagh, N. (2020, June 2). The Concept of Sustainability in Smart City Definitions. *Frontiers in Built Environment*. Frontiers Media S.A. DOI: 10.3389/fbuil.2020.00077

Tomorrow.City. (n.d.). *IA aplicada al urbanismo: Riesgos y beneficios de un movimiento que parece imparable.* The biggest platform about urban innovation. https://tomorrow.city

United Nations. (n.d.). Sustainable Development. United Nations. https://www.un.org/sustainabledevelopment/es/

van den Berg, A. C., Giest, S. N., Groeneveld, S. M., & Kraaij, W. (2020). Inclusivity in Online Platforms: Recruitment Strategies for Improving Participation of Diverse Sociodemographic Groups. *Public Administration Review, 80*(6), 989–1000. DOI: 10.1111/puar.13215

Vestbro, D. U., & Horelli, L. (2012). *Design for gender equality: The history of co-housing ideas and realities. Built Environment.* Alexandrine Press. DOI: 10.2148/benv.38.3.315

Wang, M., & Zhou, T. (2023). Does smart city implementation improve the subjective quality of life? Evidence from China. *Technology in Society, 72*. Advance online publication. DOI: 10.1016/j.techsoc.2022.102161

Weber, R. H. (2010). Internet of Things - New security and privacy challenges. *Computer Law & Security Review*, *26*(1), 23–30. DOI: 10.1016/j.clsr.2009.11.008

Wu, W., Wang, B., & Yang, S. (2019). Research on the simulation design of humanistic landscape optimization in urban residential area based on computer technology. *Informatica (Slovenia)*, *43*(3), 325–330. DOI: 10.31449/inf.v43i3.2802

Yigitcanlar, T., Desouza, K. C., Butler, L., & Roozkhosh, F. (2020). *Contributions and risks of artificial intelligence (AI) in building smarter cities: Insights from a systematic review of the literature. Energies*. MDPI AG. DOI: 10.3390/en13061473

Yigitcanlar, T., Han, H., Kamruzzaman, M., Ioppolo, G., & Sabatini-Marques, J. (2019). The making of smart cities: Are Songdo, Masdar, Amsterdam, San Francisco and Brisbane the best we could build? *Land Use Policy*, *88*. Advance online publication. DOI: 10.1016/j.landusepol.2019.104187

Zivkovic, M., Stamenkovic, M., & Petrovic, V. (2021). Towards flexible housing: Basic design principles. *Facta Universitatis - Series. Architecture and Civil Engineering*, *19*(2), 183–192. DOI: 10.2298/fuace211129014z

Zucaro, A., Maselli, G., & Ulgiati, S. (2022). Insights in Urban Resource Management: A Comprehensive Understanding of Unexplored Patterns. *Frontiers in Sustainable Cities*, *3*. Advance online publication. DOI: 10.3389/frsc.2021.807735

Chapter 12
Towards Inclusive and Sustainable Urban Streetscapes:
Lessons From Chandni Chowk's Revitalization

Lhousileno Shuya
https://orcid.org/0009-0005-9015-6379
School of Planning and Architecture, India

Manika Goel
https://orcid.org/0000-0002-8738-9084
School of Planning and Architecture, India

ABSTRACT

Streets are vital to a city's social, cultural, and economic life, serving as public spaces that enable movement and interaction. However, in many cities, including Delhi, streets have been overtaken by vehicular traffic, losing their public and social function. Chandni Chowk, once a vibrant heritage street, is now facing congestion, overcrowding, and a decline in its historic character. Increased commercial activity has further worsened traffic issues and reduced pedestrian safety. This study investigates the challenges faced by Chandni Chowk's streetscape and explores what urban streets should ideally offer. Using 25 evaluation parameters, supported by literature on street design and standards, the research involves on-site observations, interviews, and the application of design guidelines. The study concludes with an analysis of potential design benefits and offers recommendations to revive the street's livability and heritage value.

DOI: 10.4018/979-8-3373-2555-2.ch012

INTRODUCTION

A livable urban environment encompasses access to green spaces like parks, gardens, and open areas that provide opportunities for relaxation, recreation, and connecting with nature. Streets are considered important public spaces for people-environment interactions (Liu et al., 2021). It emphasizes safe, affordable, and socially cohesive communities, supported by pedestrian and cyclist-friendly infrastructure, including well-designed walkways, bike lanes, and safe crossings. Additionally, access to essential amenities such as schools, healthcare facilities, and cultural institutions plays a vital role in enhancing urban living (Lee, 2023). The streets of cities are the connecting elements that make urban life possible (Chauhan, 2022). Streetscapes should be designed as a series of orchestrated visual events, guiding (Cullen, 1961). Many cities have employed livability-enhancing urban planning methods and policies, such as mixed-use development, compact city design, and regeneration of underutilized spaces. These initiatives aim to promote lively, inclusive, and sustainable urban environments that satisfy the different needs and ambitions of their residents (Mouratidis, 2017). In terms of livable urban space, attempts have been made in Chandni Chowk to increase people's quality of life while maintaining the area's legacy. The attempts were to improve Chandni Chowk as a living urban space focus on building a more structured, pedestrian-friendly, and culturally rich environment that benefits both inhabitants and visitors (Kamal, 2024).

Every city in the past had a tiny beginning as a colony near or around a body of water and increased in population and wealth through time. It then continued to grow until it became a distinct neighborhood rich in customs and culture. Similar changes have occurred in Delhi with its comparable progression all throughout its development. According to the evidence, seven historical cities have combined to eventually create the megalopolis that is today known as Delhi. Shahjahanabad is one such Delhi city that has withstood the test of time and prosperity. A diverse population and a vibrant economy are still found there. One can see a change in the land's usage from mostly residential to commercial, making Old Delhi one of Asia's biggest wholesale marketplaces, increasing population and poor infrastructure have made the city feel like a slum (Dupont, 2004).

Chandni Chowk, considered as one of the oldest and widely known streets in Shahjahanabad, is renowned for its commercial activities and has long been a popular travel destination for tourists. It is a lane to several significant structures and highlights the historical elements and the distinctive architectural features of the past. Both tourists and residents are drawn to Chandni Chowk by its inherent attractiveness and appeal of the streets, the buildings, and the place (Sarkar, 2022). Figure 1 and Figure 2 elaborates the same.

By history, the Chowk used to be an important thoroughfare and was a street for elaborating royal processions which were formerly a familiar sight (Commission, 2017) . The planning of Chandni Chowk was initially designed and structured as a square with a central pool within the city's core. This planning also influenced the shops around the streets and the bazaars which were arranged in a half-moon pattern around the square. Additionally, producing a street planning with a network of four bazaars (Urdu Bazaar, Johri Bazaar, Ashrafi Bazaar and Fatehpur Bazaar), havelis (mansions) and the katras (houses) around the Chowk (Dokras, 2021; Liddle, 2017).

Even to this day, Chandni Chowk is still regarded as one of the city's most significant commercial areas as it is home to several streets, markets, and alleyways. Each of these areas allows walkability space for humans and vehicles, sells items, and has its own specialization of marketing areas. Some still refer to Chandni Chowk's main street as "Silver Street," as a testament to the market's historical reputation for its silverwork. However, it is increasingly becoming congested with trade and commerce (Fachrudin et al., 2025). The street took several courses of change all throughout the multiple eras and has continued to exist so unabatedly even after its development. Chandni Chowk serves as the best example to cater to pedestrians' needs. Such public spaces connect various hot spots in the city and maintain both formal and informal pedestrian flow in a city (Kamal, 2024).

While walking through the street, one would be able to observe the integration of such a built structure. A transition from an old haveli to a newly developed modern building to religious mosques to a hybrid-type building of the old and new structure. There are 325 Havelis, 783 heritage buildings, and roughly 18 designated heritage buildings on Chandni Chowk's main street (Omaxe, 2017) . More recently temples and mosques have also appeared within the vicinity, adding to the increase in human visitations.

Regarding this growth and development, the streets in Chandni Chowk needed to be taken into consideration over the connectivity and circulation of the humans, vehicles, and other public transportation towards the Chowk. There have been many approaches and solutions done to tackle the situation faced in Chandni Chowk. It is a continuous cycle of research and data analysis, new proposals for re-development, and the possible outcome of the aftermath of the street spaces. It is a given that Chandni Chowk has always been a highly populated area, such that it leads to a multiple number of issues addressed. Densely over-crowded spaces, traffic congestion, area/ parking encroachment, no proper allotment of the public squares and spaces, etc.

However, despite its rich urban and cultural fabric, the quality of spatial experience in Chandni Chowk has deteriorated due to excessive commercialization, haphazard urbanization, and insufficient pedestrian infrastructure. These changes have negatively affected street comfort, accessibility, and environmental quality raising serious concerns about urban livability in historic cores.

Hence to have a better understanding and perception of the highly, already populated street, a study of these streets based on the primary data collection, reviews and discussions have been done, refer Figure 3.

The primary objectives are:

i. To assess existing street conditions using a multi-parameter framework;
ii. To identify user perceptions and functional gaps through surveys and fieldworks;
iii. To propose context-sensitive interventions for inclusive and sustainable streetscapes.

It thoroughly emphasizes the three streets done in Chandni Chowk which are Shyama Prasad Mukherjee Street, the HC Sen Marg Street, and the Metro-Mandir Street. Each of these streets are closely interconnected with one another located at the centre of the city within the historical landmarks. All three include varied typologies: pedestrian-based, vehicular-based, and mixed land-use streets, where commercial activity dominates ground floors while residences occupy upper levels.

The study explores how various urban design parameters relate to their spatial context, how the existing street fabric could evolve through strategic redesign, and how interventions might improve circulation, pedestrian experience, and public safety.

This chapter is structured as follows:

- Section 2 outlines the methodological framework used to assess livability and spatial quality.
- Section 3 presents literature review and case studies.
- Section 4 and 5 suggest key finding and highlights intervention strategies and design recommendations.
- Section 6 and 7 concludes with a critical discussion comparing outcomes with objectives, literature, and policy implications.

TOOLS AND METHODOLOGIES

The methodology followed two modes of data collection, theoretical work, and practical field work. Under the theoretical work, it involved the basic introduction of an urban street, the analysis review, an idea, and an overview of data collected online. Whereas, under the practical field work, it involved the process of observations of the street with the people, vehicles and the buildings, the on-site mapping, the survey questionnaires, site/street pictures and the overall data collection.

Theoretical Work started with the literature review on what makes up a street, the study of street aspects, street classifications and standards, and the street parameters. This helped to analyze the street. The literature research mainly consisted of a thorough read-through of the books, including the "Life between the buildings – using public spaces" by Jan Gehl (Gehl, 1971), "Cities for People" by Jan Gehl (Gehl, 1971), "Cities for People" by Jan Gehl (Gehl, 2010) "Image of the City" by Kevin Lynch (Lynch, 1960), "A Pattern Language" by Christopher Alexander (Alexander, 1977), "A Theory of good city form" by Kevin Lynch (Lynch, 1981), "Hidden Dimension" by Edward Hall (Hall, 1966) and "Livable Street" by Donald Appleyard (Appleyard, 1981) . Furthermore, references from the existing report and new paradigm for Chandni Chowk Redevelopment proposals, which focuses on transforming the streetscapes to balance heritage conservation with modern urban design needs (Gupta, 2021) had been done. In addition, the study of all the 25 parameters on the streets in Chandni Chowk were also done, such as: Approach, Connectivity, Imageability, Legibility, Permeability, Accessibility, Landmark, Memory, Symbolism, Association, Pro-social Behavior, Urban Fabric, Urban, Grain, Urban Scale, Land Use, Building Use, Built Typology, Building Function, Building Condition, Building Height, Infrastructure Services such as - waster, drainage, sewage, electric, & other services, Movement Circulation, Adaptation, Socio-Economic Studies and Ecology.

These 25 parameters were used to understand the holistic analysis of the street and address the challenges the street faces. The parameters also encompass the critical aspects of the connectivity, the urban design, the socio-economic dynamics, cultural heritage, and much more to ensure that the redevelopment respects the street's heritage and culture identity while also addressing the urban challenges.

Field work involved the on-site observations and documentation of all the activities of the people, vehicles and the buildings involved. Other work included the on-site pictures with manual sketches of the streets and the neighbourhoods, the survey questionnaires asked to different public users, shopkeepers and the vendors with several different questions, the approximate measurements and height of the buildings, the marking of the residential/commercial, historical buildings, landmarks, location of the public areas and the car parking areas, and the placement of the existing vegetation.

By the end, discussions and suggestions were derived based on the data and the information collected on the three streets. It was a debate and an open conversation about how to facilitate a better street for the public, transportation infrastructure, and the existing streets. Results and recommendations were also integrated based on the primary analysis of the data collection, the street standards, the discussions, and the inferences put together through the street parameters.

Operationalization of Parameters

To bridge the gap between theoretical understanding and empirical investigation, the study adopted a structured framework of 25 parameters drawn from extensive literature review and urban design best practices. The parameters encompass a multi-dimensional view of streetscape analysis, covering physical, perceptual, socio-economic, and ecological dimensions.

Each parameter was derived from established theoretical sources such as Kevin Lynch's elements of city imageability, Jan Gehl's principles of public life, and Christopher Alexander's pattern language and adapted for practical field-level assessment in Chandni Chowk. The structured assessment method also resonates with the open-data driven accessibility framework proposed by a researcher, which utilizes layered parameters and survey indicators to capture pedestrian-level quality, legibility, and walkability in diverse urban environments (Liu et al., 2021).

The operationalization process involved converting abstract concepts into measurable indicators through a combination of tools including:

1. Observation checklists
2. Likert scale-based rating forms (1–5)
3. User perception surveys (100 respondents)
4. Sketches, photography, and physical measurements
5. Land use and movement mapping

Table 1 outlines how each parameter was applied on-site.

Survey Design and Demographic Profile

To better understand user perceptions of street quality, accessibility, and livability, a structured survey was conducted as part of the primary data collection. A total of 100 respondents were surveyed across the three selected streets in Chandni Chowk-Shyama Prasad Mukherjee Street, HC Sen Marg, and Metro-Mandir Street. The respondents included local residents, shopkeepers, street vendors, and visiting users.

The survey was divided into three sections: demographic information, perception-based questions on urban street quality (using a 5-point Likert scale), and open-ended suggestions for improvement. The demographic profile of the participants showed diversity in age and gender: 20% of the respondents were between 18–25 years, 35% between 26–40 years, 30% between 41–60 years, and 15% were above 60 years. In terms of gender, 65% were male, 33% female, and 2% preferred not to specify. The occupational spread included 42% shopkeepers, 28% residents, 10%

vendors, and 20% visitors. About 55% reported visiting the street daily, 25% weekly, and 20% occasionally.

These surveys helped establish stakeholder needs and perceptions regarding walkability, safety, aesthetics, functionality, accessibility, and socio-cultural attachment to the streets. Both qualitative and quantitative data were recorded, coded, and interpreted to inform the findings and proposed interventions.

LITERATURE REVIEW/CASE STUDIES

Books

a. "Life between the buildings – using public spaces" (Gehl, 1971), Jan Gehl wrote about his concerns for the people who were to move from one building to the other, or from a place to the other. It meant for the users to take care of life between buildings, to raise awareness among the architects of how much buildings can be affected by the influence of life and by the influence of people around it. The main subject of the book is the interaction of spaces between the physical environment, the activities, and the public spaces with the people been the main focal point, where social interaction, perception and experience of city life takes place. It explains from the history of the human settlements to the problems arising after the achieved inventions, to the solutions thought after and to the livability of the cities and residential areas. The ideas and tools that Jan Gehl proposes could be used for places where the spaces seemed abandoned, unwanted, accidental, where a few interactions occur, altering the spaces where overcrowding occurs as well. Knowing how people interact helps understand how we use the city. It is all the care and the consideration in the design process for the users that makes all the difference. Living Cities are very stimulating because they are rich in experience and in contrast to the lifeless and dull cities. According to Jan Gehl there are about three types of outdoor activities under the physical environment which influences life in the public spaces and the buildings around it. "Physical Environment - Necessary activities such as going to school, shopping, waiting for bus, running errands, distributing mails, etc., Optional Activities - which are under favorable exterior conditions such as taking a walk, enjoying life, sitting, sunbathing, etc. These participants are usually dependent on exterior conditions, and Social Activities - which are passive contacts such as seeing or hearing of other people, pass by one another, merely a view) These activities are supported when necessary and optional activities are given better conditions" (Gehl, 1971) . The human emotions and the interactions with their surroundings are what bring life to the city. It is a simple yet complicated and prolonged obstacle that architects tend to be overlooked for the past decades. An example Jan mentioned was the sidewalks been higher than the street, as most people prefer walking over a pathway meant for pedestrian walking, which assures them the protection of not having to worry about cars running over them. There is also direct and indirect interaction among the passer byes which brings more life to the streets between buildings and the streets which are essential in making the cities more livelier and fun. Jan proposed a prime concept that everyday life, ordinary situations, and spaces in which daily life is lived must form the centre of the attention and effort, based on the following requirements: Desirable conditions for the necessary outdoor activities, Desirable conditions for the optional, recreational activities, and Desirable conditions for the social activities.

In Context to Chandni Chowk: In "Life Between Buildings" by Jan Gehl, Chandni Chowk is mentioned as a popular street. Gehl emphasizes the importance of public space and how it can be used to enhance the quality of life in urban areas. He suggests that streets should be designed with the intention of encouraging social interaction and activities, rather than just for cars. Chandni Chowk is an example of a street that has evolved over time and is now a bustling hub of activity with markets, shops, and restaurants. Nai Sarak and Jan-path are two other streets in Delhi that are mentioned in the book as examples of streets that have successfully incorporated public space and social activities into their design.

b. "Cities for People" (Gehl, 2010), Jan Gehl discusses his most recent efforts to create or recreate human-scale cityscapes. He also explains in detail the techniques and equipment he employed to transform the urban environments into what he considers to be: cities for people. He further stresses out the four human themes that he believes as crucial, leading to a successful city design, taking into consideration changing populations and lifestyles. In the book, he provided guidance on how to create livable, secure, sustainable, and healthy cities. Gehl believed that to really appreciate the urban environment, one must experience it while walking as opposed to traveling by bus, automobile, or rail. He argues that contemporary developments frequently fail to take this up-close perspective into account. In the final chapter, Gehl urges developing countries to plan their rapidly growing cities on a human scale. The book is completed with a "Toolbox," which provides a summary of each approach, strategy, fundamental ideas, and keyword lists. Examples of over 700 images and sketches from Gehl's work throughout the world were used to illustrate the book in detail.

In Context to Chandni Chowk: Jan Gehl believes that cities should be designed for people, not just for cars. The bazaars and street of Chandni Chowk is a perfect example to validate this statement. To apply the concept of "Cities for People" to the streets in Chandni Chowk, the focus should be on creating a more pedestrian-friendly environment. This could include wider sidewalks, more green spaces, and improved lighting. By implementing these few changes, Chandni Chowk Streets could become a more vibrant and attractive place for people to visit and enjoy.

c. "Image of the City" (Lynch, 1960), in this book, Kevin Lynch discusses about the appearance of the cities, and how the people remember and interpret the characteristics of the urban environments. In addition to its various functions, Lynch further implies that the urban landscape is something to be observed, enjoyed, and remembered for. And the idea of giving a visual form to the city is a unique kind of design challenge. The legibility of the city provides an understanding of the city's physical structure as the people navigate through it. The imageability of the city are the physical qualities which is in relationship to the attributes of one's identity and structure in mental images. According to Kevin Lynch, there are five elements that makes up the city, they are divided into pathways, nodes, paths, edges, districts, and landmarks. Each of these elements are further interrelated with one another. Districts are structured with nodes, defined by edges, penetrated by paths, and sprinkled with landmarks (Lynch, 1981). These elements aid in the formation of mental images in the minds of the people and enable them to develop their own interpretations of the city's qualities. It also enables the residents to conjure up images, give them significance, and form an emotional bond with them. Lynch concludes that each person forms a distinctive impression of the city and takes these factors into consideration.

In context to Chandni Chowk: Kevin Lynch's concept of "Image of the City" can be applied to the streets in Chandni Chowk by examining the components and qualities of modern urban public space in the area. The concept emphasizes the importance of understanding the interrelationship between the physical design of public places and the behaviors of people in such spaces. By analyzing the edges, landmarks, paths, nodes, and districts of Chandni Chowk, one can identify the imageability of the area and how it affects people's behaviors. Moreover, it is essential to identify the groups of users, purposes, built environment, infrastructure, and people communicative traits in public realm.

d. "A Pattern Language" (Alexander, 1977), under this book, the contents are mainly divided into three sections called Towns, Buildings, and Construction. Under each of topics contains around 253 patterns that are described as "problems which occur repeatedly in our environment". The initial 94 patterns address the challenge on a larger scale issue including the countries, cities, town, and communities. The set from 95-205 addresses the issues relating to the different building patterns, while the patterns set 206-252 are construction patterns in context to the building. Each of these patterns together forms a language. As a result, there is a general scale that the patterns naturally follow that makes it easier to select and switch between designs.

In Context to Chandni Chowk: Chandni Chowk streetscape is mentioned in the book "Pattern Language". The primary street in Chandni Chowk is wide and extremely congested, known for its processional use. The width-to-length ratio of the streets encourages interaction between pedestrians and drivers, making it an excellent example of how specific design choices in urban planning can enhance community relationships. This also highlights the importance of transition spaces in urban environments, as discussed in the book. The presence of people and local communities is crucial in shaping a space, and the marketplace and vendors in Chandni Chowk are essential components of the social life in urban areas. Design can play a significant role in reorienting physical solutions towards more humane goals and challenging programmatic assumptions, as mentioned. Therefore, the streetscape of Chandni Chowk is a testament to urban design in fostering social cohesion and community engagement.

e. "A Theory of good city form" (Lynch, 1981), Kevin begins by analyzing the three normative theories presented in this book: those that see cities as either machines, living things, or representations of the universe. In the end, it was found that these theories weren't strong enough. The aim of these ideas was not to analyze how settlements should function, but to simply describe how they do. These theories portray cities as ecological systems, conflict zones, or webs of interconnected options. Lynch outlined his own notion of a healthy city shape, which he believed may lead to thriving communities and contains characteristics that promote development, within continuity, via openness and connection.

Additionally, he talked about the five performance components of vitality, sense, fit, access, and control. He also included the meta-criteria of fairness and effectiveness. The two meta-criteria that have an impact on the other five characteristics are effectiveness and fairness.

In context to Chandni Chowk: According to Kevin Lynch's book, "A Theory of Good City Form," the city is not just a collection of streets, but rather a network of relationships between the measurements of its spaces. Additionally, an article mentions Chandni Chowk in Delhi as an example of a public space, accessible to all people, that faces challenges in designing and maintaining due to urbanization.

f. "Hidden Dimension" (Hall, 1966), this book featured a wide variety of people like the anthropologists, ethnographers, architects, psychologists, linguists, and philosophers. It aids in educating people about the understanding about humans and their importance in architecture. Hall discussed the hidden dimensions and said that what is frequently not said is assumed to be true. He takes all the ideas concerning culture, language, perception, and senses and breaks them down by thoroughly explaining them. It effectively offers up a fresh and different way of looking at things that were previously overlooked or disregarded. Additionally, it provides a more comprehensive understanding of the role that signifies in art, culture, linguistics, senses, perception, notions of spaces and how these rotate around the proxemics. By the end, the author ultimately criticizes technology and the broad acceptance of the technocratic environment. Despite all its positive effects, technology subtly alters how people live and exist, which also meant that we end up losing touch with what it is to be truly human.

In context to Chandni Chowk: it exemplifies how cultural norms shape the spatial behavior, relying on the communication and shared understanding amid its dense crowds and sensory-rich atmosphere. Chandni Chowk showcases how built environments embody cultural values, creating a dynamic interplay between the physical space and the social interaction.

g. "Livable Street" (Appleyard, 1981), the book provides an in-depth analysis of the conflict that exists between vehicles, residents, pedestrians, and other street users. It also includes evidence-based and doable redesign ideas for city street networks that promote urban livability. During the time when Donald Appleyard's Livable Streets was released, it was hailed as a ground-breaking work. It was considered one of the most important works on urban design at the time. Later, his son Bruce updated the topic with the most recent research, fresh case studies, and best human-centered practices for building more livable streets for everyone. It uses the most recent empirical evidence on urban transportation and land use to highlight the need for more livable cities, current case studies on fruitful initiatives, campaigns, programs, and other endeavors from around the globe and includes updated coverage of population at risk.

In context to Chandni Chowk: Chandni Chowk's narrow, bustling streets reflects Appleyard's idea of streets as social spaces rather than mere traffic conduits. The interaction of pedestrians, vendors, and vehicles showcases the co-existence of diverse uses. However, the environmental strain highlights the challenges to livability that aligns with Appleyard's call for balancing accessibility, safety and social vitality in urban street design.

Standards

i. NACTO – National Association of City Transportation Officials, is an urban street design guide which provides certain standards, rules, and design consideration. It offers transportation-related concepts, perceptions, and techniques and takes a collaborative approach to problems with national transportation. The goal of NACTO is to create cities that are places for people, with transportation options that are safe, sustainable, equitable, accessible, and supportive of a healthy quality of life. "Building the Movement: Communicating a bold vision for 21st century urban mobility and building strong leadership capacity among city transportation officials, advancing a Unified City Voice: Empowering a coalition of cities to lead the way on transportation policy at the local, state, and national levels and Accelerating Change: Raising the state of the practice for street design that prioritizes people walking, biking, and taking transit" (NACTO, 1996). Figure 4, Figure 5 and Figure 6 illustrates the standards.
ii. Urban Street Design Guidelines Pune (ITDP) ((ITDP), 2016) – the contents under this street design, consisted of the vision and goals of the Urban Street Design, the applicability and implantation, scopes, and the transportation studies. The content is further chaptered to Street Typology, Street elements, Safety Elements, Universal Accessibility and Barrier Free Design Elements, Intersection, USDG Reference Templates and Street Design Process. A thorough study under the Street Element was done such that it involves the Pedestrian elements, cycle elements, Roadside elements, Carriageway elements and the intersection. This guidebook only provides recommendations for creating a street, including standards for space distribution and general design requirements for various street features.
iii. WRI – World resource Institute, "Cities Safer by design" was to understand the urban design recommendations for the fewer traffic fatalities. The goal of Cities Safer by Design is to assist cities prevent traffic fatalities through better street design and urban planning. Two strategies to increase traffic safety in cities are emphasized in Cities Safer by Design. First, by designing and modifying metropolitan areas to minimize the need for individual vehicle trips, and second, by lowering vehicle speeds in zones where automobiles, pedestrians, and bikes coexist. The report emphasizes enhancing mass transit, bicycle infrastructure, and pedestrian infrastructure. Illustrated recommendations for specific design

elements were proven to improve traffic safety, including: "Urban design that includes smaller block sizes, frequent street connections, narrower streets, and access to destinations in compact urban environments that alleviate the need for vehicle travel, Traffic calming measures such as speed humps, chicanes, curb extensions, raised pedestrian crossings and other elements, Arterials and intersections that reduce conflicts between road users by providing clear crossings, medians and refuge islands, Pedestrian facilities ranging from pedestrian-only areas to basic, consistent sidewalks, Bicycling networks that feature protected Bicycle lanes and special attention to design at intersections and Safety improvements around mass transport stations and corridors" (WRI, 2015). Figure 7 is illustrates the standards of the organization.

FINDINGS

Based on the study of the 25 parameters.

a. Collecting data on the connectivity, approach, landmarks and documenting them in a map: Connectivity and Approach in context to Chandni Chowk area – the nearest airport is Indira Gandhi International Airport - 30 mins, 15.9 km, nearest ISBT Station is Kashmiri Gate ISBT - 11 mins, 3.9 km and Anand Vihar ISBT - 28 mins, 12.9 km, nearest railway station is the Old Delhi Railway Station, DLI - 10 mins, 2.9 km, and the nearest metro station is the Chandni Chowk Metro Station (5 mins, 1.3 km), Chawari Bazaar Metro - 2 mins, 450 metre and Lal Quila Metro Station - 6 mins, 1.7 km. A quick walkthrough towards all the streets were observed. Each of the streets led to a different destination, either to the market areas, or towards the buildings, or a green area, or towards a dead end. Understanding the circulation of these network of streets, helped to better understand the approach towards the site or a street and how each street are perceived through the use of landmarks or a significant market or area point.
b. Marking of the religious landmarks and monuments in each street signifies how the landmarks influence the street and how it helps the public have a clear legibility towards the Chandni Chowk area. The major landmarks on a larger scale in Chandni Chowk are the Fatehpur Masjid towards the West, Old Delhi Railway Station towards the North, Red Fort towards the East, and Jama Masjid towards the South. Each of the streets has a landmark of its own, either a religious monument, Madir, Malls, or a public station. HC Sen Marg Street: Sri Shiv Navgrah Mandir, Omaxe Mall, Dharamshala, Hardayal Municipal Public Library; Shyama Prasad Mukherjee Marg Street: Delhi Public Library, Old Delhi Railway Station, Presentation Convent Sr. Secondary School, and

St. Mary Catholic Church; and the Metro Mandir Road: Chandni Chowk Metro Station, the Omaxe Mall, Shani Dev Temple Chandni Chowk and Shri Shiv Navgrah Mandir. Figure 8 and Figure 9 represents landmarks nearby.

c. After identifying the landmarks in the areas, the street network of the three streets with the other primary streets was done. The use of geo-location photographs was also taken of the identified areas. Within the 500-metre radius taken from the Chandni Chowk area, the three streets (HC Sen Marg Street, Metro-mandir Street and Shyama Prasad Mukherjee Street) cover an area of 194 Acres, with the neighbourhood buildings and other primary streets such as Nai Sarak Road, Rai Kedarnath Marg Road, Dariba Kalan Road, Esplande road, Diwan Hall Road, and Netaji Subhash Marg Road, refer Figure 10.

d. About the Three Streets: HC Sen Marg Street: is a 35 Metre width street stretching at .5 km consisting of two-way road with parking areas at the side of both the road and tree planted in between as a segregation. It connects the Chandni Chowk Street from the South towards the Red Fort and Fatehpur Masjid, with the Shyama Prasad Mukherjee Marg Street from the North towards Old Delhi Railway Station and connected with Metro-Mandir Street leading towards the Chandni Chowk Metro Station. This street runs along the proposed Omaxe Mall and consists of food stalls, informal vendors, and the public's entry/exit through the Chandni Chowk Metro Station. This street has signage boards making it easy for the public and visitors to identify their location.

Metro-Mandir Street: It is a two-way pedestrian-based street which has direct access from the Chandni Chowk Metro station towards the HC Sen Street and is welcomed by a Mandir right at the entrance before the Metro Station.

Shyama Prasad Mukherjee Marg Street: is a 30 Metre width street stretching at .55 km. It is a two-way road, with pedestrian walkway, parking slots and landscape in between segregating the road. Shyama Prasad Mukherjee Marg Street has legibility access from the Netaji Subhash Marg from the East and Northern Railway Hospital area from the West. The permeability access to this street is from Old Delhi Railway Station, which is also an important landmark for this street, the HC Sen Marg Street and from both end-to-end streets.

e. Interpreting the building uses, building height, building conditions, building function, and building typology: For the building uses most of the building use consist of commercial (70%). It is a mixed used areas of commercial, residential, religious, public buildings and green areas. The building height of the buildings here are mostly G+3 and each floor are about 3 Metres from floor to floor where ground floor is for the commercial, while the other floors are for warehouses and residential. The building height and the narrowness of the street makes it very proportional to the human scale. It is to compare the heights of each building and the building's relation with the scale of the street to the other building, people, plants, and trees. The building conditions, about 80% structures in Chandni Chowk is old and dilapidated, the sanitary conditions are poor. Few of the houses have underground sewerage connections. For the building function, on the bases of the case study the Chandni chowk market has been segregated into various functions such as monuments, library, garment shops, food court, green area, and boutique, etc. The building typology found in Chandni Chowk are that most of the block are linear blocks, parallel blocks, L-shaped blocks, t-shaped blocks, u-shaped blocks, linear blocks with Courtyard and superposed blocks. Figure 11, Figure 12, Figure 13, and Figure 14 illustrates all the layers.

f. The study of ecology: Ecological evaluation of the study area focused on the type, distribution, and functionality of tree species along the selected streets of Chandni Chowk. Vegetation plays a critical role in enhancing urban micro-climates, reducing heat stress, supporting biodiversity, and improving visual quality. This observation aligns with recent comparative research in Southeast Asia which emphasizes how integrated ecological and functional design elements in commercial streetscapes can improve thermal comfort, biodiversity support, and urban identity in dense city cores (Fachrudin et al., 2025). The analysis emphasizes both ecological performance and design suitability of trees planted along the three streets; Metro-Mandir Street, HC Sen Marg, and Shyama Prasad Mukherjee Marg.

Vegetation cover across the three corridors was observed to be sparse and unevenly distributed. On Metro-Mandir Street, vegetation is minimal due to space constraints imposed by dense built-up edges and unregulated vending zones. This significantly limits the ecological benefits typically offered by street trees, such as shade provision, thermal comfort, and pollution mitigation.

HC Sen Marg, which accommodates mixed traffic, exhibits isolated tree plantings functioning as green buffers, primarily aligned along the road divider or property boundaries. However, the density and placement are insufficient to provide effective thermal regulation or pedestrian comfort.

In contrast, Shyama Prasad Mukherjee Marg a high-speed vehicular corridor lacks any continuous green strip or canopy cover. There is an absence of green islands or vegetative medians at critical intersections, particularly between the junction of HC Sen Marg and Mukherjee Marg, reducing both ecological functionality and aesthetic coherence.

The key tree species documented in the study area include Kaim (Mitragyna parvifolia), Peepal (Ficus religiosa), Frangipani (Plumeria spp.), Kalpavriksha (Adansonia digitata or symbolic), and Palm (Phoenix spp./Washingtonia spp.).

These were assessed for their microclimatic benefits, canopy structure, cultural significance, and spatial suitability.

Table 1. ecological functions and design implications of the identified species

Tree Species	Ecological Functions	Design Implications
Kaim Tree (Mitragyna parvifolia)	**Moderate canopy, drought-resistant, helps regulate microclimate**	**Suitable for narrow streets due to compact form; provides light shade**
Peepal Tree (Ficus religiosa)	**Wide canopy, air purification, biodiversity host, high oxygen release**	**Excellent for plazas and courtyards; sacred value; provides deep shade**
Frangipani (Plumeria spp.)	**Seasonal flowering, low water need, supports pollinators**	**Ornamental tree suitable for aesthetic enhancement and cultural landscapes**
Kalpavriksha (Adansonia digitata or symbolic term)	**Cultural and religious significance; hardy and drought-tolerant**	**Planted as focal or symbolic tree; not ideal for dense planting**
Palm Tree (Washingtonia spp. / Phoenix spp.)	**Low canopy cover, minimal litter, low root damage risk**	**Adds verticality, low-maintenance species for narrow medians**

Source: (Salbitano et al., 2016; Raj & Lal, 2013)

In summary, the current street ecology of Chandni Chowk demonstrates a critical underutilization of greenery, with limited emphasis on ecological design principles. The findings suggest a need for integrated greening strategies that combine species selection with functional and spatial planning. Species like Peepal and Kaim, when strategically planted along sidewalks, intersections, and transition nodes, can substantially improve the thermal comfort, visual coherence, and heritage-sensitive landscape character of these historic streets.

g. Identifying the user groups and interviewing the users and stakeholders about their experiences and their perception of the street: To produce a demographic information and background analysis of the people, their purpose of their visitation, about their age and gender. To understand the user analysis and their objectives, questionnaires were done with different users such as: The customers / tourists who travel to and return by vehicles, public transport, and other means of transport.

The questions in relation to where they come from, what their purposes are for visiting such markets, why choose this destination and not the other market destination, how much time does it take for them to reach the place from their home, etc. The shopkeepers and street vendors who has been running their business for centuries or more, about how they view the current street with the previous situation, asking them for an overview of the comparative analysis and experience through their opinions and suggestions, to be able to understand the number of footfalls and the activity of the crowd population during the weekdays and weekends. The neighbourhoods or the residents living nearby, the questions are in relation to how long they've been living here, how the new development have changed the area, if the re-development would be a positive or negative impact on the current existing scenario, if the area has ever been open to any kind of threats, etc.

h. The questionnaires done were based on different age groups from young people around 18 - 25 years old to old aged around 40 - 50 years old. A total of 100 surveys were conducted under each questionnaire for the vendors, shopkeepers, and the customers/publics. Most of the answers under the vendors were the problems regarding the heavy traffic congestion and how they always had to compromise the space where they're needed to sell their products. It also included the issue of no equity distribution of spaces among the vendors, car parking areas and the public areas. Similarly, for the shopkeepers, some of them had the problem of the street utilities such as the electric infrastructure been placed haphazardly in front of their shops. Congested roads for vehicles having not enough space, scattered and tangled utilities, more shops leading to the chaotic situation for the existing market without any proper planning. As for the customers, the streets were not user-friendly according to some people. Safety concerns for the people were also mentioned, no pedestrian priority as the existing roads were inaccessible for the pedestrian, unobstructed footpaths existed as well and limited public spaces.

The customers could not get a space to breathe or get a break from the crowded markets and street.

i. Collecting quantitative data: Gather quantitative data such as the traffic analysis of the crowd population and movements analysis: Observations of the Traffic Analysis for the people, vehicle, and activities – observing the movement analysis of vehicles, the people, and the activities from morning, afternoon, and night duration and for the weekdays and weekends period. It is the comparison analysis of when the Chandni Chowk receives more people or have more gatherings during festive seasons. It is also to understand the pro-social study of the human behaviour, their daily movements with the street, and their activities. Where it is the most interactive or less interactive and if the space allotment would be needed more on the areas where it's more crowded, these analysis gives an idea of how the area around the streets influences the crowd population. For example, along the Chandni Chowk Street, which is directly connected through the HC

Sen Marg Street, there is a food restaurant corner namely the "Paranthe- Wali Gali" or "Street of the Fried Bread Makers," is a narrow lane lined with shops selling a variety of stuffed parathas and a lot of visitors and customers come to this area to try the food delicacies. This narrow lane pathway directly provides entry towards the restaurant is highly congested with public as this lane is accessible for the public from the Chandni Chowk Metro Station. These densely populated spaces obstruct the street spaces with overcrowded population and restricted movements, hence causing traffic congestion to the street areas and it also obstruct the overall circulation.

RESULT

The findings from the study, derived through on-site documentation, structured surveys, and multi-parameter analysis, reveal critical gaps in the pedestrian infrastructure, spatial quality, and livability of the three selected streets in Chandni Chowk. These include Shyama Prasad Mukherjee Street, HC Sen Marg, and Metro-Mandir Street. The outcomes validate earlier observations made in urban design research, notably by Gehl (1971) and Appleyard (1981), highlighting how vehicle dominance, inadequate pedestrian space, and limited street aesthetics diminish human-scale urban experiences.

Spatial Observations and Parameter Findings

Field observations and mapping revealed that all three streets suffer from encroachments, lack of dedicated pedestrian walkways, and inefficient use of street frontage. Narrow circulation zones, poor signage, informal vending, and inadequate lighting and sanitation infrastructure create hostile conditions for pedestrians. HC Sen Marg and Metro-Mandir Street are particularly congested, lacking clear distinctions between pedestrian, rickshaw, and vehicular movement. Meanwhile, Shyama Prasad Mukherjee Street, due to its greater width, presents opportunities for multimodal interventions.

Each street was assessed using 25 urban design parameters. Streets scored low on walkability, safety, permeability, ecological comfort, and imageability, highlighting the pressing need for intervention. Built typology was mixed and largely unregulated, contributing to visual clutter and inconsistency.

Survey-based Perception Analysis

The structured survey of 100 stakeholders (residents, vendors, shopkeepers, and visitors) confirmed the physical observations. Around 72% of respondents expressed dissatisfaction with pedestrian experience due to crowding, encroachment, and unsafe crossings. 60% felt the streets lacked sufficient lighting and security, especially after dark. Only 25% considered the streets clean, and over 48% reported difficulties for elderly and differently-abled users in accessing public space. The visual quality of HC Sen Marg was rated highest due to preserved facades and lesser encroachment, while Metro-Mandir Street scored lowest.

Respondents emphasized the need for improved waste management, pedestrian-only zones, shaded areas, and more green cover. Additionally, stakeholders called for uniform signage, accessible seating, and regulated vending areas. These perspectives reinforce the importance of integrating user feedback into urban planning for inclusive and context-sensitive redevelopment.

Key Findings and Opportunities

The data points to a clear need for differentiated design interventions:

- Pedestrianization: Two streets (HC Sen Marg and Metro-Mandir) should be converted to pedestrian-priority zones.
- Multimodal Design: Shyama Prasad Mukherjee Street offers scope for integrating dedicated rickshaw, bicycle, and pedestrian lanes.
- Vegetation and Shade: Strategic planting and green buffers are essential for ecological comfort and visual relief.
- Heritage-sensitive Approach: With over 300 havelis and several Mughal-era buildings, street improvements must protect historical continuity and local identity.
- Street Standards Application: Proposed design directions follow global and national street guidelines, including NACTO and Indian Urban Street Design Standards.

Figures 16–19 illustrate the observed site conditions and preliminary redesign concepts that respond to both stakeholder inputs and technical evaluations.

DISCUSSION

The study presents a comprehensive understanding of the challenges and opportunities for redesigning historic urban streets within dense, mixed-use settings. Drawing upon theoretical frameworks and empirical data, the findings demonstrate how street livability, pedestrian comfort, and visual quality are critically compromised in Chandni Chowk due to unregulated commercialization, spatial congestion, and lack of inclusive infrastructure.

The application of 25 urban design parameters provided a holistic evaluation framework, enabling both spatial and perceptual assessment. This approach aligns with contemporary urban theories that emphasize the importance of walkability, legibility, imageability, and social function in street environments (Gehl, 2010; Lynch, 1960). The results validate that these parameters remain relevant in Indian contexts, where overlapping layers of commerce, heritage, and human activity define the street character.

The survey findings further reinforce these observations, highlighting the discrepancy between user expectations and existing street conditions. Despite the cultural and historical significance of the area evident in landmarks such as the Red Fort, Jama Masjid, and Fatehpuri Masjid basic public infrastructure remains inadequate. Issues such as poor waste management, absence of street lighting, and encroachments were frequently cited by stakeholders, reflecting both infrastructural and managerial gaps.

A critical observation emerging from the study is the functional differentiation among the three streets. While Shyama Prasad Mukherjee Street accommodates higher vehicle throughput and wider road width, HC Sen Marg and Metro-Mandir Street demonstrate stronger pedestrian activity but suffer from unmanaged spatial fragmentation. This diversity necessitates context-sensitive street redesign strategies rather than a uniform approach.

The findings emphasize the relevance of adopting Complete Streets principles, as promoted by NACTO and Indian Street Design Guidelines, to facilitate safer, more inclusive, and multi-functional urban streets. Incorporating elements such as segregated pedestrian and cycle lanes, designated vending areas, green buffers, and wayfinding signage could significantly enhance user experience while ensuring safety and heritage conservation.

Furthermore, the integration of green infrastructure through tree-lined walkways, pocket parks, and bioswales not only enhances visual appeal but also addresses environmental concerns like heat mitigation, dust reduction, and improved air quality. These findings align with global research on green streetscapes contributing to urban sustainability and improved mental well-being.

Heritage sensitivity remains a central theme in this discussion. The architectural continuity and visual rhythm of Chandni Chowk must be retained through façade restoration, regulated signage, and controlled building heights. Rather than imposing a modernist redesign, the street proposals should adapt and enhance the existing character, supporting heritage-led regeneration.

In summary, this discussion illustrates how a multi-parameter framework, combined with empirical perception analysis, can inform meaningful urban design interventions that respond to functional, environmental, and socio-cultural demands. The study contributes to the growing body of knowledge on revitalizing historic streetscapes while supporting policy-level discourse on sustainable urban development in Indian cities.

CONCLUSION

This chapter presents a structured assessment of livability, spatial quality, and user experience across three key streets within Chandni Chowk, Old Delhi. Through the application of a multi-parameter framework and integration of stakeholder perceptions, the study highlights the critical challenges facing heritage-rich yet functionally strained urban corridors. These include inadequate pedestrian infrastructure, mixed traffic congestion, unregulated vending zones, and a disconnect between built heritage and current urban use.

The chapter demonstrates that livability in such historic urban settings cannot be improved solely through physical upgrades but requires a balanced approach integrating user needs, cultural sensitivity, and policy frameworks. By operationalizing 25 design parameters and conducting 100 user surveys, the analysis confirms that pedestrian-first design, regulated land use, and inclusive mobility infrastructure are foundational to street revitalization.

Recommendations emphasize pedestrianization of key corridors, multimodal street design, incorporation of green infrastructure, and restoration of architectural identity. These strategies aim to enhance walkability, safety, visual coherence, and environmental comfort, while respecting the historical essence of the area.

The study affirms the utility of adopting Complete Streets and Context-Sensitive Design principles tailored to Indian urban conditions. It also offers a replicable model for evaluating and redesigning similar heritage-intensive urban environments across other Indian cities.

Future efforts must involve interdepartmental coordination, public engagement, and adaptive implementation strategies to ensure that revitalization is both sustainable and culturally grounded. In doing so, urban streets like those in Chandni Chowk can once again become vibrant, accessible, and inclusive public spaces

TABLES AND FIGURES

Figure 1. (left to right) Chronology view of the Chandni Chowk in 1638, 1864, 2014 and 2023 (Urban Design Lab, 2021)

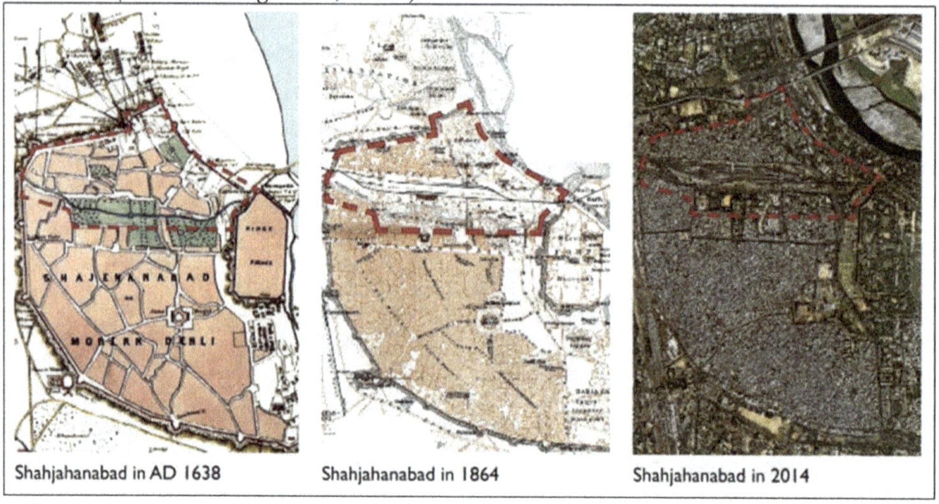

Figure 2. Three streets S1 – Shyama Prasad Mukherjee Street, S2 – HC Sen Marg and S3 – Metro-Mandir Street (Urban Design Lab, 2021)

Figure 3. The Three Streets (S1) - Shyama Prasad Mukherjee Marg, (S2) - HC Sen Marg and (S3) - Metro-Mandir Street (Author)

Figure 4. The image shows an example of a street, before and after changes while incorporating with NACTO standards. The segregation of the carriageways, the parking lanes, biking lane, and the transit wider lane (NACTO, 1996)

Figure 5. Existing conditions demonstrate how traditional design elements, such as wide travel lanes and undifferentiated street space, have had an adverse impact on how people experience the streetscape. Striping and low-cost materials can realize the benefits of a full reconstruction in the short term, while allowing a city to test and adjust a proposed design (NACTO, 1996)

Figure 6. Conventional Crosswalks - High visibility ladder, zebra, and continental crosswalk markings are preferable to standard parallel or dashed pavement markings. Corner Radii and curb extensions (NACTO, 1996)

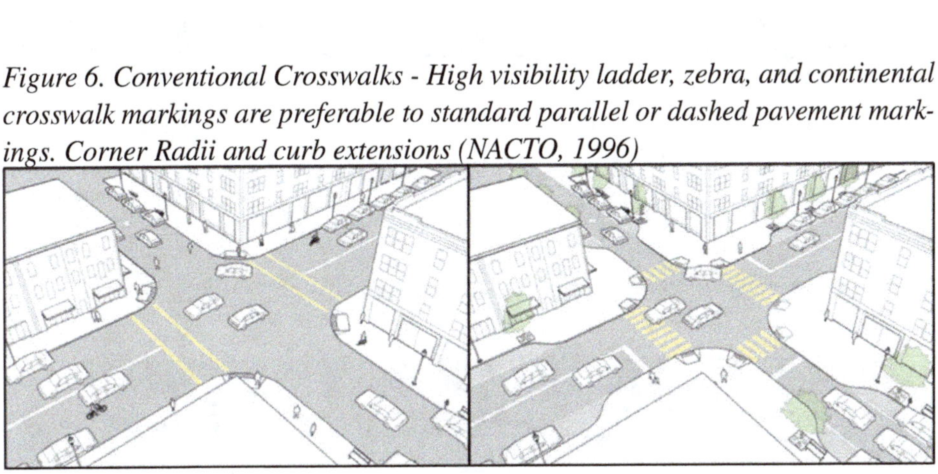

Figure 7. The following street guidelines consist of a design proposal or suggestion for a 30-metre-wide road. It was used as a reference for the 30-metre-wide HC Sen Marg Street (NACTO, 1996)

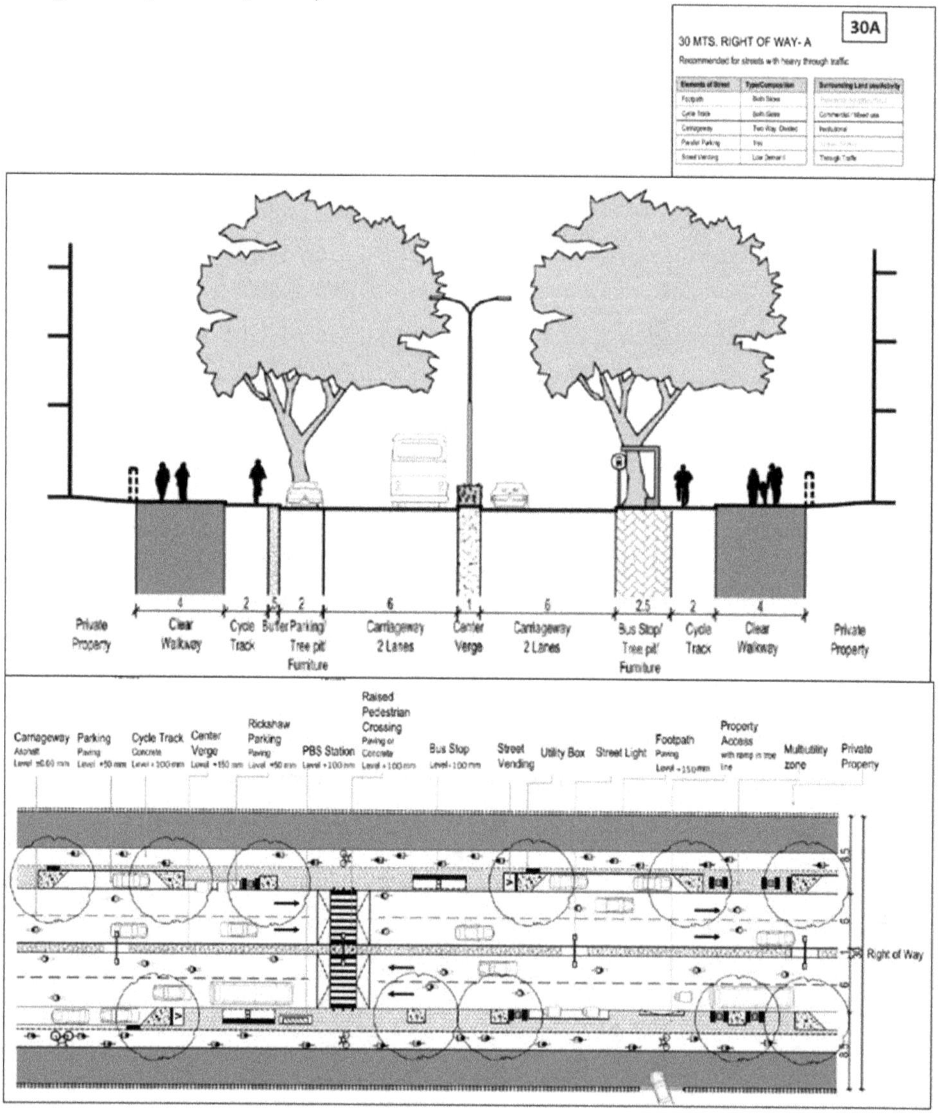

Figure 8. Approach the three streets, the landmark, the approach, and the built areas. graphical representation (Author)

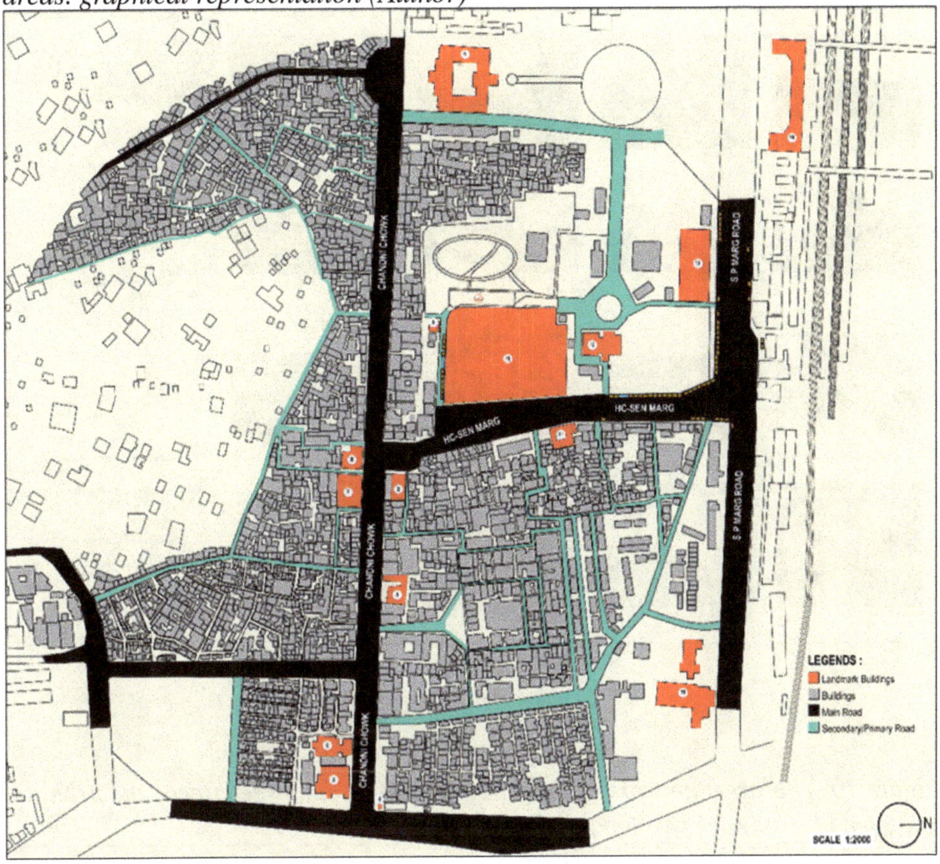

Figure 9. Images of the landmarks on site context, the street landmarks, and the approach the three streets, the landmark, the approach and the built areas (Author)

Figure 10. The street network of the main highways, the three streets, the primary and secondary streets (Author)

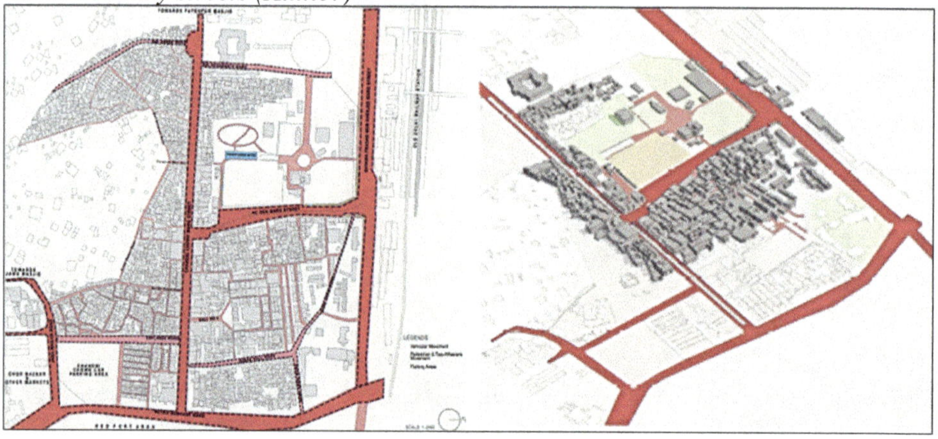

Figure 11. Interpretation of building use with pie chart presentation (Author)

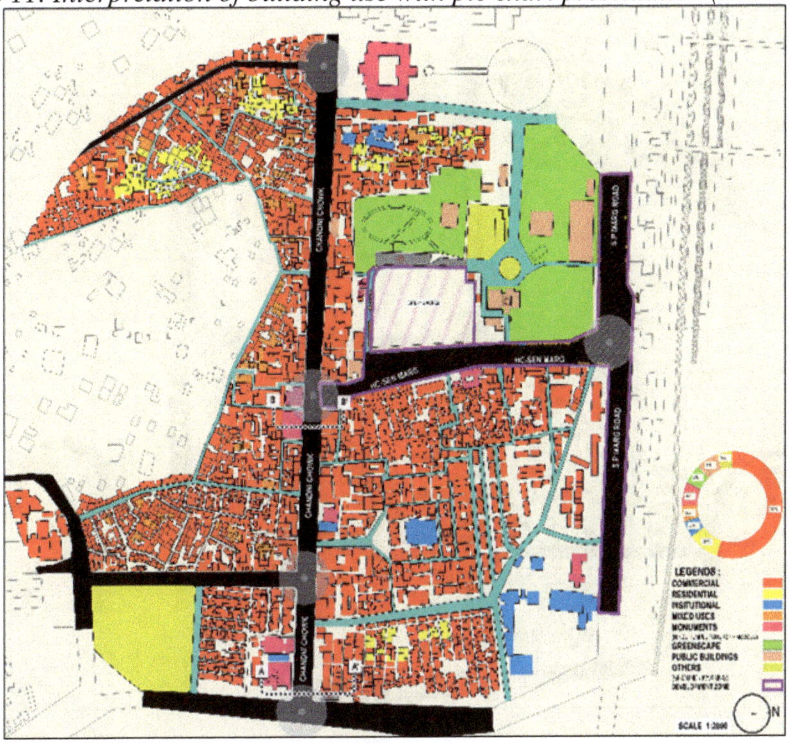

Figure 12. Interpretation of building condition with pie chart presentation (Author)

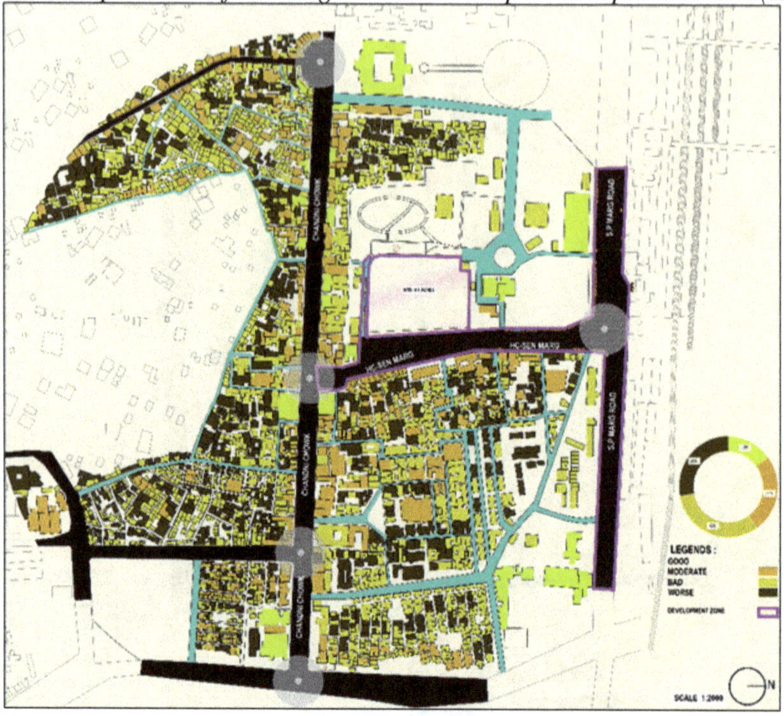

Figure 13. Interpretation of building height with pie chart presentation (Author)

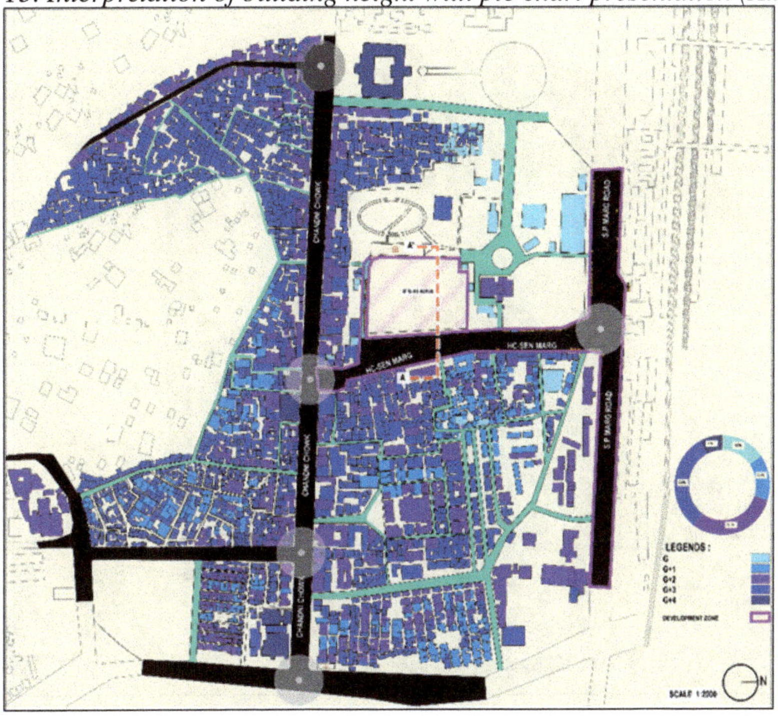

Figure 14. Interpretation of building typology with pie chart presentation (Author)

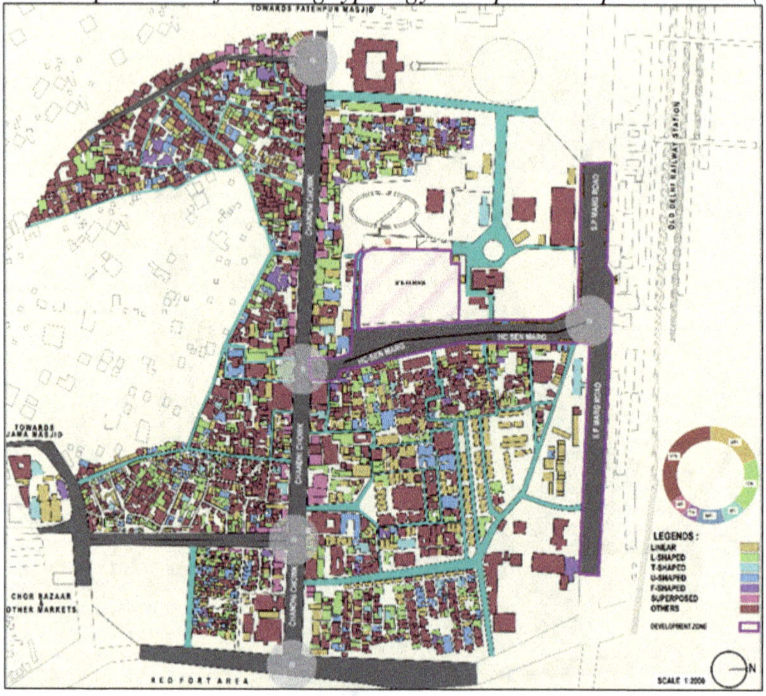

Figure 15. Interpretation of the green areas with the existing vegetation (Author)

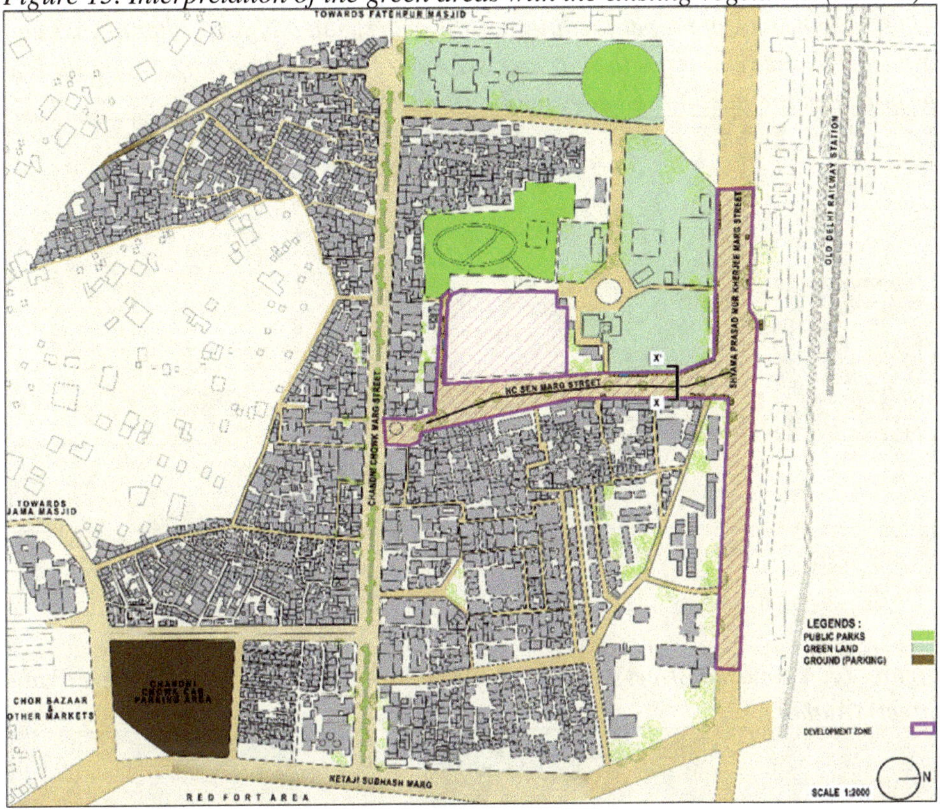

Figure 16. Existing Road Section of HC Sen Marg (30-Metre-Wide Road). Issue-1 (I1) shows the lack of proper facilities for the pedestrian and was either always encroached by the vendors or the garbage's. Issue -2 (I2) shows a narrow circulation despite been provided for a heavy public street (Author)

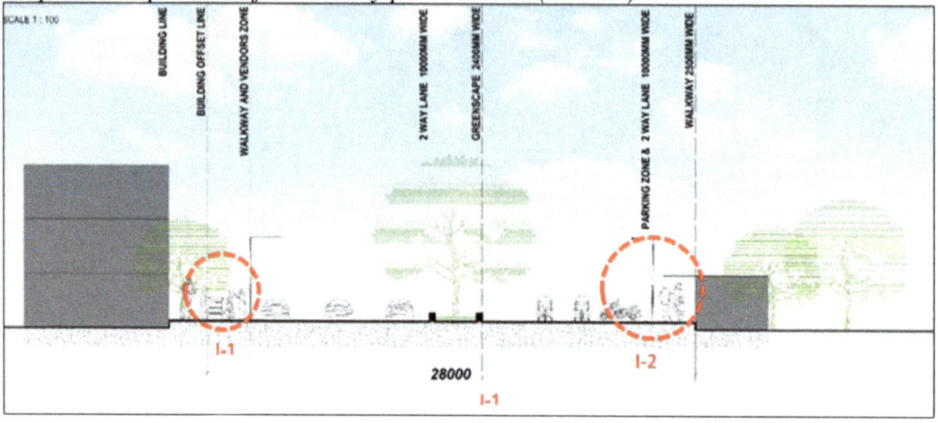

Figure 17. Proposed street section of the Metro-Mandir street (18-Metre-Wide Street) (Author)

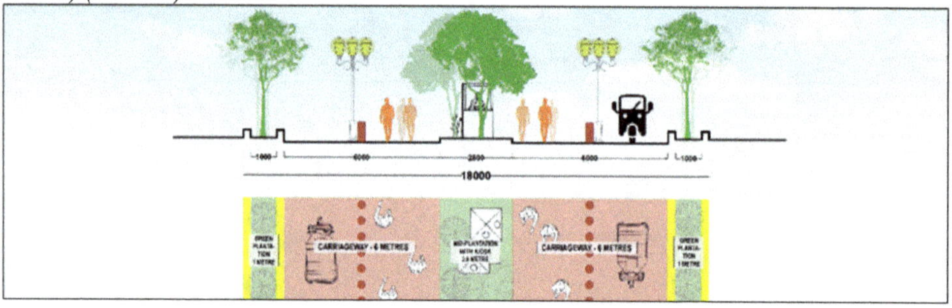

Figure 18. Proposed street section of the HC Sen Marg street (30-Metre-Wide Street) (Author)

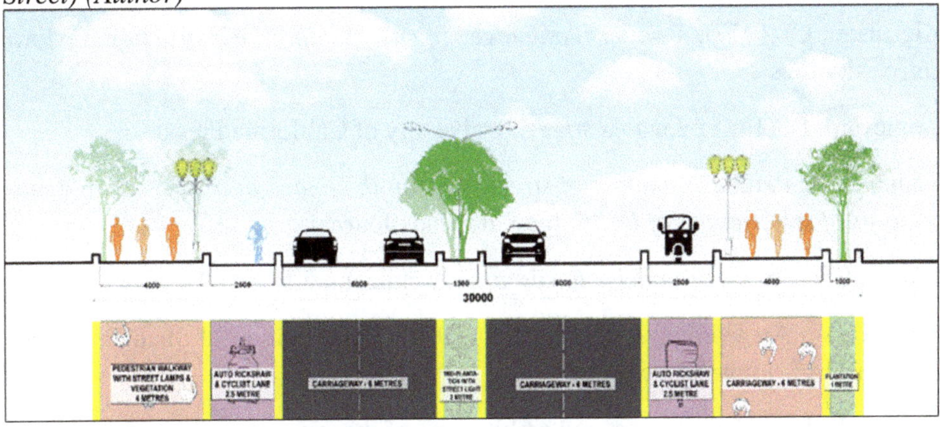

Figure 19. Proposed Street section of the Shyama Prasad Mukherjee Marg (45-Metre-Wide Street) (Author)

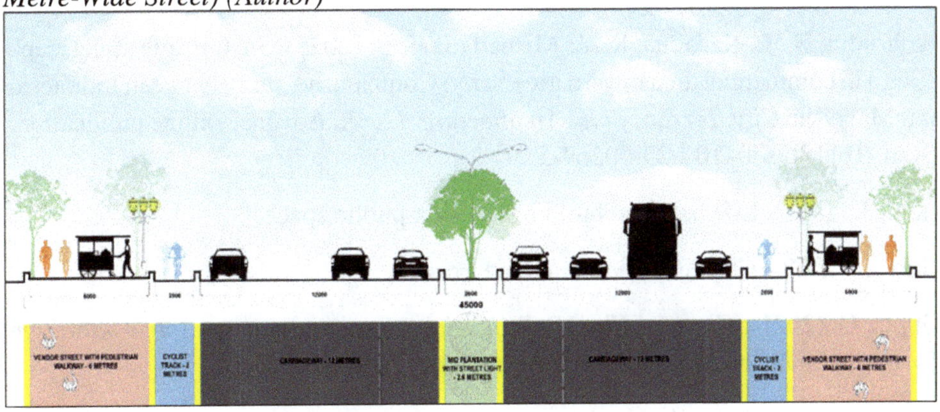

REFERENCES

Alexander, C. (1977). *A pattern language: Towns, buildings, construction*. Oxford University Press.

Appleyard, D. (1981). *Livable streets*. University of California Press.

Chauhan, A. P. (2022). *Impact of streetscape on the image of a city: A study and analysis of streetscapes of Delhi*. Ignited Minds Journals.

Cullen, G. (1961). *The concise townscape*. Architectural Press.

Delhi Urban Art Commission (DUAC). (2017). Rejuvenation of Shahjahanabad. Delhi: DUAC. https://duac.org/

Dokras, D. U. (2021). The art and architecture of the town of Shahjahanabad (Old Delhi). Academia.edu. https://www.academia.edu/47537921/

Dupont, V. (2004). Urban development and population redistribution in Delhi: Implications for categorizing population. In Hugo, T. C. (Ed.), *New forms of urbanization: Beyond the urban-rural dichotomy* (pp. 172–191). Ashgate Publishing.

Fachrudin, H. T., Karolina, R., & Misnan, S. H. B. (2025). Sustainable streetscape design in commercial area toward green city: Comparative study between Indonesia and Malaysia. *City. Territory and Architecture*, *12*, 14. Advance online publication. DOI: 10.1186/s40410-025-00269-5

Gehl, J. (1971). Life between buildings: Using public space.

Gehl, J. (2010). *Cities for people*. Island Press.

Gupta, G. (2021). *Chandni Chowk: Redeveloping urban spaces – A new paradigm for India*. Shahjahanabad Redevelopment Corporation.

Hall, E. T. (1966). The hidden dimension.

Institute for Transportation and Development Policy (ITDP). (2016). Urban street design guidelines, Pune. ITDP. https://india.uitp.org/wp-content/uploads/2022/11/USDG_Pune.pdf

Kamal, D. M. (2024). Urban regeneration of public spaces and pedestrian streets in commercial area: Case of Chandni Chowk at Old Delhi, India. *American Journal of Civil Engineering and Architecture*, *12*(1), 22–32. DOI: 10.12691/ajcea-12-1-4

Lee, C. (2023). Access to amenities such as schools, healthcare facilities, and cultural institutions is also necessary for a livable urban environment. Applied Research in Quality of Life, 18, 45–62. Springer Nature. DOI: 10.1007/s11482-023-10112-3

Liddle, S. (2017). *Chandni Chowk: The Mughal city of Old Delhi.* Speaking Tiger Publishing.

Liu, S., Higgs, C., Arundel, J., Boeing, G., Cerin, E., Moctezuma, D., & Giles-Corti, B. (2021). A generalized framework for measuring pedestrian accessibility around the world using open data. *Transportation Research Part A, Policy and Practice, 146*, 72–88. DOI: 10.1016/j.tra.2021.05.014

Lynch, K. (1960). *The image of the city.* MIT Press.

Lynch, K. (1981). *A theory of good city form.* MIT Press.

Mouratidis, K. (2018). Is compact city livable? The impact of compact versus sprawled neighbourhoods on neighbourhood satisfaction. *Urban Studies (Edinburgh, Scotland), 55*(11), 2408–2430. DOI: 10.1177/0042098017729109

National Association of City Transportation Officials (NACTO). (1996). About NACTO. https://nacto.org/about/

Omaxe. (2017, December). Rediscovering Chandni Chowk: Beyond the space. Omaxe: Turning Dreams into Reality. https://www.omaxe.com/blog/rediscovering-chandni-chowk-beyond-the-space/

Raj, A. J., & Lal, S. B. (Eds.). (2013). Forestry principles and applications (Hardback & Paperback ed.). Scientific Publishers. DOI: 10.13140/2.1.2452.3524

Salbitano, F., Borelli, S., Conigliaro, M., & Chen, Y. (2016). *Guidelines on urban and peri-urban forestry (FAO).* Food and Agriculture Organization of the United Nations.

Sarkar, A. (2022). Re-creating Chandni Chowk: Place-making and place-marketing of 'Old' Delhi's historic city centre. *Indian Journal of Spatial Science, 13*(2).

Urban Design Lab. (2021). Urban conservation plan for Chandni Chowk streetscape. Author.

World Resources Institute (WRI). (2015). Cities safer by design: Guidance and examples to promote traffic safety through urban and street design. https://www.wri.org/research/cities-safer-design

Chapter 13
Designing for All: A Road Space Equity Study of Vikas Marg, Delhi

Deeksha Khati
https://orcid.org/0009-0009-7238-3235
KR Mangalam University, India

Manika Goel
https://orcid.org/0000-0002-8738-9084
School of Planning and Architecture, India

ABSTRACT

Urban street design plays a vital role in shaping sustainable and inclusive cities by promoting walkability, connectivity, and placemaking. This study examines Vikas Marg in Delhi as a case of equitable road space allocation and effective urban design. Drawing on principles from The Image of the City, Theory of a Good City, and A Hidden Dimension, the research highlights the street's transformation toward legibility, inclusivity, and environmental quality. Through spatial analysis, stakeholder observations, and design evaluation, the study identifies initial challenges such as traffic congestion, inadequate pedestrian facilities, and lack of greenery. The redesigned street demonstrates how integrated design interventions can create vibrant, people-centric public spaces that enhance mobility, social interaction, and urban livability.

INTRODUCTION

Sustainable urban planning aims to balance environmental, social, and economic considerations, ensuring equitable access to resources for all residents (Zhang et

DOI: 10.4018/979-8-3373-2555-2.ch013

al., 2020). It addresses social inequities by providing affordable housing, improving healthcare and education, and fostering inclusive communities (González et al., 2019). By tackling these disparities, cities can enhance the well-being and quality of life for their populations. Engaging local communities in the urban planning process is crucial for achieving sustainable and equitable outcomes. Promoting mixed-income and mixed-use developments reduces spatial segregation, fosters economic diversity, and supports social integration, which contributes to improved urban resilience (Mohammad et al., 2019; Smith et al., 2021). By combining affordable and market-rate housing with commercial, recreational, and institutional spaces, cities can create diverse and inclusive neighbourhoods. Economic resilience in urban planning is achieved by diversifying local economies and fostering entrepreneurship, which is essential for long-term sustainability. This involves diversifying local economies, promoting entrepreneurship, supporting small and local businesses, and investing in job training and education programs. By creating economic opportunities and reducing income disparities, cities can enhance the economic well-being and resilience of their residents.

Vikas Marg is a major road in Delhi, India, connecting East Delhi with Central Delhi. This thoroughfare runs through several busy commercial and residential areas, linking the city to the National Capital Region (NCR). Studying a street like Vikas Marg offers valuable insights into various aspects of urban life and urban planning. Contrary to popular belief, the road is not named after Swami Vivekananda; its name translates to "Development Path" in Hindi. Vikas Marg begins at the junction of Laxmi Nagar and Shakarpur and extends to Ghaziabad in Uttar Pradesh, spanning approximately 12 kilometers. It is one of the busiest and most vital roads in East Delhi, lined with numerous landmarks, including shopping malls, restaurants, educational institutions, hospitals, and residential areas.The vibrant markets along Vikas Marg are famous for offering a wide range of products, such as clothes, electronics, jewelry, and handicrafts, attracting shoppers from across the city. The street food in the area is equally popular, drawing food enthusiasts to its bustling stalls.The road is also home to several prestigious educational institutions, including the Institute of Chartered Accountants of India, the Institute of Company Secretaries of India, and the Institute of Cost Accountants of India. These institutions attract students nationwide and have significantly contributed to the area's development. Additionally, the Karkardooma District Court Complex, one of the largest court complexes in Delhi, is located on Vikas Marg. This hub of legal administration houses several courts and plays a crucial role in the region's justice system.Vikas Marg is well-connected to other parts of Delhi and the NCR. It intersects with major roads such as NH-24, which connects Delhi with Lucknow, and the Delhi-Noida Direct (DND) Flyway, which links Delhi to Noida. The road is also served by multiple metro stations, including the Laxmi Nagar Metro Station on the Delhi Metro's Blue Line, further enhancing its connectivity (MoHUA, n.d.), refer Figure 1.

Figure 1. Vikas Marg Stretch well-connected and passes through various educational institutions, making it an ideal location for studying. Also connected to the Delhi Metro's blue line (Author)

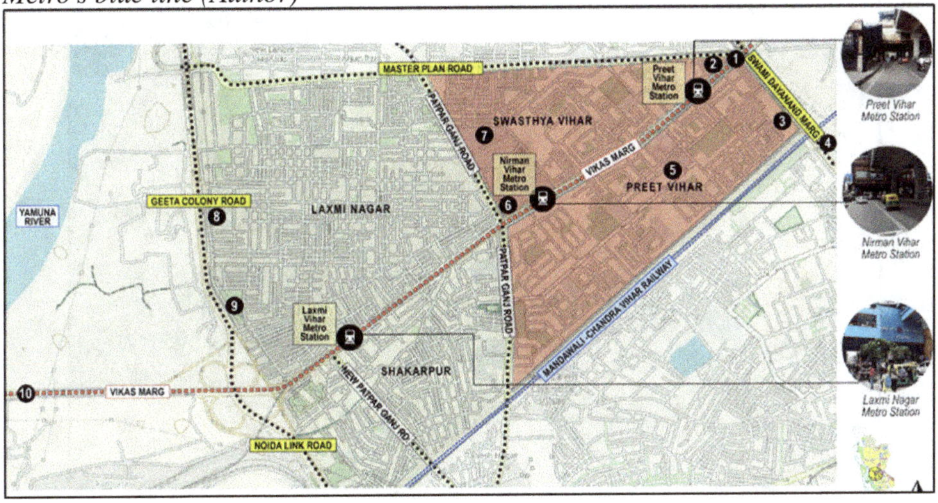

The history of Vikas Marg dates to the 1970s when the area was primarily a residential locality. At that time, the road was known as the Shahdara-Surajmal Vihar Road, a narrow two-lane road connecting Shahdara, a suburban area in East Delhi, to Surajmal Vihar, a residential colony in the eastern part of the city. In the 1980s, as Delhi experienced rapid expansion, the government recognized the need to improve infrastructure in East Delhi. One of the key projects undertaken during this period was the widening and extension of the Shahdara-Surajmal Vihar Road. The road was expanded to four lanes and extended all the way to Ghaziabad in Uttar Pradesh. Upon completion, the newly upgraded and extended road was renamed Vikas Marg (Delhipedia, 2022). Figure 2 elaborate the evolution of the context.

Figure 2. The history of the Vikas Marg with neighbouring region (Author)

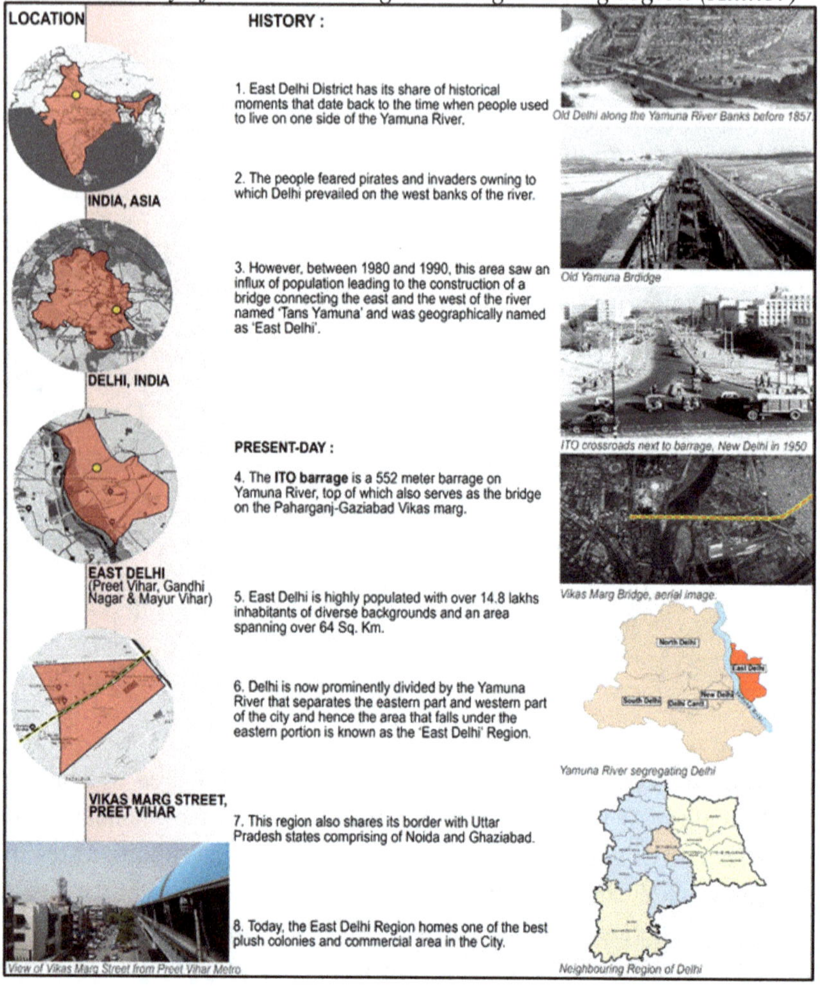

The renaming of the road was symbolic of the government's commitment to developing the area and improving the quality of life for its residents. In the following years, several landmark institutions were established along Vikas Marg. These institutions brought a wave of development to the area, attracting students and professionals from across the country. Over time, Vikas Marg has continued to evolve and grow. Today, it serves as a major commercial and residential hub, with shopping malls, restaurants, hospitals, and educational institutions lining the road (MoHUA, n.d.).

Figure 3. Street layout: Nirwan Vihar Metro station (Existing) (MoHUA, n.d.)

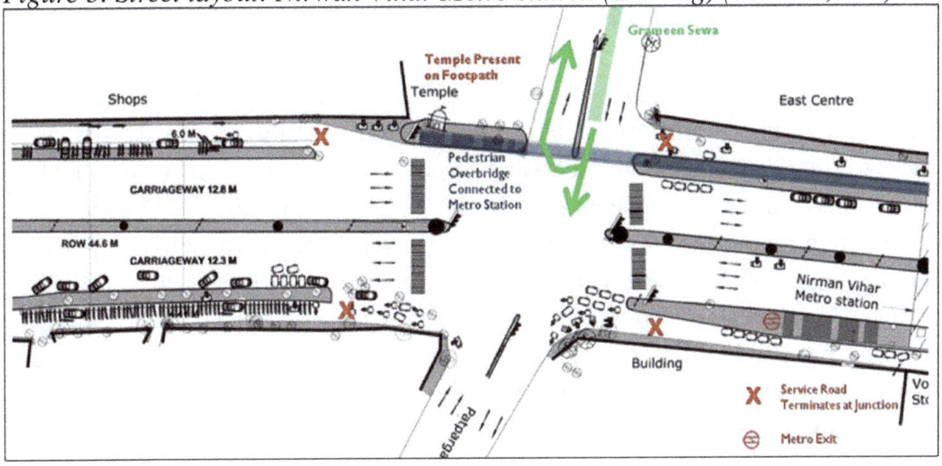

Figure 4. Street layout: between Preet Vihar and nirvana Vihar metro station (Mo-HUA, n.d.)

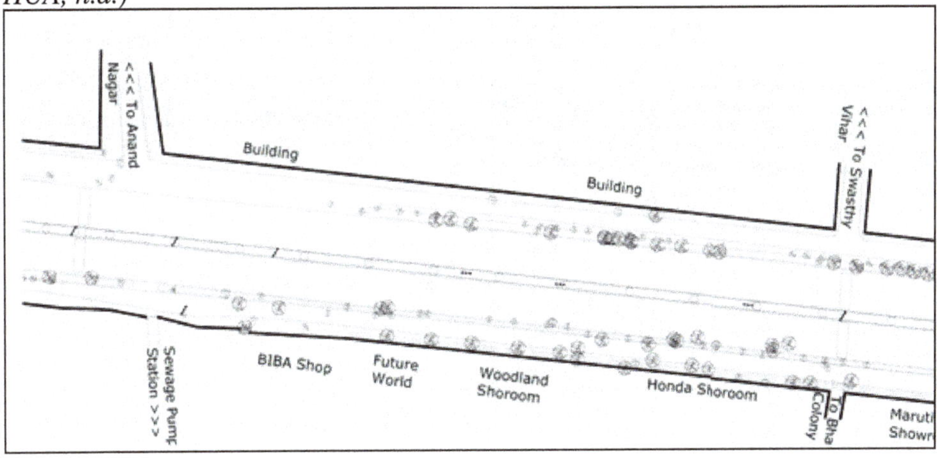

Vikas Marg is known for its high traffic volume and congestion, especially during peak hours. It is a vital transportation route for commuters, commercial vehicles, and public transportation, acting as a crucial link for people traveling between different parts of Delhi. The road's heavy traffic flow is further exacerbated by the presence of various commercial establishments, markets, educational institutions, and residential areas. Contributing factors to congestion include numerous traffic signals, narrow stretches, encroachments, and inadequate parking facilities. The increasing population, urbanization, and vehicular growth in Delhi over the years have compounded these issues, making traffic congestion a persistent problem on Vikas Marg as shown in Fgure 3 and 4 (MoHUA, n.d.).

The Delhi government has taken several initiatives to address these challenges and promote sustainable transport modes. Dedicated bus lanes have been introduced to encourage public transport use, along with the construction of footpaths and cycle tracks to create safer and more accessible spaces for pedestrians and cyclists. Traffic signals and pedestrian crossings have been strategically installed to improve road safety for non-motorized users. Other measures include promoting electric vehicles and implementing car-free days to reduce traffic volumes. Efforts such as widening the road, constructing flyovers, and implementing traffic management strategies have also been undertaken. Despite these efforts, the complex nature of the problem continues to make congestion on Vikas Marg a significant challenge for Delhi's transportation infrastructure as illustrated in Figure 5, 6, and 7 (MoHUA, n.d.).

Figure 5. Wider footpath and pedestrian crossing (MoHUA, n.d.)

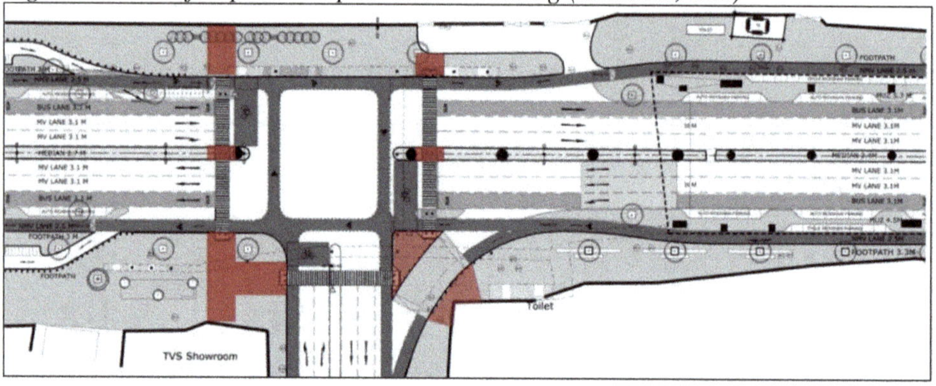

Figure 6. Separation of MV and NMV traffic (MoHUA, n.d.)

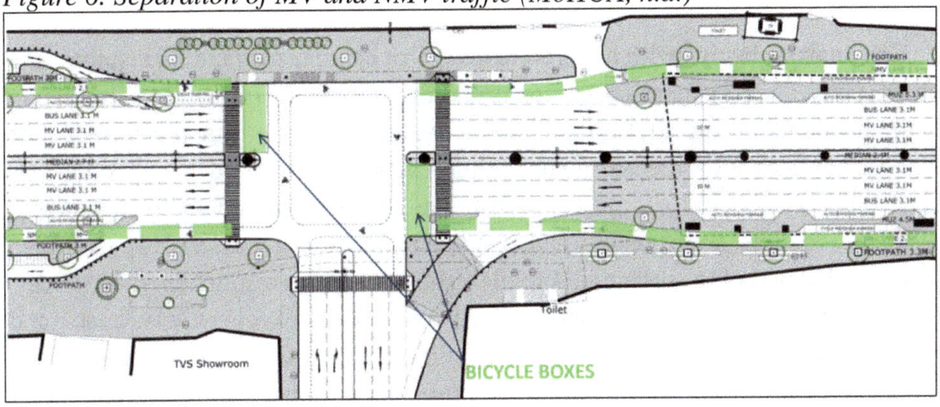

Figure 7. Alignment of lane at junctions (MoHUA, n.d.)

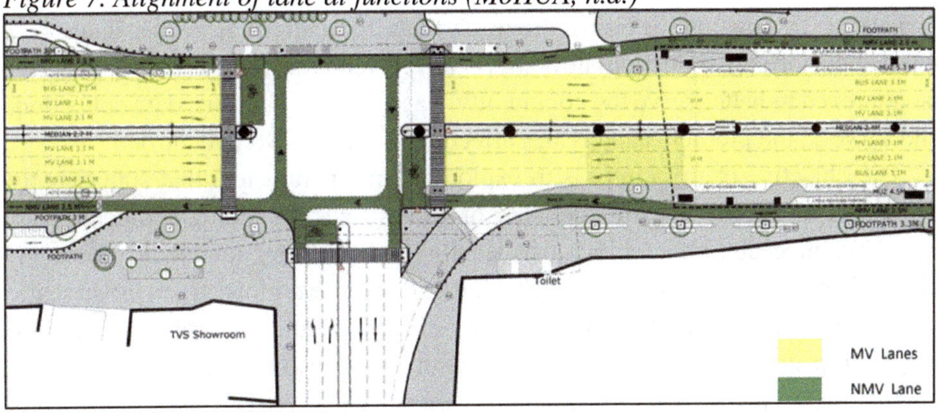

Abundant options for public buses, e-rickshaws, cabs, and metro feeder vehicles make commuting within and around the Delhi NCR region more convenient (MoHUA, n.d.).

The main objective of this study is to better understand street-related issues and propose effective solutions. The street under analysis is characterized by mixed land use (commercial and residential), a road width of 50 meters, transit-oriented infrastructure (metro, bus, subways, autos, taxis, etc.), and vegetation planted along the streets. The study focuses on a 1.8-kilometer stretch of Vikas Marg, running from Preet Vihar Metro Station to Nirman Vihar.

METHODOLOGY

The analysis begins with a literature review of case studies and books, focusing on the physical characteristics of streets and the existing infrastructure. In relation to the surrounding context, the data provides valuable insights into the current state of the streets and helps identify areas for improvement. Primary data collection was carried out to prepare the report. A questionnaire was designed and distributed to residents, vendors, shopkeepers, and customers. Face-to-face interviews were also conducted as part of the research. A total of 100 individuals responded to the survey, including a diverse group of users such as vendors, shopkeepers, customers, and residents. Surveys were conducted to gather information from various street users, including pedestrians, cyclists, motorists, and public transit users. This data helps in understanding the preferences, needs, and concerns of different users, providing critical insights into how streets can be better designed to accommodate their requirements. Understanding the activities that take place on streets is crucial for designing vibrant and functional urban spaces. Observations and analysis of activities such as walking, cycling, public transit usage, parking, commercial activities, and social interactions offer insights into how streets are currently utilized and how they can be improved to better serve the community. Based on site analysis, data collection, and surveys, the problems and issues affecting the streets were identified.

The objective of this study is to explore 25 key parameters that significantly influence the quality and functionality of streetscapes, including approach, connectivity, imageability, legibility, permeability, accessibility, landmarks, memory, symbolism, association, prosocial behaviour, urban fabric, urban grain, urban scale, land use, building use, built typology, building function, building condition, building height, infrastructure services (e.g., water, drainage, sewage, electricity), movement circulation, adaptation, socio-economic factors, and ecology. These parameters are interconnected and collectively contribute to shaping the physical, social, and ecological character of urban streets.

LITERATURE REVIEW

Understanding these parameters is essential as they serve as the foundation for assessing the design and performance of streetscapes. For instance, connectivity and permeability ensure seamless movement and integration within the urban fabric, promoting accessibility for pedestrians and vehicles (Lynch, 1960). Imageability and legibility contribute to the visual identity of a street, enabling users to form a mental image and navigate effectively, which enhances the cognitive experience of urban environments (Carmona et al., 2010). Similarly, parameters like landmarks, memory, and symbolism enrich the cultural and emotional attachment to a place, fostering a sense of belonging (Rapoport, 1990).

Moreover, urban fabric, grain, and scale are critical in defining the physical composition of streets and their harmony with the surrounding environment, while land use, building typology, and function determine the socio-economic vibrancy and functionality of streetscapes (Jacobs, 1961). The condition and height of buildings also influence perceptions of safety and aesthetics. The role of infrastructure services such as water supply, drainage, and electricity cannot be overlooked, as they ensure the street's functionality and resilience to urban challenges (Gehl, 2010). In addition, parameters like movement circulation and adaptation are vital for accommodating diverse user needs and ensuring flexibility in response to changing urban dynamics. Finally, studying socio-economic aspects and ecological considerations enables an understanding of broader implications for sustainability and inclusivity in urban streetscapes (Marshall, 2005).

By analysing these parameters, this study aims to develop a clear framework for assessing and improving the design, functionality, and sustainability of urban streets. These parameters work together and are essential for creating streets that are inclusive, vibrant, and environmentally sustainable. Additionally, this section reviews key books to provide a detailed understanding of parameters. This literature-based analysis offers both theoretical insights and practical knowledge to better understand how these elements contribute to well-designed streetscapes.

Book_01, Kevin Lynch, The image of the city

Lynch highlights the critical role of understanding how individuals perceive and navigate urban environments. The book introduces the concept of "imageability," which refers to a city's ability to create a vivid and coherent mental image for its residents and visitors. A city with high imageability is easy to navigate and possesses distinctive features that leave a lasting impression. Lynch identifies five fundamental elements that shape the image of a city: paths, edges, districts, nodes, and landmarks. Paths are the routes people use to move through the city, forming the framework of navigation. Edges act as boundaries or barriers, defining transitions between different areas. Districts represent distinct neighbourhoods or zones, each with unique functions and characteristics. Nodes serve as focal points or hubs of activity that draw people together. Lastly, landmarks are iconic and recognizable features that act as reference points, helping orient individuals within the city. Lynch emphasizes that cities succeed when these elements are effectively combined to create a cohesive, legible, and memorable urban environment. Through case studies of cities like Boston, Jersey City, and Los Angeles, he demonstrates how these elements influence people's perceptions and interactions with the built environment.

It is a groundbreaking work in urban planning and design, offering profound insights into the interplay between the physical structure of cities and human perception. Lynch underscores the importance of designing cities that are not only functional but also visually engaging and navigable. His work remains a cornerstone of urban design, providing a timeless framework for analysing and improving the visual and spatial organization of cities (Lynch, 1960).

Book_02, Kevin Lynch, Good City Form

"A Theory of the Good City" explores the principles and philosophies that define what makes an urban environment desirable, functional, and beneficial for its residents. This concept emphasizes creating cities that prioritize well-being, sustainability, equity, and quality of life. Theories in this domain serve as a foundation for guiding urban planning, design, and development (Lynch, 1981).

1. Liveability: A "good city" ensures a high standard of living for its inhabitants by fostering safe, vibrant, and inclusive spaces. It prioritizes access to essential amenities, such as housing, healthcare, education, and recreational facilities.
2. Sustainability: Environmental sustainability is at the core of a good city. This involves promoting efficient resource use, reducing pollution and waste, and integrating green practices into urban design. Cities should adopt measures to combat climate change and preserve natural ecosystems.
3. Equity and Social Justice: A good city actively addresses inequality and discrimination. It ensures equal access to opportunities and resources for all, focusing on affordable housing, accessible public spaces, and pathways for economic and social mobility. This requires tackling poverty, marginalization, and systemic barriers (Harvey, 2012).
4. Resilience: A good city is adaptable and prepared to face challenges, whether they arise from natural disasters, climate change, economic shifts, or other crises. It invests in infrastructure, policies, and community support systems to foster long-term resilience.
5. Participatory Governance: Inclusive decision-making processes are fundamental to the good city. Citizens, government bodies, community organizations, and other stakeholders must work collaboratively to shape urban policies and development. This ensures diverse perspectives are considered and that urban plans align with the aspirations of the community (Healey, 1997).

The concept of a good city highlights the importance of balancing environmental, social, and economic priorities to create spaces that serve the needs of all residents while promoting long-term sustainability and inclusivity.

Book_03, Edward T. Hall, The Hidden Dimension

In *"The Hidden Dimension,"* Edward T. Hall argues that space is more than a mere physical entity; it holds significant cultural and social meaning. He explores how different cultures use and interpret space in distinct ways, often leading to misunderstandings and conflicts when individuals from diverse cultural backgrounds interact. Hall introduces four primary categories of spatial behaviour: intimate distance, personal distance, social distance, and public distance. He delves into the culturally shaped norms and expectations tied to these categories, including acceptable physical proximity, the use of gestures and body language, and the concept of territoriality. The book also investigates how architecture and design shape spatial experiences. Hall examines the influence of architectural styles, layouts, and materials on human perception and behaviour. He highlights how the design of public spaces, such as parks, plazas, and streets, can encourage or hinder social interactions and community engagement. Through numerous examples and case studies drawn from diverse cultures and societies, Hall illustrates his theories and concepts. He underscores the importance of understanding and respecting cultural differences in spatial behaviour to foster better communication and collaboration (Hall, 1966).

In Comparison to the Books (Vikas Marg street)

The five elements—paths, edges, districts, nodes, and landmarks—together form the image of a city, influencing how people perceive, navigate, and experience urban environments. Understanding these elements can inform urban planning and design decisions to create more legible, navigable, and liveable cities (Lynch, 1960).

1. Paths include roads, streets, and pedestrian walkways that connect different parts of Laxmi Nagar, enabling movement and transportation within the locality.
2. Edges refer to the boundaries of buildings, parks, or other structures that define the shape and structure of Laxmi Nagar.
3. Districts represent distinct zones or areas within Laxmi Nagar, such as residential, commercial, or industrial zones, each with unique visual and functional characteristics.
4. Nodes are central points or hubs in Laxmi Nagar, such as intersections, junctions, or plazas, where multiple paths converge and serve as focal points of activity.

5. Landmarks include prominent features like monuments, buildings, or unique natural elements that serve as reference points or attract interest in Laxmi Nagar.

Streetscaping Vikas Marg

The streetscaping of Vikas Marg focuses on enhancing the street's visual appeal and functionality, making it more attractive, pedestrian-friendly, and sustainable. Possible improvements include:

1. Sidewalks: Wider, well-maintained sidewalks with safe pedestrian crossings, wheelchair accessibility, and designated walking areas.
2. Street Furniture: Installing benches, bus shelters, trash cans, and bike racks to enhance comfort and functionality for pedestrians.
3. Landscaping: Adding greenery, such as trees, shrubs, and flowers, to improve air quality and aesthetics.
4. Lighting: Installing adequate lighting fixtures to improve safety and security, especially at night.
5. Signage: Providing clear wayfinding, street names, and informational signs for better navigation.
6. Public Art: Incorporating murals, sculptures, or other art installations to enrich the street's cultural and artistic appeal.

Insights from "Good City Form"

In *Good City Form*, Kevin Lynch explores principles that guide the design of well-functioning and appealing urban environments. These principles can be applied to streets like Vikas Marg:

1. Urban Design Principles: Lynch emphasizes the importance of public spaces, streets, and networks. Vikas Marg should prioritize pedestrian-friendliness, efficient traffic flow, aesthetic appeal, and integration with surrounding areas.
2. Functionality: Well-designed streets support multiple activities and accommodate various modes of transport. Vikas Marg could feature defined sidewalks, safe crossings, and effective traffic management to serve pedestrians, cyclists, and vehicles.
3. Social Interaction: Urban spaces should foster community engagement. Vikas Marg can include seating, gathering spaces, and recreational amenities while integrating commercial and cultural hubs.

4. Aesthetic Appeal: Lynch highlights the role of visual harmony in urban design. Thoughtful landscaping, architectural features, and art installations can make Vikas Marg visually attractive.
5. Cultural Context: Urban spaces should reflect local culture. Vikas Marg's design could incorporate elements like public art and architectural styles that resonate with the locality's cultural identity.

Insights from "The Hidden Dimension"

In *The Hidden Dimension*, Edward T. Hall examines the cultural and social meanings of space. He discusses concepts like proxemics (the study of personal space), nonverbal communication, and the role of spatial design in human behavior. Social behaviour on Vikas Marg is shaped by its diverse mix of people, activities, and businesses.

1. Interaction and Communication: Vikas Marg is a busy area with lots of people walking, driving, shopping, and working, leading to frequent interactions among residents, commuters, shoppers, and business owners.
2. Commercial Interactions: The street has many shops and restaurants where people often bargain, shop, and interact with staff and business owners.
3. Cultural Diversity: Being part of Delhi, Vikas Marg is home to people from different backgrounds, making it a place where cultures and communities mix.
4. Road Etiquette: The heavy traffic on Vikas Marg means people need to follow road rules, respect pedestrians, and stay patient during traffic jams.

STANDARDS

The National Association of City Transportation Officials (NACTO) has developed a set of guidelines and design principles known as the "NACTO Urban Street Design Guide." These guidelines provide cities with innovative and practical strategies for designing streets that prioritize safety, efficiency, and the needs of all users. Below is an overview of key principles and recommendations outlined in the guide (NACTO, 2013) and supported by figure 8, 9 and 10:

1. Context-Sensitive Design: Streets should be tailored to the specific context of the surrounding area.
2. Active Transportation: The integration of walking, cycling, and public transit into street design is encouraged to support active transportation modes.

3. Complete Streets: Streets should accommodate the needs of all users, including pedestrians, cyclists, transit riders, and motorists. The design should prioritize the movement of people over vehicles by ensuring safe and comfortable spaces for each mode of transportation.
4. Safety and Vision Zero: The Vision Zero approach aims to eliminate traffic fatalities and serious injuries through safer street designs.
5. Transit Priority: Efficient and reliable public transit is critical. Strategies such as bus rapid transit (BRT) lanes, transit signal priority, and bus boarding infrastructure improve speed and efficiency.
6. Placemaking and Public Spaces: Creating vibrant, inviting public spaces along streets is emphasized.
7. Green Infrastructure: Incorporating green elements like trees, bioswales, and permeable pavements enhances sustainability and improves stormwater management.

Figure 8. The image shows an example of a street, before and after changes while incorporating with NACTO standards. The segregation of the carriageways, the parking lanes, biking lane, and the transit wider lane (NACTO, n.d.)

Figure 9. Existing conditions demonstrate how traditional design elements, such as wide travel lanes and undifferentiated street space, have had an adverse impact on how people experience the streetscape (NACTO, n.d.)

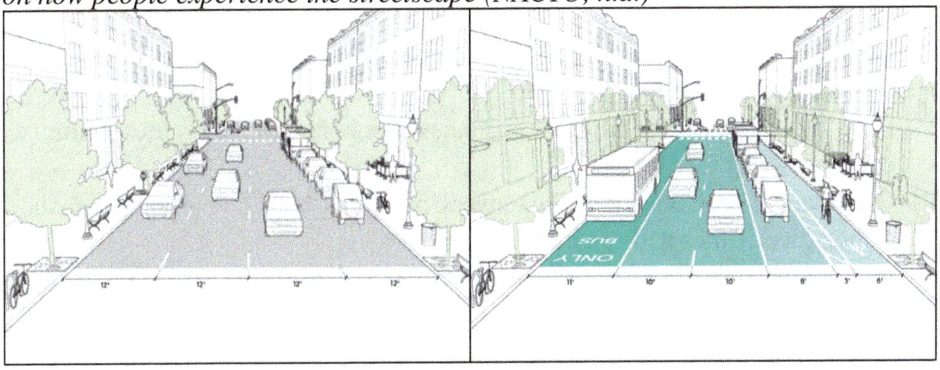

Figure 10. Reducing travel-lane capacity resulted in a decrease in vehicle traffic (NACTO, n.d.)

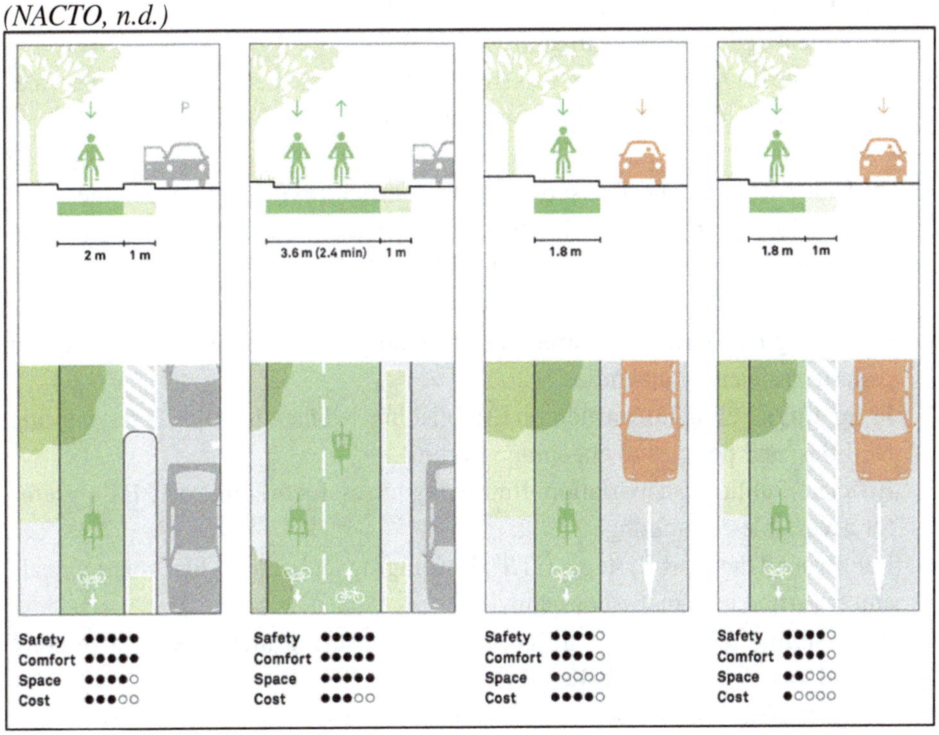

The Institute for Transportation and Development Policy (ITDP) has developed the "BRT (Bus Rapid Transit) Standard" to guide the planning, design, and implementation of high-quality BRT systems. Key elements include (ITDP, 2024):

1. Dedicated Right-of-Way: Buses should operate in dedicated lanes or corridors for efficient service.
2. Busway Alignment: Segregated bus lanes or median-aligned routes minimize conflicts with other traffic.
3. Off-Board Fare Collection: Passengers pay fares before boarding to reduce station dwell times and enhance efficiency.
4. Intersection Treatments: Priority measures like signal priority or grade separation minimize delays at intersections.
5. Bus Stations with Level Boarding Platforms: Platforms that align with bus floors enable quick and accessible boarding.
6. High-Quality Stations: Stations should have clear signage, passenger amenities, and comfortable shelters.
7. Frequent Service: High-frequency buses reduce waiting times.
8. Integration with Land Use: BRT corridors should support transit-oriented development and promote walkability.
9. Excellence in Operations: Systems should be well-managed with reliable service, effective maintenance, and high customer satisfaction.

The Unified Traffic and Transportation Infrastructure (Planning & Engineering) Centre (UTTIPEC) in Delhi, India, focuses on creating equitable road space and integrated streets. Its goals include, (UTTIPEC, 2011):

1. Enhancing Pedestrian Infrastructure: Providing safe sidewalks, crossings, and pedestrian-friendly amenities.
2. Promoting Cycling Infrastructure: Establishing dedicated cycling lanes, parking facilities, and policies to encourage cycling.
3. Efficient Public Transportation: Improving buses, metro rail, and BRT systems for accessible commuting.
4. Intelligent Transport Systems (ITS): Leveraging technology like smart signals and real-time information systems.
5. Accessibility for Persons with Disabilities: Ensuring inclusive infrastructure that addresses mobility challenges.
6. Green and Sustainable Design: Integrating green spaces, trees, and stormwater management systems to improve sustainability.

The Urban and Regional Development Plans Formulation and Implementation (URDPFI) framework, introduced by India's Ministry of Urban Development, provides guidelines for sustainable and balanced urban development. It emphasizes (URDPFI, 2015):

1. Integrated Development: Coordination across sectors to ensure comprehensive urban planning.
2. Sustainable Development: Promoting practices like resource conservation, green transportation, and energy efficiency.
3. Participatory Planning: Involving stakeholders and communities in decision-making processes.
4. Equity and Social Inclusion: Ensuring access to services and amenities for all, including marginalized groups.
5. Institutional Strengthening: Enhancing governance, capacity-building, and monitoring mechanisms.

FINDINGS

Identifying the street, while taking pictures of the street and the neighbourhood, including the geo-location picture of the street; The Approach and Connectivity of the street with other national highways/ local roads/ pedestrian/ etc.; Marking out the Landmarks, Nodes & Symbolism of the street - Neighbourhood Analysis, such as the Preet Vihar Metro Station, CBSE Building and Ex-Minister Congress House as landmark, Masjid and the Yamuna River as symbolism, at every intersection of the roads, either with cafes or malls are identified as nodes; Observing the movement Analysis of vehicles, the people, and the activities for morning, afternoon, and night duration and for the weekdays and weekends period; Demographic information and Background Analysis of the people, their purpose of their visitation, about their age, gender, etc.; Pro-Social study of the human behaviour, their daily movements with the street, and their activities; Studying the types of building surrounding the street and it's functions - the types of shops, vendors, and retail shops/ malls; Comparing the heights of each building and the relation with the scale of the street to the building, people, plants/trees, etc.; The study of ecology – types of trees planted along the road, and if they prove to be useful to the users by providing shade to the people during dire weather circumstances or if they are used for the beautification of the street, etc.; Physical Infrastructure of the drainage, electricity, and other services.

The comparative analysis involves comparing the data and findings from different streets to identify patterns, trends, and best practices. This can help in understanding the strengths and weaknesses of different streets and identifying strategies that have been successful in other contexts. The live survey study for Vikas Marg Street was conducted over the course of one month in September. The study focused on 25 parameters of the urban street, and a thorough analysis was conducted for each parameter. The survey team visited the site of the street and collected data based on the predetermined parameters. The data collected was then analysed to draw inferences and conclusions.

Connectivity

Vikas Marg is connected to other major roads and highways in Delhi, such as the Ring Road, Loni Road, and NH-24. There are several metro stations located along Vikas Marg, including Nirman Vihar, Preet Vihar, Karkardooma, Laxmi Nagar. This makes it easy for people to travel to different parts of the city and beyond. Nearest Airport: Indira Gandhi International Airport (26.3km); Nearest Railway: Delhi Junction Railway (10.9 km); Nearest Metro Station: Preet Vihar Metro Station; Connected Major Roads: Noida-Link Road & NE3.

Landmarks

It has several street features, landmarks, and notable places along its length. Some of these include Laxmi Nagar District Center, a commercial hub located on Vikas Marg, and it is one of the busiest areas in Delhi.

It features several restaurants, shops, and offices. Nirman Vihar Metro Station is a major metro station located on Vikas Marg, and it connects the area to other parts of Delhi. Swasthya Vihar Club is a popular recreational center located near Vikas Marg, and it offers a range of facilities and activities such as swimming, tennis, and yoga. The Yamuna Sports Complex is a large sports complex located near Vikas Marg, and it features several sports facilities, including an athletic stadium, a cricket stadium, and a tennis complex. Shakarpur is a neighborhood Area located on Vikas Marg, and it features several residential buildings, shops, and restaurants. East Delhi Mall is a large shopping mall located on Vikas Marg, and it features several shops, restaurants, and a cinema. The Karkardooma District Court is a major court complex located on Vikas Marg, and it serves as a hub for legal proceedings in the area. The Sai Temple is a popular religious site located near Vikas Marg, and it attracts many devotees from all over Delhi. CBSE (Central Board of Secondary Education) office is located on Vikas Marg in Delhi, India. The office is in Preet Vihar, which is a neighborhood that is situated along Vikas Marg. The CBSE office in Preet Vihar is a major hub for the administration of CBSE schools in India and is responsible for conducting examinations, issuing certificates, and providing other support services to CBSE-affiliated schools.

The "imageability" of a place refers to its ability to form a strong and coherent mental image in the minds of people who experience it. Vikas Marg has several features that contribute to its imageability. Some aspects of Vikas Marg that make it memorable and easily recognizable: Landmarks: Vikas Marg has several landmarks and notable places that help people orient themselves and remember the road. For example, the CBSE office, Yamuna Sports Complex, and East Delhi Mall are all major landmarks that can be easily identified and associated with Vikas Marg. Street Features: The street features of Vikas Marg, such as its wide sidewalks, streetlights, and tree-lined median, contribute to its visual appeal and make it more memorable. Diversity of Uses: Vikas Marg is a diverse road with a mix of commercial, residential, and institutional uses. This diversity of uses creates a rich and dynamic urban environment that people are more likely to remember and identify with. Human Scale: Despite being a major road, Vikas Marg has a human scale that makes it more accessible and inviting. The sidewalk cafes, street vendors, and small shops along the road create a sense of intimacy and connection that is often lacking in larger, more impersonal roads.

Land-Use Plan

The mainland-use findings along the Vikas Marg are Commercial: Vikas Marg has many commercial establishments, including shops, markets, malls, restaurants, and hotels. Many small shops and street vendors sell a wide variety of goods, as well as larger commercial complexes such as the V3S Mall, which is a major shopping destination in the area.

Residential: There are several residential areas located along Vikas Marg, including Preet Vihar, Karkardooma, and Anand Vihar. These areas have a mix of apartment buildings, houses, and gated communities.

Institutional: Vikas Marg is home to several important institutional buildings, such as the Central Board of Secondary Education (CBSE) headquarters, the Yamuna Sports Complex, and several schools and colleges.

Recreational: There are several parks and recreational areas located near Vikas Marg, such as the Sanjay Lake Park and the District Park in Preet Vihar. These parks provide green spaces for residents to relax and enjoy outdoor activities. Industrial: There are some industrial areas located near Vikas Marg, particularly on the eastern side of the road. These areas are home to factories and warehouses that produce a variety of goods.

Figure 11. Land -use plan with sections (Author)

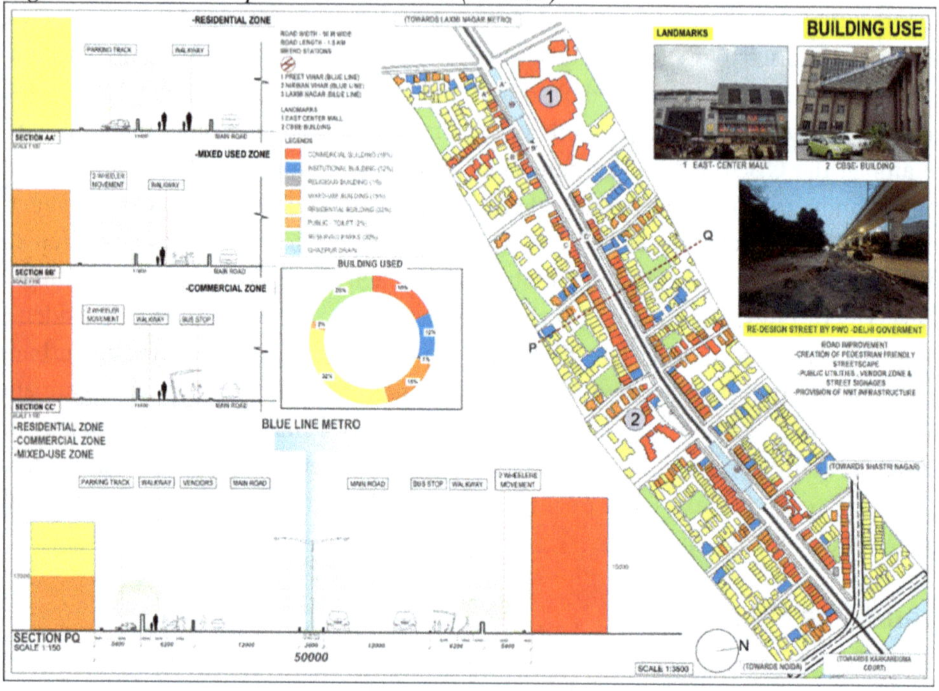

Vikas Marg is dotted with numerous commercial buildings that house offices, shops, showrooms, and other commercial establishments. These buildings are typically multistore and vary in size, design, and age. Many commercial buildings along Vikas Marg have modern facades, large display windows, and prominent signage to attract customers. They also have facilities such as elevators, parking areas, and security systems. Vikas Marg also has several residential buildings, including apartment complexes and independent houses. These buildings can range from mid-rise to high-rise structures, providing accommodation to a diverse range of residents. Residential buildings along Vikas Marg have amenities such as gated entrances, landscaped gardens, parking facilities, and community spaces for recreational activities.

Some buildings along Vikas Marg have Mixed-use characteristics, combining both commercial and residential spaces. These buildings often have commercial establishments on the lower floors, such as shops or offices, and residential units on the upper floors. They also have separate entrances and facilities for both commercial and residential sections.

The buildings along Vikas Marg designs to more traditional styles. The architectural styles depend on the age of the building, the preferences of the developers, and the local building regulations. Some buildings have sleek glass facades, while others feature more traditional brick or concrete exteriors. The age and condition of buildings along Vikas Marg can vary. Some buildings may be newly constructed with modern amenities and in excellent condition, while others may be older and require maintenance or renovation.

The condition of the buildings can impact their functionality, aesthetics, and many buildings along Vikas Marg house commercial activities such as retail stores, restaurants, banks, and offices. The type of commercial activity can vary depending on the location and zoning regulations. Buildings closer to major intersections or commercial hubs may have more retail shops and showrooms, while those further away have a mix of offices and other commercial establishments.

The accessibility of buildings along Vikas Marg may depend on their location and proximity to major roads, public transport, and other amenities. Buildings located in prime areas with good connectivity may be more sought after for commercial activities, while residential buildings may be preferred for their convenience in terms of accessibility to transportation and other amenities. The surrounding environment of buildings along Vikas Marg can also influence their analysis. Buildings that are situated in well-maintained and clean surroundings with ample greenery and public spaces be more attractive to residents and commercial.

Circulation

The circulation system around Vikas Marg faces several issues that affect pedestrian and vehicular movement. Some of the problems are: Pedestrian walkway encroachment: One of the primary issues with pedestrian circulation near Vikas Marg is the encroachment of footpaths by vendors, and residents building ramps on their properties. This obstructs pedestrian movement and forces them to walk on the main road, which poses a safety risk; Vehicle parking on footpaths: Another issue with pedestrian circulation is the unorganized parking system, particularly in markets, which leads to cars being parked on the pavement, blocking the walkway; Lack of pick-up and drop-off points: There is no designated pick- up and drop-off point on Vikas Marg, which causes vehicular congestion and chaos. To address these issues, the local authorities can take the following steps. Strict enforcement of pedestrian walkways: The authorities can strictly enforce the pedestrian walkways and remove any encroachment by vendors or residents.

Figure 12. Circulation plans with existing and proposed sections (Author)

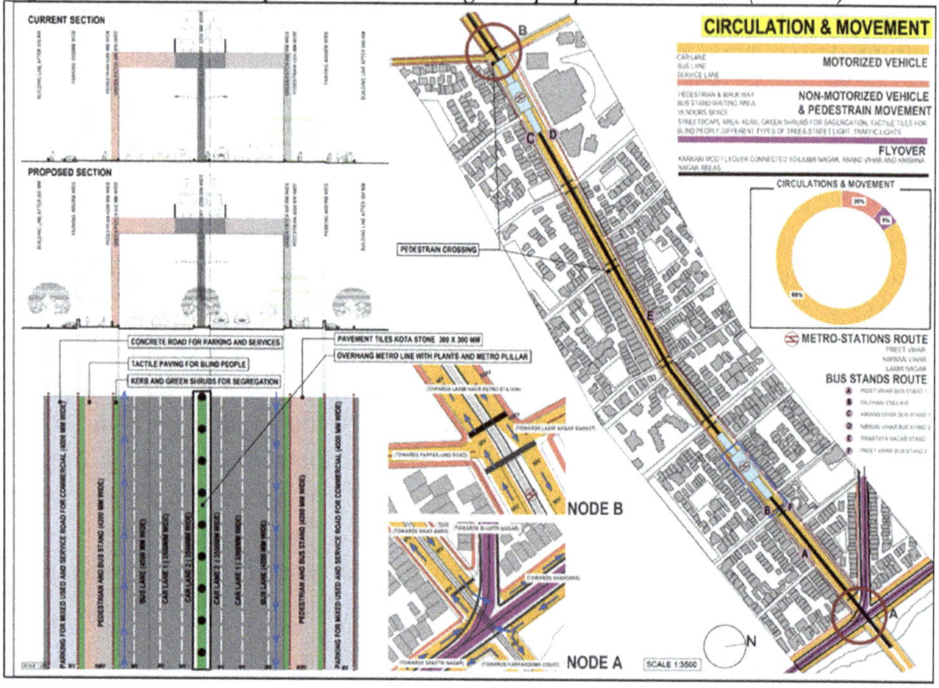

This would ensure that pedestrians can use the footpath without obstruction; Organized parking system: The local authorities can introduce an organized parking system in the markets, with designated parking areas for shoppers, refer Figure 12. This would ensure that the road space is not used for parking, and it would also help in reducing congestion; Designated pick-up and drop-off points: The authorities can identify and designate pick-up and drop-off points on Vikas Marg to reduce congestion and ensure smooth vehicular movement.

Landscape

The landscape along Vikas Marg can vary depending on the specific location, but in general, there are some common features. In the commercial areas along Vikas Marg, the landscape is dominated by buildings and shops. There are some trees and greenery on the sidewalks or in nearby parks, but these are often limited in space and number. In the residential areas, the landscape tends to be greener and more natural. Many of the housing societies and apartments along Vikas Marg have well-maintained gardens and parks with trees, flowers, and other plants. These areas provide a welcome respite from the busy road and can be enjoyed by both residents and visitors. Here are some of the common trees and shrubs that are found in the area: - Neem trees; Gulmohar trees; Peepal trees; Ashoka trees; Bougainvillea shrubs and Hibiscus shrubs etc.

RESULT

1. Traffic Congestion: Laxmi Nagar Street is often plagued by traffic congestion due to the high volume of vehicles, including cars, buses, auto-rickshaws, and cycle-rickshaws. Narrow roads, encroachments, and illegal parking often exacerbate the congestion, making it difficult for vehicles and pedestrians to move freely. This can lead to traffic jams, delays, and inconvenience for commuters and businesses.
2. Pedestrian Safety: - With the high density of vehicles, pedestrian safety can be a concern on Laxmi Nagar Street. Sidewalks may be encroached upon, forcing pedestrians to walk on the road and navigate through moving vehicles. Lack of proper pedestrian crossings, footbridges, or underpasses can also make it unsafe for pedestrians to cross the road, especially during peak hours. This poses risks to the safety and well-being of pedestrians, including children, the elderly, and differently abled individuals.
3. Air pollution: Due to heavy traffic and industrial activities in the surrounding areas, air pollution can be a problem along Vikas Marg. This can have adverse effects on the health and well-being of residents and commuters.
4. Insufficient green spaces: While there may be some parks and green spaces along Vikas Marg, the overall availability of green areas may be limited. This can impact the aesthetic appeal of the area and reduce opportunities for recreational activities.
5. Noise pollution: The busy road can contribute to high levels of noise pollution, affecting the quality of life for residents and workers in the area.
6. Garbage accumulation: Improper waste disposal and collection can lead to garbage accumulation along the road, which can cause unpleasant Odors and attract pests.
7. Parking Issues: - Parking can be a major issue on Laxmi Nagar Street, with limited parking spaces and illegal parking practices. Vehicles parked haphazardly on the road can further narrow down the already congested streets, causing inconvenience to other road users. The lack of proper parking infrastructure and enforcement of parking regulations can result in chaotic parking situations, affecting the overall flow of traffic and pedestrian movement.
8. Encroachments: -Encroachments by vendors, shops, and other establishments on sidewalks and public spaces can hinder the smooth circulation of the street in Laxmi Nagar. These encroachments may block sidewalks, create obstacles for

pedestrians, and contribute to congestion. Proper regulation and management of encroachments are required to ensure that sidewalks and public spaces are kept clear and accessible for pedestrians and other road users.

9. Waste Management: - Waste management can be an issue in Laxmi Nagar Street, with littering and improper disposal of waste leading to unclean surroundings. Improperly managed waste can accumulate on the streets, causing hygiene and sanitation issues, and contributing to environmental pollution. Proper waste collection and disposal mechanisms, along with awareness campaigns, are needed to address this issue.
10. Infrastructure Maintenance: - Maintenance of roads, sidewalks, street lighting, and other infrastructure can be a challenge in Laxmi Nagar Street. Poorly maintained roads and sidewalks can lead to potholes, cracks, and uneven surfaces, posing risks to road users. Timely repair and maintenance of infrastructure are essential to ensure safe and smooth circulation of the street.

PROPOSAL

1. Traffic congestion: Conduct a traffic flow analysis and propose alternative routes or lane configurations to ease congestion; Design pedestrian-friendly streetscapes with ample sidewalks, crosswalks, and pedestrian signals; Incorporate transit-oriented development (TOD) principles by locating mixed-use developments near transit stations to reduce dependence on private vehicles.
2. Lack of parking space: Design parking structures that blend with the surrounding urban fabric and incorporate green roofs or facades to improve aesthetics; Propose off-site parking facilities connected to transit stations, shopping centres, or other public amenities to encourage shared parking.
3. Noise pollution: Design noise barriers with materials that complement the surrounding landscape and enhance visual aesthetics; Consider acoustical treatments on building facades to mitigate noise impacts on occupants; Incorporate green spaces and water features that absorb sound and provide a buffer between the road and adjacent buildings.
4. Poor air quality: Incorporate green infrastructure elements such as bioswales, rain gardens, and street trees to filter pollutants and improve air quality; Design pedestrian-friendly streetscapes that encourage walking and cycling and reduce vehicle emissions; Propose green roofs facades on buildings to reduce the urban heat island effect and improve air quality.
5. Garbage accumulation: Design waste management systems that blend with the urban fabric and are accessible to pedestrians and waste collectors; Incorporate

composting and recycling facilities in public parks and open spaces to encourage waste diversion and reduce landfill waste.

6. Lack of green spaces: Identify potential sites for green spaces and propose design concepts that integrate ecological and recreational functions; Incorporate sustainable landscape practices such as rainwater harvesting, native plantings, and urban agriculture to promote biodiversity and community engagement; Integrate green spaces with transit-oriented developments and public amenities to create vibrant and walkable neighbourhoods.
7. Encroachment and illegal constructions: Conduct a site analysis and propose zoning and land use regulations that prevent encroachment and illegal constructions; Design public spaces that discourage encroachment by providing clear boundaries and visual cues; Propose community engagement initiatives that involve residents in maintaining and improving public spaces.
8. Garbage accumulation: Design waste management systems that blend with the urban fabric and are accessible to pedestrians and waste collectors; Incorporate composting and recycling facilities in public parks and open spaces to encourage waste diversion and reduce landfill waste.
9. Lack of green spaces: Identify potential sites for green spaces and propose design concepts that integrate ecological and recreational functions; Incorporate sustainable landscape practices such as rainwater harvesting, native plantings, and urban agriculture to promote biodiversity and community engagement; Integrate green spaces with transit-oriented developments and public amenities to create vibrant and walkable neighbourhoods.
10. Encroachment and illegal constructions: Conduct a site analysis and propose zoning and land use regulations that prevent encroachment and illegal constructions; Design public spaces that discourage encroachment by providing clear boundaries and visual cues; Propose community engagement initiatives that involve residents in maintaining and improving public spaces.

CONCLUSION

The case study of Vikas Marg highlights the importance of this urban corridor in Delhi. As a major transportation artery and a vibrant commercial hub, it attracts residents, businesses, and visitors alike. This dynamic area presents both challenges and opportunities in urban planning, design, and management. Enhancing the pedestrian experience along Vikas Marg is a key priority. Measures such as wider sidewalks, well-designed pedestrian crossings, and improved connectivity to nearby areas can create a more walkable and pedestrian-friendly environment. These improvements would not only enhance safety but also encourage people to explore the area on foot. Addressing traffic congestion and improving transportation infrastructure is crucial for ensuring smooth vehicular movement. Strategies like dedicated bus lanes, advanced traffic management systems, and promoting sustainable transportation modes, such as cycling and public transit, can alleviate congestion and enhance overall efficiency. Preserving the cultural and historical significance of the area is equally important. Incorporating elements of heritage conservation and architectural preservation can help retain the unique character of Vikas Marg while accommodating modern development. The case study also emphasizes the need for active stakeholder engagement and collaboration. Involving local communities, businesses, and governmental bodies in decision-making processes ensures diverse perspectives are considered. This approach helps align the development of Vikas Marg with the needs and aspirations of its stakeholders.

REFERENCES

Carmona, M., Tiesdell, S., Heath, T., & Oc, T. (2010). *Public Places, Urban Spaces: The Dimensions of Urban Design*. Routledge.

Delhipedia. (2022). *East Delhi's development thoroughfare: Vikas Marg.*

Gehl, J. (2010). *Cities for People*. Island Press.

González, G., Rodríguez, F., & García, M. (2019). Inclusive Urban Development for Resilient Communities: A Framework for Equitable Design. *Urban Studies (Edinburgh, Scotland)*, *56*(7), 1350–1367. DOI: 10.1177/0042098018787671

Hall, E. T. (1966). *A hidden dimension*. Penguin Random House.

Harvey, D. (2012). *Rebel Cities: From the Right to the City to the Urban Revolution*. Verso Books.

Institute for Transportation and Development Policy (ITDP). (2024). *The BRT Standard.*

Jacobs, J. (1961). *The Death and Life of Great American Cities*. Random House.

Lynch, K. (1960). *The image of the city*. The MIT Press.

Lynch, K. (1981). *Theory of a good city form*. The MIT Press.

Marshall, S. (2005). *Streets and Patterns*. Routledge.

Ministry of Housing and Urban Affairs (MOHUA). (2015). *URDPFI guidelines* (Vol. I). Government of India.

Ministry of Housing and Urban Affairs (MOHUA). (n.d.). *Equitable road space: Case study Vikas Marg*. Government of India.

Mohammad, S., Dey, A., & Saha, A. (2019). Mixed-Use Development and Its Impact on Urban Sustainability and Equity. *Journal of Urban Affairs*, *41*(4), 496–513. DOI: 10.1080/07352166.2018.1479930

National Association of City Transportation Officials (NACTO). (2013). *Urban street design guide.*

Rapoport, A. (1990). *The Meaning of the Built Environment: A Nonverbal Communication Approach*. University of Arizona Press.

Sim, D. (2019). *Soft City: Building Density for Everyday Life*. Island Press.

Smith, N., Davies, R., & Owen, G. (2021). Integrating Commercial and Residential Spaces for Sustainable Urban Futures. *Sustainable Cities and Society*, *61*, 102302. DOI: 10.1016/j.scs.2020.102302

Unified Traffic and Transportation Infrastructure (Planning & Engineering) Centre (UTTIPEC). (2011). *Street guidelines.*

Zhang, L., Wang, X., & Liu, H. (2020). Sustainable Urban Planning and Social Equity: A Systematic Approach. *Journal of Urban Planning and Development*, *146*(3), 04020034.

Chapter 14

Integrating BGI Into Urban Development: A Case of Indian Cities

Aditi Nag
https://orcid.org/0000-0002-0604-6945
Universidad Manipal, Jaipur, India

Anurag Singh Rathore
https://orcid.org/0009-0004-0112-5478
Indira Gandhi National Open University, India

Apurv Ashish
https://orcid.org/0009-0005-9076-1405
Birla Institute of Technology, India

ABSTRACT

This study examines the function of BGI in urban development, emphasizing its importance for Indian cities dealing with environmental challenges made worse by dense, rapidly growing urbanization. The study focuses on effective BGI programs that promote sustainable development, urban resilience, and public health using case studies from Bengaluru, Mumbai, London, and Copenhagen. According to the findings, implementing BGI solutions requires flexibility with regard to local context, sustainability outcomes, and stakeholder participation. Encourage BGI in Indian cities by creating thorough regulations, raising public awareness, fostering stake-holder collaborations, putting creative design principles into practice, and putting in place monitoring measures. Indian cities can create a vibrant urban environment where development and conservation can coexist because of BGI.

DOI: 10.4018/979-8-3373-2555-2.ch014

INTRODUCTION

India's rapid urbanization has bequeathed a legacy of ecological destruction. These patterns are observed in the staggering declines in green and blue areas (Brown & Mijic, 2019). India, which is among the fastest urbanizing nations globally, is expected to host approximately 814 million urban residents by 2050, posing huge challenges to sustainable urbanization (Bansal & Haridasan, 2024). This heterogeneity is followed by increasing air and water pollution, loss of biodiversity, and increased vulnerability to climate change impact, such as heatwaves and floods (Ahmad & Hassan, 2022). These need instant, innovative solutions that would improve the living standards and urban resilience of Indian cities (Chaurasia et al., 2024; Patel et al., 2024; Dhanya, Ramananda, & Dhyani, 2022). Blue-Green Infrastructure (BGI) has to be one of them. BGI is the integration of green and blue networks composed of natural elements such as parks, green roofs, urban forests and wetlands, and water bodies like lakes, rivers, and ponds (Ahmed, Meenar, & Alam, 2019; Brown & Mijic, 2019). This type of dual approach provides ecological services while playing a very critical role in climate adaptation and mitigation (Chien, Saito, & Fukushi, 2022). For instance, the green spaces purify air by filtering out pollutants, and the water bodies reduce stormwater runoff hence ensuring protection of urban places from floods (Ahmad & Hassan, 2022).

Thirdly, BGI could be aggressively promoting biodiversity in cities given that it constructs habitats for several species. This would ease a balance between nature and the huge concrete-dominated landscapes (Gupta & De, 2024). The classical approach towards grey infrastructure—relying on concrete and steel solutions—has proved to be inadequate in solving the complex issues brought about by rapid urbanization and variability within climates (Bansal & Haridasan, 2024). In comparison, BGI offers nature-based solutions that would address these issues while advancing social equity (Ahmed, Meenar, & Alam, 2019). Typical green and blue areas in cities are allocated with bias towards disadvantageous outcomes for the major sections, especially the marginalized groups (Dhyani, Majumdar, & Santhanam, 2021). By integrating BGI into the urban planning processes, such inequalities can be rectified so that the greater access to nature will benefit all users. Socio-economic inequality is large-scale throughout the country of India, but BGI can cultivate social cohesion by providing public spaces, provoking the involvement of the community (Chaurasia et al., 2024; Patel et al., 2024). In addition, BGI is in harmony with the set international goals and frameworks for sustainability by the United Nations Sustainable Development Goals (SDGs) (Ahmed, Meenar, & Alam, 2019). Indeed, these are particularly related to such issues like SDG 11-sustainable cities and communities and SDG 13-climate action (Puchol-Salort et al., 2021). Following BGI can be very instrumental for Indian cities in engaging both globally and locally in taking the lead on environmental matters (Chaurasia et al., 2024; Sehrawat & Shekhar, 2024; Rajan & Murali, 2022).

The research will discuss all the aspects of BGI, explore the problems it creates in its implementation in the Indian cities, critically review some case studies from the home and overseas contexts, and propose an entire framework for integrating BGI into the processes followed at the level of urban planning. Finally, this study aims to highlight the need for BGI as an aspect of sustainable development of Indian cities that must balance ecological soundness with social justice and economic practicality.

Need for BGI in India Cities

Indian cities are facing acute challenges due to unplanned growth in cities, which has greatly reduced green areas and water bodies by great margins (Chaurasia et al., 2024; Patel et al., 2024). The unplanned spree of rapid urbanization has resulted in severe environmental damage through increased air and water pollution, reduction of biodiversity, and making it even more susceptible to the impact of climate change (Ahmed, Meenar, & Alam, 2019). For example, the constructed area in Bengaluru increased by an astonishing 925% during the period from 1973 to 2013, while the green cover that once accounted for 68% of its landscape came down to 14% over the same time period. Similarly, the uncontrolled urban sprawl has resulted in the loss of 60% of its vegetation since 1977 in Mumbai, which is one of the worst impacts. The impacts of the uncontrolled urban sprawl have multifaceted impacts. First, the loss of green and blue spaces causes an imbalance to local ecosystems, alters the hydrological cycle, and increases the influence of the urban heat island (UHI) (Brown & Mijic, 2019). Demographic growth enhances population pressure and increases infrastructure demand. This growth in cities increases the density of cities, thus demanding new infrastructure. At the expense of natural environments, infrastructure development develops land use (Gupta & De, 2024). This causes disconnection to nature to not only reduce the quality of life within cities but also be a form of what would be termed public health issues, as can be seen from such instances of respiratory ailments and higher risks of heat stress from more extreme temperatures during incidents (Czyża & Kowalczyk, 2024; Battemarco et al., 2022).

With all these challenges, the inclusion of BGI into restoration of ecological balance and enrichment of urban livability becomes essential. In a BGI system, natural components include aspects such as trees, parks, and wetlands integrated with water bodies including lakes and rivers (Ergler, 2023; Chien, Saito, & Fukushi, 2022). This results in multifunctional landscapes that provide essential ecosystem services such as stormwater management, improvements in air quality, carbon sequestration, and provision of biodiversity habitats (Czyża & Kowalczyk, 2024; Siehr, Sun, & Aranda Nucamendi, 2022). Apart from this, BGI offers an alternative sustainable source which replaces traditional grey infrastructure toward resolving challenges within cities by nature-based solutions (Dhyani, Majumdar, & Santhanam, 2021). More than that, while the existing strategies applied in reality adapt to the challenges in cities majorly through concrete and steel, BGI utilizes natural processes to increase effectiveness and efficiency at a lower cost for managing cities (Norman, Surjan, & Booth, 2021). For example, green roofs could decrease the impact of an UHI while simultaneously making it feasible to preserve buildings warmer or cooler; permeable pavements could help elevate groundwater recharge as the stormwater discharge is regulated at source (Brown & Mijic, 2019).

Deeper still, the utilization of BGI achieves the requirement on how it would allow and make easy the accomplishment of social equity in urban areas. Such citywide access is generally not very uniform in cities and subtly discriminates the most marginalized groups (Ferreira, Dos Santos, & Campos, 2024). Incorporation of BGI into urban planning would redress such imbalances by providing every citizen with nature and its benefits (Sörensen, Persson, & Olsson, 2021). The alarming rate at which India is urbanizing makes the necessity to include BGI in its strategies for urban development glaring, as by 2050, one in every two Indians would be living in cities (Chaurasia et al., 2024). Restored and improved green and blue spaces must be at the top of the lists in city planning frameworks to bring about healthier, thriving communities, not only resilient to climate change, but thriving Indian cities (Chaurasia et al., 2024). From this point of view, Indian cities can address the issues of environmental degradation due to high-speed urbanization and BGI. Cities can build ecological integrity and quality of life for its citizens by developing natural systems that counterbalance urban problems (Chien, Saito, & Fukushi, 2022). This revolutionized approach, as found in BGI, is a revolutionary method in terms of global goals toward sustainability and concomitant local needs (Nag & Rathore, 2025a; Nag & Rathore, 2025b; Nag, Ashish, & Rathore, 2025).

Objectives of Integrating BGI

The integration of BGI into urban development is essential for addressing the myriad challenges faced by Indian cities. This section outlines the primary objectives of incorporating BGI into urban planning processes, focusing on understanding its elements, examining implementation challenges, analyzing global best practices, and developing a comprehensive framework (refer Table 1).

Table 1. Overview of research objectives, associated keywords, and descriptions for Blue-Green Infrastructure (BGI) integration in urban planning

Objective	Keywords	Description
Understanding BGI Elements	BGI components, urban planning	Identify and define the various components of BGI relevant to urban planning, including green and blue elements such as parks, wetlands, and green roofs.
Challenges in Implementation	Implementation obstacles, governance	Examine the barriers faced by Indian cities in integrating BGI, including financial constraints, lack of technical expertise, and governance issues.
Global Best Practices	International case studies, lessons learned	Analyze successful BGI implementations from other countries and assess their applicability to Indian contexts.
Framework Development	Planning framework, guidelines	Create a comprehensive framework for integrating BGI into urban planning processes, including measurable indicators and community engagement strategies.

Understanding BGI Elements

The objective relates to which components of BGI are to be considered to be used in city planning. Hence all kinds of green and blue components have to be evaluated which are parks, green roofs, wetlands, rivers, and lakes (Chien, Saito, & Fukushi, 2022). The understanding of the said components is vital for the development of efficient landscapes in the urban settings making use of the natural systems in the delivery of ecosystem services (Harlis & Seo, 2024). This will help planners design multifunctional spaces that not only bring biodiversity and better air and water quality into the urban area but also enhance the resilience of cities at large (Czyża & Kowalczyk, 2024; Siehr, Sun, & Aranda Nucamendi, 2022).

Challenges in Implementation

The second intent is to identify the barriers Indian cities are facing in bringing BGI into their dominion. In the case of BGI, systemic barriers dominate, from a paucity of funds to a technological one, besides inter-ministerial coordination, that denies proper implementation (Chaurasia et al., 2024; Patel et al., 2024). Besides, governance issues often make planning more complicated and, therefore, the efforts are fragmented, and cannot tap in into the optimum advantage delivered from the benefits accrued by the BGI. It will identify such challenges so that targeted strategies can be drawn to overcome them for integrating BGI in the already inbuilt framework of cities in an effective manner (Thammadi et al., 2022).

Global Best Practices

The third is to study the success cases of BGI adoption in other countries and the appropriateness of their applications in Indian contexts. International case studies where BGI has been merged into an urban planning process offer valuable learning opportunities (Jagadisan & Sen, 2024) - for instance, Vancouver's Rain City Strategy or Singapore's wide network of park connectors. These cases can also assimilate best practices on design principles, stakeholder engagement and financing mechanisms that would apply in Indian cities which share similar complexities.

Framework Development

The ultimate aim is to develop a comprehensive set of guidelines for incorporating BGI into urban planning practice. More precisely, developing guidelines for decision-makers and urban planners on how to systematically apply parts of BGI to their projects. This will comprise quantifiable indicators: existing conditions of blue-green spaces, standards of the new developments, and strategies on community involvement (Harlis & Seo, 2024). This would enable Indian cities to have a more aligned agenda on urban development with the sustainability goals and attempts at climate resilience.

These objectives would address the components of BGI, resolve the implementation issues of the same, learn the best practices across the globe, and ensure a robust framework for integration. The objectives achieved will enhance the livability of the city and contribute to the fulfillment of sustainable development goals within Indian cities while negotiating the complexities brought about by rapid urbanization and environmental degradation.

METHODOLOGY

The study uses systematic reviews of literature and case study analysis to extract the information about current practices in BGI among various states in India and across the world. The above methodology is used to understand BGI components, its effectiveness, and challenges within the urban context all around. The first component of methodology is the systematic literature review-an approach following established guidelines, such as Preferred Reporting Items for Systematic Reviews and Meta-Analyses (PRISMA). It enables a structured analysis of previously published literature on BGI that may give a view of publications reporting on the implementation, benefits, and challenges across various urban contexts.

1. Search Strategy: A well-developed systematic search strategy was followed while using the phrase of string for the in-depth searches: "Blue-Green Infrastructure" OR "Green Infrastructure" OR "Nature-Based Solutions" AND "Urban Planning" AND "India." This phrase was typed to filter the relevant studies arising during the period from 2013 to 2022. The relevant articles for studying BGI in both the Indian and international settings were downloaded from online databases, such as Scopus, Web of Science, and Google Scholar.
2. Included studies were based on BGI component studies, which provide ecosystem services of BGI or case studies of the implementation of BGI. The review excluded all publications based on empirical research that could be strictly labeled as theoretical discussions. The review excluded grey literature, reports, policy documents, and unpublished studies.
3. Data Extraction: The relevant papers that were identified through inclusion and exclusion criteria were used for extracting all the significant indicators to be used in determining whether BGI is effective or not. These included aspects about environmental effects, social benefits, and economic feasibility.
4. Analysis: Then the literature gathered was analyzed so as to give trends in BGI research, gaps within the existing knowledge, and variance in implementation among different geographical regions.

Besides this systematic literature review, the study uses case study analysis to provide practical insights for application of BGI. Case studies are selected from both Indian cities and from other international examples, which have been successful in integrating BGI. The cases reflect different ways BGI could be used and the outcomes. Case studies shall then be examined using some of these criteria on stakeholder engagement, sustainability-related outcomes, and flexibility towards local conditions. All these insights can then be synthesized into the best practices integrated with the urban planning process of India.

This study has, therefore been based on a systematic literature review coupled with case study analysis so as to help play a part in ensuring that BGI plays its part in enhancing urban resilience and livability. Findings from this study considerably add up to findings in further improving integration of BGI in the urban development frameworks for Indian cities and provides much-needed advice in the establishment of sustainable urban environments.

BENEFITS OF BGI

This is because BGI presents a revolutionary way to planning cities, and in that, there is a myriad of advantages it gives to environmental sustainability as well as to community welfare. Within these aims, natural systems become essential to the urban landscape at all levels in challenges related to rapid urbanization and climate change (Jagadisan & Sen, 2024). These can broadly be summarized into three primary categories: ecosystem services, climate adaptation, and social equity. BGI improves or enhances ecosystem services through air quality improvement, stormwater management, and biodiversity enhancement (Czyża & Kowalczyk, 2024; Ferreira, Dos Santos, & Campos, 2024). Such in-organic components work synergistically to create healthier urban environments (Sanchez & Govindarajulu, 2023). In addition, BGI provides essential climate adaptation solutions, including natural flood management and temperature regulation, which are increasingly vital in the face of climate variability (Ahmad & Hassan, 2022). In addition, BGI is a social equity concept because green spaces are made available to disadvantaged groups; therefore, it strengthens social cohesion and overall quality of life (Czyża & Kowalczyk, 2024). On this basis, BGI is an example of an integrated approach that fosters the perception of natural system inherent values within the urban ecosystems as a central strategy within sustainable urban development (Ergler, 2023; Liao, Deng, & Tan, 2017).

Ecosystem Services

BGI greatly enhances ecosystem services that are very vital in maintaining ecological integrity and human prosperity in urban areas (Chien, Saito, & Fukushi, 2022). All the benefits that come with incorporating natural systems into the urban environments tackle crucial environmental issues (refer Table 2).

Table 2. Ecosystem services delivered by Blue-Green Infrastructure (BGI) — descriptions, urban benefits, and key literature sources

Ecosystem Service	Description	Benefits	References
Air Quality Improvement	BGI contributes to better air quality by filtering pollutants and reducing particulate matter. Trees and vegetation absorb carbon dioxide and harmful gases.	Mitigates health risks associated with air pollution and enhances overall urban atmosphere.	Bansal & Haridasan, 2024; Czyża & Kowalczyk, 2024; Ferreira, Dos Santos, & Campos, 2024; Ahmad & Hassan, 2022; Battemarco et al., 2022; Chien, Saito, & Fukushi, 2022; Brown & Mijic, 2019; Drosou et al., 2019
Stormwater Management	BGI effectively manages stormwater through natural processes such as infiltration and evaporation, reducing the risk of flooding.	Maintains groundwater levels and mitigates impacts of heavy rainfall events, promoting sustainability.	
Biodiversity Promotion	By creating interconnected green and blue spaces, BGI supports diverse habitats that foster urban biodiversity.	Enhances ecological resilience and supports various species, enriching the urban ecosystem.	

Air Quality Improvement

One of the best advantages of BGI is air quality improvement. Filtering pollutants and particulate matter creates BGI as a significant agent for improving the urban air quality (Czyża & Kowalczyk, 2024). Additionally, the trees and plants consume carbon dioxide with all dangerous gases that result in healthier atmosphere. Natural filtration decreases health risks from air pollution at the same time and creates a more pleasing environment in urban areas (Battemarco et al., 2022).

Stormwater Management

BGI performs by managing rainwater using natural processes including infiltration and evaporation appropriately. It reduces the risk of floods and provides safe and stable levels of ground water, as well as reducing the effects of heavy downpours (Ahmad & Hassan, 2022; Battemarco et al., 2022). BGI, being incorporated in features such as green roofs and permeable pavements among others, minimizes surface runoff from downpours, hence enhancing water retention and quality (Czyża & Kowalczyk, 2024). With research already set, it has been seen that BGI can lead to the boosting of sustainability in the management of water resources within an urban location, thus increasing resilience against extreme weather conditions (Ferreira, Dos Santos, & Campos, 2024).

Biodiversity Promotion

Third, BGI focuses on biodiversity promotion. BGI disperses yet interconnects green and blue spaces to sustain diverse habitats that build up urban biodiversity. This will also help in the development of ecological resilience and the survival of different species, especially in high human population areas with fragmented natural habitats (Chien, Saito, & Fukushi, 2022). Flora and fauna diversity contribute to the richness of the urban ecosystem and enhance the health of the environment generally (Kaur & Gupta, 2022).

BGI's ecosystem services are diverse and critical to the mission of building a resilient urban space. It responds to two principal challenges: air quality improvement while managing stormwater and enhancing biodiversity, hence upgrading the quality of life of urban dwellers (Czyża & Kowalczyk, 2024).

Climate Adaptation

It addresses natural solutions for climate adaptation to effectively face the challenges imposed by climate change in urban areas. BGI achieves this by utilizing several elements such as wetlands, green roofs, and urban forests in enhancing resilience to extreme weather events and reducing the effects of climate variability (Kaur & Gupta, 2022) (refer Table 3).

Table 3. Climate adaptation benefits of Blue-Green Infrastructure (BGI) — mechanisms, key impacts, and supporting references

Climate Adaptation Benefit	Description	Key Impacts	References
Flood Management	BGI elements like wetlands and green roofs absorb excess rainwater, reducing surface runoff and minimizing flood risks. This is particularly important in urban areas prone to flooding due to inadequate drainage systems.	Decreases the likelihood of urban flooding, protecting infrastructure and communities from damage.	Gupta et al., 2024; Ahmad & Hassan, 2022; Battemarco et al., 2022; Kozak et al., 2020; Drosou et al., 2019
Temperature Regulation	BGI helps mitigate the UHI effect by providing shade and cooling through evapotranspiration, thus regulating urban temperatures.	Lowers ambient temperatures in densely populated areas, reducing energy demand for cooling and improving comfort levels.	

Flood Management

The most significant advantage of BGI is that it can ideally handle flood-related problems (Ahmad & Hassan, 2022). Wetlands, green roofs, among others in BGI can absorb extra rainwater during heavier rainfall events significantly reducing surface runoff. This means that the capacity of such BGI elements is pretty important where flooding mainly comes from poor storm drainage systems present mostly in urban areas. By allowing rainwater infiltration into the ground rather than overloading stormwater systems, BGI reduces the risk of floods while keeping the groundwater levels on course (Ahmad & Hassan, 2022; Battemarco et al., 2022).

Temperature Regulation

A critical function of BGI is temperature control (Brown & Mijic, 2019). The UHIs are problematic for urban areas characterized by dense populations because the built environments tend to retain heat. BGI, however is against this impact, in terms of shading and intensification of transpiration processes, which have natural cooling effects, thus allowing for lowered ambient temperatures (Gupta et al., 2024). This way, the cities can eliminate much of their demand for cooling energy, just what they need at a time when the world is warming fast.

Among the adaptation measures BGI has provided is fitting to enhance urban resilience: these include Flood management and temperature regulation (Dhanya, Ramananda, & Dhyani, 2022; Ahmad & Hassan, 2022; Dhyani et al., 2022). BGI, in its ways of addressing urgent environmental challenges, boosts sustainable urban development. Therefore, with cities struggling with the effect of climate change, BGI will play an instrumental role in building adaptive and resilient communities.

Social Equity

BGI is indeed a true facilitator of social equity in their urban environments, providing easier access to green spaces and better public health achievements. In line with the needs of the poor, the thrusts and initiatives spearheaded by BGI further enhance social cohesion and contribute to community wellbeing at large (Kaur & Gupta, 2022) (refer Table 4).

Table 4. Social equity benefits of Blue-Green Infrastructure (BGI) — descriptions, community impacts, and relevant literature

Social Equity Benefit	Description	Key Impacts	References
Improved Access to Green Spaces	BGI initiatives often prioritize enhancing access to parks and recreational areas, particularly for marginalized communities. This fosters social cohesion by providing spaces for community interaction and engagement.	Strengthens community ties and promotes inclusivity, ensuring that all residents benefit from green spaces.	Keita & Kourouma, 2024; Kaur & Gupta, 2022; Kozak et al., 2020; Drosou et al., 2019
Health Benefits	Access to green spaces has been linked to improved physical and mental health outcomes. BGI encourages active lifestyles by offering facilities for walking, jogging, and outdoor activities.	Enhances overall well-being, reduces stress levels, and encourages healthier lifestyles among urban residents.	

Improved Access to Green Spaces

The greatest advantage that BGI provides is an increase in access to green spaces. Many BGI projects are aimed to upgrade or build community gardens and parks while recreating recreational spaces in communities that have lacked them in the past (Kaur & Gupta, 2022). In this way, social cohesion is promoted when BGI targets renovation for the poorest communities and provides community-use sites where people can come together and interact. Quality of life rises by having such green spaces, but that empowerment of people who live there will also be achieved through a sense of ownership, as well as their engaging the local community with its environment (Czyża & Kowalczyk, 2024).

Health Benefits

Health is another key aspect of BGI. Many studies have been conducted to note the fact that spending time in green spaces is positively linked to improved physical and mental health (Drosou et al., 2019). BGI promotes healthy lifestyles, such as jogging and cycling as well as other outdoor exercises; these allow general fitness and mental wellness. This is in a way that reduces stress levels and overall qualifies the quality of city life (Czyża & Kowalczyk, 2024). On the other hand, obesity and chronic diseases are negatively correlated with higher levels of greenery, and thus BGI needs to be appropriately integrated with the design of urban areas (Ergler, 2023).

BGI can serve as a driving force in bringing social equity to cities. It builds better public health outcomes and greening access, solves environmental issues, and creates an inclusive neighborhood where all the residents can thrive amidst organic and intentional social change (Kaur & Gupta, 2022). As cities grow, there will be an increasing need to work through BGI to center social equity in creating sustainable and resilient urban landscapes (Li et al., 2022).

The inclusion of BGI in the urban plan introduces numerous advantages toward creating sustainable and resilient cities. Through the incorporation of ecosystem services, BGI enhances the quality of air and effectively manages storm water while maintaining diversity contributes to a healthier environment for the city (Bansal & Haridasan, 2024). BGI offers immense climate adaptation measures that facilitate the adaptation of the city to flood events as well as moderate temperatures associated with climate change (Ahmad & Hassan, 2022). In the process, BGI will increase social equity by expanding entrance and experience to the green spaces; this will support the overall health outcomes of the groups being viewed as disadvantaged. With rapid urban development and urbanization going on, the development of BGI will help in integrating spaces for ecological integrity and wellbeing of all residing in such areas developed (Chien, Saito, & Fukushi, 2022). In the long term, BGI adoption symbolises an approach that leads towards holistic development in cities on track with sustainability objectives and builds a quality of life in cities (Czyża & Kowalczyk, 2024).

CHALLENGES IN IMPLEMENTING BGI

Many problems may come up in the way of BGI operationalization, which mostly tends to negate any potential efficacy in urban settings. Basing our understanding on BGI offers many advantages, which include greater social justice, better public health, and climate change resilience, but to fully grasp the benefits, several impediments must be cleared away first. These impediments get classified under three broad categories: awareness of the public, budget constraints, and gap in policy (Kaur & Gupta, 2022). Each of those and the collective interaction they create form the backdrop for the whole BGI implementation and integration into urban planning processes (Willems, Kuitert, & Van Buuren, 2023). To attain sustainable urban environments employing BGI, an understanding and willingness to remove those hurdles pose a prerequisite (Keita & Kourouma, 2024).

Policy Gaps

Having huge policy gaps is one of the primary hindrances to implementation of BGI. And many urban areas lack the laws and statutory instruments that would facilitate the incorporation of BGI into the planning processes (Keita & Kourouma,

2024). The absence does allow fragmentation to occur and BGI is treated as an afterthought rather than a main tenet of urban development (refer Table 5).

Table 5. Policy gaps hindering Blue-Green Infrastructure (BGI) implementation — descriptions and urban impact implications

Policy Gap	Description	Impacts
Lack of Supportive Policies	Many cities do not have comprehensive policies that promote the integration of BGI into urban planning and development.	This results in disjointed efforts and missed opportunities for effective BGI implementation.
Inadequate Regulatory Frameworks	Existing regulations often fail to address the long-term management and maintenance of BGI, leading to neglect after initial implementation.	Without proper guidelines, BGI projects may deteriorate over time, undermining their intended benefits.
Fragmented Planning Processes	BGI is frequently overlooked in favor of traditional infrastructure projects, which are prioritized for their immediate financial returns.	This can lead to underinvestment in essential sustainable infrastructure that enhances urban resilience.

A lack of proper policies denies cities the opportunity to leverage the long-term benefits BGI can confer, such as higher climate-change resilience and a better quality of life for residents (Czyża & Kowalczyk, 2024). Effective policy pathways, thus, must be developed to ensure that BGI is considered critical infrastructure and receives an equal amount of financial, professional, and political support as do conventional infrastructure projects. On the other hand, BGI might fall into neglect after initial implementation, as existing policies often gloss over long-term maintenance issues associated with it (Keita & Kourouma, 2024). Several studies emphasize the pressing need to develop a more robust and well-defined policy framework that can be translated into statutory plans to successfully generate, protect, and enhance BGI. Then, the financial aspects related to BGI would be an issue; funding for these undertakings is almost always neglected as the benefits of BGI are externalized: those who provide the financing for its implementation may not benefit directly. This issue again widens the gap in policies, as decision-makers might not fully understand how to embed BGI into more comprehensive urban planning schemes (Keita & Kourouma, 2024). Certain policy gaps must thus be addressed if there are to be effective implementations of BGI. By developing a comprehensive and enabling regulatory framework, cities can more effectively incorporate BGI into planning procedures and further create resilient and sustainable urban environments (Kozak et al., 2020).

Financial Constraints

Financial limitations indeed present the biggest barriers for BGI's success. Since cities find it difficult to allocate funds to projects that may not bring an immediate profit, the development of larger scale BGI projects is often affected (Keita & Kourouma, 2024). The benefits of BGI are sometimes externalized and otherwise not easy to put a price on, in comparison to conventional infra investments that may bring immediate cash flow (refer Table 6).

Table 6. Financial constraints affecting Blue-Green Infrastructure (BGI) implementation — descriptions and associated urban impacts

Financial Constraint	Description	Impacts
Limited Funding	Many cities face budgetary constraints that limit their ability to invest in large-scale BGI projects. This often leads to prioritization of short-term financial gains over long-term sustainability.	Underinvestment in essential projects can exacerbate environmental challenges and reduce urban resilience.
Externalized Benefits	The advantages of BGI, such as improved air quality and enhanced biodiversity, are often not reflected in immediate financial returns, making it challenging to justify investments.	Decision-makers may overlook BGI in favor of more traditional infrastructure that promises quicker returns.
Lack of Innovative Financing Mechanisms	Cities may lack access to innovative financing options, such as public-private partnerships or green bonds, which could help fund BGI initiatives.	This limits the scope for developing comprehensive BGI projects that require significant upfront investment.

Since there are no quick incentives for money to flow, a city may sometimes choose short-term profits over long-term investments in environmentally friendly infrastructure. This, in turn, results in less funding or no funding for crucial projects instead of solving environmental-related problems such as flooding, heat management, and air quality better (Ahmad & Hassan, 2022). Another interesting fact standard BGI benefits are externalized: increased habitat diversity, flood mitigation, climate change adaptation, and human health gains (Ahmad & Hassan, 2022). Since these benefits are not always monetizable in common terms, makes it difficult for the decision-makers to put forth a policy for funding BGI projects as opposed to traditional infrastructure projects. In addition, BGI's financial elements pose challenges; support for such efforts is typically overlooked because the advantages of BGI are externalized, that is, those paying for its delivery may not indirectly benefit (Bansal & Haridasan, 2024). Since policymakers may not be fully aware of how to incorporate BGI into larger urban planning initiatives, this further exacerbates policy gaps. To be effectively implemented, some policy gaps have to be addressed (Keita & Kourouma, 2024). Cities can better integrate BGI into their planning processes and develop more sustainable and resilient cities through the development of comprehensive and supportive regulatory regimes.

Public Awareness

Lastly, the success of BGI depends, to a large extent, on public knowledge. Adoption of BGI can be greatly undermined by lack of awareness among stakeholders, from common individuals to policymakers, regarding its advantages (Kozak et al., 2020). Stakeholders will be resistant to incorporating these measures into planning procedures if they are unaware of how BGI can promote social justice, public health, and city resilience (refer Table 7).

Table 7. Public awareness challenges in Advancing Blue-Green Infrastructure (BGI) — key descriptions and societal impacts

Public Awareness Challenge	Description	Impacts
Insufficient Understanding	Many stakeholders lack comprehensive knowledge about BGI and its benefits, leading to misconceptions and skepticism.	This can result in resistance to adopting BGI solutions and hinder collaborative efforts.
Limited Stakeholder Engagement	Without effective communication strategies, community members may feel excluded from decision-making processes related to BGI initiatives.	This can diminish public support for projects and reduce the likelihood of successful implementation.
Lack of Educational Resources	There is often a shortage of accessible educational materials that explain the advantages and applications of BGI in urban settings.	This limits opportunities for informed discussions and reduces community involvement in planning.

Stakeholders are prone to be cynical and make assumptions because of inadequate knowledge about BGI. For example, policymakers may be neglecting the long-term advantage that BGI can offer in comparison to traditional infrastructure solutions promising instant results. Locals can also not understand the benefits of green spaces to their health and wellbeing and end up being uninterested in or resistant to suggested projects. Moreover, these problems have the chance to be escalated by a lack of stakeholder engagement. BGI initiatives can become viewed as mandates from above and not community participatory endeavors unless effective communication tactics are used to engage members of the community within the decision-making process (Kozak et al., 2020). Public support for projects declines as a result of this perception that also reduces the likelihood they will be executed effectively. Public awareness is therefore also hampered by the unavailability of easily accessible educational materials (Saha et al., 2024). It is probable that it is not common in most communities to be exposed to information that explains the uses and benefits of BGI in urban environments. The gap lowers participation levels of the public in planning efforts and reduces chances of intelligent debate (Kozak et al., 2020). Public awareness problems should be resolved in order to ensure effective use of BGI. Stakeholder education enhances the awareness of BGI benefits leading to support and collective action to energize and sustain cities (Saha et al., 2024). Engaging people in discussions on BGI increases community comprehension, enabling citizens to lobby for policies that benefit their area and improve their overall quality of life (Czyża & Kowalczyk, 2024). Implementing BGI creates issues requiring solutions from a collective viewpoint.

Budgetary constraints generally curb the availability of funds for sundry large projects on the outset, while other limitations of policy may cause piecemeal efforts to be taken. To make matters worse, the low profile of BGI may have the stakeholders' capacity to initiate and participate in such initiatives limited (Krishnan & Priya, 2022). Therefore, it is imperative that integrated policies be developed, innovative funding sources be identified, and publicity and educational campaigns be intensified to overcome these barriers (Ergler, 2023). Resolution of these issues will best allow urban cities to use BGI for the construction of sustainable, equitable, and resilient urban landscapes from the common good of all citizens.

CASE STUDIES

Bansal & Haridasan (2024) articulated various opportunities and events for enhancing urban resilience and sustainability through BGI. The following section imparts a few case studies from Indian cities along with international examples to demonstrate various forms of BGI integration. The case studies exemplify best practices that implemented BGI to address some of the urban problems considered: social justice, air pollution, and flooding (Dey et al., 2024; Ahmad & Hassan, 2022). Groundbreaking valuation factors for each of the case studies are adaptability to local context, sustainability advantage, and involvement of stakeholders.

Case Study 1: Bengaluru, India

Numerous projects involving urban park development and lake restoration in Bengaluru have shown the highest potential for climate resilience in the area (refer Table 8). One such project is the Lake Improvement Works Project, under which the nearby lakes are to be desilted and improved in order to enhance their water-holding capacity and encourage groundwater recharge. Nine acres of lake area have been reclaimed due to the successful removal of over 120,000 tons of silt from lakes like Ittgalpura and Sadenahalli. The increased water storage capacity, extending now up to 17 acres, aids in local agriculture and biodiversity and provides alternative livelihoods to local fishers and farmers. Another case in point is the rehabilitation of Puttenahalli Lake, which has been transformed into a bird conservation sanctuary with active community support. To find solutions to the lake problem, locals founded the Yelahanka Puttenahalli Lake and Bird Conservation Trust, which has been playing an active role (Varshini et al., 2022; Dhyani et al., 2022). Since then, bird diversity and abundance have steadily increased, especially after adding a passive water-conditioning plant to treat wastewater effectively before being discharged into the lake (Suzuki, Cervero, & Iuchi, 2013). This is a perfect example of BGI outcomes, where community pride and ownership are fostered through grassroots participation.

Table 8. Key Blue-Green Infrastructure (BGI) initiatives in Bengaluru

Initiative	Description	Outcomes
Lake Improvement Works Project	Focuses on desilting Ittgalpura and Sadenahalli lakes to enhance water-holding capacity and facilitate groundwater recharge.	Recovery of nine acres of lake land; increased water capacity supports agriculture and local livelihoods.
Puttenahalli Lake Restoration	Transformed into a bird conservation reserve with community involvement; includes a passive water conditioning plant to treat wastewater.	Improved water quality; increased bird species; serves as a model for community-led conservation efforts.

Apart from being a habitat for lake ecology, it has some benefits-the improvement of agriculture and fishery yields-thus enhancing the standard of living of persons residing near the lakes. The restoration of Puttenahalli Lake is yet another example of how environmental stewardship depends on community participation and how synergies create lasting impacts.

The Bengaluru case studies illustrate how specialized BGI programs can mediate the optimization of community engagement and urban resilience. The Lake Improvement Works Project and restoration at Puttenahalli Lake are worthy exemplars of the kind of community-based engagement that fosters sustainable environmental impacts. Drawing lessons from such examples within the framework of sustainability impacts, stakeholder interactions, and adaptability to local context would be germane for the mainstreaming of BGI across Indian urban planning (Dey et al., 2024). These illustrations demonstrate how important community cooperation and responsive management are to creating sustainable city spaces that benefit both humans and nature (Everett, Adekola, & Lamond, 2021).

Case Study 2: Mumbai, India

In response to a severe potable water shortage and localized flooding issues in Mumbai, a number of initiatives pertaining to rainwater harvesting (RWH) and green infrastructure have been undertaken. These schemes are geared toward making cities resilient through effective stormwater management and promotion of sustainable water use, but they need to be incorporated much more with city planning if they are to create the impact they envision (Dey et al., 2024). Following this trend, the BMC is going all out on rainwater harvesting as an initiative to reduce water scarcity compounded by urbanization. The Mumbai study cases shed light on the prospective benefits and bottlenecks faced in the implementation of BGI interventions such as rainwater harvesting and green roofs. The BMC has initiated various measures, including mandating RWH systems in new buildings and digging percolation pits in public gardens. Approximately 1,500 percolation pits have been dug so far in around 300 gardens, and another 700 pits are on the anvil. In the monsoon season, the systems are aimed at recharging groundwater levels, minimizing the effects of flooding and improving water availability overall. Rainwater harvesting remains not very common in residential complexes, even with these efforts. Less than 3,000 of Mumbai's 20,000 housing societies constructed RWH systems between 2007 and 2021. The perception that reservoirs offer a sufficient amount of water and lack of economic incentives for residents to install RWH systems are two reasons why it is not more commonly adopted (refer Table 9).

Table 9. Blue-Green Infrastructure (BGI) interventions in Mumbai – rainwater harvesting initiatives and their implementation challenges

Rainwater Harvesting Initiative	Description	Challenges
Mandatory Implementation	RWH systems are required for new constructions to recharge groundwater and reduce stormwater runoff.	Many existing housing societies lack motivation to install RWH due to perceived easy access to water.
Percolation Pits in Gardens	The BMC has constructed percolation pits in public gardens to enhance groundwater recharge during rainfall.	Maintenance issues and limited awareness among residents hinder effectiveness.

Along with rainwater harvesting, Mumbai is looking to unlock the possibilities of green roofs as an eco-friendly urban intervention. Green roofs are capable of mitigating stormwater runoff, the UHI effect, and fostering biodiversity (Gupta et al., 2024). As with RWH efforts, however, the roll-out of green roofs is challenged by factors such as prohibitive up-front expenses and a paucity of policies endorsing widespread application (refer Tables 10 and 11).

Table 10. Green roof initiatives in Mumbai – description and constraints in scaling nature-based urban solutions

Green Roof Initiative	Description	Challenges
Urban Heat Mitigation	Green roofs provide insulation and cooling effects while absorbing rainwater to reduce runoff.	High installation costs deter many property owners from adopting green roofs.
Biodiversity Promotion	They create habitats for various species, enhancing urban biodiversity.	Limited awareness of ecological benefits among stakeholders hampers broader acceptance.

Table 11. Summary of key Blue-Green Infrastructure strategies in Mumbai – objectives and realized outcomes

Initiative	Description	Outcomes
Rainwater Harvesting Initiatives	Promotes the installation of RWH systems in new buildings and public spaces; includes construction of percolation pits for groundwater recharge.	Aims to improve groundwater levels; however, uptake remains low due to perceived water abundance.
Green Roof Projects	Encourages the installation of green roofs on buildings to manage stormwater and reduce heat; focuses on integrating nature into urban environments.	Potential benefits for biodiversity and climate resilience are not fully realized due to limited adoption.

Although the focus of such projects is to make the urban areas more sustainable and resilient, factors like insufficient public awareness, budget limitations, and a lack of support from legislatures tend to come in the way of their effectiveness (Parkhi & Chandra, 2024). Best practices to guide future BGI projects in Mumbai and other comparable metropolitan cities can be revealed by assessing the projects on their impacts towards sustainability, stakeholder interactions, and local context flexibility (Misra & Hussain, 2022). The discoveries reaffirm how critical thoughtful planning, stakeholder participation, and adaptive management are in enabling cities to attain a more sustainable future amidst the adversities of climate change.

Case Study 3: Queen Elizabeth Olympic Park, London, UK

Among the most well-done examples of BGI deployment and the most used is Queen Elizabeth Olympic Park. Initially constructed to host the 2012 Summer Olympic and Paralympic Games, the park evolved to become a multifunctional zone that balances economic and social benefits with environmental implications (Siehr, Sun, & Aranda Nucamendi, 2022). Through solutions to specific planning, environmental, and socioeconomic concerns, professionals across disciplines worked together on a multidisciplinary front to achieve the transformation (refer Table 12).

Multifunctional Design and Community Impact

To create a livable urban space, the park's design has been integrated with green areas, streams, and recreational facilities in terms of sustainability and resilience (Roy, 2022). As the project had a budget of approximately £9 billion, it was able to carry out extensive construction that not only met the objectives of the Olympics but also laid the foundation for the use of the area in the years to come. Some of the interesting aspects of the park are:

- **Restoration of the River Lea**: The river was revitalized as a central design element, enhancing biodiversity and providing natural flood management.
- **Ecological Amenities**: The park includes diverse habitats and green spaces that promote wildlife and offer recreational opportunities for residents and visitors.
- **Community Facilities**: The park houses various sports venues, including the London Stadium, which now hosts major sporting events and concerts, ensuring continued community engagement.

Table 12. Key initiatives and outcomes of Queen Elizabeth Olympic Park – A model of multifunctional BGI design

Initiative	Description	Outcomes
River Lea Restoration	Revitalization of the River Lea as a key design component to enhance biodiversity and manage flood risks.	Improved water quality; increased habitat for wildlife; enhanced flood management capabilities.
Multifunctional Park Design	Integration of green spaces, recreational facilities, and community amenities to create a vibrant urban environment.	Attracted over **34 million visitors** in the first seven years post-reopening; created 110,000 jobs in East London.
Sustainable Infrastructure	Development of sustainable buildings and public transport systems to support ongoing community use beyond the Olympic Games.	Enhanced accessibility; promoted eco-friendly transportation options; fostered a sense of community ownership.

Queen Elizabeth Olympic Park is a wonderful proof of what can be achieved by focused BGI projects in fostering community involvement and making cities more resilient. Besides being a legacy of the 2012 Olympics, the park is also a model for urban cities of the future globally because of its multi-purpose planning and commitment to sustainability. The study reveals that it is indeed possible with BGI to design successful urban landscapes for humans and the environment surrounding the integration of ecological benefits and socioeconomic benefits. As cities continue to face the impacts of urbanization and climate change, they will equally be looking at the experience of Queen Elizabeth Olympic Park for the best practices in BGI implementation within their urban planning strategies.

Case Study 4: Copenhagen, Denmark

As part of its comprehensive Cloudburst Plan, Copenhagen has integrated blue and green measures to retain flooding of the city (Liu, Fryd, & Zhang, 2019). This novel plan came to light after the catastrophic flooding of July 2011, which caused huge destruction and highlighted the need to build greater climate resilience. The Cloudburst Management Plan covers interventions such as floodable parks, retention alleys, and tunnel systems, all intended to manage rainwater efficiently during times of heavy rainfall (Liu, Fryd, & Zhang, 2019) (refer Table 13).

Key Features of the Cloudburst Management Plan

The key ambition with the Cloudburst Management Plan has been to combine nature-based solutions with traditional infrastructure to make Copenhagen a resilient city. Some of the outstanding features include:

- Floodable Parks: Parks are planned to store temporary rainwater overflows, decreasing surface runoff and easing drainage system pressure. For instance, the old Enghavepark has been rebuilt to work as a water reservoir with a capacity to store 22,600 cubic meters of water during a storm.
- Retention Alleys: These are streets specifically designed to have permeable surfaces that enable rainwater to be absorbed into the ground instead of clogging the drainage system. This reduces not only flooding but also increases urban green space.
- Underground Tunnel System: A new system of tunnels, about 18 kilometers long, has been built to carry stormwater from the different sections of the city to the harbor. The network plays a double role: it stores rainwater during storms and releases it slowly into the harbor in periods of suitable conditions so that water quality can be kept at a high level for swimming recreation.

Table 13. Key initiatives and outcomes of Copenhagen's cloudburst management plan

Initiative	Description	Outcomes
Floodable Parks	Parks designed to temporarily hold excess rainwater, reducing surface runoff and alleviating pressure on drainage systems.	Enhanced green spaces that double as flood management tools; improved biodiversity and recreation opportunities.
Retention Alleys	Streets with permeable surfaces that facilitate rainwater infiltration, minimizing flooding risks.	Reduced stormwater runoff; increased urban greenery; improved aesthetics and community engagement.
Underground Tunnel Network	A network of tunnels that transports stormwater from various areas to the harbor while serving as a storage reservoir during heavy rains.	Improved stormwater management; maintained water quality in recreational areas; reduced urban flooding incidents.

One example of how the concentrated BGI projects can strengthen urban resilience and mitigate climate change issues is Copenhagen's Cloudburst Management Plan. This city reduces flooding risk and improves the lives of its people by combining green options with traditional infrastructural elements for lively public spaces. Coordinating city administrations, civil engineers, utility corporations, and citizens becomes imperative in building climate-resilient cities, featured suitably in the successful implementation of this project as a model for other cities facing similar problems. The innovative solution provided by Copenhagen has rich lessons to teach about the application of nature-based solutions for sustainable urban development while cities all over the world are grappling with increasing rainfall and flood calamities.

Evaluation Criteria

Certain evaluation criteria had to be put in place to decide whether or not a BGI project working in the urban setting has been successful. These criteria focus on certain main points that decide whether a BGI project can continue (refer Table 14).

Table 14. Evaluation criteria for Blue-Green Infrastructure (BGI) projects

Evaluation Criterion	Description
Stakeholder Engagement	This standard looked at the level of community involvement and cooperate with various stakeholders during the planning and implementation stages. Good stakeholder engagement is vital to ensure that BGI activities meet the needs of local communities and are supported by the members of the public. The presence of higher participation levels can improve project outcomes since stakeholders provide crucial information and induce a sense of ownership of these activities.
Sustainability Outcomes	The ecological restoration dimension looks at what environmental benefits have arisen from the implementation of BGIs such as improvements in water quality, enhancements in biodiversity, and increased climate resilience. The evaluation of sustainability outcomes allows assessing whether BGI projects are addressing issues faced by urban environments and in promoting long-term ecological integrity.
Adaptability to Local Contexts	The adaptability dimension assesses to what level BGI solutions are adapted to the specific needs and conditions of the urban setting where they are implemented. Adaptability is important to ensure BGI projects remain relevant and effective in different contexts taking into account the local geography, climate, and community dynamics. Projects that possess adaptability will more likely succeed and make long-term beneficial impacts on their environment.

With these evaluation criteria, we come to recognize who the best practitioners are, gain more knowledge about BGI projects in the urban area, and lead those processes into the future of our urban planning. Structured in this manner, the procedure allows us to understand how BGI may be used in our cities to boost sustainability, resilience, and community health in the urban environment.

Case Studies' Evaluation Criteria Insights

Evaluation of BGI interventions in a set of case studies will provide rich understanding of their strengths and weaknesses. An overview of the key findings of each case study's assessment—Bengaluru, Mumbai, Queen Elizabeth Olympic Park in London, and Copenhagen—is provided in Table 15.

Table 15. Insights from Blue-Green Infrastructure (BGI) projects in selected global cities

Evaluation Criterion	Bengaluru Insights	Mumbai Insights	Queen Elizabeth Olympic Park Insights	Copenhagen Insights
Stakeholder Engagement	High community involvement in the restoration of Puttenahalli Lake through local trusts led to successful outcomes.	Limited community participation in rainwater harvesting initiatives; residents often lack motivation.	Multidisciplinary collaboration among planners, environmentalists, and community members ensured diverse input and support.	Strong engagement with citizens and stakeholders in developing the Cloudburst Management Plan increased acceptance and support for BGI solutions.
Sustainability Outcomes	Significant improvements in water quality and biodiversity at restored lakes; enhanced agricultural productivity.	Rainwater harvesting systems show potential for improving groundwater levels but have low adoption rates.	The park's design promotes biodiversity, enhances urban green spaces, and supports recreational activities, attracting millions of visitors.	Floodable parks and retention alleys have effectively reduced urban flooding risks while enhancing green spaces and biodiversity.
Adaptability to Local Contexts	Projects are tailored to address specific water management issues in Bengaluru, enhancing community access to green spaces.	RWH solutions need customization based on local geological conditions; a one-size-fits-all approach is ineffective.	The park's multifunctional design integrates ecological needs with socio-economic benefits, making it adaptable to various community needs.	The Cloudburst Management Plan incorporates localized solutions like floodable parks that cater to Copenhagen's unique climate challenges.

Summary of Insights

Stakeholder Engagement:

- Effective community involvement is crucial for the success of BGI projects, as seen in Bengaluru's Puttenahalli Lake restoration.
- In Mumbai, low participation in rainwater harvesting highlights the need for greater public awareness and incentives.

Sustainability Outcomes:

- Successful BGI initiatives lead to improved environmental conditions, such as enhanced biodiversity and water quality.
- While projects like those in Copenhagen demonstrate effective stormwater management, Mumbai's rainwater harvesting efforts require increased adoption to realize similar benefits.

Adaptability to Local Contexts:

- Tailoring BGI solutions to local conditions is essential for effectiveness; generic approaches may not yield desired results.
- The diverse designs of projects like the Queen Elizabeth Olympic Park showcase how adaptability can enhance both ecological and social outcomes.

The lessons learned from the BGI assessments of these case examples stress the engaging stakeholder participation, the sustainability impacts, and the contextual flexibility as crucial attributes for urban resilience planning. Cities can devise BGI plans that are far more practical and that tackle environmental concerns alongside sustainable development strategies and enhancing childcare.

CONCLUSION

For Indian cities facing a gamut of environmental problems beset by rapid urbanization, BGI integration into city planning is of the utmost importance. City dwellers are increasing, thus raising the demand for sustainable means that promote resilience, health, and fair development. These are some of the features that make BGI an ecosystem-based model of urban planning aimed at assisting urban areas in enhancing biodiversity, managing stormwater, and developing green spaces to support the communities therein.

Enhancing Uran Resilience

Integrating BGI would increase the resilience of cities against flooding, heatwaves, and air pollution induced by climate change. For instance, rainwater-harvesting infrastructure and floodable parks can lessen the consequences of extreme weather events, thereby relieving pressure on conventional drainage systems. Climate adaptation can thus be enhanced for better protection of cities and their inhabitants through multifunctional green spaces that balance ecological and social functions.

Improving Public Health

BGI plays other roles in improving public health. Diminished air pollution, reduced UHI effects, and opportunities for physical activities are all features of urban green spaces. These parks and natural spaces improve mental health and quality of life; they continue social interactions and community building. Cities can therefore enhance the living standards of their citizens with BGI-based urban planning.

Fostering Sustainable Development

Furthermore, through encouraging resource efficiency and a minimization of environmental harm, BGI supports the principles of sustainable development. Apart from aiding biodiversity locally, landscaping with native plants minimizes water use and maintenance costs. BGI integration will allow urban areas to create more sustainable systems that meet service delivery needs and provide aesthetic enhancements.

Collaborative Efforts for Success

For BGI to work, it will have to combine governmental and local community stakeholders with business stakeholders. Inclusion of such diverse groups makes it possible to customize BGI projects to local circumstances and needs for optimized outcomes. For instance, in the case of Bengaluru's Puttenahalli Lake rejuvenation, local involvement has shown that ownership at the community level fosters a sense of responsibility among locals, leading to a positive environmental impact. Furthermore, public-private partnerships can inject finance and technical know-how into BGI projects and ensure that the projects remain both feasible and sustainable. By enabling cross-sector partnerships, cities harness the benefits of BGI through an array of resources and competences.

Learning From Global Examples

The urban case studies of Bengaluru, Mumbai, Queen Elizabeth Olympic Park, and Copenhagen's Cloudburst Management Plan have shown how BGI can meaningfully offset climate change while providing a much-needed platform for social justice. These case studies have highlighted the importance of innovations in design and public engagement in building urban resilience.

Indian cities can adapt these international best practices in BGI to formulate improved plans for implementation that suit their peculiar conditions. This flexibility becomes necessary in order to better respond to peculiar challenges triggered by fast urbanization across diverse geographic and socioeconomic conditions.

A Path Forward

Simply put, with increasing environmental constraints, Indian cities, to remain sustainable, must put BGI at the top of the city planning agenda; no compromises. Cities assimilating BGI into their core infrastructure can thus carve out a niche for themselves in influencing inclusive development that builds resilience and enriches the quality of life of all persons. Cultivating viable communities that blend respectfully with their natural environments will require regional-level ecological considerations across urban designs, with India fast emerging as among the most urbanizing nations. The ongoing concerted efforts and commitment toward sustainable development will soon turn Indian cities into world models for sustainability and resilience.

RECOMMENDATIONS FOR INTEGRATING BGI IN INDIAN CITIES

To make BGI effective in urban planning, a few options must be on the minds of Indian towns such as making functional policies in favor of BGI a priority-the real basis for infrastructure to support sustainable urbanization. The said policies must contain the provisions for planning, execution, and maintenance of BGI projects so that such projects can be embedded into larger urban development plans. Policymakers must also deal with environmental and social justice to ensure equitable access to blue and green spaces for all communities, especially those who are disadvantaged. Various researches have proven the positive effects of exposure to blue and green spaces on physical as well as mental health, therefore policy discourse should give prominence to such activities. Local education regarding the advantages of BGI, such as better mental health, increased cohesion at school and in the community, and sustainability, depends significantly on public information campaigns. Public information campaigns must educate the public regarding the advantages of green and blue spaces in stormwater management, reducing urban heat, and enhancing biodiversity. Participation in a BGI activity can be motivated and possession created by attracting neighboring communities with the help of public events, social media, and workshops. The people living closer to green spaces have proven to be healthier based on the evidence, therefore raising awareness will gain public backing for the program. Effective practice of BGI programs needs to come from cooperative partnership. To apply human resources, skills, and native wisdom, state and government authorities must collaborate with NGOs and associations. Partnerships will help secure project funds, promote greater levels of community engagement, and lead to programs having greater relevance in addressing local requirements. Restoration by local volunteers in the Indian state of Karnataka, at the city's Puttenahalli Lake, demonstrates community efforts' strong effect on their natural surroundings. Cities can develop a more inclusive form of urban planning that benefits all by encouraging such collaborations. To achieve the full multi-purpose potential of BGI, cities must embrace new design approaches. This includes developing green roofs that lower building insulation and save on energy expenses, or designing flood-risk parks that function as recreational areas while controlling runoff. Copenhagen's Cloudburst Management Plan is a good example of ecologically grounded solutions coupled with urban infrastructure to promote resilience to climate change and bring social benefits. It is essential to inspire urban planners to investigate new ways to incorporate nature into the urban space. To find out how BGI activities contribute to well-being in society and urban resilience, there should be a monitoring and measuring system of their efficacy. Periodic monitoring can enable areas of improvement to be found and effective strategies as well. The cities will be able to change their processes and policies based on empirical data as a result of this evidence-based practice, ensuring ongoing improvement in the implementation of BGI.

Indian cities can transition to a greener future that is sustainable and harmonizes ecological conservation and urbanization by solving these challenges. Incorporating BGI improves environmental conditions as well as healthier communities and resilient urban environments that can endure the effects of urbanization and climate change.

REFERENCES

Ahmad, N., & Hassan, Q. (2022). Enhancing Blue-Green Infrastructures for Flood and Water Stress Management: A Case Study of Chennai. In *International Conference on Trends and Recent Advances in Civil Engineering* (pp. 97-117). Singapore: Springer Nature Singapore.

Ahmed, S., Meenar, M., & Alam, A. (2019). Designing a Blue-Green Infrastructure (BGI) network: Toward water-sensitive urban growth planning in Dhaka, Bangladesh. *Land (Basel)*, *8*(9), 138.

Bansal, V., & Haridasan, H. (2024). Blue–Green Infrastructure as a Stimulant of Growth: Case of Ahmedabad. In *Blue-Green Infrastructure for Sustainable Urban Settlements: Implications for Developing Countries Under Climate Change* (pp. 365–385). Springer Nature Switzerland.

Battemarco, B. P., Tardin-Coelho, R., Veról, A. P., de Sousa, M. M., da Fontoura, C. V. T., Figueiredo-Cunha, J., & Miguez, M. G. (2022). Water dynamics and blue-green infrastructure (BGI): Towards risk management and strategic spatial planning guidelines. *Journal of Cleaner Production*, *333*, 129993.

Brown, K., & Mijic, A. (2019). *Integrating green and blue spaces into our cities: Making it happen*. Grantham Institute.

Chaurasia, M., Patel, K., Singh, R., Bhadouria, R., & Rao, K. S. (2024). Integrating Blue-Green as Next-Generation Urban Infrastructure in Developing Countries. *Blue-Green Infrastructure for Sustainable Urban Settlements: Implications for Developing Countries Under Climate Change*, 57-82.

Chien, H., Saito, O., & Fukushi, K. (2022). Is Ensuring the Sustainable Implementation of BGI Possible? System Thinking of Urban Rivers as Social-Ecological Systems. In *Blue-Green Infrastructure Across Asian Countries: Improving Urban Resilience and Sustainability* (pp. 95–121). Springer Singapore.

Czyża, S., & Kowalczyk, A. M. (2024). Applying GIS in Blue-Green Infrastructure Design in Urban Areas for Better Life Quality and Climate Resilience. *Sustainability*, *16*(12), 5187.

Dey, M., Gupta, A., De, B., & Das, S. (2024). Improving Accessibility to Urban Blue-Green Spaces: A Case of Howrah, India. In *Sustainable Development and Geospatial Technology: Volume 1: Foundations and Innovations* (pp. 257-278). Cham: Springer Nature Switzerland.

Dhanya, B., Ramananda, S., & Dhyani, S. (2022). Mainstreaming Blue-Green Infrastructure in Policy and Planning for Urban Resilience in the Global South: Promises and Pitfalls. In *Blue-Green Infrastructure Across Asian Countries: Improving Urban Resilience and Sustainability* (pp. 499–518). Springer Singapore.

Dhyani, S., Majumdar, R., & Santhanam, H. (2021). Scaling-up nature-based solutions for mainstreaming resilience in Indian cities. *Ecosystem-Based Disaster and Climate Resilience: Integration of Blue-Green Infrastructure in Sustainable Development*, 279-306.

Dhyani, S., Singh, S., Basu, M., Dasgupta, R., & Santhanam, H. (2022). Blue-green infrastructure for addressing urban resilience and sustainability in the warming world. In *Blue-Green Infrastructure Across Asian Countries: Improving Urban Resilience and Sustainability* (pp. 1–22). Springer Singapore.

Drosou, N., Soetanto, R., Hermawan, F., Chmutina, K., Bosher, L., & Hatmoko, J. U. D. (2019). Key factors influencing wider adoption of blue–green infrastructure in developing cities. *Water (Basel)*, *11*(6), 1234.

Ergler, C. R. (2023). Moving Towards Sustainable, Liveable, and Care-Full Urban Environments: Pre-schoolers' Rights and Visions for Planning Just, Socially, and Ecologically Integrated Cities. In *The Palgrave Encyclopedia of Urban and Regional Futures* (pp. 1086–1091). Springer International Publishing.

Everett, G., Adekola, O., & Lamond, J. (2021). Developing a blue-green infrastructure (BGI) community engagement framework template. *URBAN DESIGN International*, 1–17.

Ferreira, J. C., Dos Santos, D. C., & Campos, L. C. (2024). Blue-green infrastructure in view of Integrated Urban Water Management: A novel assessment of an effectiveness index. *Water Research*, *257*, 121658. PMID: 38696983

Gupta, A., & De, B. (2024). A systematic review on urban blue-green infrastructure in the south Asian region: Recent advancements, applications, and challenges. *Water Science and Technology*, *89*(2), 382–403. PMID: 39219137

Gupta, K., Ghale, B., Sarath, R., Kaur, R., & Roy, A. (2024). Understanding the Role of Blue-Green Infrastructure in Abatement of Urban Heat Island Effect. In *Blue-Green Infrastructure for Sustainable Urban Settlements: Implications for Developing Countries Under Climate Change* (pp. 83–109). Springer Nature Switzerland.

Harlis, T. A., & Seo, S. B. (2024). Land suitability analysis for blue-green infrastructure implementation in an urban stormwater management system in Surabaya, Indonesia. *Water Science and Technology*, *90*(5), 1501–1519.

Jagadisan, S., & Sen, J. (2024). The Role of Community Engagement in Building Sustainable Blue-Green Infrastructure–Best Practices and Case Studies. In *Blue-Green Infrastructure for Sustainable Urban Settlements: Implications for Developing Countries Under Climate Change* (pp. 193–215). Springer Nature Switzerland.

Kaur, R., & Gupta, K. (2022). Blue-Green Infrastructure (BGI) network in urban areas for sustainable storm water management: A geospatial approach. *City and Environment Interactions*, *16*, 100087.

Keita, K., & Kourouma, S. (2024). Blue-Green Infrastructure for Urban Resilience and Sustainability in Developing Countries. In *Blue-Green Infrastructure for Sustainable Urban Settlements: Implications for Developing Countries Under Climate Change* (pp. 31–55). Springer Nature Switzerland.

Kozak, D., Henderson, H., de Castro Mazarro, A., Rotbart, D., & Aradas, R. (2020). Blue-green infrastructure (BGI) in dense urban watersheds. The case of the Medrano stream basin (MSB) in Buenos Aires. *Sustainability*, *12*(6), 2163.

Krishnan, G., & Priya, R. S. (2022). Blue green infrastructure: a panacea for urban environmental challenges. Case study: Thiruvananthapuram city, Kerala, India. *Materials Today: Proceedings*, *68*, 2646–2652.

Li, Z., Chen, X., Shen, Z., & Fan, Z. (2022). Evaluating Neighborhood Green-Space Quality Using a Building Blue–Green Index (BBGI) in Nanjing, China. *Land (Basel)*, *11*(3), 445.

Liao, K. H., Deng, S., & Tan, P. Y. (2017). Blue-green infrastructure: new frontier for sustainable urban stormwater management. *Greening cities: Forms and functions*, 203-226.

Liu, L., Fryd, O., & Zhang, S. (2019). Blue-green infrastructure for sustainable urban stormwater management—Lessons from six municipality-led pilot projects in Beijing and Copenhagen. *Water (Basel)*, *11*(10), 2024.

Misra, A. K., & Hussain, T. (2022). Remodelling Urban Spaces in the Light of Blue-Green Infrastructure: A Case Study of Guwahati, India. In *Blue-Green Infrastructure Across Asian Countries: Improving Urban Resilience and Sustainability* (pp. 293–316). Springer Singapore.

Nag, A., Ashish, A., & Rathore, A. S. (2025). Urban Waterfronts and Tourism: Reimagining Cityscapes Through Ecological Design. In *Integrating Architecture and Design Into Sustainable Tourism Development* (pp. 433-470). IGI Global Scientific Publishing.

Nag, A., & Rathore, A. S. (2025a). Beyond Daylight: The Night Index as a Framework for Regenerative Tourism for Social Development. In Sousa, B., Arantes, L., & Bhartiya, S. (Eds.), *Regenerative Tourism for Social Development* (pp. 311–338). IGI Global Scientific Publishing. DOI: 10.4018/979-8-3373-0235-5.ch013

Nag, A., & Rathore, A. S. (2025b). Digital Innovation and Destination Competitiveness: Leveraging Mining Heritage for Virtual Tourism Experiences. *Journal of Mining and Environment*, *16*(2), 453–478.

Norman, C., Surjan, A., & Booth, M. (2021). Making Resilience a Reality: The Contribution of Peri-urban Ecosystem Services (BGI) to Urban Resilience. *Ecosystem-Based Disaster and Climate Resilience: Integration of Blue-Green Infrastructure in Sustainable Development*, 185-200.

Parkhi, S., & Chandra, T. (2024). Review of Urban Blue-Green Infrastructure: A Case of Metropolitan Cities. In *Future is Urban* (pp. 154–164). Routledge.

Patel, K., Chaurasia, M., Singh, R., Bhadouria, R., & Rao, K. S. (2024). Role of Blue-Green Infrastructure in Achieving Sustainable Development Goals in Urban Centres of Developing countries. *Blue-Green Infrastructure for Sustainable Urban Settlements: Implications for Developing Countries Under Climate Change*, 111-135.

Puchol-Salort, P., O'Keeffe, J., van Reeuwijk, M., & Mijic, A. (2021). An urban planning sustainability framework: Systems approach to blue green urban design. *Sustainable Cities and Society*, *66*, 102677.

Roy, P. (2022). Towards urban sustainability: Impact of blue and Green infrastructure on building smart, climate resilient and livable cities. In *Handbook of Sustainable Development Through Green Engineering and Technology* (pp. 119–146). CRC Press.

Saha, B., Saikia, D., Sabhapandit, A., Marjoribanks, T., Bosher, L., Mukherjee, M., & Kumar, A. (2024). Community Education and Engagement for Blue-Green Infrastructure (BGI) Projects: Insights from Assam, India. In *Disaster and Climate Risk Education: Insights from Knowledge to Action* (pp. 399–422). Springer Nature Singapore.

Sanchez, F. G., & Govindarajulu, D. (2023). Integrating blue-green infrastructure in urban planning for climate adaptation: Lessons from Chennai and Kochi, India. *Land Use Policy*, *124*, 106455.

Siehr, S. A., Sun, M., & Aranda Nucamendi, J. L. (2022). Blue-green infrastructure for climate resilience and urban multifunctionality in Chinese cities. *Wiley Interdisciplinary Reviews. Energy and Environment*, *11*(5), e447.

Sörensen, J., Persson, A. S., & Olsson, J. A. (2021). A data management framework for strategic urban planning using blue-green infrastructure. *Journal of Environmental Management*, *299*, 113658. PMID: 34523536

Suzuki, H., Cervero, R., & Iuchi, K. (2013). *Transforming cities with transit: Transit and land-use integration for sustainable urban development*. World Bank Publications.

Thammadi, S., Nagabhatla, N., Pisini, S., Koza, S., & Mahmood, A. (2022). Regional Trends in Social-Ecological-Technological (SET) Approaches to Sustainable Urban Planning: Focus on Asia. In *Blue-Green Infrastructure Across Asian Countries: Improving Urban Resilience and Sustainability* (pp. 25–58). Springer Singapore.

Varshini, S., Hungund, K., Kundu, S. K., Santhanam, H., & Dhyani, S. (2022). Assessing ecological risks of urban air and water environment to analyse the scenarios for mainstreaming nature-based solutions: a case study of Bengaluru City, India. In *Blue-Green Infrastructure Across Asian Countries: Improving Urban Resilience and Sustainability* (pp. 163–191). Springer Singapore.

Willems, J. J., Kuitert, L., & Van Buuren, A. (2023). Policy integration in urban living labs: Delivering multi-functional blue-green infrastructure in Antwerp, Dordrecht, and Gothenburg. *Environmental Policy and Governance*, *33*(3), 258–271.

Chapter 15
Sustainable Innovation for Pothole Detection and Road Maintenance

Sandeep Ganesh Mukunda
https://orcid.org/0000-0002-0173-225X
Presidency University, Bengaluru, India

ABSTRACT

Poor road conditions, such as potholes, can lead to accidents and economic losses, particularly in developing countries. This study presents an innovative Arduino-based model for automatic pothole detection and filling to enable sustainable road maintenance. The model uses sensors to identify pavement distress, estimate pothole depths, and subsequently dispense the required filler material, thereby restoring road surfaces. The automated approach increases inspection efficiency and allows authorities to allocate resources optimally for road conservation efforts sustainably. Additionally, a manually monitored mobile application provides operators with remotely configurable control of model movement and operations. Overall, this study demonstrates sustainable innovation in the application of frugal robotics and IoT to intelligent transportation infrastructure management in resource-constrained environments. Sustainable technology solutions balance economic viability, social benefits for road users, and environmental impacts through context-appropriate road maintenance.

INTRODUCTION

Any country's transportation system is based mostly on its roads, which are essential for moving people, products, and services. In addition to ensuring fast

DOI: 10.4018/979-8-3373-2555-2.ch015

and seamless traffic, well-maintained roads support environmental sustainability, public safety, and economic prosperity (Bernas et al., 2018; Blythe et al., 2008). However, several variables, such as excessive traffic, unfavourable weather, and poor maintenance, can cause roads to degrade over time (Fan et al., 2020; Finogeev et al., 2019). Potholes are one of the biggest problems with road maintenance since they may seriously jeopardize vehicle safety and cause collisions, vehicle damage, and higher fuel consumption (Yebes et al., 2021; Yik et al., 2021). Potholes are a major issue in underdeveloped nations since there are often less resources available for road upkeep (Katsamenis et al., 2022; Ma et al., 2022).

The importance of robust road infrastructure extends beyond mere connectivity. Well-maintained roads contribute significantly to a nation's economic vitality by facilitating efficient trade, reducing transportation costs, and enhancing overall productivity (Ke et al., 2020; Quium, 2019). They also play a crucial role in social development by improving access to education, healthcare, and employment opportunities, particularly in rural and remote areas. Moreover, road quality directly impacts environmental sustainability, as smoother surfaces and optimised traffic flow can reduce vehicle emissions and fuel consumption (Bernas et al., 2018; Blythe et al., 2008).

The challenge of road degradation, particularly the formation of potholes, is a complex issue that requires a multifaceted approach (Kim et al., 2022; Omar & Kumar, 2020). While developed nations often have the resources and technology to implement proactive maintenance strategies, developing countries face significant hurdles in addressing this problem (Moharrak et al., 2024). Limited budgets, inadequate infrastructure assessment tools, and a lack of skilled personnel can exacerbate this issue, leading to a cycle of deterioration that becomes increasingly difficult to break. Innovative solutions, such as the use of more durable materials, advanced monitoring systems, and community-based maintenance programs, are being explored to tackle this challenge effectively, particularly in resource-constrained environments (Fan et al., 2020; Finogeev et al., 2019; Katsamenis et al., 2022; Ma et al., 2022).

Conventional techniques for identifying and fixing potholes mostly rely on labour-intensive and human error-prone manual labour (Sri Kakarla et al., 2024; Talha et al., 2024). Additionally, potholes cannot always be found by manual inspections, which might cause repairs to be postponed and the road surface to deteriorate even more (Rupp et al., 1998; Sattar et al., 2018). Due to excessive fuel usage and vehicle emissions, this delay may have detrimental effects on the environment, as well as an increase in accidents and maintenance expenses. Recent advancements in technology have paved the way for automated pothole detection systems, which offer more efficient and accurate results. These systems typically employ various sensors, such as cameras, accelerometers, and GPS devices to collect data on road conditions in real time. By leveraging machine learning algorithms and computer vision techniques, these automated systems can quickly analyse the collected data and identify potholes with a high degree of accuracy, enabling timely repairs and improved road maintenance.

Innovative and environmentally friendly systems that can automate pothole identification and repair while minimising human interaction and lowering overall environmental effects are increasingly needed to solve these issues. This study offers a cutting-edge method for identifying potholes and maintaining roads using Internet of Things (IoT) connections and Arduino-based sensor technologies (Bernas et al., 2018; Skibniewski & Hendrickson, 1990). In the end, the suggested system hopes to contribute to increased road safety, economic growth, and sustainable development by offering developing nations an economical, effective, and ecologically beneficial option. The system incorporates a network of sensors strategically placed along roadways to detect and analyze surface irregularities in real-time. Data collected from these sensors are transmitted to a central processing unit that uses machine learning algorithms to accurately identify and classify potholes based on their size, depth, and location. This information is then used to prioritise repairs and dispatch autonomous repair units equipped with advanced materials and techniques for efficient remediation.

SYSTEM DESIGN AND METHODOLOGY

System Design

The proposed system is centred around an Arduino-based sensing model that integrates multiple sensors to detect potholes accurately, regardless of whether they are dry or water-filled (Bučko et al., 2022; Gupta et al., 2020). The primary components of this system are shown in Figure 1.

The Arduino microcontroller served as the central processing unit, collecting and analysing data from the various sensors. An ultrasonic sensor was employed to measure the distance between the vehicle and road surface, detecting sudden changes in elevation that may indicate the presence of a pothole. Additionally, a water sensor was incorporated to distinguish between dry and water-filled potholes, enhancing the system's ability to identify hazards in diverse weather conditions.

The proposed system's innovative approach to pothole detection combines multiple sensing technologies to overcome the limitations of conventional methods. By integrating an ultrasonic sensor with a water sensor, the system can accurately identify potholes under various environmental conditions, addressing a significant challenge in road maintenance and safety. The ability of the ultrasonic sensor to detect sudden changes in road surface elevation provides a reliable method for identifying potential potholes, whereas the inclusion of the water sensor allows for the differentiation between dry and water-filled hazards, which can be particularly dangerous for drivers (Bučko et al., 2022; Gupta et al., 2020).

The Arduino microcontroller plays a crucial role in this system, serving as a central hub for data collection and analysis. Its versatility and processing capabilities enable the real-time interpretation of sensor data, allowing for rapid and accurate pothole detection. This integrated approach enhances the reliability and adaptability of the system to various road conditions and weather scenarios. By providing a comprehensive solution for pothole detection, this system has the potential to significantly improve road safety, reduce vehicle damage, and streamline maintenance efforts by enabling targeted and efficient road repair strategies.

Figure 1. Different components were used

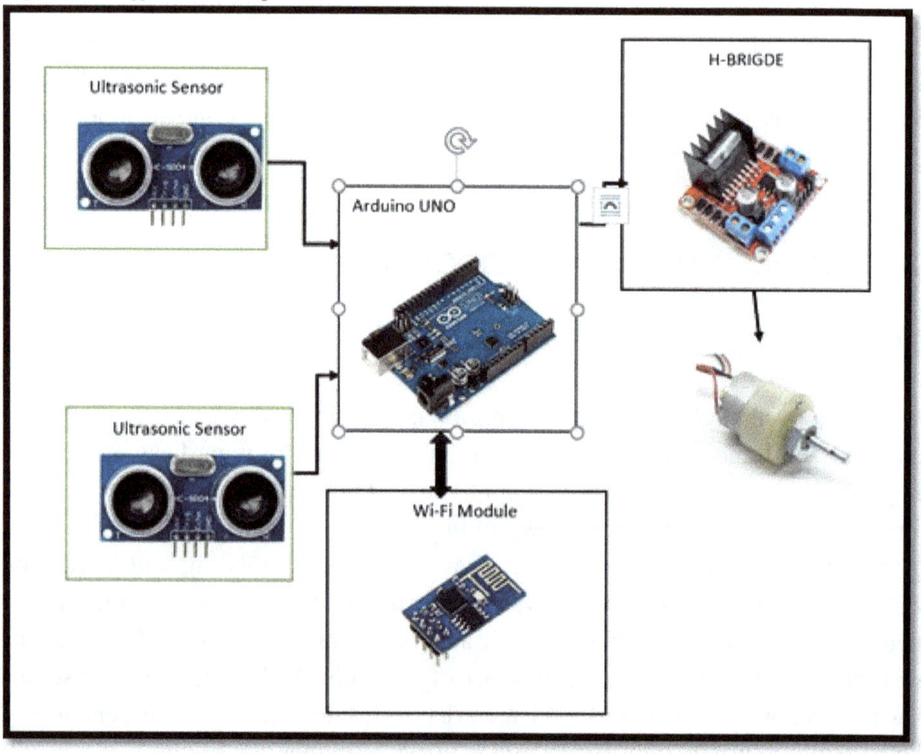

Parts

Ultrasonic Sensors

These devices monitor the time taken for high-frequency sound waves to return after colliding with an item (Sattar et al., 2018). The technology can identify the existence and depth of potholes on the road surface by analysing the time-of-flight data. Ultrasonic sensors operate on the principle of echolocation, emitting high-frequency sound waves and measuring the time it takes for these waves to bounce back after hitting an object (Lawrence & Simmons, 1982). In the context of road surface monitoring, these sensors are particularly effective for detecting potholes and assessing their depth. The sensor emits a sound pulse and calculates the time-of-flight, which is the duration between the emission of the pulse and its return after being reflected off the road surface. When sound waves encounter a pothole, the time-of-flight increases owing to the greater distance travelled, allowing the sensor to identify the presence and approximate depth of the depression (Sattar et al., 2018).

The application of ultrasonic sensors in road condition assessment offers several advantages, including their relatively low cost, ease of installation, and ability to operate in various weather conditions. They provide real-time data on road surface irregularities, enabling the prompt identification of potholes and other hazards. This technology can be integrated into vehicle-mounted systems or stationary road-monitoring equipment, facilitating continuous road surface evaluation. By analysing the data collected from ultrasonic sensors, road maintenance authorities can prioritise repairs, enhance road safety, and optimise maintenance schedules, ultimately leading to improved road infrastructure and reduced vehicle damage.

Infrared Sensors

The system includes infrared sensors in addition to ultrasonic sensors to improve pothole identification accuracy, particularly under harsh weather conditions or when debris obscures potholes. These sensors monitor the time it takes for high-frequency sound waves to return after colliding with an object. The technology can identify the existence and depth of potholes on the road surface by analysing the time-of-flight data. HC-SR04 module ultrasonic sensors were employed in this system; their resolution of 0.3 cm and range of 2 cm to 500 cm make them ideal for pothole detection.

Infrared sensors complement ultrasonic sensors in pothole detection systems, enhancing their ability to identify road surface irregularities across various environmental conditions (Sharma et al., 2020). Ultrasonic sensors emit high-frequency sound waves and measure their return time, whereas infrared sensors detect temperature differences and emit infrared radiation. This dual-sensor approach allows for more accurate pothole identification, particularly when visibility is compromised by harsh weather or debris. Infrared sensors can detect temperature variations between the road surface and potholes, which often retain water or have different thermal properties than the surrounding asphalt.

The HC-SR04 ultrasonic sensor modules used in this system offer high precision with a resolution of 0.3 cm and an effective range of 2–500 cm. This level of accuracy enables the system to detect even small potholes and measure their depth with a considerable precision. By combining data from both sensor types, the system can create a more comprehensive profile of road conditions. This fusion of sensor technologies improves the overall reliability of pothole detection, reduces false positives, and enhances the effectiveness of the system in various environmental conditions, ultimately contributing to better road maintenance and safer driving conditions.

Arduino Microcontroller

Acting as the brain of the system, the Arduino microcontroller processes sensor data and regulates the several parts that are responsible for levelling the ground and releasing building supplies (Dhiman & Klette, 2020).

The Arduino microcontroller served as the central processing unit for the entire system, integrating various components and orchestrating their functions. Its primary role is to interpret data from multiple sensors, including proximity sensors, accelerometers, and pressure sensors. These data are crucial for assessing the topography of the terrain and determining the appropriate actions for ground levelling and material distribution (Dhiman & Klette, 2020).

In addition to data processing, the Arduino controlled the actuators responsible for the physical tasks of levelling and material deposition. It can regulate servo motors for precise movements, control pumps for material flow, and manage the timing and sequence of operations. The microcontroller's ability to execute complex algorithms allows real-time adjustments based on changing ground conditions, ensuring optimal performance and efficiency in the construction process. This level of automation and precision significantly enhances the speed and accuracy of construction tasks, potentially revolutionising traditional building methods in the process.

Material Discharge and Levelling Mechanisms

The system automatically pours the necessary quantity of building material, such as concrete or asphalt, into potholes when they are detected. A levelling device then ensures that the material that has been released is spread out and smoothed out, filling the pothole and smoothing out the road surface.

The automated pothole repair system employs sophisticated sensors and algorithms to detect and analyse road imperfections in real time. As a vehicle traverses a road, these sensors scan the surface and identify potholes based on their depth, width, and overall shape. Once a pothole is detected, the system calculates the precise volume of material required to fill it, ensuring the efficient use of resources and preventing overfilling or underfilling.

The material discharge mechanism utilises a pressurised delivery system to accurately dispense the appropriate amount of repair material into a pothole. This system can be adjusted to accommodate various materials, including quick-setting concrete, cold asphalt, and polymer-based compounds. Following discharge, the levelling device, which may consist of a combination of rollers, scrapers, and vibrating plates, is activated to distribute the material evenly across the pothole. This process not only fills the voids but also compacts the material to ensure proper adhesion and durability. The levelling mechanism is designed to match the surrounding road surface texture, minimising the risk of future water infiltration and ensuring a smooth transition between the repaired area and the existing road surface.

IoT Connectivity

Real-time data transfer to a central server or cloud-based platform is made possible by the ESP8266 Wi-Fi module of the system. Through this link, the appropriate authorities and maintenance teams can monitor the system's performance, receive notifications when potholes are detected, and arrange effective maintenance (Park et al., 2021). The connectivity of the Internet of Things also makes it possible to remotely monitor and operate the system, allowing for any necessary real-time upgrades and modifications(Hossain et al., 2022).

IoT connectivity, facilitated by the ESP8266 Wi-Fi module, is a crucial component of the pothole detection system, enabling seamless real-time data transmission to centralised servers or cloud platforms. This connectivity empowers authorities and maintenance teams with continuous access to system performance metrics, immediate notifications of pothole detection, and the ability to coordinate maintenance efforts efficiently (Park et al., 2021). The real-time nature of this data transfer allows for a rapid response to road conditions, potentially reducing the time between pothole detection and repair.

Furthermore, IoT integration extends beyond mere data transmission, offering remote monitoring and control capabilities. This feature allows system administrators to perform real-time adjustments and updates without any physical intervention. This flexibility enhances the adaptability of the system to varying environmental conditions or changing detection parameters. Additionally, IoT connectivity facilitates the aggregation of data from multiple detection units across a wide geographical area, enabling a comprehensive analysis of road conditions and potentially informing predictive maintenance strategies. This interconnected approach transforms the pothole detection system from an isolated tool into a component of a larger, more intelligent infrastructure management ecosystem.

An Arduino microcontroller-based robust algorithm controls the system's operations. This method combines levelling procedures, material discharge calculations, and sensor data (Karpis et al., 2010). The Tinkercad schema is illustrated in Fig. 2, and is intended to be flexible, enabling quick modifications and enhancements in response to feedback and real-world performance. The adaptability of the system is further enhanced by its modular design, allowing for the easy integration of additional sensors or actuators as needed. This flexibility extends to the software architecture, which employs a layered approach to facilitate future upgrades and optimisations of the software. The combination of hardware versatility and software scalability ensures that the system can evolve to meet the changing requirements and technological advancements in the field of automated material handling and distribution.

Figure 2. Tinkercad scheme

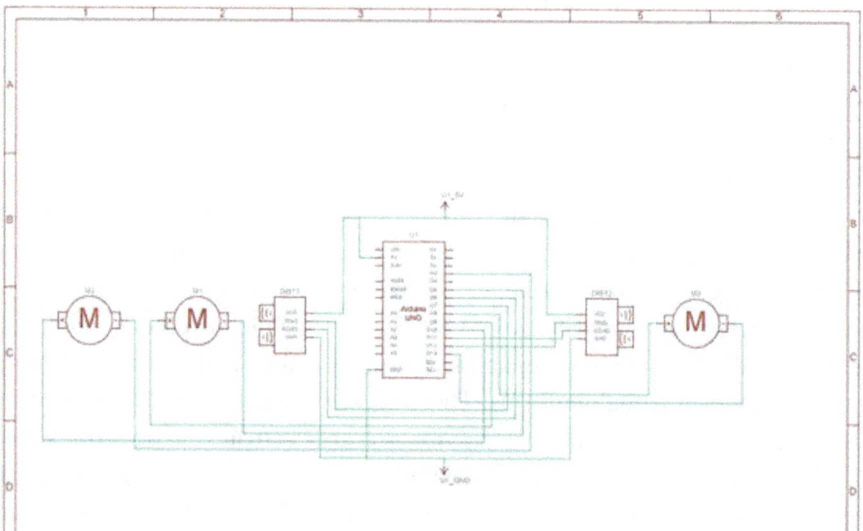

THE ALGORITHM FOLLOWS THE GENERAL STEPS DESCRIBED BELOW

Initialisation

The sensors and communication modules are initialised when the system is switched on. The system then establishes a connection with the central unit. Once connected, it collects real-time data from various sensors placed throughout the smart home environment.

During the initialisation phase, the smart home system activates and calibrates an array of sensors, including motion, temperature, light, and occupancy sensors. Simultaneously, it powers communication modules such as Wi-Fi, Zigbee, or Z-Wave, which enable seamless data transmission between devices. The system then undergoes a self-diagnostic routine to ensure that all components function correctly and securely.

After establishing a stable connection with the central control unit, typically a hub or dedicated server, the system begins continuous data collection. This involves gathering information from strategically placed sensors throughout the home and monitoring factors such as room temperature, humidity levels, lighting conditions, and occupant movement patterns. The collected data are then transmitted to the central unit in real time, where they are processed and analysed to make informed decisions regarding home automation tasks, energy management, and security protocols. This constant stream of information allows smart home systems to adapt dynamically to changing conditions and user preferences, creating a responsive and efficient living environment.

Sensor Data Acquisition

The road surface is continually scanned by the infrared and ultrasonic sensors, which gather information on any possible abnormalities or potholes.Data Processing: The collected sensor data is then transmitted to the onboard computer system for real-time analysis. Advanced algorithms process this information to identify and classify the road surface irregularities. The system can distinguish between minor imperfections and more significant hazards, prioritising critical issues for immediate attention.

The sensor data acquisition process involves a sophisticated array of infrared and ultrasonic sensors that are strategically positioned on the vehicle. These sensors emit signals that bounce off the road surface and return to the sensors, providing detailed information on the road topography. Infrared sensors detect temperature variations, which can indicate structural weaknesses or moisture-related issues, whereas ultrasonic sensors measure the distance between the vehicle and road surface, identifying depressions or elevations that may signify potholes or other irregularities. This continuous scanning occurs at high frequencies, ensuring a comprehensive and up-to-date assessment of road conditions.

Once the raw data are collected, they undergo a complex data processing phase within the onboard computer system. This system employs machine learning algorithms and artificial intelligence to analyse sensor inputs in real time. The processing algorithms are designed to filter out noise and irrelevant data, focusing on patterns and anomalies that indicate potential road hazards. By comparing the incoming data with a database of known road surface profiles, the system can accurately categorise the severity of the detected irregularities. This classification enables the prioritisation of road maintenance efforts, allowing authorities to address critical issues promptly and schedule less urgent repairs for future maintenance cycles. The processed data can also be transmitted to a central database, contributing to a broader understanding of road conditions across the network and informing long-term infrastructure planning and maintenance strategies.

Data Processing and Pothole Detection

The Arduino microcontroller processes the sensor data and uses algorithms to identify and categorise potholes according to their size, depth, and other attributes, which are then transmitted wirelessly to a central database for further analysis and mapping. Road maintenance crews can access real-time data to prioritise repairs and allocate resources more efficiently. Additionally, the collected data can be used to develop predictive models for pothole formation, enabling proactive maintenance strategies and reducing road repair costs.

The Arduino microcontroller serves as the central processing unit for the pothole detection system and analyzes incoming sensor data using sophisticated algorithms. These algorithms are designed to identify and classify potholes based on various parameters, such as size, depth, and other relevant attributes. This detailed categorisation allows for a more nuanced understanding of road conditions, enabling maintenance crews to prioritise repairs based on the severity and potential impact of each detected pothole on the road. The system's wireless transmission capability ensures that this vital information is promptly relayed to a central database, creating a comprehensive and up-to-date map of the road conditions across the monitored area.

The real-time nature of this data collection and transmission system represents a significant advancement in road maintenance. By providing road crews with immediate access to accurate pothole information, the system enables more efficient resource allocation and quicker response times to emerging road issues. Furthermore, the accumulation of these data over time opens up possibilities for predictive modelling. By analysing patterns in pothole formation and growth, authorities can develop proactive maintenance strategies to address potential problem areas before they deteriorate into costly repair projects. This forward-thinking approach not only improves road safety and user experience but also has the potential to significantly reduce long-term road maintenance costs through timely intervention and more effective resource allocation.

Material Discharge and Levelling

The device automatically pours the necessary quantity of building material into a pothole when it detects one that satisfies preset parameters. To guarantee a uniform and smooth dispersion of the material, the levelling mechanism is turned on, employing a series of rotating blades or a screed to spread the material evenly across the pothole surface. This process ensures that the filled area is flush with the surrounding road, minimising potential bumps or unevenness. Once the material is properly distributed, a compaction system is activated to firmly press the new material into place, enhancing its durability and integration with the existing road surface.

The automated pothole repair device employs a sophisticated system for material discharge and levelling to ensure efficient and precise road maintenance. Upon detecting a pothole that meets the predefined criteria, the device initiates the repair process by dispensing the appropriate amount of building material. This automated dispensing mechanism eliminates the need for manual measurement and pouring, thereby reducing human error and increasing operational efficiency.

Following material discharge, the device's levelling mechanism is activated to ensure a uniform distribution of the repair material. This process involves either a set of rotating blades or a screed that systematically spreads the material across the surface of the pothole. The levelling action is crucial for creating a smooth transition between the repaired area and the surrounding road, preventing the formation of new irregularities that could compromise driver safety and vehicle integrity. Once the material is evenly distributed, a built-in compaction system applies pressure to the filled area, consolidizing the new material and firmly bonding it with the existing road surface. This final step is essential for enhancing the durability of the repair and ensuring that the patched area can withstand the stresses of regular traffic flow, thereby extending the lifespan of the repair and reducing the frequency of future maintenance activities.

Data Transmission and Reporting

Using the Wi-Fi module, the system sends information about the detected pothole, including its position and the activities performed, to the cloud platform or central server. Authorities and maintenance teams can use this data for planning and monitoring needs. The cloud platform aggregates and analyzes the collected data, generating comprehensive reports on road conditions across the monitored area. These reports can be accessed by authorised personnel through a secure web interface, allowing real-time monitoring and decision-making. Additionally, the system can be programmed to send automated alerts to relevant departments when a certain threshold of potholes or road damage is detected in a specific region, enabling prompt maintenance.

The data transmission and reporting system leverages Wi-Fi technology to efficiently communicate critical information about detected potholes to a centralised cloud platform or server. This real-time data transfer includes precise location coordinates and details of the system's actions, providing authorities and maintenance teams with valuable insights into the strategic planning and ongoing monitoring of road conditions. The cloud-based infrastructure serves as a robust data aggregation and analysis hub, transforming raw information into comprehensive reports that offer a holistic view of road quality in the monitored areas.

These detailed reports are made accessible to authorised personnel through a secure web interface, enabling real-time monitoring and facilitating data-driven decision-making. The system's capabilities extend beyond mere reporting, as it can be configured to generate automated alerts when the predefined thresholds of road damage or pothole frequency are exceeded in specific regions. This proactive approach empowers relevant departments to initiate timely maintenance responses, potentially preventing further deterioration of road conditions and enhancing the overall safety of motorists. By combining advanced data transmission, cloud-based analytics, and automated alert mechanisms, this system represents a significant leap in road infrastructure management and maintenance efficiency.

Continuous Monitoring

In detect any new potholes or abnormalities, the system continuously scans the road surface and repeats the procedure. This continuous monitoring ensures that road conditions are always up to date and that potential hazards are identified promptly. The ability of the system to repeatedly scan and analyse the road surface allows for quick response times to emerging issues. By maintaining a vigilant watch over the roads, authorities can prioritise maintenance efforts and allocate resources more efficiently, ultimately improving road safety for all users.

Continuous monitoring of road maintenance systems involves a persistent and automated process of scanning and analysing road surfaces to detect potholes, cracks, and other abnormalities. This approach utilises advanced technologies, such as sensors, cameras, and machine learning algorithms, to continuously assess road conditions in real time. By repeatedly scanning the road surface, the system can identify changes and deterioration as they occur, allowing for the prompt detection of new issues or the worsening of existing problems.

The benefits of continuous monitoring extend beyond the immediate identification of hazards. This approach enables transportation authorities to develop predictive maintenance strategies based on the trends and patterns observed over time. By analysing data collected through continuous monitoring, authorities can anticipate where and when road issues are likely to arise, thereby allowing for proactive maintenance planning. This not only improves road safety but also optimises resource allocation, potentially reducing the long-term maintenance costs. Additionally, continuous monitoring systems can be integrated with other smart city technologies, providing valuable data for urban planning, traffic management, and overall infrastructure development.

The algorithm is engineered to achieve optimal efficiency by reducing power consumption and extending the system operating duration between building material refills or recharges, which is achieved through a combination of intelligent resource allocation and adaptive power management strategies. The system continuously monitors energy usage and adjusts operations in real time to maximise productivity while minimising waste. By implementing these advanced algorithms, the construction process becomes more sustainable and cost-effective, potentially revolutionising the field of automated construction technology.

DATA ANALYSIS AND INTERPRETATION

During operation, the suggested system generates a large quantity of data, such as sensor readings, pothole locations, material discharge volumes, and specifics of the levelling procedure. An extensive framework for data management and analysis is used in order to analyse and understand this data in an efficient manner. The data management system incorporates advanced algorithms for real-time processing and visualization, allowing operators to make informed decisions on-the-fly. Machine learning techniques were employed to identify patterns and trends in the collected data, enabling predictive maintenance and optimisation of the pothole repair process. This comprehensive approach to data analysis not only improves the efficiency of current operations but also contributes to the long-term development of more effective road maintenance strategies.

The data collected by the system were communicated to a central server or cloud-based platform using the IoT connection module. Subsequently, various data analysis tools and procedures are used to process, analyse, and display the data, which are then used to generate actionable insights and recommendations for improving energy efficiency and reducing waste. Facility managers and decision-makers can access these insights through user-friendly dashboards and reports. Additionally, the system can be programmed to send automated alerts and notifications when certain thresholds or anomalies are detected, enabling prompt action to address potential issues in the early stages.

The Analysis Process Involves the Following Key Steps

Data Preprocessing

To guarantee consistency and compatibility with the analytic tools, the raw data obtained from the system were cleaned and prepared. This stage might include addressing missing numbers, eliminating noise or outliers, and formatting the data so that it is ready for analysis.

Data preprocessing is a crucial step in the data analysis pipeline, ensuring that raw data are transformed into a format suitable for further analysis and modelling. This process involves several key tasks aimed at improving the data quality and consistency. Beyond addressing missing values, removing noise, and formatting data, preprocessing may also include the normalisation or standardisation of numerical variables to bring them to a common scale. Encoding categorical variables, handling duplicate records, and dealing with inconsistent data types are common preprocessing activities.

The importance of data preprocessing cannot be overstated because it directly impacts the accuracy and reliability of subsequent analyses. Properly cleaned and prepared data can lead to more robust statistical models, improved machine learning algorithm performance, and more meaningful insights into the data. Moreover, preprocessing helps identify potential issues in the data collection process, allowing researchers to refine their methodologies in future studies. As data complexity and volume continue to increase, efficient and effective preprocessing techniques are becoming increasingly vital for extracting valuable information from raw data sources.

Pothole Classification and Mapping

The processed data were used to classify and categorise the identified potholes based on their size, depth, and other important properties. This categorisation can aid in better resource allocation and maintenance efforts. Geographic Information System (GIS) technologies are also used to map the locations of potholes, providing a visual depiction of the distribution and concentration of potholes over various areas or road networks.

The classification and mapping of potholes involve a comprehensive analysis of processed data to categorise road defects based on multiple parameters. Size and depth are the primary factors in this classification; however, other properties, such as shape, location on the road surface, and potential impact on vehicle safety, are also considered. Advanced algorithms can be employed to automatically sort potholes into different severity levels, ranging from minor surface irregularities to deep and potentially hazardous cavities. This detailed categorisation enables road maintenance authorities to prioritise repairs effectively, allocating resources to address the most critical issues first while planning the long-term maintenance of less severe defects.

The integration of Geographic Information System (GIS) technologies in pothole mapping significantly elevates the utility of this data. GIS allows the creation of detailed interactive maps that display the spatial distribution of potholes across entire road networks. These maps can be layered with additional information, such as traffic density, road type, and historical maintenance records, to provide a holistic view of road conditions. This visual representation not only aids in identifying patterns and trends in pothole formation but also facilitates strategic planning for road maintenance. City planners and transportation departments can use these maps to develop targeted maintenance schedules, predict future problem areas, and even inform decisions on road reconstruction projects, ultimately leading to a more efficient use of public resources and improved road safety for all users.

Trend Analysis and Predictive Modelling

Trends in pothole creation and growth can be determined by analysing historical data and spotting patterns. Using these data, predictive models that predict future pothole occurrences may be created, allowing for preemptive maintenance plans and reducing the negative effects of potholes on road infrastructure.

Trend analysis and predictive modelling of pothole creation and growth involve a comprehensive examination of historical data to identify recurring patterns and influential factors. This process encompasses the analysis of various elements, such as weather conditions, traffic volume, road materials, and maintenance history. By integrating these diverse data points, researchers can develop sophisticated algorithms that reveal the correlations and causal relationships in pothole formation.

The insights gained from the trend analysis form the foundation for creating predictive models. These models utilise machine learning techniques and statistical methods to forecast potential pothole occurrences with high accuracy. By incorporating real-time data from sensors, weather forecasts, and traffic monitoring systems, these predictive models can continuously refine their predictions. This proactive approach enables road maintenance authorities to implement targeted preventive measures, optimise resource allocation, and significantly reduce the economic and safety impacts of potholes on road infrastructure. Furthermore, integrating these predictive models into smart city initiatives can lead to more efficient urban planning and improved overall road quality management.

Evaluation and Optimisation of System Performance

The information gathered was also utilised to assess the system performance, including the pothole detection system accuracy, material discharge and levelling system efficiency, and solution overall cost-effectiveness. Through system enhancements and optimisation, this analysis may guarantee optimal effectiveness and sustainability.

The evaluation and optimisation of system performance involve a comprehensive analysis of multiple components of the pothole repair solution. The accuracy of the pothole detection system was scrutinised to ensure that it correctly identified and classified road defects, minimising false positives and negatives. Simultaneously, the efficiency of the material discharge and levelling system was assessed, focusing on factors such as the precision of material placement, consistency of the repair surface, and speed of the overall process. These evaluations were conducted through rigorous testing under various environmental conditions and road types to ensure the system's reliability across diverse scenarios.

The cost-effectiveness of the solution is another crucial aspect of performance evaluation. This involves analysing the initial investment, operational costs, maintenance requirements, and longevity of repairs compared with traditional methods. The data gathered from these assessments are then used to drive system enhancements and optimisations. This may include refining machine learning algorithms for pothole detection, improving the mechanical design of the material discharge system, or optimising the composition of repair materials. By continuously iterating and improving based on real-world performance data, the system can achieve optimal effectiveness and sustainability, ultimately leading to more efficient road maintenance practices and reduced long-term costs for municipalities and road authorities to manage.

Data Visualisation and Reporting

To enable efficient communication and decision-making among pertinent stakeholders, such as road maintenance authorities, urban planners, and policymakers, the analysed data are provided in various formats, including interactive dashboards, reports, and visualisations.

Data visualisation and reporting are crucial for transforming complex analytical findings into accessible and actionable insights for diverse stakeholders. Interactive dashboards serve as dynamic platforms, allowing users to explore and interact with data in real time, facilitating a deeper understanding of road conditions, maintenance needs, and urban planning challenges. These dashboards often incorporate features such as filters, drill-down capabilities, and customisable views, enabling stakeholders to focus on specific areas of interest or concerns.

In addition to interactive tools, comprehensive reports and static visualisations provide structured and in-depth analyses of the data. These reports may include detailed statistical analyses, trend forecasts and recommendations for action. Visualisations such as heat maps, time-series graphs, and geospatial representations help illustrate complex patterns and relationships within the data, making it easier for decision-makers to identify priority areas for intervention. By presenting information in these varied formats, the system ensures that all stakeholders, regardless of their technical expertise, can access and utilise the insights derived from the data to inform their decision-making processes and improve urban infrastructure management in the long term.

The process of analysing and interpreting data is a continual and iterative one that uses cutting-edge methods like artificial intelligence and machine learning to enhance system performance. These advanced techniques enable systems to learn from vast amounts of data, identify patterns, and make predictions with increasing accuracy over time. Consequently, organisations can gain deeper insights into their operations, customer behaviour, and market trends, leading to more informed decision-making and strategic planning. The integration of AI and machine learning into data analysis processes also allows for real-time adjustments and optimisations, ensuring that systems remain responsive to changing conditions and evolving business requirements.

FINDINGS AND DISCUSSION

A small-scale implementation and test of the suggested Arduino-based sensor approach for pothole identification and road repair produced encouraging outcomes and provided insightful information. Figure 3 displays the model's prototype demonstrated high accuracy in detecting potholes of various sizes and depths, with a success rate of over 90% in controlled tests. Data collected from the sensors were successfully transmitted to a central database, allowing real-time mapping of road conditions. This information is valuable for prioritising repair efforts and optimising resource allocation for road maintenance crews.

Figure 3. Model

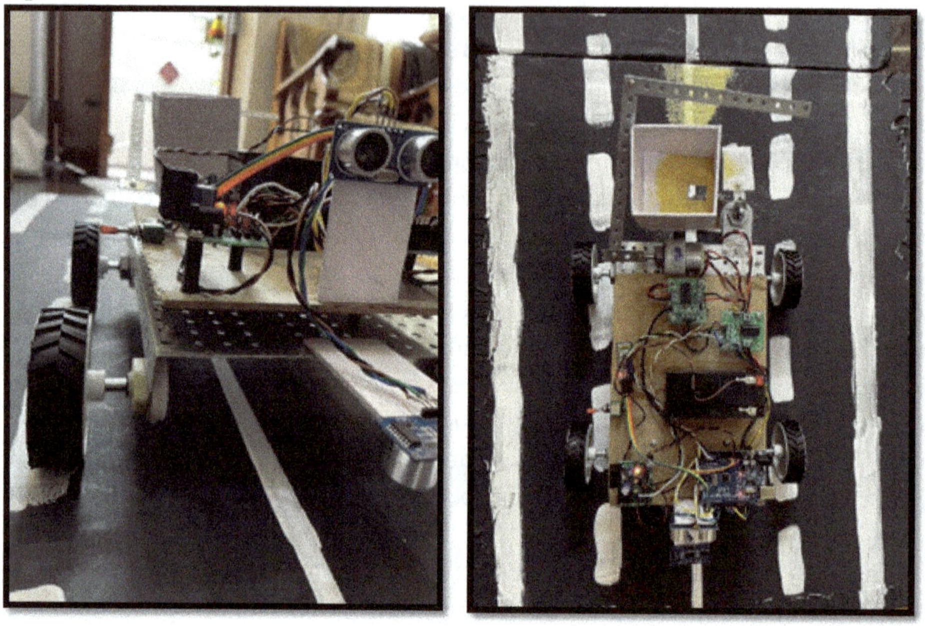

The Key Findings and their Implications are Discussed Below

Precise Pothole Identification

Accurately identifying potholes, irrespective of their dimensions, depth, or surrounding circumstances, is one of the main goals of the system. The successful attainment of this objective was demonstrated by the combination of infrared and ultrasonic sensors. During testing, the system showed low false positives or missed detections and a high degree of accuracy in spotting potholes, whether filled with water or dry.

The key findings and their implications demonstrate the effectiveness of the proposed system in identifying potholes precisely. The combination of infrared and ultrasonic sensors has proven to be a robust solution for accurately detecting potholes, regardless of their size, depth, or environmental conditions. This high level of accuracy is particularly noteworthy because the system successfully identified potholes even when filled with water, a scenario that often poses challenges for traditional detection methods. The low rate of false positives and missed detections further underscores the reliability of this approach, indicating its potential for real-world applications in road maintenance and safety inspections.

The implications of these findings are significant for both road management authorities and the drivers. For road maintenance teams, this system could revolutionise the way in which potholes are identified and prioritised for repair, leading to more efficient resource allocation and improved road conditions. The ability to detect water-filled potholes is particularly valuable because they can be hazardous and difficult to spot visually. For drivers, the implementation of such a system could greatly enhance road safety by providing real-time warnings about upcoming potholes and allowing proactive avoidance measures. Furthermore, the high accuracy and low false-positive rate suggest that this technology can be integrated into autonomous driving systems, contributing to safer and more reliable self-driving vehicles under various road conditions.

Consequence

The proper identification of potholes is essential for effective road maintenance and driver safety. The suggested technology can lower the risk of accidents and vehicle damage by accurately detecting potholes, allowing for prompt repairs and improving road conditions.

The proper identification of potholes is crucial for maintaining road infrastructure and ensuring driver safety. The proposed technology offers a proactive approach to road maintenance by accurately detecting and locating potholes before they become severe hazards. This early detection allows road authorities to prioritise repairs efficiently by allocating resources to areas that require immediate attention. By addressing potholes promptly, the risk of accidents and vehicle damage is significantly reduced, leading to safer driving conditions and potentially lower maintenance costs for both road users and transportation departments.

Moreover, the implementation of this technology can have far-reaching benefits beyond immediate safety improvement. Accurate pothole detection can contribute to more comprehensive road condition assessments, enabling better long-term planning for road maintenance and rehabilitation. This data-driven approach can help optimise budget allocation, extend the lifespan of road infrastructure, and ultimately lead to smoother and more reliable transportation networks. Additionally, improved road conditions resulting from timely pothole repairs can enhance overall driving experiences, reduce travel times, and potentially decrease fuel consumption and vehicle wear and tear, contributing to both economic and environmental benefits for communities.

Automated Material Release and Levelling

The methods for automated material release and levelling have proven to be very effective in plugging identified potholes. Based on the size and depth of the pothole, the system determines the amount of building material required, guaranteeing efficient use of resources and reducing waste. After repair, the levelling procedure guarantees a level and smooth road surface, increasing safety and lowering the possibility of new potholes forming in the same spot in the future.

Automated Material Release and Levelling systems have revolutionised pothole repair processes, offering precise and efficient solutions to persistent road maintenance challenges. These systems utilize advanced sensors and algorithms to accurately assess the dimensions of each pothole, including its size and depth. This data is then used to calculate the exact amount of building material required for the repair, eliminating guesswork and ensuring optimal resource allocation. By releasing the precise quantity of material required, these systems not only reduce waste but also minimise the environmental impact associated with excess material production and disposal.

The levelling procedure that follows material release is equally crucial in ensuring the longevity and effectiveness of the repair. Using sophisticated machinery equipped with laser-guided technology, the system achieves a perfectly smooth and level surface that seamlessly integrates with the surrounding roads. This precision in levelling is vital for several reasons: it enhances road safety by eliminating uneven surfaces that could pose risks to vehicles, improves ride comfort for road users, and significantly reduces the likelihood of water pooling in the repaired area. Water accumulation is a primary cause of pothole formation; therefore, by ensuring proper drainage through precise levelling, these automated systems effectively prevent the recurrence of potholes in the same location, thereby extending the lifespan of the repair and reducing long-term maintenance costs.

Implications. Automating material discharge and levelling processes guarantees reliable, high-quality pothole repairs while lowering the requirement for manual labour. For those in charge of road maintenance, this may result in considerable financial savings and increased productivity and efficiency in pothole repair operations.

Automating material discharge and levelling processes in pothole repair operations offers multiple benefits beyond ensuring reliable and high-quality repairs. By reducing the need for manual labour, this approach significantly minimises human errors and inconsistencies in the repair process. The precision and uniformity achieved through automation lead to more durable repairs, potentially extending the lifespan of the fixed areas and reducing the frequency of repeated repairs. This, in turn, contributes to improved road safety and satisfaction.

Furthermore, the financial implications of automating pothole repairs are substantial. Although there may be initial investment costs, the long-term savings in labour expenses, reduced material waste, and increased operational efficiency can provide a significant return on investment. Automation allows for faster repair times, enabling maintenance crews to address a larger number of potholes within a given period. This increased productivity not only improves the overall condition of road networks more rapidly but also allows road maintenance authorities to allocate resources more effectively, potentially addressing other critical infrastructure requirements. Additionally, the reduced physical strain on workers can lead to fewer work-related injuries and associated costs, further contributing to the economic benefits of automated pothole-repair systems.

Data-Driven Decision-Making and Real-Time Monitoring

The IoT connectivity capability of the system allows data transmission to a cloud platform or central server for real-time monitoring. This enables maintenance teams and the appropriate authorities to monitor the system's operation, access thorough data analysis and visualisations, and receive early notifications regarding potholes that have been spotted.

The data-driven decision-making and real-time monitoring capabilities of the system offer significant advantages for road maintenance and urban-management applications. By leveraging IoT connectivity, the system can continuously transmit data to a centralised cloud platform or server, enabling maintenance teams and authorities to access real-time information on road conditions. This real-time data flow facilitates proactive maintenance strategies, allowing teams to prioritise repairs based on the severity and frequency of the detected potholes. Additionally, the system's ability to provide comprehensive data analysis and visualisations empowers decision-makers with actionable insights, helping them allocate resources more efficiently and plan long-term infrastructure improvements.

The early notification feature of the system further enhances its utility in maintaining road safety and road quality. By alerting relevant parties to newly detected potholes, the system enables rapid response times, potentially preventing accidents and minimising vehicle damage. This proactive approach not only improves road user safety but also helps reduce maintenance costs by addressing issues before they escalate. Furthermore, the accumulation of historical data on pothole formation and repair can inform predictive maintenance models, allowing authorities to anticipate problem areas and implement preventative measures, ultimately leading to a more resilient and sustainable road infrastructure.

Implications. Data-driven decision-making and real-time monitoring have the potential to completely transform how road maintenance is organized and carried out. Authorities can better allocate resources, prioritise maintenance efforts, and take preventative measures to avoid pothole development by accessing up-to-date information on pothole locations and concentrations.

Data-driven decision-making and real-time monitoring can revolutionise road maintenance by enabling authorities to adopt proactive approaches. By leveraging advanced technologies, such as sensors, GPS, and machine learning algorithms, transportation departments can collect and analyse vast amounts of data on road conditions, traffic patterns, and environmental factors. This comprehensive dataset allows the identification of high-risk areas prone to pothole formation, enabling targeted interventions before significant damage occurs. Moreover, real-time monitoring systems can alert maintenance crews to emerging issues, facilitating rapid responses and minimising the impact on road users.

The implementation of these technologies can lead to significant improvements in efficiency and cost effectiveness. By prioritising maintenance efforts based on data-driven insights, authorities can optimise resource allocation, ensuring that limited budgets are utilised where they are most required. This approach not only enhances the overall quality of road infrastructure but also contributes to increased safety for motorists and reduces vehicle damage. Additionally, the ability to predict and prevent pothole formation through data analysis can result in longer-lasting road surfaces, ultimately reducing the frequency and cost of repairs. As these systems become more sophisticated and widely adopted, they have the potential to create a paradigm shift in road maintenance strategies, moving from reactive to predictive and preventive approaches.

Impact on the Environment and Sustainability

Compared with conventional pothole repair techniques, the suggested technology may considerably lessen the environmental impact. The detection and repair procedures may be automated to limit the need for heavy machinery and manual labour, which, in turn, reduces greenhouse gas emissions and the carbon footprint. Furthermore, the effective use of materials and levelling systems in the system helps reduce waste and conserve resources.

The proposed technology for pothole repair offers significant environmental benefits beyond emission reduction and resource conservation. By automating the detection and repair processes, traffic disruptions and congestion typically associated with road maintenance are minimised, further decreasing overall vehicle emissions. The precision of the system in material application ensures that repairs are more durable, reducing the frequency of maintenance cycles and the associated environmental costs of repeated interventions.

Moreover, this innovative approach aligns with the broader sustainability goals of urban infrastructure management. By integrating smart technologies and data-driven decision-making, cities can optimise their road maintenance strategies, leading to more efficient resource allocation and long-term savings. This proactive maintenance model not only extends the lifespan of road infrastructure but also contributes to improved road safety and reduced vehicle damage, indirectly supporting sustainable transportation. As cities worldwide strive for smarter and more sustainable urban environments, such technologies play a crucial role in creating resilient and eco-friendly transport networks.

Implications. The suggested system's advantages for sustainability and the environment complement international initiatives to slow down climate change and advance sustainable development. Developing nations may contribute to resource conservation and environmental preservation while simultaneously improving their road infrastructure using this creative idea.

The advantages of the proposed system for sustainability and the environment extend beyond immediate resource conservation. By utilising recycled materials and reducing the need for new asphalt production, this approach significantly decreases the carbon emissions associated with road construction. This aligns with global efforts to mitigate the impacts of climate change and promote circular economy principles. Moreover, the system's potential to enhance road durability could lead to less frequent maintenance and reconstruction, further reducing long-term environmental impact and resource consumption.

In developing countries, implementing this innovative road construction method could yield multiple benefits. It not only addresses the pressing need for improved infrastructure but also provides an opportunity to leapfrog traditional resource-intensive construction techniques. By adopting this eco-friendly approach, these countries can demonstrate leadership in sustainable development practices and potentially attract international support and investment. Additionally, the system can create new local industries focused on plastic waste collection and processing, contributing to both environmental cleanup efforts and economic growth. This holistic approach to infrastructure development aligns with the United Nations Sustainable Development Goals, particularly those related to sustainable cities, responsible consumption, and climate action.

Cost-Effectiveness and Scalability

Pothole identification and road repair have been shown to be economically feasible tasks for the Arduino-based sensor paradigm. The system is economical and accessible because of its use of open-source technology and widely available components, especially for developing nations with low resources.

The Arduino-based sensor paradigm for pothole identification and road repair is significantly cost-effective and scalable, making it an attractive solution for various regions, particularly in developing nations with limited resources. By leveraging open-source technology and readily available components, this system minimises the implementation costs while maintaining efficiency. The accessibility of Arduino platforms and associated sensors allows for easy adoption and modification, enabling local engineers and technicians to customise the system to suit specific regional needs and road conditions of the region.

Furthermore, the scalability of this approach is significant. As the system relies on modular components, it can be easily expanded or replicated across different areas without substantial investments. This scalability factor is crucial for widespread implementation, allowing municipalities or road maintenance authorities to gradually increase their coverage and improve road quality on a larger scale. The combination of low-cost hardware and open-source software also facilitates ongoing improvements and updates, ensuring that the system remains relevant and effective in the long term, even as technology advances and new challenges in road maintenance emerge.

Implications. The proposed system is a feasible choice for broad acceptance and deployment throughout several areas and road networks owing to its affordability and scalability. This can result in major advancements in the general state of road networks and support the growth and development of the economies of emerging nations.

The affordability and scalability of the proposed system make it a promising solution for widespread adoption across diverse regions and road networks. By offering a cost-effective approach to road monitoring and maintenance, this system can be implemented by governments and organisations with different budgets and resources. This accessibility allows for a more comprehensive and uniform approach to road management, potentially bridging the gap between urban and rural areas in terms of infrastructure quality and road safety.

The broad deployment of this system can lead to significant improvements in the overall road network conditions, which can have far-reaching effects on emerging economies. Well-maintained roads facilitate the more efficient transportation of goods and services, enhance connectivity between communities, and improve access to education and healthcare. These factors contribute to increased economic activity, reduced transportation costs, and improved residents' quality of life. Moreover, the system's ability to provide real-time data and predictive maintenance capabilities can help optimise resource allocation, reduce long-term infrastructure costs, and enable more strategic planning for future development projects.

The results of the system testing and execution demonstrate its potential as an inventive and sustainable pothole detection and road maintenance solution. To exploit the benefits of the system and handle any potential obstacles or limitations that may occur, further studies, development, and widespread deployment are required.

The testing and execution results of the proposed system demonstrated its promise as an innovative and sustainable approach to pothole detection and road maintenance. By leveraging advanced technologies, such as machine learning algorithms, sensor networks, and real-time data processing, the system offers a more efficient and cost-effective method for identifying and addressing road surface issues. This proactive approach to infrastructure management has the potential to significantly reduce maintenance costs, improve road safety and enhance overall transportation efficiency.

However, to fully realise the potential of the system, further research and development are necessary. This includes refining detection algorithms to improve accuracy across diverse road conditions and weather scenarios, optimising the sensor network for broader coverage and energy efficiency, and integrating the system with existing road maintenance workflows. Additionally, large-scale deployment and real-world testing in various urban and rural environments are crucial for validating the system's effectiveness and identifying any unforeseen challenges. Addressing these aspects will not only enhance the reliability of the system but also facilitate its adoption by transportation authorities and pave the way for smarter and more responsive road maintenance practices in the future.

IMPLEMENTATION AND RESULTS

The proposed technique was implemented on a modest scale and evaluated using a prototype model made of easily obtainable parts. The outcomes showed how well the system could identify dry and water-filled potholes of various sizes and depths.

Compared to conventional human techniques, the automated material discharge and levelling systems proved to be dependable and effective, greatly reducing the time and effort required for pothole repair. Furthermore, real-time data transfer made possible by the Internet of Things connectivity feature allows for timely maintenance planning and reaction by the appropriate authorities.

The prototype model demonstrated promising results in pothole detection and repair, showcasing the system's ability to accurately identify both dry and water-filled potholes across a range of dimensions. This versatility in detection is crucial for addressing the diverse nature of road damage in real-world scenarios. The automated material discharge and levelling systems proved to be a significant improvement over traditional manual methods, offering enhanced reliability and efficiency. By reducing the time and labour required for pothole repair, this automation has the potential to revolutionise road maintenance practices, allowing more frequent and widespread repairs to be conducted with fewer resources.

The integration of Internet of Things (IoT) connectivity further amplifies the utility of the system by enabling real-time data transmission. This feature provides maintenance teams with up-to-date information on road conditions, facilitating strategic and responsive maintenance planning. By allowing for the immediate notification of newly detected potholes or completed repairs, the IoT component enhances the overall efficiency of road maintenance operations. This real-time capability not only improves the speed of repairs but also enables better resource allocation and prioritisation of maintenance tasks based on the severity and location of the road damage.

The sustainability of the proposed system is one of its main benefits. The pothole identification and repair procedure may be automated to minimise its related environmental effects and carbon footprint by reducing the requirement for physical labour. Additionally, the system's capacity to quickly fix potholes leads to increased fuel economy, decreased vehicle damage, and better road safety, all of which have favourable effects on the environment.

The suggested Arduino-based sensor paradigm for pothole identification and road repair creatively and sustainably solves a long-standing problem faced by poor nations. Through the utilisation of cutting-edge sensor technologies, automation, and Internet of Things connections, the system provides an economical and effective method for maintaining road infrastructure.

Prospective advancements in this field may encompass the incorporation of machine learning algorithms to enhance the precision of pothole identification and prioritisation, along with the investigation of substitute building materials and methodologies for more enduring and eco-friendly pothole restorations.

The proposed Arduino-based sensor system for pothole detection and road repair offers a comprehensive solution to the persistent infrastructure challenges in developing countries. By leveraging advanced sensor technologies, automation, and Internet of Things (IoT) connectivity, this system presents a cost-effective and efficient approach for maintaining road infrastructure. The automation of pothole detection and repair processes significantly reduces the need for manual labour, thereby minimising the associated environmental impact and carbon footprint. This sustainability aspect is further enhanced by the system's ability to swiftly address road damage, leading to improved fuel efficiency, reduced vehicle wear and tear, and enhanced road safety, all of which contribute positively to environmental conservation.

The potential for future advancements in this field is substantial and promising. The integration of machine learning algorithms can significantly improve the accuracy of pothole detection and prioritisation, allowing for more targeted and efficient repairs.

Additionally, exploring alternative construction materials and techniques could lead to more durable and environmentally friendly road surfaces than those currently used. These innovations could further enhance the effectiveness and sustainability of the system, potentially revolutionising road maintenance practices in developing nations. By addressing the critical issue of road infrastructure maintenance in a creative and sustainable manner, this Arduino-based sensor paradigm represents a significant step forward in improving transportation networks and quality of life in resource-constrained environments.

REFERENCES

Bernas, M., Płaczek, B., Korski, W., Loska, P., Smyła, J., & Szymała, P. (2018). A Survey and Comparison of Low-Cost Sensing Technologies for Road Traffic Monitoring. *Sensors (Basel), 18*(10), 3243. DOI: 10.3390/s18103243 PMID: 30261665

Blythe, P., Bryan, H., Watson, P., Sharif, B., Neasham, J., Edwards, S., Wagner, J., Bell, M. C., & Suresh, V. (2008). An environmental sensor system for pervasively monitoring road networks. *IET Road Transport Information and Control Conference and the ITS United Kingdom Members' Conference (RTIC 2008)*, 91–91. DOI: 10.1049/ic.2008.0805

Bučko, B., Lieskovská, E., Zábovská, K., & Zábovský, M. (2022). Computer Vision Based Pothole Detection under Challenging Conditions. *Sensors (Basel), 22*(22), 8878. DOI: 10.3390/s22228878 PMID: 36433474

Dhiman, A., & Klette, R. (2020). Pothole Detection Using Computer Vision and Learning. *IEEE Transactions on Intelligent Transportation Systems, 21*(8), 3536–3550. DOI: 10.1109/TITS.2019.2931297

Fan, R., Ozgunalp, U., Hosking, B., Liu, M., & Pitas, I. (2020). Pothole Detection Based on Disparity Transformation and Road Surface Modeling. *IEEE Transactions on Image Processing, 29*, 897–908. DOI: 10.1109/TIP.2019.2933750 PMID: 31449016

Finogeev, A., Finogeev, A., Fionova, L., Lyapin, A., & Lychagin, K. A. (2019). Intelligent monitoring system for smart road environment. *Journal of Industrial Information Integration, 15*, 15–20. DOI: 10.1016/j.jii.2019.05.003

Gupta, B. D., Semwal, V., & Pathak, A. (2020). Nanotechnology-based fiber-optic chemical and biosensors. In *Nano-Optics* (pp. 163–195). Elsevier. DOI: 10.1016/B978-0-12-818392-2.00007-X

Hossain, M. J., & Bari, Md. A., & Khan, M. M. (2022). Development of an IoT Based Health Monitoring System for e-Health. *2022 IEEE 12th Annual Computing and Communication Workshop and Conference (CCWC)*, 31–37. DOI: 10.1109/CCWC54503.2022.9720825

Karpis, O., Juricek, J., & Micek, J. (2010). Application of Wireless Sensor Networks for Road Monitoring. *IFAC Proceedings Volumes, 43*(24), 183–188. https://doi.org/DOI: 10.3182/20101006-2-PL-4019.00035

Katsamenis, I., Bimpas, M., Protopapadakis, E., Zafeiropoulos, C., Kalogeras, D., Doulamis, A., Doulamis, N., Martín-Portugués Montoliu, C., Handanos, Y., Schmidt, F., Ott, L., Cantero, M., & Lopez, R. (2022). Robotic Maintenance of Road Infrastructures: The HERON Project. *Proceedings of the 15th International Conference on PErvasive Technologies Related to Assistive Environments*, 628–635. DOI: 10.1145/3529190.3534746

Ke, X., Lin, J. Y., Fu, C., & Wang, Y. (2020). Transport Infrastructure Development and Economic Growth in China: Recent Evidence from Dynamic Panel System-GMM Analysis. *Sustainability, 12*(14), 5618. DOI: 10.3390/su12145618

Kim, Y.-M., Kim, Y.-G., Son, S.-Y., Lim, S.-Y., Choi, B.-Y., & Choi, D.-H. (2022). Review of Recent Automated Pothole-Detection Methods. *Applied Sciences (Basel, Switzerland), 12*(11), 5320. DOI: 10.3390/app12115320

Lawrence, B. D., & Simmons, J. A. (1982). Measurements of atmospheric attenuation at ultrasonic frequencies and the significance for echolocation by bats. *The Journal of the Acoustical Society of America, 71*(3), 585–590. DOI: 10.1121/1.387529 PMID: 7085967

Ma, N., Fan, J., Wang, W., Wu, J., Jiang, Y., Xie, L., & Fan, R. (2022). Computer vision for road imaging and pothole detection: A state-of-the-art review of systems and algorithms. *Transportation Safety and Environment, 4*(4), tdac026. Advance online publication. DOI: 10.1093/tse/tdac026

Moharrak, M., Nguyen, N. P., & Mogaji, E. (2024). Business environment and adoption of AI: Navigation for internationalization by new ventures in emerging markets. *Thunderbird International Business Review, 66*(4), 355–372. DOI: 10.1002/tie.22384

Omar, M., & Kumar, P. (2020). Detection of Roads Potholes using YOLOv4. *2020 International Conference on Information Science and Communications Technologies (ICISCT)*, 1–6. DOI: 10.1109/ICISCT50599.2020.9351373

Park, S.-S., Tran, V.-T., & Lee, D.-E. (2021). Application of Various YOLO Models for Computer Vision-Based Real-Time Pothole Detection. *Applied Sciences (Basel, Switzerland), 11*(23), 11229. DOI: 10.3390/app112311229

Quium, A. (2019). Transport Corridors for Wider Socio–Economic Development. *Sustainability, 11*(19), 5248. DOI: 10.3390/su11195248

Rupp, T., Volz, H., & Cord, T. (1998). Automated and Robotics-Based Techniques for Road Construction. *IFAC Proceedings Volumes, 31*(3), 515–520. DOI: 10.1016/S1474-6670(17)44137-1

Sattar, S., Li, S., & Chapman, M. (2018). Road Surface Monitoring Using Smartphone Sensors: A Review. *Sensors (Basel), 18*(11), 3845. DOI: 10.3390/s18113845 PMID: 30423962

Sharma, S. K., Phan, H., & Lee, J. (2020). An Application Study on Road Surface Monitoring Using DTW Based Image Processing and Ultrasonic Sensors. *Applied Sciences (Basel, Switzerland), 10*(13), 4490. DOI: 10.3390/app10134490

Skibniewski, M., & Hendrickson, C. (1990). Automation and Robotics for Road Construction and Maintenance. *Journal of Transportation Engineering, 116*(3), 261–271. DOI: 10.1061/(ASCE)0733-947X(1990)116:3(261)

Sri Kakarla, U., Venkataramana, A., Gopinath, S., Shivaram, K., Yugendhar, K., Harika, K. V. S. L., & Dixit, S. (2024). Revolutionizing Agricultural Harvesting with IoT Application. *MATEC Web of Conferences, 392*, 01014. DOI: 10.1051/matecconf/202439201014

Talha, S. A., Manasreh, D., & Nazzal, M. D. (2024). The Use of Lidar and Artificial Intelligence Algorithms for Detection and Size Estimation of Potholes. *Buildings (Basel, Switzerland), 14*(4), 1078. DOI: 10.3390/buildings14041078

Yebes, J. J., Montero, D., & Arriola, I. (2021). Learning to Automatically Catch Potholes in Worldwide Road Scene Images. *IEEE Intelligent Transportation Systems Magazine, 13*(3), 192–205. DOI: 10.1109/MITS.2019.2926370

Yik, Y. K., Alias, N. E., Yusof, Y., & Isaak, S. (2021). A Real-time Pothole Detection Based on Deep Learning Approach. *Journal of Physics: Conference Series, 1828*(1), 012001. DOI: 10.1088/1742-6596/1828/1/012001

APPENDIX

Code used:

```
Program
const int pingPin_1 = 2; // Trigger Pin of Ultrasonic Sensor front uv_sensor
const int echoPin_1 = 3; // Echo Pin of Ultrasonic Sensor
const int pingPin = 5; // Trigger Pin of Ultrasonic Sensor
const int echoPin = 4; // Echo Pin of Ultrasonic Sensor // pothole sensor
int IN1=14; //FORWARD BACKWARD
int IN2=15;
int IN3=16;
int IN4=17;
Program
int IN5=10; //FILLING
int IN6=11;
int IN7=8; // LEVELING
int IN8=9;
void FORWARD();
void REVERSE();
void LEFT();
void RIGHT();
void STOP();
void OPEN_MUD();
void LEVEL_DOWN();
void LEVEL_UP();
long microsecondsToCentimeters(long microseconds);
//char ch;
char ch;
int a=0;
void setup()
{
pinMode(IN1,OUTPUT);
pinMode(IN2,OUTPUT);
pinMode(IN3,OUTPUT);
pinMode(IN4,OUTPUT);
pinMode(IN5,OUTPUT);
pinMode(IN6,OUTPUT);
```

```
pinMode(IN7,OUTPUT);
pinMode(IN8,OUTPUT
```

Compilation of References

Abasova, A., & Mamedov, E. (2025). *Analysis of modern development trends in the construction sector*. Academic Press.

Abbas, M., Akhai, S., Abbas, U., Jafri, R., & Arif, S. M. (2025). Ai-enabled sustainable urban planning and management. In *Real-World Applications of AI Innovation* (pp. 233–260). IGI Global Scientific Publishing.

Abbass, M., Abbas, U., Jafri, R., Arif, S. M., & Akhai, S. (2025). AI and Machine Learning Applications in Sustainable Smart Cities. In *Sustainable Smart Cities and the Future of Urban Development* (pp. 1–32). IGI Global Scientific Publishing.

Abdel-Rahman, S., Dokhan, S., & El-Dash, K. (2023). Outsourcing engineering tasks by construction firms: State-of-the-art. *Engineering Research Journal, 52*(4), 157–168.

Abidin, H., Andreas, H., Gumilar, I., & Brinkman, J. (2015). Study on the risk and impacts of land subsidence in Jakarta. *Proceedings of the International Association of Hydrological Sciences, 372*(372), 115–120.

Abioye, S. O., Oyedele, L. O., Akanbi, L., Ajayi, A., Delgado, J. M. D., Bilal, M., Akinade, O. O., & Ahmed, A. (2021). Artificial intelligence in the construction industry: A review of present status, opportunities and future challenges. *Journal of Building Engineering, 44*, 103299.

Adam, A., & Lindahl, G. (2017). Applying the dynamic capabilities framework in the case of a large public construction client. *Construction Management and Economics, 35*(7), 420–431. DOI: 10.1080/01446193.2016.1274452

Adams, R., Jeanrenaud, S., Bessant, J., Denyer, D., & Overy, P. (2016). Sustainability-oriented innovation: A systematic review. *International Journal of Management Reviews, 18*(2), 180–205.

Adeyemi, A. B., Ohakawa, T. C., Okwandu, A. C., Iwuanyanwu, O., & Ifechukwu, G.-O. (2024). Integrating modular and prefabricated construction techniques in affordable housing: Architectural design considerations and benefits. *Journal of Construction Innovation, 18*(2), 67–82.

Agee, P., Nikdel, L., & Roberts, S. (2021). A measured energy use, solar production, and building air leakage dataset for a zero energy commercial building. *Scientific Data, 8*(1). Advance online publication. DOI: 10.1038/s41597-021-01082-8

Aghimien, D., Aigbavboa, C., & Matabane, K. (2023). Dynamic capabilities for construction organizations in the Fourth Industrial Revolution era. *International Journal of Construction Management, 23*(5), 855–864. DOI: 10.1080/15623599.2020.1813832

Agwu, K., Oraefo, O., Jude, B., Onuorah, I., Onyemaechi, C., & Chinenye, C. (2024). Exploring the impacts of sustainable design practices on construction cost. *International Journal of Progressive Research in Engineering Management and Science, 4*, 1531–1537.

Agyemang, E., Ofori-Dua, K., Dwumah, P., & Forkuor, J. B. (2024). Towards responsible resource utilization: A review of sustainable vs. Unsustainable reuse of wood waste. *PLoS One, 19*(12), e0312527. PMID: 39715227

Ahmad, N., & Hassan, Q. (2022). Enhancing Blue-Green Infrastructures for Flood and Water Stress Management: A Case Study of Chennai. In *International Conference on Trends and Recent Advances in Civil Engineering* (pp. 97-117). Singapore: Springer Nature Singapore.

Ahmed, A., Ge, T., Peng, J., Yan, W.-C., Tee, B. T., & You, S. (2022). Assessment of the renewable energy generation towards net-zero energy buildings: A review. *Energy and Building, 256*, 111755.

Ahmed, S., Meenar, M., & Alam, A. (2019). Designing a Blue-Green Infrastructure (BGI) network: Toward water-sensitive urban growth planning in Dhaka, Bangladesh. *Land (Basel), 8*(9), 138.

Aina, L. B. &. H. M. a. &. Y. A. (2023). Developing an urban computing framework for smart and sustainable neighborhoods: a case study of Alkhaledia in Jizan City, Saudi Arabia. ideas.repec.org. http://ideas.repec.org/a/gam/jsusta/v15y2023i5p4057-d1077795.html

Ajaj, R. (2024). A case study on the barriers towards achieving sustainable smart city for Abu Dhabi. *IET Smart Cities, 6*(2), 112–128. DOI: 10.1049/smc2.12077

Ajirotutu, R. O., Adeyemi, A. B., Ifechukwu, G.-O., Iwuanyanwu, O., Ohakawa, T. C., & Garba, B. M. P. (2024). Future cities and sustainable development: Integrating renewable energy, advanced materials, and civil engineering for urban resilience. *International Journal of Sustainable Urban Development, 3*.

Akhai, S., & Abbass, M. (2025). Toward Resilient Futures: The Role of AI-Driven Strategies in Climate Adaptation. In *Nexus of AI, Climatology, and Urbanism for Smart Cities* (pp. 325-340). IGI Global Scientific Publishing.

Akhai, S., & Rana, M. (2022). Taguchi-based grey relational analysis of abrasive water jet machining of Al-6061. *Materials Today: Proceedings*, *65*, 3165–3169.

Akhai, S., Srivastava, P., Sharma, V., & Bhatia, A. (2021, August). Investigating weld strength of AA8011-6062 alloys joined via friction-stir welding using the RSM approach. [). IOP Publishing.]. *Journal of Physics: Conference Series*, *1950*(1), 012016.

Akhai, S., & Taneja, T. (2025). The Critical Role of Water Quality. In *Smart Water Technology for Sustainable Management in Modern Cities* (pp. 101–116). IGI Global. DOI: 10.4018/979-8-3693-8074-1.ch005

Akhil, U. V., Radhika, N., Saleh, B., Aravind Krishna, S., Noble, N., & Rajeshkumar, L. (2023). A comprehensive review on plant-based natural fiber reinforced polymer composites: Fabrication, properties, and applications. *Polymer Composites*, *44*(5), 2598–2633.

Akhimien, N. G., Latif, E., & Hou, S. S. (2021, June 1). Application of circular economy principles in buildings: A systematic review. *Journal of Building Engineering*. Elsevier Ltd. DOI: 10.1016/j.jobe.2020.102041

Alam Bhuiyan, M. M., & Hammad, A. (2024). Engineering and Design for Sustainable Construction: A Bibliometric Analysis of Current Status and Future Trends. *Sustainability*, *16*(7), 2959.

Aldossari, M. Q., & Sidorova, A. (2020). Consumer Acceptance of Internet of Things (IoT): Smart Home Context. *Journal of Computer Information Systems*, *60*(6), 507–517. DOI: 10.1080/08874417.2018.1543000

Alexander, C. (1977). *A pattern language: Towns, buildings, construction*. Oxford University Press.

Alfano, D. F. R., Dell'Isola, M., Ficco, G., & Tassini, F. (2012). Experimental analysis of air tightness in Mediterranean buildings using the fan pressurization method. *Building and Environment*, *53*, 16–25. DOI: 10.1016/J.BUILDENV.2011.12.017

Alfano, F. R. d., Dell'Isola, M., Ficco, G., Palella, B. I., & Riccio, G. (2016). Experimental air-tightness analysis in mediterranean buildings after windows retrofit. *Sustainability*, *8*, 991.

Alhasnawi, B. N., Almutoki, S. M. M., Hussain, F. F. K., Harrison, A., Bazooyar, B., Zanker, M., & Bureš, V. (2024). A new methodology for reducing carbon emissions using multi-renewable energy systems and artificial intelligence. *Sustainable Cities and Society*, *114*, 105721. DOI: 10.1016/j.scs.2024.105721

Alijoyo, F. A. (2024). AI-powered deep learning for sustainable industry 4.0 and internet of things: Enhancing energy management in smart buildings. *Alexandria Engineering Journal*, *104*, 409–422.

Almarzouq, A., & Sakhrieh, A. (2018). Effects of glazing design and infiltration rate on energy consumption and thermal comfort in residential buildings. *Thermal Science*, *2018*, 2951–2960. DOI: 10.2298/TSCI170910073A

Almeida, F. (2023). Prospects of Cybersecurity in Smart Cities. *Future Internet*, *15*(9). Advance online publication. DOI: 10.3390/fi15090285

Almihat, M. G. M., & Munda, J. L. (2025). The role of smart grid technologies in urban and sustainable energy planning. *Energies*, *18*(7), 1618. DOI: 10.3390/en18071618

Alnoman, A. (2021). Edge computing services for smart Cities: A review and case study. *2022 International Symposium on Networks, Computers and Communications (ISNCC)*, 1–6. DOI: 10.1109/isncc52172.2021.9615785

Al-Numan, B. S. O. (2024). Construction Industry Role in Natural Resources Depletion and How to Reduce It. In *Natural Resources Deterioration in MENA Region: Land Degradation, Soil Erosion, and Desertification* (pp. 93-109). Springer.

Alshehri, K., Chen, I.-C., Rugani, B., Sapsford, D., Harbottle, M., & Cleall, P. (2024). A novel uncertainty assessment protocol for integrated ecosystem services-life cycle assessments: A comparative case of nature-based solutions. *Sustainable Production and Consumption*, *47*, 499–515.

Alshuwaikhat, H. M., Adenle, Y. A., & Almuhaidib, T. (2022). A Lifecycle-Based Smart Sustainable City Strategic Framework for Realizing Smart and Sustainability Initiatives in Riyadh City. *Sustainability (Switzerland)*, *14*(14). Advance online publication. DOI: 10.3390/su14148240

Altomonte, S. (2008). Daylight for energy savings and psycho-physiological well-being in sustainable built environments. *Journal of Sustainable Development*, *1*(3), 3–16.

Aly, M. (2012). *Development of an eco-friendly composite material for engineering applications* (Doctoral dissertation, Dublin City University).

An, Z., Chen, Q., & Li, J. (2020). Ecological strategies of urban ecological parks–a case of Bishan Ang Mo Kio park and Kallang river in Singapore. E3S Web of Conferences.

Andrade, R. O., Yoo, S. G., Tello-Oquendo, L., & Ortiz-Garces, I. (2020). A Comprehensive Study of the IoT Cybersecurity in Smart Cities. *IEEE Access : Practical Innovations, Open Solutions, 8*, 228922–228941. DOI: 10.1109/ACCESS.2020.3046442

Anguelovski, I., & Frumkin, H. (2025). Safeguarding nature and cities in a burning country. *Nature Cities*, 1-2.

Annan-Diab, F., & Molinari, C. (2017). Interdisciplinarity: Practical approach to advancing education for sustainability and for the Sustainable Development Goals. *International Journal of Management Education, 15*(2), 73–83.

Annex – 2014 Guidelines on the Method of Calculation of the Attained Energy Efficiency Design Index (EEDI) for New Ships. (n.d.). https://imorules.com/GUID-507FC12C-452D-4925-8D8E-052333B40148.html

Anthopoulos, L. (2015). Defining Smart City Architecture for Sustainability. *Electronic Government and Electronic Participation: Joint Proceedings of Ongoing Research and Projects of IFIP WG 8.5 EGOV and EPart 2015*, 140–147.

Appleyard, D. (1981). *Livable streets*. University of California Press.

Argote, L., & Miron-Spektor, E. (2011). Organizational learning: From experience to knowledge. *Organization Science, 22*(5), 1123–1137.

Arif, S. M., Akhai, S. S., & Dadhwal, R. (2025). High-Performance Aluminium Alloys for Shipbuilding. In *Innovative Materials for Next-Generation Defense Applications* (pp. 107–144). IGI Global. DOI: 10.4018/979-8-3373-0933-0.ch004

Arora, N. K., & Mishra, I. (2019). United Nations Sustainable Development Goals 2030 and environmental sustainability: Race against time. *Environmental Sustainability, 2*(4), 339–342. DOI: 10.1007/s42398-019-00092-y

Ashour, T. (2017). Composites using agricultural wastes. *Handbook of composites from renewable materials*. Wiley.

Asim, M., Uddin, G. M., Jamshaid, H., Raza, A., Hussain, U., Satti, A. N., & Arafat, S. M. (2020). Comparative experimental investigation of natural fibers reinforced light weight concrete as thermally efficient building materials. *Journal of Building Engineering*, *31*, 101411.

Ataş, M., & Talay, A. C. (2022). Development of automatic tree counting software from UAV based aerial images with machine learning. arXiv preprint . 2201. 02698. DOI: 10. 48550/ arXiv

Athira, G., Bahurudeen, A., & Appari, S. (2021). Rice-straw ash as a potential supplementary cementitious material: Influence of thermochemical conversion on its properties. *Journal of Materials in Civil Engineering*, *33*(6), 04021123.

Ayub, M. A., Usman, M., Faiz, T., Umair, M., ul Haq, M. A., Rizwan, M., & Zia ur Rehman, M. (2020). Restoration of degraded soil for sustainable agriculture. *Soil health restoration and management*, 31-81.

Babalola, A., & Harinarain, N. (2024). Policy barriers to sustainable construction practice in the Nigerian construction industry: An exploratory factor analysis. *Journal of Engineering. Defence Technology*, *22*(1), 214–234.

Baduge, S. K., Thilakarathna, S., Perera, J. S., Arashpour, M., Sharafi, P., Teodosio, B., Shringi, A., & Mendis, P. (2022). Artificial intelligence and smart vision for building and construction 4.0: Machine and deep learning methods and applications. *Automation in Construction*, *141*, 104440.

Bakıcı, T., Almirall, E., & Wareham, J. (2013). A Smart City Initiative: The Case of Barcelona. *Journal of the Knowledge Economy*, *4*(2), 135–148. DOI: 10.1007/s13132-012-0084-9

Banerjee, K., Vats, I., Akhtar, N., Harsha, K. G., Vinooth, P., Kumar, A., . . . Gautam, P. (2022, December). A review on artificial intelligence based sign language recognition techniques. In *2022 5th International Conference on Contemporary Computing and Informatics (IC3I)* (pp. 2195-2201). IEEE.

Bansal, V., & Haridasan, H. (2024). Blue–Green Infrastructure as a Stimulant of Growth: Case of Ahmedabad. In *Blue-Green Infrastructure for Sustainable Urban Settlements: Implications for Developing Countries Under Climate Change* (pp. 365–385). Springer Nature Switzerland.

Barczewski, M., Matykiewicz, D., Piasecki, A., & Szostak, M. (2018). Polyethylene green composites modified with post agricultural waste filler: Thermo-mechanical and damping properties. *Composite Interfaces*, *25*(4), 287–299.

Barles, S. (2010). Society, energy and materials: The contribution of urban metabolism studies to sustainable urban development issues. *Journal of Environmental Planning and Management, 53*(4), 439–455. DOI: 10.1080/09640561003703772

Barney, J. (1991). Firm resources and sustained competitive advantage. *Journal of Management, 17*(1), 99–120. DOI: 10.1177/014920639101700108

Barreira, A. P., Andraz, J., Ferreira, V., & Panagopoulos, T. (2023). Perceptions and preferences of urban residents for green infrastructure to help cities adapt to climate change threats. *Cities (London, England), 141*. Advance online publication. DOI: 10.1016/j.cities.2023.104478

Barreiro, J., Zaragoza, S., & Diaz-Casas, V. (2022). Review of ship energy efficiency. *Ocean Engineering, 257*, 111594. DOI: 10.1016/j.oceaneng.2022.111594

Bartosiewicz, S. (1992). *Econometrics with elements of mathematical programming and comparative analysis*. Wrocław University of Economics Press.

Battemarco, B. P., Tardin-Coelho, R., Veról, A. P., de Sousa, M. M., da Fontoura, C. V. T., Figueiredo-Cunha, J., & Miguez, M. G. (2022). Water dynamics and blue-green infrastructure (BGI): Towards risk management and strategic spatial planning guidelines. *Journal of Cleaner Production, 333*, 129993.

Batty, M. (2018). Digital twins. *Environment and Planning. B, Urban Analytics and City Science, 45*(5), 817–820. DOI: 10.1177/2399808318796416

Batty, M., Axhausen, K. W., Giannotti, F., Pozdnoukhov, A., Bazzani, A., Wachowicz, M., & Portugali, Y. (2012). Smart cities of the future. *The European Physical Journal. Special Topics, 214*(1), 481–518. DOI: 10.1140/epjst/e2012-01703-3

Bayraktar, M., & Yuksel, O. (2023). A scenario-based assessment of the energy efficiency existing ship index (EEXI) and carbon intensity indicator (CII) regulations. *Ocean Engineering, 278*, 114295. DOI: 10.1016/j.oceaneng.2023.114295

Bechtel Corporation. (2025). *Dream, Design, Deliver.* Retrieved from https://www.bechtel.com/approach/

Behar, C. (2014). Utilising resident feedback to inform energy-saving interventions at the Barbican. *Local Environment, 19*(5), 539–559.

Benites, A. J., & Simões, A. F. (2021). Assessing the urban sustainable development strategy: An application of a smart city services sustainability taxonomy. *Ecological Indicators, 127*. Advance online publication. DOI: 10.1016/j.ecolind.2021.107734

Bernas, M., Płaczek, B., Korski, W., Loska, P., Smyła, J., & Szymała, P. (2018). A Survey and Comparison of Low-Cost Sensing Technologies for Road Traffic Monitoring. *Sensors (Basel), 18*(10), 3243. DOI: 10.3390/s18103243 PMID: 30261665

Bernstein, M. S., Levi, M., Magnus, D., Rajala, B., Satz, D., & Waeiss, C. (2021). ESR: ethics and society review of artificial intelligence research. arXiv preprint https:// arxiv. org/ pdf/ 2106. 11521.pdf

Bertino, G., Kisser, J., Zeilinger, J., Langergraber, G., Fischer, T., & Österreicher, D. (2021). Fundamentals of building deconstruction as a circular economy strategy for the reuse of construction materials. *Applied Sciences (Switzerland), 11*(3), 1–31. DOI: 10.3390/app11030939

Bhandari, M., Hun, D., Shrestha, S., Pallin, S., & Lapsa, M. (2018). A simplified methodology to estimate energy savings in commercial buildings from improvements in airtightness. *Energies, 11*(12). Advance online publication. DOI: 10.3390/en11123322

Bhuyan, N., Narzari, R., Gogoi, L., Bordoloi, N., Hiloidhari, M., Palsaniya, D. R., & Kataki, R. (2020). Valorization of agricultural wastes for multidimensional use. In *Current developments in biotechnology and bioengineering* (pp. 41–78). Elsevier.

Bibri, S. E., & Krogstie, J. (2017, May 1). Smart sustainable cities of the future: An extensive interdisciplinary literature review. *Sustainable Cities and Society*. Elsevier Ltd. DOI: 10.1016/j.scs.2017.02.016

Bilal, M., Khan, K. I. A., Thaheem, M. J., & Nasir, A. R. (2020). Current state and barriers to the circular economy in the building sector: Towards a mitigation framework. *Journal of Cleaner Production, 276*. Advance online publication. DOI: 10.1016/j.jclepro.2020.123250

Binici, H., Eken, M., Kara, M., & Dolaz, M. (2013, October). An environment-friendly thermal insulation material from sunflower stalk, textile waste and stubble fibers. In *2013 International conference on renewable energy research and applications (ICRERA)* (pp. 833-846). IEEE.

Biswal, B. K., Bolan, N., Zhu, Y.-G., & Balasubramanian, R. (2022). Nature-based Systems (NbS) for mitigation of stormwater and air pollution in urban areas: A review. *Resources, Conservation and Recycling, 186*, 106578.

Blayse, A. M., & Manley, K. (2004). Key influences on construction innovation. *Construction Innovation, 4*(3), 143–154.

Blythe, P., Bryan, H., Watson, P., Sharif, B., Neasham, J., Edwards, S., Wagner, J., Bell, M. C., & Suresh, V. (2008). An environmental sensor system for pervasively monitoring road networks. *IET Road Transport Information and Control Conference and the ITS United Kingdom Members' Conference (RTIC 2008)*, 91–91. DOI: 10.1049/ic.2008.0805

Boafo, F. E., Kim, J.-H., & Kim, J.-T. (2016). Performance of modular prefabricated architecture: Case study-based review and future pathways. *Sustainability*, *8*(6), 558.

Boopathi, S., Myilsamy, S., & Sukkasamy, S. (2021). *Experimental Investigation and Multi-Objective Optimization of Cryogenically Cooled Near-Dry Wire-Cut EDM Using TOPSIS Technique*. IJAMT PREPRINT.

Boopathi, S., & Sivakumar, K. (2013). Experimental investigation and parameter optimization of near-dry wire-cut electrical discharge machining using multi-objective evolutionary algorithm. *Springer: The International Journal of Advanced Manufacturing Technology*, *67*(9–12), 2639–2655.

Boopathi, S., Umareddy, M., & Elangovan, M. (2023). Applications of Nano-Cutting Fluids in Advanced Machining Processes. In *Sustainable Utilization of Nanoparticles and Nanofluids in Engineering Applications* (pp. 211–234). IGI Global.

Boyer, R., Peterson, N., Arora, P., & Caldwell, K. (2016). Five approaches to social sustainability and an integrated way forward. *Sustainability*. Advance online publication. DOI: 10. 3390/ su809 0878

Bradley, C. A., & Altizer, S. (2007, February). Urbanization and the ecology of wildlife diseases. *Trends in Ecology & Evolution*. Advance online publication. DOI: 10.1016/j.tree.2006.11.001 PMID: 17113678

Bramiana, C. N., Entrop, A. G., & Halman, J. I. M. (2016). Relationships between Building Characteristics and Airtightness of Dutch Dwellings. In *Energy Procedia* (Vol. 96, pp. 580–591). Elsevier Ltd. DOI: 10.1016/j.egypro.2016.09.103

Brevini, B. (2020). Black boxes, not green: Mythologizing artificial intelligence and omitting the environment. *Big Data & Society*, *7*(2). Advance online publication. DOI: 10. 1177/ 20539 51720 935141

Brown, K., & Mijic, A. (2019). *Integrating green and blue spaces into our cities: Making it happen*. Grantham Institute.

Bučko, B., Lieskovská, E., Zábovská, K., & Zábovský, M. (2022). Computer Vision Based Pothole Detection under Challenging Conditions. *Sensors (Basel)*, *22*(22), 8878. DOI: 10.3390/s22228878 PMID: 36433474

Bulkeley, H., & Castán Broto, V. (2013). Government by experiment? Global cities and the governing of climate change. *Transactions of the Institute of British Geographers, 38*(3), 361–375. DOI: 10.1111/j.1475-5661.2012.00535.x

Burgos, M. J. M., de Toledo, H. S. Á., Lezcano, R. A. G., & de Mera, A. G. (2020). The sedentary process and the evolution of energy consumption in eight native American dwellings: Analyzing sustainability in traditional architecture. *Sustainability (Switzerland), 12*(5), 1–28. DOI: 10.3390/su12051810

Burry, M. (2022). A new agenda for AI-based urban design and planning. In *Artificial Intelligence in Urban Planning and Design: Technologies, Implementation, and Impacts* (pp. 3–20). Elsevier. DOI: 10.1016/B978-0-12-823941-4.00005-6

Canton, H. (2021). United Nations Environment Programme—UNEP. In *The Europa directory of international organizations 2021* (pp. 188–214). Routledge.

Carbajal-Pina, C., & Acur, N. (2024). From principles to practice: Responsible implementation of emerging technologies through innovation dilemmas. *IEEE Transactions on Engineering Management.*

Carmona, M., Tiesdell, S., Heath, T., & Oc, T. (2010). *Public Places, Urban Spaces: The Dimensions of Urban Design.* Routledge.

Castaldo, N. F. (2022). *Buildings That Breathe: Greening the World's Cities.* Lerner+ ORM.

Cavagnoli, S., Fabiani, C., de Albuquerque Landi, F. F., & Pisello, A. L. (2024). Advancing sustainable construction through comprehensive analysis of thermal, acoustic, and environmental properties in prefabricated panels with recycled PET materials. *Energy and Building, 312*, 114218. Advance online publication. DOI: 10.1016/j.enbuild.2024.114218

Chaganti, V. N. (2008). *Reclamation potential of amendments for soils irrigated with saline-sodic drainage water.* California State University.

Challoumis, C. (2024, October). Building a sustainable economy-how ai can optimize resource allocation. In *XVI International Scientific Conference* (pp. 190-224).

Chan, , W. R., Joh, J., & Sherman, M. H. (2013). Analysis of air leakage measurements of US houses. *Energy and Building, 66*, 616–625. DOI: 10.1016/j.enbuild.2013.07.047

Chan, W. R., Nazaroff, W. W., Price, P. N., Sohn, M. D., & Gadgil, A. J. (2005). Analyzing a database of residential air leakage in the United States. *Atmospheric Environment, 39*(19), 3445–3455. DOI: 10.1016/j.atmosenv.2005.01.062

Charef, R., Lu, W., & Hall, D. (2022). The transition to the circular economy of the construction industry: Insights into sustainable approaches to improve the understanding. *Journal of Cleaner Production, 364*, 132421.

Charnock, G., March, H., & Ribera-Fumaz, R. (2021). From smart to rebel city? Worlding, provincialising and the Barcelona Model. *Urban Studies (Edinburgh, Scotland), 58*(3), 581–600. DOI: 10.1177/0042098019872119

Chatzitheodoridis, F., Melfou, K., Kontogeorgos, A., & Kalogiannidis, S. (2023). Exploring Key Aspects of an Integrated Sustainable Urban Development Strategy in Greece: The Case of Thessaloniki City. *Smart Cities, 6*(1), 19–39. DOI: 10.3390/smartcities6010002

Chauhan, A. P. (2022). *Impact of streetscape on the image of a city: A study and analysis of streetscapes of Delhi*. Ignited Minds Journals.

Chaurasia, M., Patel, K., Singh, R., Bhadouria, R., & Rao, K. S. (2024). Integrating Blue-Green as Next-Generation Urban Infrastructure in Developing Countries. *Blue-Green Infrastructure for Sustainable Urban Settlements: Implications for Developing Countries Under Climate Change*, 57-82.

Chavan, L., Kumbhakar, R., Agrawal, P., Rakesh, N., Parvathi, R., & Shahid, M. (2024, November). AI in Agriculture: Yield Optimization and Food Security. In *2024 4th International Conference on Technological Advancements in Computational Sciences (ICTACS)* (pp. 536-541). IEEE.

Chen, L., Chen, Z., Zhang, Y., Liu, Y., Osman, A. I., Farghali, M., & Yap, P. S. (2023). Artificial intelligence-based solutions for climate change: A review. *Environmental Chemistry Letters, 21*(5), 2525–2557.

Chen, L., Hu, Y., Wang, R., Li, X., Chen, Z., Hua, J., Osman, A. I., Farghali, M., Huang, L., & Li, J. (2024). Green building practices to integrate renewable energy in the construction sector: A review. *Environmental Chemistry Letters, 22*(2), 751–784.

Chen, L., Yang, M., Chen, Z., Xie, Z., Huang, L., Osman, A. I., Farghali, M., Sandanayake, M., Liu, E., Ahn, Y. H., Al-Muhtaseb, A. H., Rooney, D. W., & Yap, P. S. (2024). Conversion of waste into sustainable construction materials: A review of recent developments and prospects. *Materials Today Sustainability, 27*, 100930. DOI: 10.1016/j.mtsust.2024.100930

Chen, L., Yang, Y., Liu, J., & Xie, P. (2024). Advancing eco-friendly cementitious composites: Integrating carbon capture and sustainable resource utilization. *Cement and Concrete Research, 170*, 106302. DOI: 10.1016/j.cemconres.2024.106302

Chien, H., Saito, O., & Fukushi, K. (2022). Is Ensuring the Sustainable Implementation of BGI Possible? System Thinking of Urban Rivers as Social-Ecological Systems. In *Blue-Green Infrastructure Across Asian Countries: Improving Urban Resilience and Sustainability* (pp. 95–121). Springer Singapore.

Chiesa, G., Di Vita, D., Ghadirzadeh, A., Herrera, A. H. M., & Rodriguez, J. C. L. (2020). A fuzzy-logic IoT lighting and shading control system for smart buildings. *Automation in Construction*, *120*, 103397.

Chunmei, W. (2017). Light art design of public architecture based on the urban humanistic environment. *Agro Food Industry Hi-Tech*, *28*(1), 322–326. https://blog.iaac.net/course/moec-23-24-urbanismo-humanista/

Cidell, J. (2009). Building green: The emerging geography of LEED-certified buildings and professionals. *The Professional Geographer*, *61*(2), 200–215.

ClassNK. (n.d.). https://www.classnk.com/hp/pdf/tech_info/tech_img/T1245e

Cohen, W. M., & Levinthal, D. A. (1990). Absorptive capacity: A new perspective on learning and innovation. *Administrative Science Quarterly*, *35*(1), 128–152.

Congedo, P. M., Baglivo, C., D'Agostino, D., & Albanese, P. M. (2024). Overview of EU building envelope energy requirement for climate neutrality. *Renewable & Sustainable Energy Reviews*, *202*, 114712.

Construction, P. C. L. (2024). *Company website*. Retrieved from https://www.pcl.com

Costa, E. M. (2021). From smart cities to more Humane and Sustainable Smart Cities: the world in transformation. In *Humane and Sustainable Smart Cities: A Personal Roadmap to Transform Your City after the Pandemic* (pp. 41–63). Elsevier. DOI: 10.1016/B978-0-12-819186-6.00005-1

Crilly, M., & Lemon, M. (2018). The misalignment of policy and practice in sustainable urban design. *Smart Futures, Challenges of Urbanisation, and Social Sustainability*, 69-87.

Cullen, G. (1961). *The concise townscape*. Architectural Press.

Czermański, E., Oniszczuk-Jastrząbek, A., Spangenberg, E. F., Kozłowski, Ł., Adamowicz, M., Jankiewicz, J., & Cirella, G. T. (2022). Implementation of the Energy Efficiency Existing Ship Index: An important but costly step towards ocean protection. *Marine Policy*, *145*, 105259. DOI: 10.1016/j.marpol.2022.105259

Czyża, S., & Kowalczyk, A. M. (2024). Applying GIS in Blue-Green Infrastructure Design in Urban Areas for Better Life Quality and Climate Resilience. *Sustainability*, *16*(12), 5187.

Darvish, A., Eghbali, S. R., Eghbali, G., & Mahlabani, Y. G. (2020). The effects of building glass facade geometry on wind infiltration and heating and cooling energy consumption. *International Journal of Technology, 11*(2), 235–247. DOI: 10.14716/ijtech.v11i2.3201

Das, S., Rawat, A., & Singh, S. K. (2020). Impacts of temperature, precipitation and population growth on energy demands of Delhi. *International Journal of Design & Nature and Ecodynamics, 15*(4), 539–544. DOI: 10.18280/ijdne.150411

Davies, C., Chen, W. Y., Sanesi, G., & Lafortezza, R. (2021). The European Union roadmap for implementing nature-based solutions: A review. *Environmental Science & Policy, 121*, 49–67.

de la Edificación, C. T. (2019). Documento Básico Ahorro de Energía HE1: Limitación de la Demanda Energética. *Madrid. December.*

de la Edificación, C. T. (2018). *Documento Básico he Ahorro de energía. CTE.* DB-HE.

De Luca, F., Sepúlveda, A., & Varjas, T. (2022). Multi-performance optimization of static shading devices for glare, daylight, view and energy consideration. *Building and Environment, 217*, 109110.

Delhi Urban Art Commission (DUAC). (2017). Rejuvenation of Shahjahanabad. Delhi: DUAC. https://duac.org/

Delhipedia. (2022). *East Delhi's development thoroughfare: Vikas Marg.*

Deloitte. (2024). *U.S. global outsourcing survey 2024 report: Multidimensional sourcing*. Deloitte Consulting LLP.

Deraman, R., Nawi, M. N. M., Mydin, M. A. O., Ismail, M. H., Nordin, N. D. M., Sari, M. W., & Mohd-Danuri, M. S. (2022). Production of roof board insulation using agricultural wastes towards sustainable building material. *Journal of Advanced Research in Fluid Mechanics and Thermal Sciences, 99*(1), 66–89.

Dey, M., Gupta, A., De, B., & Das, S. (2024). Improving Accessibility to Urban Blue-Green Spaces: A Case of Howrah, India. In *Sustainable Development and Geospatial Technology: Volume 1: Foundations and Innovations* (pp. 257-278). Cham: Springer Nature Switzerland.

Dhiman, A., & Klette, R. (2020). Pothole Detection Using Computer Vision and Learning. *IEEE Transactions on Intelligent Transportation Systems, 21*(8), 3536–3550. DOI: 10.1109/TITS.2019.2931297

Dhyani, S., Majumdar, R., & Santhanam, H. (2021). Scaling-up nature-based solutions for mainstreaming resilience in Indian cities. *Ecosystem-Based Disaster and Climate Resilience: Integration of Blue-Green Infrastructure in Sustainable Development*, 279-306.

Di Gaetano, F., Cascone, S., & Caponetto, R. (2023). Integrating BIM processes with LEED certification: A comprehensive framework for sustainable building design. *Buildings (Basel, Switzerland), 13*(10), 2642.

Di Paola, L. (2021). Milan's private Vertical Forests vs. horizontal urban greening. In *The Green City and Social Injustice* (pp. 25–34). Routledge.

Di Sarno, L., Albuhairi, D., & Medeiros, J. M. P. (2024). Exploring innovative resilient and sustainable bio-materials for structural applications: Hemp-fibre concrete. *Energy Policy, 38*(11), 6398–6408.

Di Vaio, A., Palladino, R., Hassan, R., & Escobar, O. (2020). Artificial intelligence and business models in the sustainable development goals perspective: A systematic literature review. [CrossRef]. *Journal of Business Research, 121*, 283–314.

Diabat, A., & Govindan, K. (2011). An analysis of the drivers affecting the implementation of green supply chain management. *Resources, Conservation and Recycling, 55*(6), 659–667.

Dokras, D. U. (2021). The art and architecture of the town of Shahjahanabad (Old Delhi). Academia.edu. https://www.academia.edu/47537921/

Don't Wait. Act now. Manage your SEEMP Part III compliance today. (n.d.). https://www.lr.org/en/seemp-part-iii-ship-energy-efficiency

Donati, G. F. A., Bolliger, J., Psomas, A., Maurer, M., & Bach, P. M. (2022). Reconciling cities with nature: Identifying local Blue-Green Infrastructure interventions for regional biodiversity enhancement. *Journal of Environmental Management, 316*. Advance online publication. DOI: 10.1016/j.jenvman.2022.115254 PMID: 35576714

Dorst, H., Van der Jagt, A., Raven, R., & Runhaar, H. (2019). Urban greening through nature-based solutions–Key characteristics of an emerging concept. *Sustainable Cities and Society, 49*, 101620.

Dos Santos, L. F. B. S. (2020). *Efficient modeling strategies for performance-based building design supported by daylight and building energy simulations*. University of California.

Doub, S., Hanford, J., Sprug, M., Hellstern, C., & Misel, K. (2019). Living Buildings: The Bullitt Center. In *Architecture and Health* (pp. 260-279). Routledge.

Dreiseitl, H. (2020). Water and sustainable design. In *Sustainable Built Environments* (pp. 187–205). Springer.

Drosou, N., Soetanto, R., Hermawan, F., Chmutina, K., Bosher, L., & Hatmoko, J. U. D. (2019). Key factors influencing wider adoption of blue–green infrastructure in developing cities. *Water (Basel), 11*(6), 1234.

Dulaimi, A., Hamida, R., Naser, M., & Mawed, M. (2022). Digital twin solution implemented on energy hub to foster sustainable smart energy city: Case study of sustainable smart energy hub. *ISPRS Annals of the Photogrammetry, Remote Sensing and Spatial Information Sciences, X-4*(W3), 41–48. DOI: 10.5194/isprs-annals-X-4-W3-2022-41-2022

Dulian, M. (2024). Revision of the Energy Performance of Buildings Directive–Fit for 55 package. *EU Legislation, 19*(1).

Dupont, V. (2004). Urban development and population redistribution in Delhi: Implications for categorizing population. In Hugo, T. C. (Ed.), *New forms of urbanization: Beyond the urban-rural dichotomy* (pp. 172–191). Ashgate Publishing.

Eberhardt, L. C. M., Birkved, M., & Birgisdottir, H. (2022). Building design and construction strategies for a circular economy. *Architectural Engineering and Design Management, 18*(2), 93–113. DOI: 10.1080/17452007.2020.1781588

Edge Workspaces. (n.d.). *Integrated co-living and co-working spaces: The future of urban living and working.* https://www.edgeworkspaces.com/integrated-co-living-and-co-working-spaces-the-future-of-urban-living-and-working/

EEA. (2022). European Maritime Transport. *European Environment Agency.* https://www.eea.europa.eu/publications/maritime-transport/

Elassy, M., Al-Hattab, M., Takruri, M., & Badawi, S. (2024). Intelligent transportation systems for sustainable smart cities. *Transportation Engineering, 16*, 100252. DOI: 10.1016/j.treng.2024.100252

Eley, C., & Eley, C. (2016). Smart Building Design: Contextual Design, Energy Efficiency, and Curtailment. *Design Professional's Guide to Zero Net Energy Buildings*, 9-60.

El-Hakim, Y., & AbouZeid, M. N. (2024). Towards mitigating climate change negative impact: The role of regulations and governance in the construction industry. *Sustainability, 16*(16), 6822.

EllisDon. (2025). *EllisDon furniture, equipment and design division more than meets the eye.* Retrieved from https://www.ellisdon.com/news/ellisdon-furniture-equipment-and-design-division-more-than-meets-the-eye

Engineering.com. (2016). *Outsourcing CAD services.* Retrieved from https://www.engineering.com/outsourcing-cad-good-bad/

ENR. (2009). *State Engineers Union Claims Private Engineers "Cost Double".* Retrieved from https://www.enr.com/blogs/12-california-views/post/15255-state-engineers-union-claims-private-engineers-cost-double

Ergler, C. R. (2023). Moving Towards Sustainable, Liveable, and Care-Full Urban Environments: Pre-schoolers' Rights and Visions for Planning Just, Socially, and Ecologically Integrated Cities. In *The Palgrave Encyclopedia of Urban and Regional Futures* (pp. 1086–1091). Springer International Publishing.

European Parliament, European Directive 2018/844 Amending Directive 2010/31/EU on the Energy Performance of Buildings Directive 2012/27, EU on energy efficiency, 2018.

Everett, G., Adekola, O., & Lamond, J. (2021). Developing a blue-green infrastructure (BGI) community engagement framework template. *URBAN DESIGN International*, 1–17.

Fachrudin, H. T., Karolina, R., & Misnan, S. H. B. (2025). Sustainable streetscape design in commercial area toward green city: Comparative study between Indonesia and Malaysia. *City. Territory and Architecture*, *12*, 14. Advance online publication. DOI: 10.1186/s40410-025-00269-5

Fang, Z., Deng, W., Zhang, Y., Ding, X., Tang, M., Liu, T., & Wang, X. (2017). Open burning of rice, corn and wheat straws: Primary emissions, photochemical aging, and secondary organic aerosol formation. *Atmospheric Chemistry and Physics*, *17*(24), 14821–14839.

Fan, R., Ozgunalp, U., Hosking, B., Liu, M., & Pitas, I. (2020). Pothole Detection Based on Disparity Transformation and Road Surface Modeling. *IEEE Transactions on Image Processing*, *29*, 897–908. DOI: 10.1109/TIP.2019.2933750 PMID: 31449016

Fatima, K. (2024). Building Better from LEED to Living: An Approach to Net Zero Water Management.

Fatima, S., Desouza, K., Dawson, G., & Denford, J. (2021). *Analyzing Artificial Intelligence Plans in 34 Countries.* Available online: https://www.brookings.edu/blog/techtank/2021/05/13/analyzing-artificial-intelligence-plans-in-34-countries

Fenner, A. E., Kibert, C. J., Woo, J., Morque, S., Razkenari, M., Hakim, H., & Lu, X. (2018). The carbon footprint of buildings: A review of methodologies and applications. *Renewable & Sustainable Energy Reviews*, *94*, 1142–1152.

Ferdyn-Grygierek, J., Bartosz, D., Specjał, A., & Grygierek, K. (2018). Analysis of Accuracy Determination of the Seasonal Heat Demand in Buildings Based on Short Measurement Periods. *Energies*, *11*, 2734.

Fernández-Agüera, J., Domínguez-Amarillo, S., Sendra, J. J., & Suárez, R. (2016). An approach to modelling envelope airtightness in multi-family social housing in Mediterranean Europe based on the situation in Spain. *Energy and Building*, *128*, 236–253. DOI: 10.1016/j.enbuild.2016.06.074

Fernandez-Antolin, M. M., del Río, J. M., Costanzo, V., Nocera, F., & Gonzalez-Lezcano, R. A. (2019). Passive design strategies for residential buildings in different Spanish climate zones. *Sustainability (Switzerland)*, *11*(18). Advance online publication. DOI: 10.3390/su11184816

Fernandez-Antolin, M. M., del Río, J. M., & Gonzalez-Lezcano, R. A. (2021). The use of gamification in higher technical education: Perception of university students on innovative teaching materials. *International Journal of Technology and Design Education*, *31*(5), 1019–1038. DOI: 10.1007/s10798-020-09583-0

Fernandez-Antolin, M. M., Del-Río, J. M., Del Ama Gonzalo, F., & Gonzalez-Lezcano, R. A. (2020). The relationship between the use of building performance simulation tools by recent graduate architects and the deficiencies in architectural education. *Energies*, *13*(5). Advance online publication. DOI: 10.3390/en13051134

Fernandez-Antolin, M. M., del-Río, J. M., & Gonzalez-Lezcano, R. A. (2019). Influence of solar reflectance and renewable energies on residential heating and cooling demand in sustainable architecture: A case study in different climate zones in Spain considering their urban contexts. *Sustainability (Switzerland)*, *11*(23). Advance online publication. DOI: 10.3390/su11236782

Ferreira Silva, M., Jayasinghe, L. B., Waldmann, D., & Hertweck, F. (2020). Recyclable architecture: Prefabricated and recyclable typologies. *Sustainability*, *12*(4), 1342.

Ferreira, J. C., Dos Santos, D. C., & Campos, L. C. (2024). Blue-green infrastructure in view of Integrated Urban Water Management: A novel assessment of an effectiveness index. *Water Research*, *257*, 121658. PMID: 38696983

Fewings, P., & Henjewele, C. (2019). *Construction project management: An integrated approach* (3rd ed.). Routledge.

Finogeev, A., Finogeev, A., Fionova, L., Lyapin, A., & Lychagin, K. A. (2019). Intelligent monitoring system for smart road environment. *Journal of Industrial Information Integration*, *15*, 15–20. DOI: 10.1016/j.jii.2019.05.003

Firoozi, A. A., Firoozi, A. A., Oyejobi, D., Avudaiappan, S., & Flores, E. S. (2024). Emerging trends in sustainable building materials: Technological innovations, enhanced performance, and future directions. *Results in Engineering*, *24*, 103521.

Flammer, C. (2020). Green Bonds: Effectiveness and implications for public policy. *Environmental and Energy Policy and the Economy*, *1*, 95–128. DOI: 10.1086/706794

Fleischman, R. B., & Seeber, K. (2016). New construction for resilient cities: The argument for sustainable low damage precast/prestressed concrete building structures in the 21st century. *Scientia Iranica*, *23*(4), 1578–1593.

Fluor Corporation. (2016, October 18). *Fluor completes module fabrication for oil sands project*. Retrieved from https://newsroom.fluor.com/news-releases/news-details/2016/Fluor-Completes-Module-Fabrication-for-Oil-Sands-Project/default.aspx

Folorunso, A., Olanipekun, K., Adewumi, T., & Samuel, B. (2024). A policy framework on AI usage in developing countries and its impact. *Global Journal of Engineering and Technology Advances*, *21*(01), 154–166.

FR-504. (2022). *Final report: Transportation infrastructure evaluation and performance assessment* (Report No. FR-504). Louisiana Transportation Research Center, Louisiana State University.

Friedman, A., & Matheson, M. (2017). Selecting and Installing Energy-Efficient Windows to Improve Dwelling Sustainability. VITRUVIO - International Journal of Architectural Technology and Sustainability, 2(2), 2. https://doi.org/DOI: 10.4995/vitruvio-ijats.2017.7687

Frontczak, M., Schiavon, S., Goins, J., Arens, E., Zhang, H., & Wargocki, P. (2012). Quantitative relationships between occupant satisfaction and satisfaction aspects of indoor environmental quality and building design. *Indoor Air*, *22*(2), 119–131. DOI: 10.1111/j.1600-0668.2011.00745.x

Galvin, R., & März, S. (2025). A finance scheme to help Germany's small private landlords sharply increase their buildings' energy performance: Tapping into the banking system. *Energy Research & Social Science*, *120*, 103929.

Gann, D. M. (2001). Putting academic ideas into practice: Technological progress and the absorptive capacity of construction organizations. *Construction Management and Economics*, *19*(3), 321–330. DOI: 10.1080/01446190010022632

Gann, D. M., & Salter, A. J. (2000). Innovation in project-based, service-enhanced firms: The construction of complex products and systems. *Research Policy*, *29*(7-8), 955–972.

Garay, R. M. M., Rallo, M. D. L. B., Carmona, R. C., & Araya, J. C. (2009). Characterization of anatomical, chemical, and biodegradable properties of fibers from corn, wheat, and rice residues. *Chilean Journal of Agricultural Research*, *69*(3), 406–415.

Geels, F. W., Sovacool, B. K., Schwanen, T., & Sorrell, S. (2017). Sociotechnical transitions for deep decarbonization. *Science*, *357*(6357), 1242–1244. DOI: 10.1126/science.aao3760

Gehl, J. (1971). Life between buildings: Using public space.

Gehl, J. (2010). *Cities for people*. Island Press.

Gehl, J. (2010). *Cities for People*. Island Press.

GhaffarianHoseini, A., Dahlan, N. D., Berardi, U., GhaffarianHoseini, A., Makaremi, N., & GhaffarianHoseini, M. (2013). Sustainable energy performances of green buildings: A review of current theories, implementations, and challenges. *Renewable & Sustainable Energy Reviews*, *25*, 1–17. DOI: 10.1016/j.rser.2013.01.010

Giacomello, E. (2015). Case study: Bosco verticale, milan: A new urban forest rises in Milan. *CTBUH Journal*, (1), 12–18.

Gianni, M., Pietra, A., Coraddu, A., & Taccani, R. (2022). Impact of SOFC Power Generation Plant on Carbon Intensity Index (CII) Calculation for Cruise Ships. *Journal of Marine Science and Engineering*, *10*(10), 1478. DOI: 10.3390/jmse10101478

Gijón Mancheño, A., Jafino, B. A., Hofland, B., van Wesenbeeck, B. K., Kazi, S., & Urrutia, I. (2025). Nature Meets Infrastructure: The Role of Mangroves in Strengthening Bangladesh's Coastal Flood Defenses. *Sustainability*, *17*(4), 1567.

Gil, J., & Duarte, J. P. (2013). Tools for evaluating the sustainability of urban design: A review. *Proceedings of the Institution of Civil Engineers: Urban Design and Planning*. Thomas Telford Services Ltd. DOI: 10.1680/udap.11.00048

Giles-Corti, B., Vernez-Moudon, A., Reis, R., Turrell, G., Dannenberg, A. L., Badland, H., … Owen, N. (2016, December 10). City planning and population health: a global challenge. *The Lancet*. Lancet Publishing Group. DOI: 10.1016/S0140-6736(16)30066-6

Gil-Garcia, J. R., Chen, T., & Gasco-Hernandez, M. (2023). Smart City Results and Sustainability: Current Progress and Emergent Opportunities for Future Research. *Sustainability (Switzerland), 15*(10). Advance online publication. DOI: 10.3390/su15108082

Glady, J. B. P., D'Souza, S. M., Priya, A. P., Amuthachenthiru, K., Vikram, G., & Boopathi, S. (2024). A Study on AI-ML-Driven Optimizing Energy Distribution and Sustainable Agriculture for Environmental Conservation. In *Harnessing High-Performance Computing and AI for Environmental Sustainability* (pp. 1–27). IGI Global. DOI: 10.4018/979-8-3693-1794-5.ch001

Goh, E., & Loosemore, M. (2017). The impacts of industrialization on construction subcontractors: A resource-based view. *Construction Management and Economics, 35*(5), 288–304. DOI: 10.1080/01446193.2016.1248985

González, G., Rodríguez, F., & García, M. (2019). Inclusive Urban Development for Resilient Communities: A Framework for Equitable Design. *Urban Studies (Edinburgh, Scotland), 56*(7), 1350–1367. DOI: 10.1177/0042098018787671

González-Lezcano, R. A., & Hormigos-Jiménez, S. (2016). Energy saving due to natural ventilation in housing blocks in Madrid. In IOP Conference Series: Materials Science and Engineering (Vol. 138). Institute of Physics Publishing. DOI: 10.1088/1757-899X/138/1/012002

González-Lezcano, R. A., López-Fernández, E., Cesteros-García, S., & Sanglier-Contreras, G. (2020). Evaluation of energy-saving strategies according to spanish regulations. *International Journal of Energy and Environmental Engineering, 26*(3), 197–217.

González-Lezcano, R. A., Sanglier Contreras, G., Iglesias Sanz, C. M., Sancho Alambillaga, R., & López Fernández, E. J. (2023). *Construction parameters that affect the air leaks of the envelope in dwellings in Madrid.* Frontiers in Built Environment. DOI: 10.3389/fbuil.2023.1220559

Gonzalo, F. D. A., Griffin, M., Laskosky, J., Yost, P., & González-lezcano, R. A. (2022). Assessment of Indoor Air Quality in Residential Buildings of New England through Actual Data. *Sustainability (Switzerland), 14*(2). Advance online publication. DOI: 10.3390/su14020739

Gordon, K. S., Kyung, T., Perez, C. R., Holec, P. V., Ramos, A., Zhang, A. Q., & Birnbaum, M. E. (2022). Screening for CD19-specific chimaeric antigen receptors with enhanced signalling via a barcoded library of intracellular domains. *Nature Biomedical Engineering, 6*(7), 855–866. DOI: 10.1038/s41551-022-00896-0 PMID: 35710755

Goubran, S., Qi, D., Saleh, W. F., & Wang, L. (Leon). (2017). Comparing methods of modeling air infiltration through building entrances and their impact on building energy simulations. *Energy and Building*, *138*, 579–590. DOI: 10.1016/j.enbuild.2016.12.071

Goudarzi, A., Ghayoor, F., Waseem, M., Fahad, S., & Traore, I. (2022). A survey on IoT-enabled smart grids: Emerging, applications, challenges, and outlook. *Energies*, *15*(19), 6984.

Guillén-Lambea, S., Rodríguez-Soria, B., & Marín, J. M. (2019). Air infiltrations and energy demand for residential low energy buildings in warm climates. *Renewable & Sustainable Energy Reviews*, *116*, 109469. DOI: 10.1016/J.RSER.2019.109469

Gulev, S. K. (2023). Global Climate Change and the Oceans. *Studies on Russian Economic Development*, *34*(6), 738–745. DOI: 10.1134/S1075700723060060

Gupta, A., & De, B. (2024). A systematic review on urban blue-green infrastructure in the south Asian region: Recent advancements, applications, and challenges. *Water Science and Technology*, *89*(2), 382–403. PMID: 39219137

Gupta, B. D., Semwal, V., & Pathak, A. (2020). Nanotechnology-based fiber-optic chemical and biosensors. In *Nano-Optics* (pp. 163–195). Elsevier. DOI: 10.1016/B978-0-12-818392-2.00007-X

Gupta, G. (2021). *Chandni Chowk: Redeveloping urban spaces – A new paradigm for India*. Shahjahanabad Redevelopment Corporation.

Guy, S., & Farmer, G. (2001). Reinterpreting sustainable architecture: The place of technology. *Journal of Architectural Education*, *54*(3), 140–148.

Haleem, A., Javaid, M., Khan, I. H., & Mohan, S. (2023). Significant applications of artificial intelligence towards attaining sustainability. *Journal of Industrial Integration and Management*, *8*(04), 489–520.

Hall, E. T. (1966). The hidden dimension.

Hall, E. T. (1966). *A hidden dimension*. Penguin Random House.

Halliday, S., & Atkins, R. (2019). *Sustainability: RIBA Plan of Work 2013 Guide*. RIBA Publishing.

Hamby, D. M. (1994). A review of techniques for parameter sensitivity analysis of environmental models. *Environmental Monitoring and Assessment*, *32*(2), 135–154. DOI: 10.1007/BF00547132

Handfield, R. B., Melnyk, S. A., Calantone, R. J., & Curkovic, S. (2002). Integrating environmental concerns into the design process: The gap between theory and practice. *IEEE Transactions on Engineering Management*, *48*(2), 189–208.

Hanus, M. J., & Harris, A. T. (2013). Nanotechnology innovations for the construction industry. *Progress in Materials Science*, *58*(7), 1056–1102.

Harlis, T. A., & Seo, S. B. (2024). Land suitability analysis for blue-green infrastructure implementation in an urban stormwater management system in Surabaya, Indonesia. *Water Science and Technology*, *90*(5), 1501–1519.

Harvey, D. (2012). *Rebel Cities: From the Right to the City to the Urban Revolution*. Verso Books.

Harwood, E. (2011). The Barbican, City of London. *DASH| Delft Architectural Studies on Housing,* (4), 22-33.

Hassan, H., Rodriguez-Ubinas, E., Al Tamimi, A., Trepci, E., Mansouri, A., & Al-mehairbi, K. (2024). Towards innovative and sustainable buildings: A comprehensive review of 3D printing in construction. *Automation in Construction*, *163*, 105417.

Hayes, A. T., Jandaghian, Z., Lacasse, M. A., Gaur, A., Lu, H., Laouadi, A., Ge, H., & Wang, L. (2022). Nature-based solutions (nbss) to mitigate urban heat island (UHI) effects in Canadian cities. *Buildings (Basel, Switzerland)*, *12*(7), 925.

Hegazy, B., Khodeir, L., & Fathy, F. (2025). Achieving Socio-economic Resilience in Neighborhood Through Nature-based Solutions: A Systematic Review. *Results in Engineering*, 104266.

Heidrich, O., Kamara, J., Maltese, S., Re Cecconi, F., & Dejaco, M. C. (2017). A critical review of the developments in building adaptability. *International Journal of Building Pathology and Adaptation*, *35*(4), 284–303.

Herazo, B., & Lizarralde, G. (2015). The influence of green building certifications in collaboration and innovation processes. *Construction Management and Economics*, *33*(4), 279–298.

Hernando Castro, S. M., Rodríguez, R. C., López Arquillo, J. D., & Álvarez de Eulate, M. P. (2025). Comparative evaluation of construction insulation materials: Environmental performance across production, end-of-life, and beyond system boundaries. *Journal of Building Engineering*, *99*, 111570. DOI: 10.1016/j.jobe.2024.111570

Hirsch, P. L., Shwom, B. L., Yarnoff, C., Anderson, J. C., Kelso, D. M., Olson, G. B., & Colgate, J. E. (2001). Engineering design and communication: The case for interdisciplinary collaboration. *International Journal of Engineering Education, 17*(4/5), 343–348.

Homchick Crowe, J. (2020). Architectural advocacy: The Bullitt Center and environmental design. *Environmental Communication, 14*(2), 236–254.

Hong, G., & Kim, C. (2022). Experimental analysis of airtightness performance in high-rise residential buildings for improved code-compliant simulations. *Energy and Building, 261*. Advance online publication. DOI: 10.1016/j.enbuild.2022.111980

Hossain, M. J., & Bari, Md. A., & Khan, M. M. (2022). Development of an IoT Based Health Monitoring System for e-Health. *2022 IEEE 12th Annual Computing and Communication Workshop and Conference (CCWC)*, 31–37. DOI: 10.1109/CCWC54503.2022.9720825

Hossain, M. R., Oo, A. M., & Ali, A. S. (2013). Smart grid. *Smart grids: opportunities, developments, and trends*, 23-44.

Huang, H., Lu, J., Jin, L., & Ren, H. (2024). The future of environmental engineering technology: A disruptive innovation perspective. *Engineering*, *41*(10), 153–160. DOI: 10.1016/j.eng.2024.06.009

Hui, C. X., Dan, G., Alamri, S., & Toghraie, D. (2023). Greening smart cities: An investigation of the integration of urban natural resources and smart city technologies for promoting environmental sustainability. *Sustainable Cities and Society, 99*. Advance online publication. DOI: 10.1016/j.scs.2023.104985

IEA. (2021). Buildings: a source of enormous untapped efficiency potential. International Energy Agency (IEA).

IEA. (2021). *Empowering Cities for a Net Zero Future: Unlocking Resilient, Smart, Sustainable Urban Energy Systems*. International Energy Agency. https://www.iea.org/reports/empowering-cities-for-a-net-zero-future

Ijabadeniyi, A., & Vanclay, F. (2020). Socially-tolerated practices in environmental and social impact assessment reporting: Discourses, displacement, and impoverishment. *Land (Basel), 9*(2), 33.

Ikudayisi, A. E., Chan, A. P., Darko, A., & Adedeji, Y. M. (2023). Integrated practices in the Architecture, Engineering, and Construction industry: Current scope and pathway towards Industry 5.0. *Journal of Building Engineering, 73*, 106788.

IMO Environment Committee approves amendments to cut ship emissions. (n.d.). https://www.imo.org/en/MediaCentre/PressBriefings/pages/42-MEPC-short-term-measure.aspx

IMO. (2021). *Reducing greenhouse gas emissions from ships.* https://www.imo.org/en/MediaCentre/HotTopics/Pages/Reducing-greenhouse-gas-emissions-from-ships.aspx

IMO. (2023). *IMO's work to cut GHG emissions from ships.* https://www.imo.org/en/MediaCentre/HotTopics/Pages/Cutting-GHG-emissions.aspx

Institute for Transportation and Development Policy (ITDP). (2016). Urban street design guidelines, Pune. ITDP. https://india.uitp.org/wp-content/uploads/2022/11/USDG_Pune.pdf

Institute for Transportation and Development Policy (ITDP). (2024). *The BRT Standard.*

Instituto de Arquitectura Avanzada de Cataluña (IAAC). (n.d.). *MOeC 23/24 Urbanismo Humanista.* https://blog.iaac.net/course/moec-23-24-urbanismo-humanista/

Ipsen, K. L., Pizzol, M., Birkved, M., & Amor, B. (2021). How lack of knowledge and tools hinders the eco-design of buildings—A systematic review. *Urban Science (Basel, Switzerland)*, *5*(1), 20.

Isaac, M., & van Vuuren, D. P. (2009). Modeling global residential sector energy demand for heating and air conditioning in the context of climate change. *Energy Policy*, *37*(2), 507–521. DOI: 10.1016/j.enpol.2008.09.051

Iwuanyanwu, O., Gil-Ozoudeh, I., Okwandu, A. C., & Ike, C. S. (2024). Retrofitting existing buildings for sustainability: Challenges and innovations. *Engineering Science & Technology Journal*, *5*(8), 2616–2631.

Jacobs, J. (1961). *The Death and Life of Great American Cities.* Random House.

Jagadisan, S., & Sen, J. (2024). The Role of Community Engagement in Building Sustainable Blue-Green Infrastructure–Best Practices and Case Studies. In *Blue-Green Infrastructure for Sustainable Urban Settlements: Implications for Developing Countries Under Climate Change* (pp. 193-215). Springer.

Jain, R., Singh, S. K., Palaniappan, D., & Associates. (2025). Integrating smart cyber-physical systems into urban planning: The future of smart cities. In *Smart Cyber Physical Systems: Innovations and Real-World Implications* (Ch. 5). CRC Press.

Jain, R., Singh, S. K., Palaniappan, D., & Parmar, K., & T, P. (2025). Data-Driven Civil Engineering: Applications of artificial intelligence, machine learning, and deep learning. *Turkish Journal of Engineering*, *9*(2), 354–377. DOI: 10.31127/tuje.1581564

Jamawat, J. (2019). *Redesign, redeploy, and re-envision urban corporate headquarters: Amazon's Seattle campus case study*. Massachusetts Institute of Technology.

Janda, K. B., Wilson, C., Bartiaux, F., & Moezzi, M. (2015). Improving efficiency in buildings: conventional and alternative approaches. *Global Energy: Issues, Potentials, and Policy Implications*, 163-188.

Javaid, A., Jain, D., & Kwatra, N. (2024). Pre-treatment and its effect on the thermal conductivity of natural lignocellulosic rice straw stubble waste boards: Analysis vis a vis potential for the civil engineering applications. *Construction & Building Materials*, *445*, 137903.

Jean, G. G. (2024). Green Financing and Green Product Innovation. *Green Financing and Green Product Innovation. Green Financing in India: Insights from NIFTY 100's top 5 Companies and their Role in Financial and Sustainable Growth, 285.*

Jiang, D., An, P., Cui, S., Xu, F., Tuo, T., Zhang, J., & Jiang, H. (2018). Effect of leaf fiber modification methods on mechanical and heat-insulating properties of leaf fiber cement-based composite materials. *Journal of Building Engineering*, *19*, 573–583.

Jiang, Y., Zhao, D., Wang, D., & Xing, Y. (2019). Sustainable performance of buildings through modular prefabrication in the construction phase: A comparative study. *Sustainability*, *11*(20), 5658.

Jim, C. Y. (2013). Sustainable urban greening strategies for compact cities in developing and developed economies. *Urban Ecosystems*, *16*(4), 741–761. DOI: 10.1007/s11252-012-0268-x

Ji, X., Ren, J., & Ulgiati, S. (2018). Towards urban-rural sustainable cooperation: Models and policy implication. *Journal of Cleaner Production*, *213*, 892–898. DOI: 10.1016/j.jclepro.2018.12.097

Johnson, M. T. J., & Munshi-South, J. (2017, November 3). Evolution of life in urban environments. *Science*. American Association for the Advancement of Science. DOI: 10.1126/science.aam8327

Johnson, C., Tilt, J. H., Ries, P. D., & Shindler, B. (2019). Continuing professional education for green infrastructure: Fostering collaboration through interdisciplinary trainings. *Urban Forestry & Urban Greening*, *41*, 283–291.

Jokisalo, J., Kurnitski, J., Korpi, M., Kalamees, T., & Vinha, J. (2009). Building leakage, infiltration, and energy performance analyses for Finnish detached houses. *Building and Environment*, *44*(2), 377–387. DOI: 10.1016/j.buildenv.2008.03.014

Joo, Y. M. (2023). Developmentalist smart cities? the cases of Singapore and Seoul. *International Journal of Urban Sciences*, *27*(S1), 164–182. DOI: 10.1080/12265934.2021.1925143

Jurasz, J., Canales, F., Kies, A., Guezgouz, M., & Beluco, A. (2019). A review on the complementarity of renewable energy sources: Concept, metrics, application and future research directions. *Solar Energy*, *195*, 703–724. DOI: 10.1016/j.solener.2019.11.087

Kalaiselvi, T., Saravanan, G., Haritha, T., Babu, A. V. S., Sakthivel, M., & Boopathi, S. (2024). A Study on the Landscape of Serverless Computing: Technologies and Tools for Seamless Implementation. In *Serverless Computing Concepts, Technology and Architecture* (pp. 260–282). IGI Global. DOI: 10.4018/979-8-3693-1682-5.ch016

Kamal, D. M. (2024). Urban regeneration of public spaces and pedestrian streets in commercial area: Case of Chandni Chowk at Old Delhi, India. *American Journal of Civil Engineering and Architecture*, *12*(1), 22–32. DOI: 10.12691/ajcea-12-1-4

Kamali, M., & Hewage, K. (2017). Development of performance criteria for sustainability evaluation of modular versus conventional construction methods. *Journal of Cleaner Production*, *142*, 3592–3606.

Kamali, M., Hewage, K., & Sadiq, R. (2022). Economic sustainability benchmarking of modular homes: A life cycle thinking approach. *Journal of Cleaner Production*, *348*. Advance online publication. DOI: 10.1016/j.jclepro.2022.131290

Kamal, M. A. (2012). An overview of passive cooling techniques in buildings: Design concepts and architectural interventions. *Acta Technica Napocensis: Civil Engineering & Architecture*, *55*(1), 84–97.

Kamble, Z., & Behera, B. K. (2021). Sustainable hybrid composites reinforced with textile waste for construction and building applications. *Construction & Building Materials*, *284*, 122800.

Kamrowska-Załuska, D. (2021, November 1). Impact of ai-based tools and urban big data analytics on the design and planning of cities. *Land*. MDPI. DOI: 10.3390/land10111209

Karpis, O., Juricek, J., & Micek, J. (2010). Application of Wireless Sensor Networks for Road Monitoring. *IFAC Proceedings Volumes, 43*(24), 183–188. https://doi.org/DOI: 10.3182/20101006-2-PL-4019.00035

Katsamenis, I., Bimpas, M., Protopapadakis, E., Zafeiropoulos, C., Kalogeras, D., Doulamis, A., Doulamis, N., Martín-Portugués Montoliu, C., Handanos, Y., Schmidt, F., Ott, L., Cantero, M., & Lopez, R. (2022). Robotic Maintenance of Road Infrastructures: The HERON Project. *Proceedings of the 15th International Conference on PErvasive Technologies Related to Assistive Environments*, 628–635. DOI: 10.1145/3529190.3534746

Kaur, R., & Gupta, K. (2022). Blue-Green Infrastructure (BGI) network in urban areas for sustainable storm water management: A geospatial approach. *City and Environment Interactions*, *16*, 100087.

Kaya, C. (2025). Intelligent Environmental Control in Plant Factories: Integrating Sensors, Automation, and AI for Optimal Crop Production. *Food and Energy Security*, *14*(1), e70026.

Kazeem, K. O., Olawumi, T. O., & Osunsanmi, T. (2023). Roles of artificial intelligence and machine learning in enhancing construction processes and sustainable communities. *Buildings (Basel, Switzerland)*, *13*(8), 2061.

Kellert, S. (2011). *Biophilic design: The theory, science and practice of bringing buildings to life*. John Wiley & Sons.

Kennedy, C., Pincetl, S., & Bunje, P. (2010). The study of urban metabolism and its applications to urban planning and design. *Environmental Pollution*, *159*(8–9), 1965–1973. DOI: 10.1016/j.envpol.2010.10.022

Kenny, S. (2022, February). The innovation function: An interview with Shaun Kenny of Bechtel. *McKinsey & Company*. Retrieved from https://www.mckinsey.com/industries/infrastructure/our-insights/the-innovation-function-an-interview-with-shaun-kenny-of-bechtel

Ke, X., Lin, J. Y., Fu, C., & Wang, Y. (2020). Transport Infrastructure Development and Economic Growth in China: Recent Evidence from Dynamic Panel System-GMM Analysis. *Sustainability*, *12*(14), 5618. DOI: 10.3390/su12145618

Khan, , M. S., Woo, M., Nam, K., & Chathoth, P. K. (2017). Smart city and smart tourism: A case of Dubai. *Sustainability (Switzerland)*, *9*(12). Advance online publication. DOI: 10.3390/su9122279

Khan, I. (2019). Energy-saving behaviour as a demand-side management strategy in the developing world: The case of Bangladesh. *International Journal of Energy and Environmental Engineering*, *10*(4), 493–510. DOI: 10.1007/s40095-019-0302-3

Khan, S. A., Jassim, M., Ilcan, H., Sahin, Ö., Bayer, İ. R., Sahmaran, M., & Koç, M. (2023). 3D printing of circular materials: Comparative environmental analysis of materials and construction techniques. *Case Studies in Construction Materials, 18*, e02059. DOI: 10.1016/j.cscm.2023.e02059

Kibert, C. J. (2016). *Sustainable construction: green building design and delivery*. John Wiley & Sons.

Kim, K. (2012). A model-driven workflow fragmentation framework for collaborative workflow architectures and systems. *Journal of Network and Computer Applications, 35*(1), 97–110.

Kim, Y.-M., Kim, Y.-G., Son, S.-Y., Lim, S.-Y., Choi, B.-Y., & Choi, D.-H. (2022). Review of Recent Automated Pothole-Detection Methods. *Applied Sciences (Basel, Switzerland), 12*(11), 5320. DOI: 10.3390/app12115320

Konstantoglou, M., & Tsangrassoulis, A. (2016). Dynamic operation of daylighting and shading systems: A literature review. *Renewable & Sustainable Energy Reviews, 60*, 268–283.

Kor, M., Yitmen, I., & Alizadehsalehi, S. (2023). An investigation for integration of deep learning and digital twins towards Construction 4.0. *Smart and Sustainable Built Environment, 12*(3), 461–487.

Korman, R. (2017, August 8). *Fluor stumbles on fixed-price power projects*. Engineering News-Record. Retrieved from https://www.enr.com/articles/42478-fluor-stumbles-on-fixed-price-power-projects

Kozak, D., Henderson, H., de Castro Mazarro, A., Rotbart, D., & Aradas, R. (2020). Blue-green infrastructure (BGI) in dense urban watersheds. The case of the Medrano stream basin (MSB) in Buenos Aires. *Sustainability, 12*(6), 2163.

KPMG. (2023). *Future of outsourcing*. KPMG LLP.

Krass, D., Nedorezov, T., & Ovchinnikov, A. (2013). Environmental taxes and the choice of green technology. *Production and Operations Management, 22*(5), 1035–1055.

Krishnan, G., & Priya, R. S. (2022). Blue green infrastructure: a panacea for urban environmental challenges. Case study: Thiruvananthapuram city, Kerala, India. *Materials Today: Proceedings, 68*, 2646–2652.

Kumar, H., Wadhwa, A. S., Akhai, S., & Kaushik, A. (2024). Parametric analysis, modeling and optimization of the process parameters in electric discharge machining of aluminium metal matrix composite. *Engineering Research Express, 6*(2), 025542.

Kumar, H., Wadhwa, A. S., Akhai, S., & Kaushik, A. (2024). Parametric optimization of the machining performance of Al-SiCp composite using combination of response surface methodology and desirability function. *Engineering Research Express*, *6*(2), 025505.

Kumar, P., Channi, H. K., Kumar, R., Rajiv, A., Kumari, B., Singh, G., Singh, S., Dyab, I. F., & Lozanović, J. (2024). A comprehensive review of vehicle-to-grid integration in electric vehicles: Powering the future. *Energy Conversion and Management, X*, 100864. DOI: 10.1016/j.ecmx.2024.100864

La Notte, A., & Zulian, G. (2021). An ecosystem services-based approach to frame NBS in Urban Context. In *Nature-Based Solutions for More Sustainable Cities–A Framework Approach for Planning and Evaluation* (pp. 47–65). Emerald Publishing Limited.

Lai, Z., & Chen, Y. (2024). Enhancing the mechanical and environmental performance of solidified soil using construction waste and glass micro-powder. *Heliyon*, *10*(22), e40187. DOI: 10.1016/j.heliyon.2024.e40187 PMID: 39584091

Lawrence, B. D., & Simmons, J. A. (1982). Measurements of atmospheric attenuation at ultrasonic frequencies and the significance for echolocation by bats. *The Journal of the Acoustical Society of America*, *71*(3), 585–590. DOI: 10.1121/1.387529 PMID: 7085967

Lawrence, R. (2020). *Creating built environments: Bridging knowledge and practice divides*. Routledge.

Lee, C. (2023). Access to amenities such as schools, healthcare facilities, and cultural institutions is also necessary for a livable urban environment. Applied Research in Quality of Life, 18, 45–62. Springer Nature. DOI: 10.1007/s11482-023-10112-3

Lee, D., & Lee, S. (2023). Artificial intelligence enabled energy-efficient heating, ventilation and air conditioning system: Design, analysis and necessary hardware upgrades. *Applied Thermal Engineering*, *235*, 121253. DOI: 10.1016/j.applthermaleng.2023.121253

Lee, J., Park, J., Jung, H., & Park, J. (2017). Renewable energy potential by the application of a building integrated photovoltaic and wind turbine system in global urban areas. *Energies*, *10*(12), 2158. DOI: 10.3390/en10122158

Lee, M. J., Kim, N. I., & Ryou, H. S. (2011). Air tightness measurement with transient methods using sudden expansion from a compressed chamber. *Building and Environment*, *46*(10), 1937–1945. DOI: 10.1016/j.buildenv.2011.04.001

Lehtiö, A., Hartikainen, M., Ala-Luopa, S., Olsson, T., & Väänänen, K. (2023). Understanding citizen perceptions of AI in the smart city. *AI & Society*, *38*(3), 1123–1134. DOI: 10.1007/s00146-022-01589-7

Leoto, R., & Lizarralde, G. (2019). Challenges in evaluating strategies for reducing a building's environmental impact through Integrated Design. *Building and Environment*, *155*, 34–46.

Leshinsky, R., & Schatz, L. (2018). I don0t think my landlord will find out: Airbnb and the challenges of enforcement. *Urban Policy and Research*, *36*, 417–428.

Lezcano, R. A. G., Fernández, E. J. L., García, S. C., & Contreras, G. S. (2020). Optimisation of Heat Loss in Cities in Different Climatic Zones in a Mediterranean Climate.

Lezcano, R. A. G., & Burgos, M. J. M. (2021). Airflow analysis of the haida plank house, a breathing envelope. *Energies*, *14*(16). Advance online publication. DOI: 10.3390/en14164871

Liao, K. H., Deng, S., & Tan, P. Y. (2017). Blue-green infrastructure: new frontier for sustainable urban stormwater management. *Greening cities: Forms and functions*, 203-226.

Li, D., Heimeriks, G., & Alkemade, F. (2020). The emergence of renewable energy technologies at country level: Relatedness, international knowledge spillovers and domestic energy markets. *Industry and Innovation*, *27*(9), 991–1013. DOI: 10.1080/13662716.2020.1713734

Liddle, S. (2017). *Chandni Chowk: The Mughal city of Old Delhi*. Speaking Tiger Publishing.

Likita, A. J., Jelodar, M. B., Vishnupriya, V., & Rotimi, J. O. B. (2024). Lean and BIM integration benefits construction management practices in New Zealand. *Construction Innovation*, *24*(1), 106–133.

Li, L., Yi, Z., Jiang, F., Zhang, S., & Zhou, J. (2023). Exploring the mechanism of digital transformation empowering green innovation in construction enterprises. *Developments in the Built Environment*, *15*, 100199. DOI: 10.1016/j.dibe.2023.100199

Lim, M., & Xenarios, S. (2021). Economic assessment of urban space and blue-green infrastructure in Singapore. *Journal of Urban Ecology*, *7*(1), juab020.

Lin, P. C., & Huang, Y. H. (2012). The influence factors on choice behavior regarding green products based on the theory of consumption values. *Journal of Cleaner Production*, *22*(1), 11–18.

Liu, L., Fryd, O., & Zhang, S. (2019). Blue-green infrastructure for sustainable urban stormwater management—Lessons from six municipality-led pilot projects in Beijing and Copenhagen. *Water (Basel)*, *11*(10), 2024.

Liu, L., & Jensen, M. B. (2018). Green infrastructure for sustainable urban water management: Practices of five forerunner cities. *Cities (London, England)*, *74*, 126–133.

Liu, S., Higgs, C., Arundel, J., Boeing, G., Cerin, E., Moctezuma, D., & Giles-Corti, B. (2021). A generalized framework for measuring pedestrian accessibility around the world using open data. *Transportation Research Part A, Policy and Practice*, *146*, 72–88. DOI: 10.1016/j.tra.2021.05.014

Li, X., Ruan, T., Hou, K., & Qu, R. (2022). The configuring pathways of green technology advance, organizational strategy and policy environment for realizing low-carbon manufacturing from the perspective of simmelian tie: A qualitative comparative analysis of listed companies in China. *Journal of Cleaner Production*, *382*, 135149. DOI: 10.1016/j.jclepro.2022.135149

Li, X., Wang, S., Duan, L., Hao, J., Li, C., Chen, Y., & Yang, L. (2007). Particulate and trace gas emissions from open burning of wheat straw and corn stover in China. *Environmental Science & Technology*, *41*(17), 6052–6058. PMID: 17937281

Li, Z., Chen, X., Shen, Z., & Fan, Z. (2022). Evaluating Neighborhood Green-Space Quality Using a Building Blue–Green Index (BBGI) in Nanjing, China. *Land (Basel)*, *11*(3), 445.

Logeswaran, J., Shamsuddin, A. H., Silitonga, A. S., & Mahlia, T. M. I. (2020). Prospect of using rice straw for power generation: A review. *Environmental Science and Pollution Research International*, *27*, 25956–25969. PMID: 32378100

Loorbach, D. (2017). *Urban Sustainability Transition: Retrofitting the City. Retrofitting Cities for Tomorrow's World.* John Wiley & Sons Ltd.

Lozinsky, C. H., & Touchie, M. F. (2018). Improving energy model calibration of multi-unit residential buildings through component air infiltration testing. *Building and Environment*, *134*, 218–229. DOI: 10.1016/j.buildenv.2018.02.040

Luca, M. L., Lepri, B., Gallotti, R., Paolazzi, S., Bigi, M., & Pistore, M. (2024). Towards civic digital twins: Co-design the citizen-centric future of Bologna. *arXiv*.

Lucatello, S., & Alcántara-Ayala, I. (2024). Sustainable Synergy: Strengthening disaster risk reduction in Latin America and the Caribbean through nature-based solutions. *International Journal of Disaster Risk Reduction*, *113*, 104860.

Luo, X., & Oyedele, L. O. (2021). Assessment and optimisation of life cycle environment, economy and energy for building retrofitting. *Energy for Sustainable Development, 65*, 77–100.

Luo, X., Yan, R., & Wang, S. (2024). After five years' application of the European Union monitoring, reporting, and verification (MRV) mechanism: Review and prospectives. *Journal of Cleaner Production, 434*, 140006. DOI: 10.1016/j.jclepro.2023.140006

Lv, Z., & Xie, S. (2024). Artificial intelligence in the digital twins: State of the art, challenges, and future research topics. *Digital Twin, 1*(1), 12.

Lynch, K. (1960). *The image of the city*. MIT Press.

Lynch, K. (1981). *A theory of good city form*. MIT Press.

Lynch, K. (1981). *Theory of a good city form*. The MIT Press.

Ma, C. (2021). Smart city and cyber-security; technologies used, leading challenges and future recommendations. *Energy Reports, 7*, 7999–8012. DOI: 10.1016/j.egyr.2021.08.124

Macaione, I., Raffa, A., & Andaloro, B. (2024). Climate-Adaptive Nature-Based Regenerative Urban Green Streetscapes: Design Exploration from the City of Matera. *Sustainability, 16*(16), 6811.

Madakam, S., Ramaswamy, R., & Tripathi, S. (2015). Internet of Things (IoT): A literature review. *Journal of Computer and Communications, 3*(5), 164–173. DOI: 10.4236/jcc.2015.35021

Madan, B. S., Viraj Dravid, S., Deenadayalan, G., Rathod, P., Arumugam, G., & Boopathi, S. (2024). Innovations and Applications of Engineered Cementitious Composites and Alternative Materials: Reimagining Construction. In *Advances in Chemical and Materials Engineering* (pp. 53–86). IGI Global. DOI: 10.4018/978-1-6684-8182-0.ch003

Mageswari, D. U., Kareemullah, H., Jithesh, K., Boopathi, S., Rachel, P. M. P. P., & Ramkumar, M. S. (2024). Experimental investigation of mechanical properties and multi-objective optimization of electronic, glass, and ceramic waste–mixed concrete. *Environmental Science and Pollution Research International*. Advance online publication. DOI: 10.1007/s11356-024-33751-7 PMID: 38806982

Mahizhnan, A. (1999). Smart cities: The Singapore case. *Cities (London, England), 16*(1), 13–18. DOI: 10.1016/s0264-2751(98)00050-x

Mahmoud, I., Morello, E., Bisello, A., & Kolokotsa, D. (2024). *Embedding technologies for improving Nature-Based Solutions performance and fostering social inclusion in urban greening strategies: Augmented NBS for cities* (Vol. 93). Elsevier.

Maji, S., Dwivedi, D. H., Singh, N., Kishor, S., & Gond, M. (2020). Agricultural waste: Its impact on environment and management approaches. *Emerging eco-friendly green technologies for wastewater treatment*, 329-351.

Mamatha, G., Gupta, A., & Koujalagi, A. Pooja, Pramanik, S., & Roy, A. (2025). *Applying Generative AI in a Global Language School to Improve Curriculum Planning* (pp. 151–184). IGI Global Scientific Publishing. DOI: 10.4018/979-8-3693-8744-3.ch007

Ma, N., Fan, J., Wang, W., Wu, J., Jiang, Y., Xie, L., & Fan, R. (2022). Computer vision for road imaging and pothole detection: A state-of-the-art review of systems and algorithms. *Transportation Safety and Environment*, *4*(4), tdac026. Advance online publication. DOI: 10.1093/tse/tdac026

Manchana, R. (2022). Optimizing Real Estate Project Management through Machine Learning, Deep Learning, and AI. *Journal of Scientific and Engineering Research*, *9*(4), 192–208.

Marshall, S. (2005). *Streets and Patterns*. Routledge.

Martín-Garín, A., Millán-García, J. A., Hidalgo-Betanzos, J. M., Hernández-Minguillón, R. J., & Baïri, A. (2020). Airtightness analysis of the built heritage⇓field measurements of nineteenth century buildings through blower door tests. *Energies*, *13*(24). Advance online publication. DOI: 10.3390/en13246727

Masyhur, R. T., Alias, A. H., Haron, N. A., & Zulkafli, Z. (2024). A systematic review on green practices in the Malaysian construction industry: Status, challenges, key motivations, rating systems and technology evolution. *Energy and Building*, *320*, 114550.

Mauser, W., Klepper, G., Rice, M., Schmalzbauer, B. S., Hackmann, H., Leemans, R., & Moore, H. (2013). Transdisciplinary global change research: The co-creation of knowledge for sustainability. *Current Opinion in Environmental Sustainability*, *5*(3-4), 420–431.

McClure, W. R., & Bartuska, T. J. (2011). *The built environment: a collaborative inquiry into design and planning*. John Wiley & Sons.

McNally, A. (2020). *Creating Trustworthy AI for the Environment: Transparency, Bias, and Beneficial Use*. Available online: https://www.scu.edu/environmental-ethics/resources/creating-trustworthy-ai-for-the-environment-transparency-biasand-beneficial-use

Meacham, B., Bowen, R., Traw, J., & Moore, A. (2005). Performance-based building regulation: Current situation and future needs. *Building Research and Information*, *33*(2), 91–106. DOI: 10.1080/0961321042000322780

Meire, C., Liñares, P., & Hermo, V. (2024). El Sistema constructivo modular Walluminium, análisis de la envolvente hermética y termoacústica y su sistema de producción. *Informes de la Construcción*, *76*(573), 6451. DOI: 10.3989/ic.6451

Mello Rose, F., Thiel, J., & Grabher, G. (2022). Selective inclusion: Civil society involvement in the smart city ecology of Amsterdam. *European Urban and Regional Studies*, *29*(3), 369–382. DOI: 10.1177/09697764221092587

Meng, Q., Hu, L., Li, M., & Qi, X. (2024). Assessing the environmental impact of building life cycle: A carbon reduction strategy through innovative design, intelligent construction, and secondary utilization. *Environmental Science & Technology*, *58*(4), 765–778. DOI: 10.1021/acs.est.3c07876

Menyeh, B. O., & Acheampong, T. (2024). Crowdfunding renewable energy investments: Investor perceptions and decision-making factors in an emerging market. *Energy Research & Social Science*, *114*, 103602. DOI: 10.1016/j.erss.2024.103602

MEPC 78/17/Add.1 Annex 12, page 1 I:8 78-17-Add.1.docx ANNEX 12 RESOLUTION MEPC.350(78) (adopted on 10 June 2022) 2022 GUIDELINES ON THE METHOD OF CALCULATION OF THE ATTAINED ENERGY EFFICIENCY EXISTING SHIP INDEX (EEXI). (n.d.). https://www.register-iri.com/wp-content/uploads/MEPC.35078.pdf

MEPC.328(76)-AMENDMENTS TO THE ANNEX OF THE PROTOCOL OF 1997 TO AMEND THE INTERNATIONAL CONVENTION FOR THE PREVENTION OF POLLUTION FROM SHIPS, 1973, AS MODIFIED BY THE PROTOCOL OF 1978 RELATING THERETO 2021 Revised MARPOL Annex VI. (n.d.). https://www.ccs.org.cn/ccswzen//specialDetail?id=202206220255211398

MEPC.352(78)-2022 Guidelines on Operational Carbon Intensity Indicators and the Calculation Methods (CII Guidelines, G1). (n.d.). https://www.ccs.org.cn/ccswzen//articleDetail?id=202212210561846752

MEPC.353(78)-2022 Guidelines on the Reference Lines for Use with Operational Carbon Intensity Indicators (CII Reference Lines Guidelines, G2). (n.d.). https://www.ccs.org.cn/ccswzen/articleDetail?id=202212210517258440

Merili. (2025, January 15). *Solar Energy in Urban Design: Powering Sustainable Cities.* Impact Day. https://impactday.eu/blog/solar-energy-in-urban-design-powering-sustainable-cities/

Mersin, K. (2020). A New Method for Calculating Fuel Consumption by Using Speed Loss Function. *International Journal of Environment and Geoinformatics, 7*(1), 64–67. DOI: 10.30897/ijegeo.700557

Mesjasz-Lech A. (2017), Factors Conducive to the Development of Business Cooperation in a Cluster Initiative – a Statistical Analysis, Zeszyty Naukowe. *Organizacja i Zarządzanie / Politechnika Śląska, 1984*(109), 65-75.

Mesjasz-Lech, A. (2018). Educational and research potential of regions and the innovativeness of businesses - an analysis of spatial concentration. IA TED Academy.

Ministry of Housing and Urban Affairs (MOHUA). (2015). *URDPFI guidelines* (Vol. I). Government of India.

Ministry of Housing and Urban Affairs (MOHUA). (n.d.). *Equitable road space: Case study Vikas Marg.* Government of India.

Mishra, B. K., Roy, A. I, V. V., Maity, A., Kumar, S., Sheet, A., & Dutta, D. (2024). A Review of India's Renewable Energy Transition: Policies, Achievements, and Integration Challenges. *2024 10th International Conference on Electrical Energy Systems (ICEES)*, 1–7. DOI: 10.1109/ICEES61253.2024.10776919

Miszczuk, A., & Heim, D. (2021). Parametric study of air infiltration in residential buildings—The effect of local conditions on energy demand. *Energies, 14*(1). Advance online publication. DOI: 10.3390/en14010127

Mitra, A., Bera, B., Das Ashok, K., Jamal Sajjad, S., & You, I. (2023). Impact on blockchain-based AI/ML-enabled big data analytics for Cognitive Internet of Things environment. *Computer Communications, 197*, 173–185. DOI: 10.1016/j.comcom.2022.10.010

Mlote, D. S., Budig, M., & Cheah, L. (2024). Adaptability of buildings: A systematic review of current research. *Frontiers in Built Environment, 10*, 1376759.

Mohammad, S., Dey, A., & Saha, A. (2019). Mixed-Use Development and Its Impact on Urban Sustainability and Equity. *Journal of Urban Affairs, 41*(4), 496–513. DOI: 10.1080/07352166.2018.1479930

Moharrak, M., Nguyen, N. P., & Mogaji, E. (2024). Business environment and adoption of AI: Navigation for internationalization by new ventures in emerging markets. *Thunderbird International Business Review*, *66*(4), 355–372. DOI: 10.1002/tie.22384

Moradi, S., & Sormunen, P. (2024a). Integrating lean construction with BIM and sustainability: A comparative study of challenges, enablers, techniques, and benefits. *Construction Innovation*, *24*(7), 188–203.

Mosneaga, A. (2022). Unsustainable development, disasters and displacement: Revisiting the governance challenge. *International Journal of Disaster Risk Reduction*, *79*, 103172.

Mouratidis, K. (2018). Is compact city livable? The impact of compact versus sprawled neighbourhoods on neighbourhood satisfaction. *Urban Studies (Edinburgh, Scotland)*, *55*(11), 2408–2430. DOI: 10.1177/0042098017729109

Mudunuri, L. N. R., & Attaluri, V. (2025). Urban Development Challenges and the Role of Cloud AI-Powered Blue-Green Solutions. In *Integrating Blue-Green Infrastructure Into Urban Development* (pp. 507–522). IGI Global Scientific Publishing.

Muhammed, M., & Tsakalakos, T. (2003). Nanostructured materials and nanotechnology: Overview. *Journal of the Korean Ceramic Society*, *40*(11), 1027–1046.

Murali, G., & Wong, L. S. (2024). A comprehensive review of biochar-modified concrete: Mechanical performance and microstructural insights. *Construction & Building Materials*, *425*, 135986.

Murphy, K. M. (2016). *Sustainable and energy-efficient development interventions and their application toward net-zero or net-positive energy and water building development*. Massachusetts Institute of Technology.

Murtagh, N., Roberts, A., & Hind, R. (2016). The relationship between motivations of architectural designers and environmentally sustainable construction design. *Construction Management and Economics*, *34*(1), 61–75.

Musonda, J., Ma, L., Liu, T. Y., Shahzad, M., & Ali, A. (2025). Megaprojects' environmental impact mitigation by sustainable construction mediated by sustainable practices and regulations. *Cleaner and Responsible Consumption*, 100281.

Muthukumar, R., Pratheep, G. V. G., S. J., S., Kunduru, K. R., Kumar, P., & Boopathi, S. (2024). Leveraging Fuel Cell Technology With AI and ML Integration for Next-Generation Vehicles: Empowering Electric Mobility. In *Advances in Mechatronics and Mechanical Engineering* (pp. 312–337). IGI Global. DOI: 10.4018/979-8-3693-5247-2.ch016

Mutschler, R., Rüdisüli, M., Heer, P., & Eggimann, S. (2021). Benchmarking cooling and heating energy demands considering climate change, population growth and cooling device uptake. *Applied Energy*, *288*. Advance online publication. DOI: 10.1016/j.apenergy.2021.116636

Myilsamy, S., & Sampath, B. (2017). Grey Relational Optimization of Powder Mixed Near-Dry Wire Cut Electrical Discharge Machining of Inconel 718 Alloy. *Asian Journal of Research in Social Sciences and Humanities*, *7*(3), 18–25.

Nag, A., Ashish, A., & Rathore, A. S. (2025). Urban Waterfronts and Tourism: Reimagining Cityscapes Through Ecological Design. In *Integrating Architecture and Design Into Sustainable Tourism Development* (pp. 433-470). IGI Global Scientific Publishing.

Nag, A., & Rathore, A. S. (2025a). Beyond Daylight: The Night Index as a Framework for Regenerative Tourism for Social Development. In Sousa, B., Arantes, L., & Bhartiya, S. (Eds.), *Regenerative Tourism for Social Development* (pp. 311–338). IGI Global Scientific Publishing. DOI: 10.4018/979-8-3373-0235-5.ch013

Nag, A., & Rathore, A. S. (2025b). Digital Innovation and Destination Competitiveness: Leveraging Mining Heritage for Virtual Tourism Experiences. *Journal of Mining and Environment*, *16*(2), 453–478.

Najini, H., Nour, M., Al-Zuhair, S., & Ghaith, F. (2020). Techno-economic analysis of green building codes in United Arab Emirates based on a case study office building. *Sustainability*, *12*(21), 8773.

Najjar, M., Figueiredo, K., Hammad, A. W., & Haddad, A. (2019). Integrated optimization with building information modeling and life cycle assessment for generating energy efficient buildings. *Applied Energy*, *250*, 1366–1382.

Namany, S., Govindan, R., Di Martino, M., Pistikopoulos, E. N., Linke, P., Avraamidou, S., & Al-Ansari, T. (2021). An Energy-Water-Food nexus-based decision-making framework to guide national priorities in Qatar. *Sustainable Cities and Society*, *75*, 103342. DOI: 10.1016/j.scs.2021.103342

Narra, M., Vij, M., Shreenidhi, K. S., Anandaram, H., Hari, B. S., & Boopathi, S. (2024). Redefining Sustainability in Building Practices in Circular Construction Process. In *Innovations in Energy Efficient Construction Through Sustainable Materials* (pp. 293–324). IGI Global. DOI: 10.4018/979-8-3693-3398-3.ch011

Naser, M., & Kodur, V. (2025). Vulnerability of structures and infrastructure to wildfires: A perspective into assessment and mitigation strategies. *Natural Hazards*, *121*(8), 9995–10015.

National Association of City Transportation Officials (NACTO). (1996). About NACTO. https://nacto.org/about/

National Association of City Transportation Officials (NACTO). (2013). *Urban street design guide*.

Nguyen, T. A., & Aiello, M. (2013). Energy intelligent buildings based on user activity: A survey. *Energy and Building*, *56*, 244–257.

NIST. (2014). NIST Framework and Roadmap for Smart Grid Interoperability Standards, Release 3.0.

Nonaka, I. (1994). A dynamic theory of organizational knowledge creation. *Organization Science*, *5*(1), 14–37.

Nooradzhar, F. S., Abdullah, M. N., & Asmoni, M. (2024). Strategic Pathways to Sustainable Development: A Comprehensive Review of the Benefits and Barriers of Green Buildings. *International Journal of Real Estate Studies*, *18*(2), 91–102.

Noori, N., Hoppe, T., & de Jong, M. (2020). Classifying pathways for smart city development: Comparing design, governance and implementation in Amsterdam, Barcelona, Dubai, and Abu Dhabi. *Sustainability (Switzerland)*, *12*(10). Advance online publication. DOI: 10.3390/SU12104030

Norman, C., Surjan, A., & Booth, M. (2021). Making Resilience a Reality: The Contribution of Peri-urban Ecosystem Services (BGI) to Urban Resilience. *Ecosystem-Based Disaster and Climate Resilience: Integration of Blue-Green Infrastructure in Sustainable Development*, 185-200.

Nwulu, E. O., Elete, T. Y., Omomo, K. O., Akano, O., & Erhueh, O. (2023). The importance of interdisciplinary collaboration for successful engineering project completions: A strategic framework. *World Journal of Engineering and Technology Research*, *2*(3), 48–56.

Oakland, J. S., & Marosszeky, M. (2017). *Total construction management: Lean quality in construction project delivery*. Routledge.

Octavianti, T., & Charles, K. (2018). Disaster capitalism? Examining the politicisation of land subsidence crisis in pushing Jakarta's seawall megaproject. *Water Alternatives*, *11*(2).

OECD. (2018). *Building Resilient Cities: An Assessment of Disaster Risk Management Policies in Southeast Asia, OECD Green Growth Studies*. OECD Publishing. DOI: 10.1787/9789264305397-

Oh, J., Hong, T., Hakpyeong, K., An, J., Kwangbok, J., & Koo, C. (2017). Advanced Strategies for Net-Zero Energy Building: Focused on the Early Phase and Usage Phase of a Building's Life Cycle. *Sustainability*, *9*, 2272. DOI: 10.3390/su9122272

Oke, O. S., Aliu, J. O., Oke, A. E., Ekundayo, D., & Duduyegbe, O. M. (2025). Green Building Practices: Fuzzy Synthetic Evaluation of the Drivers of Deforestation and Forest Degradation in a Developing Economy. *Sustainability*, *17*(4), 1538.

Olawade, D. B., Wada, O. Z., David-Olawade, A. C., Fapohunda, O., Ige, A. O., & Ling, J. (2024). Artificial intelligence potential for net zero sustainability: Current evidence and prospects. *Next Sustainability*, *4*, 100041.

Oliveira, R., Vicente, R., Almeida, R. M. S. F., & Figueiredo, A. (2021). The importance of in situ characterisation for the mitigation of poor indoor environmental conditions in social housing. *Sustainability (Switzerland)*, *13*(17). Advance online publication. DOI: 10.3390/su13179836

Omar, M., & Kumar, P. (2020). Detection of Roads Potholes using YOLOv4. *2020 International Conference on Information Science and Communications Technologies (ICISCT)*, 1–6. DOI: 10.1109/ICISCT50599.2020.9351373

Omaxe. (2017, December). Rediscovering Chandni Chowk: Beyond the space. Omaxe: Turning Dreams into Reality. https://www.omaxe.com/blog/rediscovering-chandni-chowk-beyond-the-space/

Omer, A. M. (2008). Renewable building energy systems and passive human comfort solutions. *Renewable & Sustainable Energy Reviews*, *12*(6), 1562–1587.

Omopariola, E. D., Olanrewaju, O. I., Albert, I., Oke, A. E., & Ibiyemi, S. B. (2024). Sustainable construction in the Nigerian construction industry: Unsustainable practices, barriers and strategies. *Journal of Engineering. Defence Technology*, *22*(4), 1158–1184.

Onyeaka, H., Tamasiga, P., Nwauzoma, U. M., Miri, T., Juliet, U. C., Nwaiwu, O., & Akinsemolu, A. A. (2023). Using artificial intelligence to tackle food waste and enhance the circular economy: Maximising resource efficiency and minimising environmental impact: A review. *Sustainability*, *15*(13), 10482.

Orieno, O. H., Ndubuisi, N. L., Eyo-Udo, N. L., Ilojianya, V. I., & Biu, P. W. (2024). Sustainability in project management: A comprehensive review. *World Journal of Advanced Research and Reviews*, *21*(1), 656–677.

Oxenswärdh, A. (2019). Knowledge sharing and sustainable development. In *Encyclopedia of sustainability in higher education* (pp. 1057–1065). Springer.

Pal, A., Maity, A., Roy, A., Goswami, S., Samanta, N., & Bera, S. (2024). A Review of Renewable Energy Systems and IMO's Advanced Strategies for Marine Pollution Mitigation. *2024 10th International Conference on Electrical Energy Systems (ICEES)*, 1–7. DOI: 10.1109/ICEES61253.2024.10776828

Pandey, M. M. (2004). Present status and future requirements of farm equipment for crop production. *Central Institute of Agricultural Engineering, Bhopal, 24*, 69-113.

Pan, Y., & Zhang, L. (2023). Integrating BIM and AI for smart construction management: Current status and future directions. *Archives of Computational Methods in Engineering, 30*(2), 1081–1110.

Parkhi, S., & Chandra, T. (2024). Review of Urban Blue-Green Infrastructure: A Case of Metropolitan Cities. In *Future is Urban* (pp. 154–164). Routledge.

Park, S.-S., Tran, V.-T., & Lee, D.-E. (2021). Application of Various YOLO Models for Computer Vision-Based Real-Time Pothole Detection. *Applied Sciences (Basel, Switzerland), 11*(23), 11229. DOI: 10.3390/app112311229

Partridge, C. C. (2019). *Green Municipal Bonds and the Financing of Green Infrastructure in the United States UCL*. University College London.

Patel, K., Chaurasia, M., Singh, R., Bhadouria, R., & Rao, K. S. (2024). Role of Blue-Green Infrastructure in Achieving Sustainable Development Goals in Urban Centres of Developing countries. *Blue-Green Infrastructure for Sustainable Urban Settlements: Implications for Developing Countries Under Climate Change*, 111-135.

Perera, A. T. D., Coccolo, S., & Scartezzini, J. (2019). The influence of urban form on the grid integration of renewable energy technologies and distributed energy systems. *Scientific Reports, 9*(1). Advance online publication. DOI: 10.1038/s41598-019-53653-w

Pérez-Cornejo, C., Rodríguez-Gutiérrez, P., & de Quevedo-Puente, E. (2023). City reputation and the role of sustainability in cities. *Sustainable Development, 31*(3), 1444–1455. DOI: 10.1002/sd.2459

Perini, K., & Sabbion, P. (2016). Green-blue infrastructure in urban areas, the case of the Bronx River (NYC) and Paillon (Nice). *Techne : Journal of Technology for Architecture and Environment*, 97–103.

Peschiera, G., & Taylor, J. E. (2012). The impact of peer network position on electricity consumption in building occupant networks utilizing energy feedback systems. *Energy and Building, 49*, 584–590. DOI: 10.1016/j.enbuild.2012.03.011

Phillips, C., Jiao, J., & Clubb, E. (2024). Testing the Capability of AI Art Tools for Urban Design. *IEEE Computer Graphics and Applications*, *44*(2), 37–45. DOI: 10.1109/MCG.2024.3356169 PMID: 38241102

Piracha, A., & Chaudhary, M. T. (2022). Urban air pollution, urban heat island and human health: A review of the literature. *Sustainability*, *14*(15), 9234.

Polli, G. H. B. (2020). A comparison about european environmental sustainability rating systems: Breeam UK, DGNB, LiderA, ITACA and HQE. *U. Porto Journal of Engineering*, *6*(2), 46–58.

Poopalam, K. D., Ismail, T. N. M. T., Hanzah, N. A., Alias, A. H., Wahab, N. A., Ibrahim, Z., Ibrahim, Z., Subramaniam, V., Armylisas, A. H. N., & Idris, Z. (2024). Utilization of oil palm biomass and polyurethanes as sustainable construction materials: A review. *Developments in the Built Environment*, *17*, 100380. DOI: 10.1016/j.dibe.2024.100380

Popescu, S. M., Mansoor, S., Wani, O. A., Kumar, S. S., Sharma, V., Sharma, A., & Chung, Y. S. (2024). Artificial intelligence and IoT driven technologies for environmental pollution monitoring and management. *Frontiers in Environmental Science*, *12*, 1336088.

Popkova, E. G. (2023). Sustainable development of agro-industrial complex and food security through digital agriculture: Contribution of robots, big data, and AI. In *Digital agriculture for food security and sustainable development of the agro-industrial complex* (pp. 3–9). Springer International Publishing.

Pramanik, S., Bag, S., & Roy, A. (2025). *Grading the Blockchain Using Sharding*. IGI Global Scientific Publishing. DOI: 10.4018/979-8-3693-3956-5.ch009

Pramanik, S., Bag, S., Roy, A., Ghosh, R., & Rakshit, P. (2025). *SMOTE vs. ADASYN*. IGI Global Scientific Publishing. DOI: 10.4018/979-8-3693-8714-6.ch004

Pramanik, S., Bag, S., Roy, A., & Rakshit, P. (2025). *Comprehensive Modernity of Fish Smoking Techniques Using the IoT*. IGI Global Scientific Publishing. DOI: 10.4018/979-8-3373-4332-7.ch006

Pramanik, S., Roy, A., & Rakshit, P. (2025). Resuming Transfer Technique for Two-Level Sharding Point of Rupture for a Single Table. In *AI and the Revival of Big Data* (pp. 221–238). IGI Global Scientific Publishing. DOI: 10.4018/979-8-3693-8472-5.ch010

Pramanik, S., Roy, A., Rakshit, P., & Bag, S. (2025). *Blockchain-Based Decentralised Management of Data for the Cellular Internet of Things*. IGI Global Scientific Publishing. DOI: 10.4018/979-8-3693-8664-4.ch011

Puchol-Salort, P., O'Keeffe, J., van Reeuwijk, M., & Mijic, A. (2021). An urban planning sustainability framework: Systems approach to blue green urban design. *Sustainable Cities and Society, 66*. Advance online publication. DOI: 10.1016/j.scs.2020.102677

Puri, V., Kumar, S., Grover, K., & Sharma, M. (2022, July). Development of Eco-Friendly Bricks for Sustainable Construction. *IOP Conference Series. Materials Science and Engineering, 1248*(1), 012109.

PwC España. (2024, febrero 26). *Los expertos señalan la falta de oferta y una legislación inadecuada como las principales causas del problema de la vivienda en España*. PwC España. https://www.pwc.es/es/sala-prensa/notas-prensa/2024/expertos-oferta-legislacion-problema-vivienda-espana.html

Qamar, F., Thomas, T., & Ali, M. (2018). Use of natural fibrous plaster for improving the out of plane lateral resistance of mortarless interlocked masonry walling. *Construction & Building Materials, 174*, 320–329.

Qian, H., Zhang, F., & Qiu, B. (2023). Deciphering the evolution, frontier, and knowledge clustering in sustainable city planning: A 60-year interdisciplinary review. *Sustainability, 15*(24), 16854.

Quium, A. (2019). Transport Corridors for Wider Socio–Economic Development. *Sustainability, 11*(19), 5248. DOI: 10.3390/su11195248

Radhakrishnan, K. (2010). *Development of a decision-making procedure for outsourcing engineering services in TxDOT*. Texas Department of Transportation.

Radicchi, A., Cevikayak Yelmi, P., Chung, A., Jordan, P., Stewart, S., Tsaligopoulos, A., & Grant, M. (2021). Sound and the healthy city. *Cities & Health, 5*(1–2), 1–13. DOI: 10.1080/23748834.2020.1821980

Radinja, M., Atanasova, N., & Zavodnik Lamovšek, A. (2021). The water-management aspect of blue-green infrastructure in cities. *Urbani Izziv, 32*(1), 98–110. DOI: 10.5379/URBANI-IZZIV-EN-2021-32-01-003

Rahla, K. M., Mateus, R., & Bragança, L. (2021). Implementing circular economy strategies in buildings—From theory to practice. *Applied System Innovation, 4*(2). Advance online publication. DOI: 10.3390/asi4020026

Rahla, K. M., Mateus, R., & Bragança, L. (2021). Selection criteria for building materials and components in line with the circular economy principles in the built environment—A review of current trends. *Infrastructures, 6*(4), 49.

Raj, A. J., & Lal, S. B. (Eds.). (2013). Forestry principles and applications (Hardback & Paperback ed.). Scientific Publishers. DOI: 10.13140/2.1.2452.3524

Ramos, N. M. M., Almeida, R. M. S. F., Curado, A., Pereira, P. F., Manuel, S., & Maia, J. (2015). Airtightness and ventilation in a mild climate country rehabilitated social housing buildings – What users want and what they get. *Building and Environment, 92*, 97–110. DOI: 10.1016/J.BUILDENV.2015.04.016

Rana, M. N., Islam, M. N., Nath, S. K., Das, A. K., Ashaduzzaman, M., & Shams, M. I. (2019). Properties of low-density cement-bonded composite panels manufactured from polystyrene and jute stick particles. *Journal of Wood Science, 65*, 1–10.

Rana, M., & Akhai, S. (2022). Multi-objective optimization of Abrasive water jet Machining parameters for Inconel 625 alloy using TGRA. *Materials Today: Proceedings, 65*, 3205–3210.

Rane, N. (2023). Integrating building information modelling (BIM) and artificial intelligence (AI) for smart construction schedule. *Cost, Quality, and Safety Management: Challenges and Opportunities.*

Rane, N. (2023). Integrating leading-edge artificial intelligence (AI), internet of things (IOT), and big data technologies for smart and sustainable architecture. *Engineering and Construction (AEC) Industry: Challenges and Future Directions.*

Rapoport, A. (1990). *The Meaning of the Built Environment: A Nonverbal Communication Approach.* University of Arizona Press.

Rashid, A., & Rasheed, R. (2022). A Paradigm for measuring sustainable performance through big data analytics–artificial intelligence in manufacturing firms. *Available at SSRN* 4087758.

Ravikumar, A. P., Baker, E., Bates, A., Nock, D., Venkataraman, D., Johnson, T., Ash, M., Attari, S. Z., Bowie, K., Carley, S., Castellanos, S., Cha, M., Clark, D. L., Deane-Ryan, D., Djokic, D., Ford, J. C., Goldstein, A., Grubert, E., Hu, L., & Tuominen, M. (2022). Enabling an equitable energy transition through inclusive research. *Nature Energy, 8*(1), 1–4. DOI: 10.1038/s41560-022-01145-z

Reddy, P. Y., Madan, B. S., Anandaram, H., Rathod, P., Vasanthaseelan, S., & Boopathi, S. (2024). Analyzing Fuel Cell Vehicles Through Intelligent Battery Management Systems (BMS): AI and ML Technologies for E-Mobility. In *Advances in Web Technologies and Engineering* (pp. 335–355). IGI Global. DOI: 10.4018/979-8-3693-1487-6.ch016

Reddy, V. J., Hariram, N., Ghazali, M. F., & Kumarasamy, S. (2024). Pathway to sustainability: An overview of renewable energy integration in building systems. *Sustainability*, *16*(2), 638.

Reuters. (2024, May 23). *How AI is arming cities in the battle for climate resilience.* Reuters.

Rodriguez-Ubinas, E., Alzarouni, S., Alhammadi, N., Alantali, M., Behzad, F., & Sgouridis, S. (2020). Sustainability through energy conservation building codes: Comparative analysis of green building regulations in the middle east. *WIT Transactions on Ecology and the Environment*, *249*, 85–100.

Rojek, I., Mikołajewski, D., Mroziński, A., & Macko, M. (2024). Green Energy Management in Manufacturing Based on Demand Prediction by Artificial Intelligence—A Review. *Electronics (Basel)*, *13*(16), 3338.

Rowe, P. G., Hee, L., Rowe, P. G., & Hee, L. (2019). Gardens, parks and green reserves. *A City in Blue and Green: The Singapore Story*, 82-113.

Roy, A., Pramanik, S., Mitra, K., & Chakraborty, M. (2024). *Application of Neural Network for Reducing Emission and Optimizing Performance of Hydrogen with Biofuel CI Engine*. DOI: 10.1007/978-981-97-7532-3_26

Roy, A., & Chakraborty, M. (2024). Effects of ship emissions on Asian haze pollution, health, and IMO strategies. *Societal Impacts*, *3*, 100055. DOI: 10.1016/j.socimp.2024.100055

Roy, A., & Chakraborty, M. (2025). A review of ship emissions impacts on environmental, health, societal impacts and IMO's mitigation policies. *Regional Studies in Marine Science*, *81*, 103964. DOI: 10.1016/j.rsma.2024.103964

Roy, A., & Pramanik, S. (2024). A review of the hydrogen fuel path to emission reduction in the surface transport industry. *International Journal of Hydrogen Energy*, *49*, 792–821. DOI: 10.1016/j.ijhydene.2023.07.010

Roy, P. (2022). Towards urban sustainability: Impact of blue and Green infrastructure on building smart, climate resilient and livable cities. In *Handbook of Sustainable Development Through Green Engineering and Technology* (pp. 119–146). CRC Press.

Ruiz Trejo, C. E. (2018). *Feasability study on maize husk as resource for a novel composite material* (Doctoral dissertation).

Rules on ship carbon intensity and rating system enter into force. (n.d.). https://www.imo.org/en/MediaCentre/PressBriefings/pages/CII-and-EEXI-entry-into-force.aspx

Rupp, T., Volz, H., & Cord, T. (1998). Automated and Robotics-Based Techniques for Road Construction. *IFAC Proceedings Volumes, 31*(3), 515–520. DOI: 10.1016/S1474-6670(17)44137-1

Saadi, S., & Azdimousa, H. (2023, October). Artificial Intelligence and Green Marketing: What Link? In *International Conference on Advanced Intelligent Systems for Sustainable Development* (pp. 104-113). Cham: Springer Nature Switzerland.

Sadineni, S. B., Madala, S., & Boehm, R. F. (2011). Passive building energy savings: A review of building envelope components. *Renewable & Sustainable Energy Reviews, 15*(8), 3617–3631.

Sadineni, S. B., Madala, S., & Boehm, R. F. (2011). *Passive building energy savings: A review of building envelope components. Renewable and Sustainable Energy Reviews*. Elsevier Ltd. DOI: 10.1016/j.rser.2011.07.014

Saeed El Etaiby, Y. (2023). Smart agriculture and future technologies for achieving food security. *International Journal of Family Studies. Food Science and Nutrition Health, 4*(1), 56–71.

Sagala, S., Murwindarti, A., Avila, B. E., Rosyidie, A., & Azhari, D. (2022). Sustainable urban drainage system (SUDS) as nature based solutions approach for flood risk management in high-density urban settlement. IOP Conference Series: earth and environmental science. *Energy and Building, 188*, 278–285.

Saha, A., Kulkarni, N. D., & Kumari, P. (2024). Development of Bambusa tulda fiber-micro particle reinforced hybrid green composite: A sustainable solution for tomorrow's challenges in construction and building engineering. *Construction & Building Materials, 441*, 137486.

Saha, B., Saikia, D., Sabhapandit, A., Marjoribanks, T., Bosher, L., Mukherjee, M., & Kumar, A. (2024). Community Education and Engagement for Blue-Green Infrastructure (BGI) Projects: Insights from Assam, India. In *Disaster and Climate Risk Education: Insights from Knowledge to Action* (pp. 399–422). Springer Nature Singapore.

Saini, K., Matsagar, V. A., & Kodur, V. R. (2024). Recent advances in the use of natural fibers in civil engineering structures. *Construction & Building Materials, 411*, 134364.

Salbitano, F., Borelli, S., Conigliaro, M., & Chen, Y. (2016). *Guidelines on urban and peri-urban forestry (FAO)*. Food and Agriculture Organization of the United Nations.

Sanchez, F. G., & Govindarajulu, D. (2023). Integrating blue-green infrastructure in urban planning for climate adaptation: Lessons from Chennai and Kochi, India. *Land Use Policy*, *124*, 106455.

Sandström, P. (2025, March 18). *Sweden's city of tomorrow: How Scandi innovation is leading climate-smart urban development.* The Guardian.

Sangmesh, B., Patil, N., Jaiswal, K. K., Gowrishankar, T. P., Selvakumar, K. K., Jyothi, M. S., & Kumar, S. (2023). Development of sustainable alternative materials for the construction of green buildings using agricultural residues: A review. *Construction & Building Materials*, *368*, 130457.

Santamaria, B. M., Del Ama Gonzalo, F., Pinette, D., Gonzalez-Lezcano, R. A., Aguirregabiria, B. L., & Ramos, J. A. H. (2020). Application and validation of a dynamic energy simulation tool: A case study with water flow glazing envelope. *Energies*, *13*(12). Advance online publication. DOI: 10.3390/en13123203

Santamouris, M., Papanikolaou, N., Livada, I., Koronakis, I., Georgakis, C., Argiriou, A., & Assimakopoulos, D. N. (2001). On the impact of urban climate on the energy consuption of building. *Solar Energy*, *70*(3), 201–216. DOI: 10.1016/S0038-092X(00)00095-5

Santos, A. K., Ferreira, V. M., & Dias, A. C. (2025). Promoting decarbonisation in the construction of new buildings: A strategy to calculate the embodied carbon footprint. *Journal of Building Engineering*, *103*, 112037. DOI: 10.1016/j.jobe.2025.112037

Sarkar, A. (2022). Re-creating Chandni Chowk: Place-making and place-marketing of 'Old' Delhi's historic city centre. *Indian Journal of Spatial Science, 13*(2).

Sattar, S., Li, S., & Chapman, M. (2018). Road Surface Monitoring Using Smartphone Sensors: A Review. *Sensors (Basel)*, *18*(11), 3845. DOI: 10.3390/s18113845 PMID: 30423962

Satyam, V., & Sharma, P. (2024). Unsustainable construction material and process leading to severe environmental degradation. *AIP Conference Proceedings*, *3100*(1), 040018.

Satyam, V., & Sharma, P. (2024). Unsustainable construction material and process leading to severe environmental degradation. *Global Change Biology*, *27*(8), 1518–1546.

Segara, S., Li, Q. J., Gallotta, A., Wang, Y., Gosling, J., & Rezgui, Y. (2024). Taxonomy of circularity indicators for the built environment: Integrating circularity through the Royal Institute of British architects (RIBA) plan of work. *Journal of Cleaner Production*, *446*, 141429.

Sendra, J. J., Domínguez-Amarillo, S., Rojas, P. B., & Leon, A. L. (2013). Energy Intervention in the residential sector in the south of Spain: Current challenges. *Informes de la Construcción*, *65*, 457–464.

Serra-Parareda, F., Julián, F., Espinosa, E., Rodríguez, A., Espinach, F. X., & Vilaseca, F. (2020). Feasibility of barley straw fibers as reinforcement in fully biobased polyethylene composites: Macro and micro mechanics of the flexural strength. *Molecules (Basel, Switzerland)*, *25*(9), 2242. PMID: 32397611

Sfakianaki, A., Pavlou, K., Santamouris, M., Livada, I., Assimakopoulos, M. N., Mantas, P., & Christakopoulos, A. (2008). Air tightness measurements of residential houses in Athens, Greece. *Building and Environment*, *43*(4), 398–405. DOI: 10.1016/J.BUILDENV.2007.01.006

Shafaghat, A., & Keyvanfar, A. (2022). Dynamic façades design typologies, technologies, measurement techniques, and physical performances across thermal, optical, ventilation, and electricity generation outlooks. *Renewable & Sustainable Energy Reviews*, *167*, 112647.

Shahraki, A. A. (2021). Urban planning for physically disabled people's needs with case studies. *Spatial Information Research*, *29*(2), 173–184. DOI: 10.1007/s41324-020-00343-9

Shah, S. F. A., Iqbal, M., Aziz, Z., Rana, T. A., Khalid, A., Cheah, Y.-N., & Arif, M. (2022). The role of machine learning and the internet of things in smart buildings for energy efficiency. *Applied Sciences (Basel, Switzerland)*, *12*(15), 7882.

Shaikh, P. H., Nor, N. B. M., Nallagownden, P., Elamvazuthi, I., & Ibrahim, T. (2014). A review on optimized control systems for building energy and comfort management of smart sustainable buildings. *Renewable & Sustainable Energy Reviews*, *34*, 409–429.

Sharma, M., Sharma, M., Sharma, N., & Boopathi, S. (2024). Building Sustainable Smart Cities Through Cloud and Intelligent Parking System. In *Handbook of Research on AI and ML for Intelligent Machines and Systems* (pp. 195–222). IGI Global.

Sharma, S. K., Phan, H., & Lee, J. (2020). An Application Study on Road Surface Monitoring Using DTW Based Image Processing and Ultrasonic Sensors. *Applied Sciences (Basel, Switzerland)*, *10*(13), 4490. DOI: 10.3390/app10134490

Sharma, V., Zaki, M., Jha, K. N., & Krishnan, N. A. (2022). Machine learning-aided cost prediction and optimization in construction operations. *Engineering, Construction, and Architectural Management*, *29*(3), 1241–1257.

Sherman, M. (1995). The Use of Blower-Door Data. *Indoor Air, 5*(3), 215–224. DOI: 10.1111/j.1600-0668.1995.t01-1-00008.x

Sherman, M. H. (1987). Estimation of infiltration from leakage and climate indicators. *Energy and Building, 10*(1), 81–86. DOI: 10.1016/0378-7788(87)90008-9

Shibani, A., Agha, A., Hassan, D., & Naomi, F. (2021). The impact of green certification BREEAM on occupancy rates of commercial buildings in the UK. *International Journal of Innovations in Engineering Research and Technology, 8*(9), 153-167.

Shibeika, A., & Harty, C. (2015). Diffusion of digital innovation in construction: A case study of a UK engineering firm. *Construction Management and Economics, 33*(5-6), 453–466.

Ship Energy Efficiency Management Plan (SEEMP). (n.d.). https://www.amsa.gov.au/marine-environment/air-pollution/ship-energy-efficiency-management-plan-seemp

Short, J. R., & Farmer, A. (2021). Cities and Climate Change. *Earth (Switzerland), 2*(4), 1038–1045. DOI: 10.3390/earth2040061

Shortt, N. K., & Hammett, D. (2013). Housing and health in an informal settlement upgrade in Cape Town, South Africa. *Journal of Housing and the Built Environment, 28*(4), 615–627.

Shrestha, P. M., Humphrey, J. L., Barton, K. E., Carlton, E. J., Adgate, J. L., Root, E. D., & Miller, S. L. (2019). Impact of low-income home energy-efficiency retrofits on building air tightness and healthy home indicators. *Sustainability (Switzerland), 11*(9). Advance online publication. DOI: 10.3390/su11092667

Shwedeh, F., Hami, N., Bakar, S. Z. A., Yamin, F. M., & Anuar, A. (2022). The Relationship between Technology Readiness and Smart City Performance in Dubai. *Journal of Advanced Research in Applied Sciences and Engineering Technology, 29*(1), 1–12. DOI: 10.37934/araset.29.1.112

Siehr, S. A., Sun, M., & Aranda Nucamendi, J. L. (2022). Blue-green infrastructure for climate resilience and urban multifunctionality in Chinese cities. *Wiley Interdisciplinary Reviews. Energy and Environment, 11*(5), e447.

Sim, D. (2019). *Soft City: Building Density for Everyday Life*. Island Press.

Singh Madan, B., Logeswaran, S., Veerapathran, M., Rai, P., Kumar Singh, S., & B., S. (2024). Engineered Cementitious Composites (ECC) Through Nano-Composite Integration: Enhancing Construction. In *Advances in Chemical and Materials Engineering* (pp. 210–243). IGI Global. https://doi.org/DOI: 10.4018/978-1-6684-8182-0.ch008

Singh, A., Rajput, V. D., Rawat, S., Sharma, R., Kumar, S. A., Kumar, S. A., & Tomar, R. S. (2021). Emerging tools for sustainable agriculture and food security: An introduction. *Emerging Tools for Sustainable Agriculture and Food Security*, *1*(1), 3–16.

Singh, L., & Brar, B. S. (2021). A review on rice straw management strategies. *Nat Environ Pollut Technol*, *20*(4), 1485–1493.

Singh, N., Singh, S., & Mall, R. (2020). Urban ecology and human health: implications of urban heat island, air pollution and climate change nexus. In *Urban ecology* (pp. 317–334). Elsevier.

Singh, S. K., & Alasali, B. (2025). *Urban Mobility and Smart Energy: Pathways for Global South Cities*. IGI Global.

Singh, S. K., & Jain, R. (2025). Future of mobility: potential challenges and ethical considerations. In *AIoT for Smart Transportation: Transforming the future of mobility* (pp. 235–253). IET., DOI: 10.1049/pbtr047e_ch12

Singh, S. K., Toryila, M., & Alasali, B. (2025). Utilization of sustainable resources for energy efficiency using smart technologies. In *Artificial Intelligence for Environmental Sustainability*. Apple Academic Press – CRC Press.

Sinha, K. K., Osiyevskyy, O., & Radnejad, A. B. (2022). Propositions for R&D governance regimes: A behavioral perspective. *International Journal of Innovation and Technology Management*, *19*(6), 2250017.

Sini, R., & Sini, R. (2020). The Singapore Playground: System of Themed Public Parks that Addresses Environmental, Social and Cultural Sustainability. *Singapore's Park System Master Planning: A Nation Building Tool to Construct Narratives in Post-Colonial Countries*, 253-328.

Sizirici, B., Fseha, Y., Cho, C.-S., Yildiz, I., & Byon, Y.-J. (2021). A review of carbon footprint reduction in construction industry, from design to operation. *Materials (Basel)*, *14*(20), 6094.

Skibniewski, M., & Hendrickson, C. (1990). Automation and Robotics for Road Construction and Maintenance. *Journal of Transportation Engineering*, *116*(3), 261–271. DOI: 10.1061/(ASCE)0733-947X(1990)116:3(261)

Slater, A. G., & Cooper, A. I. (2015). Function-led design of new porous materials. *Science*, *348*(6238), aaa8075. PMID: 26023142

Smith, N., Davies, R., & Owen, G. (2021). Integrating Commercial and Residential Spaces for Sustainable Urban Futures. *Sustainable Cities and Society, 61*, 102302. DOI: 10.1016/j.scs.2020.102302

Smith, R. E. (2010). *Prefab architecture: A guide to modular design and construction*. John Wiley & Sons.

SNC-Lavalin. (2022, May). *Investor relations overview: May 2022*. SNC-Lavalin. Retrieved from https://s29.q4cdn.com/301499865/files/doc_presentations/2022/06/SNC_Investor-Relations-Overview-May-2022_FINAL.pdf

Soares, M., Faria, L., Miranda, T., Pereira, E., Vilarinho, C., & Carvalho, J. (2025). The potential of agri-food waste to solve construction's environmental problems: A review. *Cleaner and Circular Bioeconomy, 10*, 100138. DOI: 10.1016/j.clcb.2025.100138

Sohail, A., Shen, B., Cheema, M. A., Ali, M. E., Ulhaq, A., Babar, M. A., & Qureshi, A. (2024). Beyond data, towards sustainability: A Sydney case study on urban digital twins. *arXiv*.

Son, N., Chen, C. R., & Syu, C. H. (2024). Towards artificial intelligence applications in precision and sustainable agriculture. *Agronomy (Basel), 14*(2), 239.

Soori, M., Arezoo, B., & Dastres, R. (2023). Artificial intelligence, machine learning and deep learning in advanced robotics, a review. *Cognitive Robotics, 3*, 54–70.

Sörensen, J., Persson, A. S., & Olsson, J. A. (2021). A data management framework for strategic urban planning using blue-green infrastructure. *Journal of Environmental Management, 299*, 113658. PMID: 34523536

Sorensen, N. L., Rasmussen, F. N., Oien, T. B., & Frandsen, A. K. (2020). Holistic sustainability: Advancing interdisciplinary building design through tools and data in Denmark. *Construction Economics and Building, 20*(2), 25–44.

Sovacool, B., Heffron, R., & McCauley, D. (2016). Energy decisions reframed as justice and ethical concerns. *Nature Energy, 1*, 16024. DOI: 10.1038/nenergy.2016.24

Sowińska-Świerkosz, B., & García, J. (2022). What are Nature-based solutions (NBS)? Setting core ideas for concept clarification. *Nature-Based Solutions, 2*, 100009.

Spence, R., & Mulligan, H. (1995). Sustainable development and the construction industry. *Habitat International, 19*(3), 279–292.

Sri Kakarla, U., Venkataramana, A., Gopinath, S., Shivaram, K., Yugendhar, K., Harika, K. V. S. L., & Dixit, S. (2024). Revolutionizing Agricultural Harvesting with IoT Application. *MATEC Web of Conferences, 392*, 01014. DOI: 10.1051/matecconf/202439201014

Sri Skandha Moorthy, R., & Nathiya, N. (2023). Botnet Detection Using Artificial Intelligence. *Procedia Computer Science, 218*, 1405–1413. DOI: 10.1016/j.procs.2023.01.119

Srivastava, A., & Maity, R. (2023). Assessing the potential of AI–ML in urban climate change adaptation and sustainable development. *Sustainability, 15*(23), 16461.

Sunikka-Blank, M., & Galvin, R. (2012). Introducing the prebound effect: The gap between performance and actual energy consumption. *Building Research and Information, 40*(3), 260–273. DOI: 10.1080/09613218.2012.690952

Sutkowska, M., Stefańska, A., Vaverkova, M. D., Dixit, S., & Thakur, A. (2024). Recent advances in prefabrication techniques for biobased materials towards a low-carbon future: From modules to sustainability. *Journal of Building Engineering*, 109558.

Su, W., Zhang, W., & Chang, Q. (2025). Tailored green and blue infrastructure for heat mitigation under renewal planning of urban blocks in Beijing. *The Science of the Total Environment, 967*, 178759.

Suzuki, H., Cervero, R., & Iuchi, K. (2013). *Transforming cities with transit: Transit and land-use integration for sustainable urban development*. World Bank Publications.

Synnefa, A., Santamouris, M., & Akbari, H. (2007). Estimating the effect of using cool coatings on energy loads and thermal comfort in residential buildings in various climatic conditions. *Energy and Building, 39*(11), 1167–1174. DOI: 10.1016/j.enbuild.2007.01.004

Talha, S. A., Manasreh, D., & Nazzal, M. D. (2024). The Use of Lidar and Artificial Intelligence Algorithms for Detection and Size Estimation of Potholes. *Buildings (Basel, Switzerland), 14*(4), 1078. DOI: 10.3390/buildings14041078

Talwar, R., & Koury, A. (2017). Artificial intelligence – the next frontier in IT security? *Network Security, 2017*(4), 14–17. DOI: 10.1016/S1353-4858(17)30039-9

Tam, C., Tam, V. W., Chan, J. K., & Ng, W. C. (2005). Use of prefabrication to minimize construction waste-a case study approach. *International Journal of Construction Management, 5*(1), 91–101.

Tapeh, A. T. G., & Naser, M. (2023). Artificial intelligence, machine learning, and deep learning in structural engineering: A scientometrics review of trends and best practices. *Archives of Computational Methods in Engineering, 30*(1), 115–159.

Terzi, D. S., Terzi, R., & Sagiroglu, S. (2017). Big data analytics for network anomaly detection from netflow data. In *2nd International Conference on Computer Science and Engineering, UBMK 2017* (pp. 592–597). Institute of Electrical and Electronics Engineers Inc. https://doi.org/DOI: 10.1109/UBMK.2017.8093473

Terzis, D. (2022). Monitoring innovation metrics in construction and civil engineering: Trends, drivers and laggards. *Developments in the Built Environment, 9*, 100064. DOI: 10.1016/j.dibe.2021.100064

Thirunavukkarasu, A., Nithya, R., Sivashankar, R., & Sathya, A. B. (2018). Bio-based building materials for a green and sustainable environment. *Bioprocess engineering for a green environment*, 47-65.

Tien, P. W., Wei, S., Darkwa, J., Wood, C., & Calautit, J. K. (2022). Machine learning and deep learning methods for enhancing building energy efficiency and indoor environmental quality–a review. *Energy and AI, 10*, 100198.

Tirado, R., Aublet, A., Laurenceau, S., & Habert, G. (2022). Challenges and Opportunities for Circular Economy Promotion in the Building Sector. *Sustainability (Switzerland), 14*(3). Advance online publication. DOI: 10.3390/su14031569

Tokhayeva, Z. O., Almukhambetova, B. Z., Keneshbayev, B., & Akhmetova, K. (2020). Innovative processes' management in agriculture and food security: Development opportunities. *Entrepreneurship and Sustainability Issues*, *7*(3), 1565.

Tokuç, A., & İnan, T. (2017). A Green Outlook to Tall Building Facades via Milan's "Vertical Forest". Proceedings of the International Advanced Technologies Symposium (IATS).

Toli, A. M., & Murtagh, N. (2020, June 2). The Concept of Sustainability in Smart City Definitions. *Frontiers in Built Environment*. Frontiers Media S.A. DOI: 10.3389/fbuil.2020.00077

Toll, D., Lindgren, I., Melin, U., & Madsen, C.Ø. (2020). Values, benefits, considerations and risks of AI in government: A study of AI policies in Sweden. *JeDEM-eJournal democracy Open Gov.*, 12, 40–60.

Tomorrow.City. (n.d.). *IA aplicada al urbanismo: Riesgos y beneficios de un movimiento que parece imparable.* The biggest platform about urban innovation. https://tomorrow.city

Tran, T. A. (2020a). Effect of ship loading on marine diesel engine fuel consumption for bulk carriers based on the fuzzy clustering method. *Ocean Engineering, 207*, 107383. DOI: 10.1016/j.oceaneng.2020.107383

Tran, T. A. (2020b). Simulation and analysis on the ship energy efficiency operational indicator for bulk carriers by Monte Carlo simulation method. *International Journal of Modeling, Simulation and Scientific Computing*, *11*(04), 2050036. DOI: 10.1142/S1793962320500361

Ukoba, K., Olatunji, K. O., Adeoye, E., Jen, T.-C., & Madyira, D. M. (2024). Optimizing renewable energy systems through artificial intelligence: Review and future prospects. *Energy & Environment*, *35*(7), 3833–3879.

Umana, A. U., Garba, B. M. P., Ologun, A., Olu, J. S., & Umar, M. O. (2024). The role of government policies in promoting social housing: A comparative study between Nigeria and other developing nations. *World Journal of Advanced Research and Reviews*, *23*(03), 371–382.

UN DESA. (2019). *World Urbanization Prospects: The 2018 Revision.* United Nations Department of Economic and Social Affairs. https://population.un.org/wup/

UN-Habitat. (2020). *World Cities Report 2020: The Value of Sustainable Urbanization.* United Nations Human Settlements Programme. https://unhabitat.org/world-cities-report-2020-the-value-of-sustainable-urbanization

Unified Traffic and Transportation Infrastructure (Planning & Engineering) Centre (UTTIPEC). (2011). *Street guidelines.*

United Nations. (n.d.). Sustainable Development. United Nations. https://www.un.org/sustainabledevelopment/es/

Urban Design Lab. (2021). Urban conservation plan for Chandni Chowk streetscape. Author.

Valzania, G. (2022). Towers once in the park: Uprooting Toronto's welfare landscapes. *Geografiska Annaler. Series B, Human Geography*, *104*(3), 227–249.

van den Berg, A. C., Giest, S. N., Groeneveld, S. M., & Kraaij, W. (2020). Inclusivity in Online Platforms: Recruitment Strategies for Improving Participation of Diverse Sociodemographic Groups. *Public Administration Review*, *80*(6), 989–1000. DOI: 10.1111/puar.13215

van der Schoor, T. (2022). Lessons from EU-projects for energy renovation. *Urban Planning*, *7*(2), 123–130.

Van Tonder, E. (2017). Passive innovation resistance–a conservative consumer perspective. *European Business Review*, *29*(6), 642–663.

Vellingiri, A., Kokila, R., Nisha, P., Kumar, M., Chinnusamy, S., & Boopathi, S. (2024). Harnessing GPS, Sensors, and Drones to Minimize Environmental Impact: Precision Agriculture. In *Designing Sustainable Internet of Things Solutions for Smart Industries* (pp. 77–108). IGI Global. https://doi.org/DOI: 10.4018/979-8-3693-5498-8.ch004

Verbeke, A., & Greidanus, N. S. (2009). The end of the opportunism versus trust debate: Bounded reliability as a new envelope concept in research on MNE governance. *Journal of International Business Studies*, *40*(9), 1471–1495. DOI: 10.1057/jibs.2009.44

Vestbro, D. U., & Horelli, L. (2012). *Design for gender equality: The history of co-housing ideas and realities. Built Environment*. Alexandrine Press. DOI: 10.2148/benv.38.3.315

Villi, G., Peretti, C., Graci, S., & De Carli, M. (2013). Building leakage analysis and infiltration modelling for an Italian multi-family building. *Journal of Building Performance Simulation*, *2*(6), 98–118. DOI: 10.1080/19401493.2012.699981

Vinuesa, R., Azizpour, H., Leite, I., Balaam, M., Dignum, V., Domisch, S., & Fuso Nerini, F. (2020). The role of artificial intelligence in achieving the Sustainable Development Goals. *Nature Communications*, *11*(1), 233. PMID: 31932590

Vision and level of ambition of the Initial IMO Strategy. (n.d.). https://www.imo.org/en/OurWork/Environment/Pages/Vision-and-level-of-ambition-of-the-Initial-IMO-Strategy.aspx

von Davier, A. A., Hao, J., Liu, L., & Kyllonen, P. (2017). Interdisciplinary research agenda in support of assessment of collaborative problem solving: Lessons learned from developing a collaborative science assessment prototype. *Computers in Human Behavior*, *76*, 631–640.

Walker, A., & Kwong Wing, C. (1999). The relationship between construction project management theory and transaction cost economics. *Engineering, Construction, and Architectural Management*, *6*(2), 166–176. DOI: 10.1108/eb021111

Walker, G., & Weber, D. (1984). A transaction cost approach to make-or-buy decisions. *Administrative Science Quarterly*, *29*(3), 373–391.

Wang, B., Gu, F., Wu, F., Ma, J., Yang, H., & Hu, Z. (2022). Analysis of influencing factors and operation quality evaluation strategy of straw crushing and scattering system. *Agriculture*, *12*(4), 508.

Wang, M., Xu, H., Zhao, J., Sun, C., Liu, Y., & Li, J. (2025). The Carbon Sequestration Potential of Skyscraper Greenery: A Bibliometric Review (2003–2023). *Sustainability*, *17*(5), 1774.

Wang, M., & Zhou, T. (2023). Does smart city implementation improve the subjective quality of life? Evidence from China. *Technology in Society*, *72*. Advance online publication. DOI: 10.1016/j.techsoc.2022.102161

Wang, X., Gard, W., Borska, H., Ursem, B., & Van De Kuilen, J. (2020). Vertical greenery systems: From plants to trees with self-growing interconnections. *European Journal of Wood and Wood Products*, *78*, 1031–1043.

Wang, X., & Li, H. (2024). Enhancing Public Education of Water Resource Conservation Through Urban Landscape: A Case Study of Bishan-Ang Mo Kio Park in Singapore. *Communications in Humanities Research*, *45*, 25–31.

Wang, X., Yu, Y., & Lin, L. (2020). Tweeting the United Nations Climate Change Conference in Paris (COP21): An analysis of a social network and factors determining the network influence. *Online Social Networks and Media*, *15*, 100059. DOI: 10.1016/j.osnem.2019.100059

Wang, X., Zhang, Y., Hu, X., Jia, X., Li, K., Wang, C., & Wang, Y. (2024). Sustainability-oriented construction materials for traditional residential buildings: From material characteristics to environmental suitability. *Building and Environment*, *234*, 109944. DOI: 10.1016/j.buildenv.2024.109944

Waqar, A. (2024). Intelligent decision support systems in construction engineering: An artificial intelligence and machine learning approaches. *Expert Systems with Applications*, *249*, 123503.

Weber, R. H. (2010). Internet of Things - New security and privacy challenges. *Computer Law & Security Review*, *26*(1), 23–30. DOI: 10.1016/j.clsr.2009.11.008

Well, W. (2014). *WELL building standard*. International Well Building Institute.

Willems, J. J., Kuitert, L., & Van Buuren, A. (2023). Policy integration in urban living labs: Delivering multi-functional blue-green infrastructure in Antwerp, Dordrecht, and Gothenburg. *Environmental Policy and Governance*, *33*(3), 258–271.

Williams, D. E. (2007). *Sustainable design: ecology, architecture, and planning*. John Wiley & Sons.

Williamson, O. E. (1996). *The mechanisms of governance*. Oxford University Press.

World Maritime Theme 2022. (n.d.). https://www.imo.org/en/About/Events/Pages/World-Maritime-Theme-2022.aspx

World Resources Institute (WRI). (2015). Cities safer by design: Guidance and examples to promote traffic safety through urban and street design. https://www.wri.org/research/cities-safer-design

Wortmann, T., & Tunçer, B. (2017). Differentiating parametric design: Digital workflows in contemporary architecture and construction. *Design Studies*, *52*, 173–197.

Woschank, M., Rauch, E., & Zsifkovits, H. (2020). A review of further directions for artificial intelligence, machine learning, and deep learning in smart logistics. *Sustainability*, *12*(9), 3760.

Wu, W., Wang, B., & Yang, S. (2019). Research on the simulation design of humanistic landscape optimization in urban residential area based on computer technology. *Informatica (Slovenia)*, *43*(3), 325–330. DOI: 10.31449/inf.v43i3.2802

Xu, X., Mumford, T., & Zou, P. X. (2021). Life-cycle building information modelling (BIM) engaged framework for improving building energy performance. *Energy and Building*, *231*, 110496.

Yadav, K., & Nikalje, G. C. (2024). Comprehensive analysis of bioplastics: Life cycle assessment, waste management, biodiversity impact, and sustainable mitigation strategies. *PeerJ*, *12*, e18013. PMID: 39282116

Yadav, M., & Agarwal, M. (2021). Biobased building materials for sustainable future: An overview. *Materials Today: Proceedings*, *43*, 2895–2902.

Yang, L., Chen, G., Zhao, J., & Rytter, N. G. M. (2020). Ship Speed Optimization Considering Ocean Currents to Enhance Environmental Sustainability in Maritime Shipping. *Sustainability*, *12*(9), 3649. DOI: 10.3390/su12093649

Yang, M., Lin, C., Lai, J., Osman, A. I., Farghali, M., Rooney, D. W., & Yap, P. S. (2024). Advancing environmental sustainability in construction through innovative low-carbon, high-performance cement-based composites: A review. *Materials Today Sustainability*, *26*, 100712. DOI: 10.1016/j.mtsust.2024.100712

Yebes, J. J., Montero, D., & Arriola, I. (2021). Learning to Automatically Catch Potholes in Worldwide Road Scene Images. *IEEE Intelligent Transportation Systems Magazine*, *13*(3), 192–205. DOI: 10.1109/MITS.2019.2926370

Yevenes, M. A., Kolok, A. S., & Araneda, A. (2025). Citizen science to promote a nature-based solution: Barriers and lessons learned from a vegetated vermifilter implementation experience in a Chilean community. *Environmental Monitoring and Assessment*, *197*(2), 1–12.

Yigitcanlar, T., Desouza, K. C., Butler, L., & Roozkhosh, F. (2020). *Contributions and risks of artificial intelligence (AI) in building smarter cities: Insights from a systematic review of the literature. Energies*. MDPI AG. DOI: 10.3390/en13061473

Yigitcanlar, T., Han, H., Kamruzzaman, M., Ioppolo, G., & Sabatini-Marques, J. (2019). The making of smart cities: Are Songdo, Masdar, Amsterdam, San Francisco and Brisbane the best we could build? *Land Use Policy, 88*. Advance online publication. DOI: 10.1016/j.landusepol.2019.104187

Yigitcanlar, T., Kamruzzaman, M., & Teriman, S. (2015). Neighborhood sustainability assessment: Evaluating residential development sustainability in a developing country context. *Sustainability, 7*(3), 2570–2602. DOI: 10.3390/su7032570

Yik, Y. K., Alias, N. E., Yusof, Y., & Isaak, S. (2021). A Real-time Pothole Detection Based on Deep Learning Approach. *Journal of Physics: Conference Series, 1828*(1), 012001. DOI: 10.1088/1742-6596/1828/1/012001

Yu, H., Zahidi, I., Chow, M. F., Liang, D., & Madsen, D. Ø. (2024). Reimagining resources policy: Synergizing mining waste utilization for sustainable construction practices. *Journal of Cleaner Production, 464*, 142795.

Yu, Z., Nurdiawati, A., Kanwal, Q., Al-Humaiqani, M. M., & Al-Ghamdi, S. G. (2024). Assessing and mitigating environmental impacts of construction materials: Insights from environmental product declarations. *Journal of Building Engineering, 98*, 110929. DOI: 10.1016/j.jobe.2024.110929

Zawada, M. (2009). Application of selected methods of multivariate statistical analysis in the investment process on the capital market in Poland. In A. Mesjasz-Lech (Ed.), *Modern management instruments* (pp. 78). Publishing Section of the Faculty of Management, Częstochowa University of Technology.

Zhang, R., Nie, L., Zhao, C., & Zhao, X. (2025). Eco efficiency as a catalyst for citizen co production: Evidence from Chinese cities. *arXiv*.

Zhang, H., Xu, J. J., Cui, H. W., Li, L., Yang, Y., Tang, C. S., & Boers, N. (2024). When Geoscience Meets Foundation Models: Toward a general geoscience artificial intelligence system. *IEEE Geoscience and Remote Sensing Magazine*.

Zhang, L., Sun, H., Pu, T., Sun, H., & Chen, Z. (2024). Do green finance and hi-tech innovation facilitate sustainable development? Evidence from the Yangtze River Economic Belt. *Economic Analysis and Policy, 81*, 1430–1442.

Zhang, L., Wang, X., & Liu, H. (2020). Sustainable Urban Planning and Social Equity: A Systematic Approach. *Journal of Urban Planning and Development, 146*(3), 04020034.

Zhang, R., Tang, Y., Liu, G., Wang, Z., & Zhang, Y. (2024). Unveiling the impact of innovation on competitiveness among construction projects: Moderating and mediating role of environmental regulation and sustainability. *Ain Shams Engineering Journal, 15*(4), 102558. DOI: 10.1016/j.asej.2023.102558

Zhang, T., Tan, Y., Yang, H., & Zhang, X. (2016). The application of air layers in building envelopes: A review. *Applied Energy, 165*, 707–734.

Zhang, W., Gu, X., Tang, L., Yin, Y., Liu, D., & Zhang, Y. (2022). Application of machine learning, deep learning and optimization algorithms in geoengineering and geoscience: Comprehensive review and future challenge. *Gondwana Research, 109*, 1–17.

Zhou, Y., & Kankan Halli, A. (2021). AI regulation for smart cities: Challenges and principles. In Estevez, E., Pardo, T., & Scholl, H. (Eds.), *Smart Cities and Smart Governance: Towards the 22nd Century Sustainable City* (pp. 101–118). Springer.

Zhu, J., Zhang, C., Zhang, M., & Yuan, J. (2024). Modeling and simulation of technological innovation diffusion of high-performance construction materials in transportation engineering. *Transportation Research Part D, Transport and Environment, 120*, 103838. Advance online publication. DOI: 10.1016/j.trd.2024.103838

Zhu, Q., Sarkis, J., & Lai, K. H. (2008). Confirmation of a measurement model for green supply chain management practices implementation. *International Journal of Production Economics, 111*(2), 261–273.

Zivkovic, M., Stamenkovic, M., & Petrovic, V. (2021). Towards flexible housing: Basic design principles. *Facta Universitatis - Series. Architecture and Civil Engineering, 19*(2), 183–192. DOI: 10.2298/fuace211129014z

Zucaro, A., Maselli, G., & Ulgiati, S. (2022). Insights in Urban Resource Management: A Comprehensive Understanding of Unexplored Patterns. *Frontiers in Sustainable Cities, 3*. Advance online publication. DOI: 10.3389/frsc.2021.807735

About the Contributors

Roberto González-Lezcano is Full Professor at the Department of Architecture and Design, area of Building Systems, within the Institute of Technology of Universidad CEU San Pablo. Coordinator of the subjects Building Systems as well as Electrical and Lighting systems. Professor Accredited by ANECA in the figures of Tenured Professor. Two six-year research period by CNEAI (last period 2016-2021). Coordinator of the post-graduate degree of Energy Efficiency and Sustainability in Buildings and Coordinator of the Laboratory of Building Systems within Universidad CEU San Pablo. Coordinator of line 2: Innovation, Construction and Technology, in the Doctoral Program of Composition, History and Techniques in Architecture and Urbanism. Extraordinary Doctorate Award. San Pablo CEU University. 57 impact articles in the last 5 years. H Index:17. Editor of 8 JCR-indexed journals (140 edited articles, indexed in the JCR). Principal Investigator of the ARIE Group: Architecture, Efficient Installations. CEU San Pablo University.

Sunil Kumar Sansaniwal is a Senior Researcher, deals in multidisciplinary research including green buildings, building energy efficiency, occupant adaptive behavior, indoor environmental quality and renewable energy. He has a rich experience of more than ten years in scientific research, teaching and academic consultancy. Dr. Sansaniwal has been involved in various national and international collaborative research projects under different capacities. He has published peer-reviewed research in reputed journals and also working as a reviewer and provides editorial services to various scientific journals and publishers. He has also secured various awards and got recognitions from different renowned organizations. Dr. Sansaniwal has excellent academic credentials and has secured a Master of Technology and Bachelor of Engineering (with distinction) both in Mechanical Engineering.

***

Atheek Ahamed is a Civil Engineering graduate from Marwadi University, India. His expertise lies in Building Information Modeling (BIM), construction project management, and architectural design, with proficiency in tools such as AutoCAD, Revit, Navisworks, and Primavera P6. He has contributed to various engineering projects, including cost management enhancements using 5D BIM technology. Passionate about sustainable infrastructure and innovation in construction, he is dedicated to advancing industry practices through research and professional collaborations.

Shalom Akhai has established himself as a prominent figure in the field of engineering, leveraging over 17+ years of academic, research, and technical experience to make significant contributions. With a robust portfolio of over 100+ publications in prestigious journals and conferences, Dr. Akhai demonstrates a commitment to advancing knowledge through scholarly work. Additionally, he has authored/co-authored several technical textbooks, further solidifying his expertise and dedication to education in the field. Beyond academia, Dr. Akhai's impact extends to real-world applications through the establishment of IGNIS Technical Solutions, a consultancy firm to translate his knowledge into real-world applications. His skills have resulted in the successful development of over 50 patents+, underscoring his innovative approach and influence in shaping the engineering landscape.

Apurv Ashish is an Assistant Professor in the Department of Architecture and Planning, Birla Institute of Technology, Mesra where he is also pursuing his PhD. His research area focuses on climate responsive design and sustainable practices.

Vijay Bhutani is currently completing his Ph.D. in Mechanical Engineering at Amity University, UP, India. Concurrently, he serves as an Assistant Professor in the Department of Mechanical Engineering at Amity University, Greater Noida, U.P., India. His research contributions include numerous publications in esteemed journals indexed in SCI/SCOPUS and other reputable indexing platforms. His expertise lies in thermal engineering, renewable and sustainable energy, green energy, interdisciplinary technology integration, and the design of solar energy systems.

Anosh Nadeem Butt is a Tutor at Glasgow International College, University of Glasgow. His major research interests include transdisciplinary collaboration, nature-based design approaches, pedagogies in architectural design and urbanism, sustainable built environment, and urban resilience. He received his Master's (MArch in Architectural Design – International) and PhD from the Department of Architecture, University of Strathclyde. His PhD focussed on developing a transdisciplinary collaboration framework (TCF) for the nature-based design

(NBD) of sustainable buildings. Information on his research outputs and professional activities is available at: https://orcid.org/0000-0002-9662-9473.

Manashi Chakraborty is a faculty member at the School of Science & Technology, The Neotia University. Her academic contributions focus on interdisciplinary research bridging science and sustainability, with recent engagements in maritime and environmental impact studies.

Manika Goel is a Doctor of Philosophy and Visiting Faculty at School of Planning and Architecture, India.

Jaishri Gothania is an academician, author, researcher, and data scientist with over 9 years of experience. She has published 12 research articles, presented 6 conference papers, and holds multiple patents. Her work bridges data science and innovation, contributing significantly to academia, research, and technological advancement.

Kalyan Mitra, who, till recently was a Faculty at IMU-KC, received an Omkarnath and Chuni Wazir outstanding teaching Award from the Institute of Marine Engineers (India) for excellence in education. This award was instituted to judge the Teacher in DG Shipping recognized MTI's. It was awarded by the Institute of Marine Engineers. Dr. Kalyan Mitra also received Excellence in Research Award from Min. of Shipping & Transport in 2007-08.

Sandeep Mukunda received his Ph. D. degree in Mechanical Engineering from Presidency University, Bengaluru, India, in 2023. His academic journey has been centered on mechanical engineering, with a strong focus on finite element analysis, structural mechanics, and computational simulations. Currently, he serves as an Assistant Professor at Presidency University, Bengaluru, where he is actively involved in teaching, mentoring students, and coordinating academic programs. He is also the Club Head of Team Alfa Zyrus, engaging students in technical and research-based projects. His research contributions span areas such as rotary friction analysis, material behavior under dynamic loads, and computational mechanics, with publications in reputed journals and conferences. Dr. Mukunda is an ANSYS-certified professional, demonstrating expertise in finite element simulations. His dedication to research and education has been recognized through [mention awards, if applicable]. He is a member of professional societies such as ISTE & IEI and actively participates in academic collaborations and committee work.

Aditi Nag is an Architect and Urban Planner, currently serving as an Assistant Professor in the Faculty of Science, Technology and Architecture at Manipal

University Jaipur. She has a Ph.D. in heritage planning and tourism competitiveness and management from the Department of Architecture and Planning at Birla Institute of Technology, Mesra, Ranchi, India. Her research focuses on urban and rural planning, and she has several publications to her credit. Notably, she is the first recipient of the DST Inspire Fellowship in Architecture and Planning, awarded by the Ministry of Science and Technology, Government of India.

Oleksiy Osiyevskyy is an Associate Professor of Entrepreneurship & Innovation at the Haskayne School of Business, University of Calgary. In his scholarship, supported by the Future Fund Fellowship, he concentrates on the problem of achieving and sustaining firm growth and prosperous longevity through successful engagement in entrepreneurial strategies. Oleksiy founded his first Internet startup back in 2003, during his undergraduate studies. Since that time, he has been actively engaged in real-world business practice across the globe as an entrepreneur, manager within large companies, consultant and educator. Oleksiy has mentored, co-founded, invested in, and served on the boards of numerous startups within a broad range of industries (education, food, medical devices, transportation, e-commerce, among others), and engaged in strategic innovation projects within large organizations.

Sabyasachi Pramanik currently works as an Associate Professor in the Department of Computer Science & Engineering, Haldia Institute of Technology. Sabyasachi does research in Steganography, Computer Communications, Data Hiding and Image Processing. Their current project is 'A new encrypted method in image steganography'.

Anurag Singh Rathore is a student of Tourism Studies, pursuing a degree course from IGNOU, New Delhi in the Department of Tourism. He has a fascination in the field of modern tourism trends and traditional branding techniques.

Atanu Roy, CEng (India), FIMarE, FIE, is an Extra First Class Marine Engineer from Kolkata, India, with extensive experience managing bulk carriers. A Chartered Engineer (India) and a Fellow of both the Institute of Marine Engineers (India) and the Institution of Engineers (India), he is committed to professional excellence and the sustainable development of the maritime industry.

Rita Sharma is currently serving as an Assistant Professor in Physics at Rayat Bahra University, where she has been teaching since 2021. With over 10 years of experience in teaching and academic mentorship, she has made significant contributions to the fields of experimental material science and physics. She earned her MSc in Applied Physics, followed by an MPhil and PhD in Physics from Guru Nanak Dev University. Her doctoral research focused on the structural, optical, and electrical properties of In-Sb-Ag-Se thin films, a quaternary chalcogenide system. Dr. Sharma has published her research in peer-reviewed journals and has collaborated on projects with leading institutions, including the Central University of Jammu. She is actively guiding PhD scholars and is a member of the Board of Studies (BoS), contributing to curriculum development and academic policy. Her work bridges theoretical understanding and experimental innovation, with ongoing efforts to integrate AI and machine learning into materials research.

Lhousileno Shuya graduated with a Bachelor's degree in Architecture in 2023 and currently in the final year of Master's program in Urban Planning.

Sitesh Kumar Singh is an accomplished academic and Chartered Engineer with a Ph.D. in Civil Engineering and over nine years of diverse experience in education, research, consultancy, and educational administration. Currently serving as an Assistant Professor in the Department of Civil and Environmental Engineering at the National University of Science and Technology, Muscat, Sultanate of Oman, he specializes in Civil and Transportation Engineering, Project Management, and Data Analytics. Dr. Singh holds multiple advanced degrees including M.Tech, B.Tech, PGDIPR, PGDUPDL, and an M.A. in Education. His career encompasses curriculum development, research supervision, quality assurance, and strategic academic planning. He has supervised over 40 postgraduate and 2 doctoral students, contributing to academic leadership through workshops, conferences, and editorial roles, and has published over 40 research articles, several patents, and authored books and book chapters.

Michael Tiza holds a Ph.D. in Civil Engineering (2024), an M.Eng. in Transportation Engineering (2017), an MBA in Construction Management (2016), and a B.Eng. in Civil Engineering (2015). He is a COREN-registered engineer, an NSE member, and a Fellow of the International Organization for Academic and Scientific Development (IOASD, India), the Chartered Institute of Management and Leadership (CIML, USA), and the International Society for Development and Sustainability (ISDS, Japan).

Lara Vélez-González is Professor in the Faculty of Humanities and Communication Sciences at the University San Pablo-CEU. Principal researcher of the consolidated group ICOIDI (Research in Communication through Image and Design). Responsible for line 3 'Communication and New Technologies' in the Social Communication program of the International Doctoral School CEINDO. She has been director of the Department of Audiovisual Communication and Advertising at USP-CEU (2015-2019), academic secretary of the same (2007-2015), director of the Informative Graphics section (2007-2015) and director of the

Research Support Service of USP-CEU SAI-CA (2015-2017). PhD in Information Sciences from the UCM, her research revolves around the design of periodicals and the typographic evolution of newspapers and magazines. Author of more than thirty articles in indexed journals specialized in Communication, six books and more than forty chapters in collective works. She has curated three exhibitions on Design and Typography. Her undergraduate teaching activity focuses on the subjects 'Fundamentals of Design', 'Journalistic Design' and 'Graphic Communication and New Narratives', in the Communication Degrees at USP-CEU. As a postgraduate lecturer, she has been linked to the PhD program in Social Communication at the CEINDO International School since 2016, having taught in the previous PhD program since 2010. She has also been and is a guest lecturer in different official masters related to publication design Master in Editorial Design (Istituto Europeo di Design. IED), Master in Fashion and Beauty Communication. Telva & Yo Dona (Unidad Editorial USP-CEU), Master in Corporate Communication Arts and Master in Graphic Design (TRACOR/USP-CEU). Similarly, and seeking to reinforce a research coherence, during the last years, she has been able to participate in some projects and contracts where she has researched and produced on issues related to the processes of graphic design of information. All this scientific production has been taken to more than one hundred national and international congresses, with the presentation of communications, papers or invited lectures. In 2018 and in 2021 he obtained the First Prize for Teaching Innovation awarded by the USP-CEU. He has two sexenios of research recognized and has directed more than a dozen doctoral theses. Member of the Research Commission of the Faculty of Humanities and Communication Sciences and of the Teaching Innovation Commission of the USP-CEU (2020-2022). Member of the SEP (Spanish Society of Journalism), AE-IC (Spanish Association of Communication Research) and DI_MAD (Madrid Designers Association). This research and teaching career has been reinforced with more than thirty years of experience as a design and graphic communication consultant in different companies and institutions.

Jagdeep Walia is a Research Scholar in the Department of Physics at Rayat Bahara University, Mohali-Punjab, India.

Mitchell Woodworth is a structural engineer with extensive experience in both consulting and contracting roles, currently working at Aecon Group Inc. in Calgary, Alberta. He holds a professional engineering designation (P.Eng) and is completing his MBA at the Haskayne School of Business, University of Calgary, with a focus on innovation and entrepreneurship. Mitchell's work centers on integrating engineering and project delivery models to drive efficiency, knowledge retention, and technological advancement in large-scale infrastructure projects. His academic interests include strategic insourcing, digital transformation in construction, and the evolving role of in-house engineering teams in fostering innovation. He is actively involved in research and publication on these topics through his directed studies and applied industry projects.

Amit Yadav is an aspiring Civil Engineer at his final year at Marwadi University, India.

Index

Symbols

A

B

C

P

R

S

T

U

W

www.ingramcontent.com/pod-product-compliance
Lightning Source LLC
LaVergne TN
LVHW081248011026
849162LV00004BA/754
* 9 7 9 8 3 3 7 3 2 5 5 5 2 *